# Floods and Droughts in the Tulare Lake

By John T. Austin

Figure 1. Flood on the Kaweah River, January 2, 1997.
Photograph by Tony Caprio

COVER PHOTO
Flood on Rock Creek, July 29, 2011. Photograph by Dave Alexander

Sequoia Natural History Association
Three Rivers, CA 93271

**COLOR:** You have purchased the black and white edition of this book. A separate edition is available on Amazon.com for a higher price in full color. All of the charts, graphs, maps, and photos in this one-color edition may be viewed for free in full color online at www.floodsdroughts.weebly.com.

**Ordering Copies:** Printed and bound versions of this book as well as Kindle eBook will be available on Amazon.com. Search the title/author on the Amazon website. For library copies or wholesale contact the publisher at snha@sequoiahistory.org.

First Edition December 2012        Printed in the United States
One color Printed version ISBN 978-1-878441-37-9

**LIBRARY OF CONGRESS DATA FOR FULL COLOR EDITION ONLY**
Library of Congress Cataloging-in-Publication Data:

Austin, John T.
  Floods and droughts in the Tulare Lake Basin / by John Austin. -- 1st ed.
     p. cm.
  ISBN 978-1-878441-32-4
  1.  Floods--California--Tulare Lake Watershed--History. 2.  Droughts--California--Tulare Lake Watershed--History. 3.  Tulare Lake Watershed (Calif.)--Environmental conditions--History.  I. Sequoia Natural History Association. II. Title.
  GB1399.4.C2A97 2013
  551.48'90979485--dc23
                        2012048707
Written by John T. Austin
Publishing Coordination and Kindle version layout by Mark Tilchen
Published by Sequoia Natural History Association
www.sequoiahistory.org

# Contents

# Figures

# Tables

# Preface

The following report began as an effort to understand the hydrologic cycles of Sequoia and Kings Canyon National Parks, but it has turned into something even more important. The pages that follow provide priceless insights into an entire region: the Tulare Lake Basin of Central California. This distinct geographic zone contains not only the southern Sierra Nevada with its twin national parks but also major cities, a significant part of the richest agricultural area in the United States, and the bed of what was less than two centuries ago the largest freshwater lake in the western half of the conterminous United States.

Water, through both its presence and its absence, affects this region profoundly and in distinctive ways. The Tulare Lake Basin has a number of characteristics that make this particularly true. First, the region is close to the Pacific Ocean and thus well within range of intense oceanic storms. It has a Mediterranean climate which sees the great majority of its precipitation fall during the winter months of November through April. A further factor is that, because the region falls within the mid-thirties latitude range, it occupies the highly variable frontier between the wet winter climate of the Pacific Northwest and the often very dry winter climate of Southern California and northwestern Mexico. Adding more interest is the presence of the high-altitude terrain of the Sierra Nevada (including Mt. Whitney), which means that when the right kind of disturbances do arrive, extremely heavy precipitation can be extracted as the storms move eastward. And finally, the region's major rivers — the Kings, Kaweah, Tule and Kern — flow into an interior basin rather than into the ocean. This last fact is true of no other significant area along the immediate Pacific Coast of the United States.

Assemble these characteristics and the complexity and significance of the region comes into focus. The Tulare Lake Basin has a highly variable climate, irregularly endures oceanic storms of ferocious magnitude, and (naturally at least) collects and holds all the runoff that occurs within its watershed. And, we must not forget, it is the home of large numbers of human beings and their institutions, everything from national parks to cities and corporate farms.

Author John T. Austin has approached the problems inherent in this report from the perspective of history: that is, he has sought out historical evidence from the many sources that document the Tulare Lake Basin's highly variable patterns of flood and drought. As the pages that follow will document, he has thrown his metaphorical net wide, taking in everything from newspaper accounts to the literature regarding geological sediment cores. In doing so, the author has brought together a large and rich body of knowledge that has simply never been looked at before as a part of a single, unified pattern.

The value of such a unified perspective is immense. In many ways, we modern humans are just coming to know the Tulare Lake Basin. In less than two centuries we have settled the region and harnessed it to our needs. Yet, as this report so clearly demonstrates, we have not been here nearly long enough to know how the basin actually works. We have yet to experience either floods or droughts of the intensity found within the 2,000-year-long period documented. The report warns us how much we have yet to learn if we are to build a sustainable civilization within the basin.

Adding importance to this study is the accelerating presence of global climate change. The management policies of everything from cities and farms to national parks assume a "normal" world, one where averages can be defined and counted on. Discerning a core of climatic normality in a place as variable as the Tulare Lake Basin is no easy feat, and now we face the challenge that even such normality as we have known is inevitably evolving into something else. It is in this final context that this report adds yet more value by giving us a longer-term context in which to consider those things that will yet occur.

*Floods and Droughts in the Tulare Lake Basin* is a report of significant long-term value to all who live in or care about this important region. In a way not seen before, it provides a historically powerful climatic overview of the region and how it works. It should be studied carefully by all who intend to manage lands or make their homes living within this dynamic region.

Wm. Tweed
Chief Park Naturalist, Sequoia and Kings Canyon National Parks
1996–2006

# Purpose and Scope

## Purpose

This document has a variety of purposes:

- To tell the story of water in the Tulare Lake Basin, to make it meaningful to the public. Why should residents of the Tulare Lake Basin care about the nearly 2,000-year history of the hydrology of this basin when considering modern day agriculture, dams, public health and safety, etc.?

- To provide a human dimension to the long-term climate record in an easily readable format. This document is meant to be read by the general public, not just by scientists and public land managers.

- To provide a context for understanding the predictions of the various climate models. Those models predict that the future will be different relative to the recent past. This document tells us what that past really looked like.

- To provide a single source for what is known about the history of floods and droughts within the Tulare Lake Basin. However, it is not intended to be a scientific treatise on that subject.

- To provide information so that the reader can better understand the risks that we face in preparing for future floods and droughts. To raise awareness of the seriousness of those risks.

- To provide context for understanding the link between storm precipitation and flooding.

- To provide a resource for interpreters and education specialists — to serve as a basic sourcebook for answering visitor questions as well as for building programs, exhibits, and other interpretive media.

- To provide a context for understanding and interpreting Sequoia and Kings Canyon National Parks' collection of flood photographs. In this document, Sequoia and Kings Canyon National Parks are generally referred to simply as the "national parks."

## Scope

The intention of this document is to present the historical record of floods and droughts that have occurred within the Tulare Lake Basin over approximately the last 2,000 years. To the extent possible, this history is based on records specific to this basin. However, it has often been useful to include records from outside the basin for one or more of the following reasons:

- Records from within the basin are sometimes inadequate to describe a particular flood or drought, particularly in the early years of Euro-American settlement.

- Including records from outside the region is useful for major flood and drought events because those are larger than regional events. Examples of floods that affected an area much larger than the Tulare Lake Basin were the floods of 1861–62, 1916, 1938, 1964, and 1969.

- It's useful for us to have an understanding of low-frequency events; that is, events that occur infrequently. For example, what does a 1,000-year storm look like? Or what happens when an 8-inch-per-hour storm hits a recently burned slope? By their nature, it takes a long time to observe such events in any given area, especially in a basin with as few gages and monitoring sites as the Tulare Lake Basin has. By looking beyond the boundaries of our basin, we can get a sense of what risks we might face in this time of climate change. Therefore, there are a number of 1,000-year events described in this document, and even some as rare as 300,000-year events.

This page intentionally left blank.

Floods and Droughts in the Tulare Lake Basin

**This page intentionally left blank**

# Summary

## Overview of the Document's Contents

This document consists of two parts:
1. Overview and background material useful in understanding our history of floods and droughts, such as:
   - Maps of the Tulare Lake Basin and the adjacent basins.
   - Description and history of Tulare Lake and the neighboring lakes.
   - The types and causes of floods, and the terminology used to describe them.
   - Description of the federal reservoirs and how the conveyance structures work below the dams.
   - Summary graphs and tables showing runoff, floods, droughts, and temperatures.
   - Description of the different types of droughts.
   - Description of the consequences of using more water than we have.
2. A history of each of the floods and droughts, over approximately the last 2,000 years, for which we were able to find records.

## Quick Start Guide

This document was intended to be read in order, starting at the beginning. That works for some readers, but others like to jump right into the history section, skipping the background material. If you like to jump right in, here is a suggested path for doing just that:

1) A good place to start is to look at the maps, especially Figure 2, Figure 4, and Figure 5. Then check out Figure 13 to get a sense of how widely runoff varies from year to year in this highly variable climate.

2) Read a flood story from the modern era. A good one to start with involves Bobbie McDowall and her dad on the North Fork of the Kaweah. This drama took place late one night during the big November 1950 flood.

3) Then read about the huge 1861-62 flood. This is usually viewed as a Central Valley flood, but it was even bigger than that. It affected a very wide area stretching from the Columbia River to the Mexican border. The atmospheric mechanisms behind the storms of 1861–62 are unknown; however, the storms were likely the result of an intense atmospheric river, or a series of atmospheric rivers. Atmospheric rivers are relatively narrow regions in the atmosphere that are responsible for most of the horizontal transport of water vapor outside of the tropics. A strong atmospheric river can create major flooding when it makes landfall. The Tulare Lake Basin has had at least five large atmospheric river floods that we know of. Be sure to check out the link to Figure 21, where a satellite caught the remains of Super Typhoon Melor sitting over Japan, while simultaneously pummeling the Southern Sierra and Sequoia and Kings Canyon National Parks with an atmospheric river on October 14, 2009.

4) A representative drought to read about from the early pioneer days is the 1862–64 drought. It was severe, especially by the third year. There was no state or federal water system, so every rancher and farmer was on his own. The saving grace in many ways was Tulare Lake. Thanks to the 1861–62 flood, the lake was brimful when the drought set in. The lake served rather

like a water hole on the Serengeti, albeit a 40-mile-long water hole. Vast herds of cattle would spread over the country for miles, traveling as far back from the lake as they could go without water in search of the scant grasses. Then they would rush back to the shore each day to quench their thirst.

5) Another good section to read is the one on California mega-floods. These floods are even bigger than the 1861–62 flood and recur on a regular cycle of approximately 200 years. If they hold to their past schedule, the next one is expected to return within the next few decades.

6) Then go back to the beginning of the document and read any of the background material that interests you.

## Using the Document Electronically

This document is all hot-linked from the Table of Contents, so there's no need to print it out, let alone read it in order. The body of the document also contains many hot-links. All figure and table references are hot-linked. There are also hot links to resources such as stream gage databases.

Citations (endnote numbers) are scattered throughout the document. The endnotes are located in the Literature Cited section at the end of the document. In the original Microsoft Word document, double-clicking on the citation number would take you to that section and then return you to where you were in the body of the document. The process of converting the Word document to an Adobe pdf file deactivated the endnotes. Therefore, you have to manually move to the entry that you are looking for in the Literature Cited section.

Many of the citations in the Literature Cited section have hot-links that will take you right to the original publication. Some citations are not available electronically and thus not hot-linked.

## Key Findings about Runoff and Floods

The 118-year average runoff (1894–2011) of the four rivers in the Tulare Lake Basin (Kings, Kaweah, Tule, and Kern) is 2,975,682 acre-feet. It can be hard to grasp just how much water that is. The four federal reservoirs aren't designed to capture the total runoff; that isn't how they operate. Water continuously enters and is more or less continuously released from reservoirs. Ideally, reservoirs are drawn down most of the way before a flood, freeing up the flood-control pool.

However, the size of the reservoirs is useful for visualizing the volume of the runoff. The four reservoirs have a combined current capacity of 1,608,073 acre-feet. It's useful to compare that capacity against the size of the historic runoff:
- The 1,608,073 acre-feet in combined current capacity can hold 54% of the 118-year average runoff (1894–2011) of the four rivers.
- The combined runoff of the four major rivers in the Tulare Lake Basin in 1983 was 8,746,222 acre-feet (see Table 66 and Figure 13). That is 5.4 times the combined current capacity of the federal reservoirs on those four rivers.

A look at Figure 13 on page 109 will show how widely runoff varies from year to year. Wet years and dry years commonly alternate, at least to some extent. The Tulare Lake Basin doesn't have normal conditions in the sense of a statistical average. What is reliable about our climate is its extreme and relentless variability. That is our real normal; that is the lesson of Figure 13.

Likewise, floods are amazingly commonplace in our area. A look at the Table of Contents or Figure 17 on page 139 will show just how commonplace: this document describes what we know about approximately 183 floods that have occurred during the last 2,000 years.

Floods occur at all manner of times. They occur in wet years, and they occur during multi-year droughts. They occur during the winter wet season, and they occur during the summer dry season. When they occur varies so widely because there are such a variety of causes for floods.

There's a surprising variety in what constitutes a flood. This document contains a definition of what constitutes a drought, but it does not have an all-encompassing definition of a flood. That has proven too messy a concept to define.

Some of our floods are obvious: a river overflows its banks or a downpour overwhelms a city's drainage system. At the other extreme, some of our floods have two components: hydrologic and socio-political. Society decides what its tolerance for natural processes is; that is, where to allow a river to flow. Thus some of our floods are the result of water appearing at the wrong place at the wrong time. They're an inconvenience. For example, farmers wanted to drain Tulare Lake so that the lakebed could be used for agricultural purposes. They viewed Tulare Lake as an inconvenience, a nuisance to be prevented. Their viewpoint has prevailed. As a result, society has defined the presence of excess water in the lakebed as a flood. Water managers go to great efforts to minimize that type of flood.

## Preparing for the Next Big Flood

Recalling our long-term flood history can be highly instructive. Historical information on floods can be used to prepare for taking future actions. That was the original impetus for preparing this document. One of the big lessons of this document is that our rivers have been relatively quiet of late. Table 17 on page 135 shows that the Tulare Lake Basin hasn't experienced any 50-year floods or 100-year floods in over 40 years. The Kaweah and Tule Rivers haven't even seen any 20-year or larger floods during that time period. This finding is based on the natural flow of the rivers without factoring in the flood control provided by the reservoirs.

The fact that our rivers have been relatively quiet during the last 40 years probably doesn't mean anything; it's just a statistical coincidence. The problem is more psychological. We have become complacent. When we don't experience a big flood for a while, we tend to forget just how big our floods can be. We have come to think of the federal reservoirs and our levees as protecting us from the effects of big floods, and that isn't necessarily realistic when we consider our flood history.

The last really big flood in the Tulare Lake Basin was the December 1966 flood. It's sobering to reflect back on the experience of that flood. Fifteen-foot waves were reported to have been common on the mainstem of the Kaweah in Three Rivers. Today we think of Dry Creek below Terminus Dam as not much more than a quiet foothills stream. However, in the 1966 flood, Dry Creek carried 44% more water than the Merced River in Yosemite Valley in the much more famous January 1997 flood.

The take-away message is that it would be prudent to prepare for big floods, floods much larger than we've been experiencing during the last 40 years. This is particularly important for those of us who live and work in areas that aren't protected by a federal reservoir. We cannot assume that emergency plans have already been prepared by the county emergency agencies charged with planning for floods. Check just to be sure, ask to see the plans. If the county has an emergency plan, ask whether it includes a flood warning system, evacuation routes, etc.

People who live below the reservoirs tend to think that they're safe, that the reservoirs are so big that they can catch and hold the floodwaters of the biggest events. Those reservoirs have been very effective at protecting downstream communities since their construction, but they do have their limits.

Authorized flood-control reservoirs are designed to provide a particular flood-control pool. That flood-control pool is used to store high inflows from a flood event so that flows downstream of a dam do not exceed the stated channel capacity. Hydrologists manage this flood-control pool to temporarily store the rain-flood runoff which would otherwise pass by a dam. The goal is to keep flows downstream of the dam within their stated channel capacity so that flooding conditions are avoided. The maximum size flood that a dam can control is termed a dam's "level of flood protection."

A reservoir's level of flood protection can change over time. An example is Terminus Dam which forms Lake Kaweah. When originally constructed in 1962, Lake Kaweah's storage capacity was estimated to be sufficient to provide a 60-year level of flood protection downstream. But as sediment accumulated in the reservoir, the level of protection had decreased by 1978 to only a 46-year level of flood protection. When fuse gates increased the flood-control pool size of the reservoir in 2004, the level of protection increased to a 70-year level of flood protection (see Table 6).

To learn more about the ins and outs of the level of flood protection, see the section of this document that describes Flood Rate and Flood-risk terminology.

In a large flood like the 1966 flood, reservoirs such as Lake Kaweah and Lake Success were pushed to their limit or beyond. In a huge flood like the 1867–68 flood, even greater flows would be passed downstream to the communities that sit below the reservoirs. The reservoirs were never designed to fully control floods of this magnitude. Sedimentation since construction has further reduced their flood control abilities.

Communities that sit below the reservoirs have to rely on levees for their fallback flood protection. And many of those levees have a history that dates back to the 19th century. Visalia is a prime example.

At the time that Visalia was founded in 1852, the flow of the Kaweah River was distributed largely along the south side of the Kaweah Delta. That all changed thanks to the huge 1861–62 flood and the even bigger 1867–68 flood. One of the legacies of those floods was the creation of the St. Johns River which rerouted the majority of the Kaweah River floodwaters along the north side of the delta, to the north of Visalia.

It didn't take Visalia long to erect a levee along the south bank of the newly formed river. That levee was built using material pulled up out of the river channel. That levee has failed numerous times since its initial construction, but none of the repairs corrected the levee's significant structural shortcomings. The Visalia Chamber of Commerce hosted a thorough after-action review after a particularly bad levee failure that occurred during the 1945 flood. That review disclosed that the ability of the levee to protect Visalia was now significantly degraded, and it wasn't obvious how the levee could be upgraded sufficiently to protect Visalia in the event of another flood of similar magnitude. The St. Johns River is subject to floods that are much larger than the 1945 flood.

Over six decades have passed since that after-action review, but there are still major concerns about the ability of the St. Johns channel to safely pass floodwaters. The 2005–06 Tulare County Civil Grand Jury investigated the St. Johns levee and found that it was not constructed to U.S. Army Corps of Engineers (USACE) certification standards, it was not being adequately maintained, and there was no adequate source of funds for its maintenance.

The Tulare County Resource Management Agency surveyed property owners in the levee district in 2002, but those owners were generally uninterested in levee maintenance and did not want to put more of their tax dollars into maintenance. Because the south levee was in such bad shape, $17 million was then needed to bring it up to USACE certification standards. However, no source for those funds has yet been found, and the levee is in approximately the same shape now that it was in when the grand jury assessed the situation.

In June 2009, the Federal Emergency Management Agency (FEMA) found that the levee was in such bad repair that it provided essentially no reliable flood protection for Visalia.

The USACE has also noted that many levees in our area were originally built to protect agricultural lands, but now protect urban development. As a result, they are under-designed for the purpose that they now serve.

The takeaway message is that, like the dams, we shouldn't assume that all of our levees are designed, constructed, and maintained to provide protection from reasonably foreseeable floods. They are not. We have become complacent. We have not planned for flood events that are relatively common from the long-term perspective.

To help make historical knowledge applicable to future catastrophic events, the USGS Multi Hazards Demonstration Project (MHDP) applies science to improve the resiliency of communities in Southern California to a variety of major natural hazards. The MHDP assembled experts from a number of agencies to design a large, but scientifically plausible, hypothetical storm scenario that would provide emergency responders, resource managers, and the public a realistic assessment of what is historically possible. One of the MDHP's full scenarios, called ARkStorm, addresses massive West Coast storms analogous to those that devastated California in 1861–62. This is a particularly reasonable assumption because storms of this magnitude are projected to become more frequent and intense as a result of climate change.

The ARkStorm scenario is patterned after the 1861–62 historical events. The ARkStorm scenario draws heat and moisture from the tropical Pacific, forming a series of atmospheric rivers that approach the ferocity of hurricanes and then slam into the West Coast over several weeks, resulting in large scale flooding. With the right alignment of conditions, a single intense atmospheric river hitting the Sierra east of Sacramento could bring devastation to the Central Valley. The U.S. Geological Survey (USGS) strongly urges risk management agencies to plan for the return of a flood as big as the 1861–62 event. That was the purpose of creating the ARkStorm scenario. The website for the USGS's Multi Hazards Demonstration Project warns that an ARkStorm is plausible, perhaps inevitable. The 1861–62 storm was not a freak event and was not the last time that California will experience such a severe storm.

## Preparing for the Next Big Landslide

Very large storm events occasionally result in not just floods, but also very large landslides. The ARkStorm scenario was based in large part on the 1861–62 flood. That was a huge flood in the Tulare Lake Basin. The force of the flood was so great that all four of our major rivers (Kings, Kaweah, Tule, and Kern) relocated and cut new channels. USGS has urged local risk management agencies to prepare for a return of a flood as big as the 1861–62 flood. We definitely do not at this time have plans in place for dealing with such an event.

However, the Tulare Lake Basin experienced a flood even bigger than the 1861–62 flood just six years later: the 1867–68 flood. It was a bigger flood on all four of our major rivers. In addition to being a major flood, the storm that brought on that flood was a deep soaking rain that lasted for upwards of six weeks. Hillslopes in the middle elevations of the Sierra are typically quite steep. Some are weathered in place, but many consist of unconsolidated colluvial debris slopes. When these hillslopes are soaked to depth, huge landslides can be triggered.

In a short intense storm like the December 2010 storm, we get many relatively small landslides and debris flows. By contrast, in an extended event like the 1867–68 storm, the mid-elevation zone can experience cataclysmic landslides. In the past, some of those have formed landslide dams across our major rivers that were up to 400 feet high. When dams such as those fail, the results downstream can be catastrophic. For example, the residents of Bakersfield woke on New Year's Day, 1868, to a 200-foot-high flood coming out of the Kern Canyon.

In the 1867–68 storm, the landslide dams on the Kaweah and Kern held the flooding rivers back long enough for the residents downstream to react and get out of the floodplain. In contrast, the dams on the San Joaquin River and Mill Flat Creek presented a less clear signal downstream, partly because those events happened at night.

The residents of Old Kernville and Weldon had about 24 hours' notice because the river stopped running. They were able to evacuate their towns before the Kern River submerged them under about 50 feet of water. The residents of Millerton, the county seat of Fresno at the time, weren't aware of what was happening. The disintegrating remnants of one or more landslide dams hit their town just before midnight on Christmas Eve, 1867, destroying it. That's why Fresno is now the county seat of Fresno County.

Just as USGS is urging us to prepare for a return of a storm similar to the 1861–62 flood, it might be prudent to prepare for a return of a storm similar to the 1867–68 flood, complete with large landslides and landslide dams. This is especially true in high landslide hazard zones.

## Preparing for the Next Mega-Flood

The website for the USGS's Multi Hazards Demonstration Project warns that the 1861–62 storm was not a freak event, was not the last time that California will experience such a severe storm, and was not the worst case. The geologic record shows that six mega-storms more severe than the 1861–62 flood have struck California in the last 1,800 years. There is absolutely no reason to believe that similar events won't occur again.

As detailed in Table 1, a huge flood strikes Southern California approximately every 200 years (for more information on these floods, see the section of this document that describes the California

6

mega-floods). The flood presumably strikes Northern and Central California as well, but the research has largely been done in Southern California. That storm cycle appears to be associated with the roughly 208-year cycle of solar activity (the Suess Cycle). The result seems to be that as solar activity decreases, the climate cools; and this shifts the prevailing wind patterns and associated storm tracks toward the Equator. Whatever the mechanism for these mega-floods, the most complete record that we have of their occurrence is from the Santa Barbara Basin off Southern California.

Table 1. Mega-floods in Southern California.

| Approximate Date of Flood |
|---|
| 212 |
| 440 |
| 603 |
| 1029 |
| 1418 |
| 1605 |

The bicentennial flooding in Southern California was skipped only three times since 212 and never twice in a row. The last skip was in the early 1800s, leading researchers in the Santa Barbara Basin to conclude that *we foresee the possibility for a historically unprecedented flooding in Southern California during the first half of the 21st century.*

The skip in the early 1800s may have only been a skip from the local perspective of the Santa Barbara Basin. The Central Valley experienced a huge flood in 1805, one that was even bigger than the huge 1861–62 flood. Perhaps the storm track that year just didn't extend far enough south to be recorded in the Santa Barbara Basin.

Based on past experience, we can expect floods resulting from huge storms like those of 1861–62 and 1867–68 to last for up to three months. That's very serious, and we certainly aren't prepared at this time for such events. However, they're similar to the type of floods that we've experienced in more recent times, just much bigger.

Judging from what we know of the 1605 flood, mega-floods appear to be a much longer-term type of event. Once such an event begins, it can last for up to 10 years. During that period, multiple episodes of flooding and extreme runoff can occur as well as other unusual climatic events. The 1605 flood was associated with a large-scale change in climate that affected the Northern Hemisphere from roughly 1600–1610.

The decade 1600–1609 stands out as the coldest in a 570-year (A.D. 1400–1970) comprehensive record of summer temperatures across the Northern Hemisphere, based on tree-ring and ice-core data. The years 1601 and 1605 produced unusually narrow tree rings in the Sierra, suggesting very cold growing seasons.

The 16-year period from 1597–1613 in the Sacramento River Basin had the maximum reconstructed riverflow for the 420-year (1560–1980) time period. There was a major flood on several Northern California rivers, including the Salmon and Klamath, in about 1600. Those rivers wouldn't see another flood that big until the December 1964 flood.

The year 1602 was the end of a 36-year drought (1566–1602) drought in the Sierra. Mono Lake filled to record levels from approximately 1600–1650. Those were the highest levels of the past millennium. The Mojave River terminates at the Silver Lake playa in the Mojave Desert. (That playa is located along Interstate 15, just north of the town of Baker.) At very infrequent times, the Mojave River delivers so much water to the playa that it forms a perennial lake. The last time that this happened was likely during approximately the 1600–1610 time period.

Along the coast of California in the Santa Barbara area, 1604 was the fourth wettest year, and 1601–1611 was the third-wettest 11-year period in a 620-year (A.D. 1366–1985) reconstruction of precipitation. There was a major flood on the Santa Ana River in Orange County in about 1600. That river wouldn't see another flood that big until the January 22, 1862 flood. Severe flooding occurred around Mexico City in 1604 and 1607.

It's tempting to think that there would also have been floods in the Tulare Lake Basin during the 1600–1610 time period since the regions to our north, east, and south were experiencing immense precipitation events at that time. The Tulare Lake Basin would have been under the same general storm tracks. However, we haven't found records of such floods.

If the mega-storms hold to their past schedule, the next one is expected to return within the next few decades. It may or may not be prudent to prepare for a return of a mega-storm similar to the 1605 or 1805 events. That's a question for risk managers to decide. The good news is that by preparing for events such as a return of the 1861–62 or 1867–68 storms, we'd be in a much better position to deal with a mega-storm should it materialize.

## Key Findings about Droughts

This document describes what we know about approximately 31 multi-year droughts.

One of the findings is that it's surprisingly common for floods to occur during droughts. That seems counter-intuitive, but it happens repeatedly.

The term "drought" is commonly used in two different ways in the Tulare Lake Basin. Traditionally, the term "drought" has been used to reflect precipitation that is significantly less than average. That is generally how the term drought is used in this document — a period of significantly less than average precipitation. Typically this condition has to last for at least two years before it is recognized as a drought. This type of drought ends when precipitation returns to approximately average or above-average conditions. The 1976–77 drought is an example of this type of drought.

The other use of the term "drought" refers to conditions when, for various reasons, we don't have all the water that we feel we are entitled to. In this kind of drought, 1) available water supply fails to meet demand, and 2) we perceive that water that is rightfully ours is being used somewhere else. If "they" would just let us have our water, then we would be better able to meet our needs. This type of drought has two components: hydrologic and socio-political. The Tulare Lake Basin has had a lot of this sort of drought in recent years. It is more challenging to mark the end of this type of drought. Precipitation can return to average or even above-average conditions, but there still isn't enough water to meet demand. The latter part of the 2007–09 drought is an example of this type of drought.

In the 2007–09 drought, water years 2007 and 2008 were drought years by the traditional definition. The runoff was so low in those years that the state's water year index rated those years as critically dry. The 2007–09 drought was California's first drought for which a statewide proclamation of drought emergency was issued.

That turned out to be critical. When precipitation returned to near-average or above-average, it was hard politically for the governor to declare an end to the drought. There clearly wasn't enough water to go around. It wasn't until March 30, 2011, after an incredibly wet winter, that the state of emergency was finally rescinded. That was long after the end of the hydrologic drought. The 2007–09 drought had morphed from the traditional type of drought (below-average precipitation) into the socio-political type (we're entitled to more water than we're getting).

When surface supplies are inadequate to meet the demand, we pump out of the groundwater aquifer. When surface supplies allow, water districts work to recharge the groundwater aquifer. In recent years, it has become increasingly apparent that we are withdrawing more than we are returning, we have a groundwater overdraft. Our demand for water is dramatically higher than the surface supply, so we are mining the underground supply. By that definition, we are effectively in drought conditions most of the time, even when precipitation is above average.

The groundwater supplies of the San Joaquin Valley are being depleted by an average of over 2.8 million acre-feet per year. For perspective, that 2.8 million acre-feet overdraft is nearly as great (80%) as the combined annual flow of the two largest rivers in the southern San Joaquin Valley. The San Joaquin River produces a long-term average annual flow (measured at Millerton Dam) of 1.8 million acre-feet per year. The Kings River has an average annual flow (measured at Pine Flat Dam) of 1.7 million acre-feet. Together, these two rivers produce an average of 3.5 million acre-feet of water.

In the San Joaquin Valley, water for agriculture, cities, rivers, wildlife refuges, etc. comes from three sources:
1. The most sustainable and local of these three is the Sierra, mainly in the form of water from the Tuolumne, Merced, San Joaquin, Kings, Kaweah, and Kern Rivers.
2. The second source of our water is exports from the Sacramento–San Joaquin River Delta, which is the immediate source of water that we import from Northern California via the Delta-Mendota Canal and California Aqueduct. That water, moved south at considerable expense, is increasingly fought over and hard to get.
3. The third source is what we pump from the groundwater aquifer, much of which is never replaced. A century ago, much of the valley had groundwater almost to the surface; artesian wells were common. Now, many areas have been mined for water to a depth of several hundred feet. The groundwater situation in Tulare County is particularly well studied and understood.

The implications of this are hard to escape. Even if our snowpack were to remain stable, the groundwater table in the San Joaquin Valley will continue to drop due to the overdraft. One of the consequences of this is that the wells will continue to go ever deeper and the cost of pumping will continue to rise. Eventually the pumping will be limited by supply and demand. Agriculture (the valley's single biggest water user) will be forced to reduce its reliance on groundwater.

The consequences of overdrawing the groundwater aquifer aren't just financial. Subsidence in the San Joaquin Valley is one of the great changes that human activity has imposed on the environment. The San Joaquin Valley has the largest vertical subsidence (a maximum of 29.7 feet, or more than 28 feet, depending on the source) and the largest areal extent (5,400 square miles, or 5,200 square miles, depending on the source) of subsidence in the world because of groundwater withdrawal.

Those measurements were as of 1991. Further significant subsidence has occurred along the western side of the Tulare Lake Basin in the ensuing years, particularly in Kern, Kings, and Fresno Counties. It has been speculated that some areas in those counties have now subsided upwards of 50 or 60 feet. USGS is currently conducting a study to measure the actual amount.

With such a huge shortage of water in the Tulare Lake Basin, it would seem that we would be anxious to hold onto all the water that we could. However, it turns out that in wet years, we go to considerable efforts to move water out of the basin. This is done in order to keep the rivers from following their natural course back to the Buena Vista and Tulare Lakebeds.

As discussed above, we as a society have decided that we would rather use those lakebeds for growing crops than for storing water as the natural reservoirs that they were until the late nineteenth century. At the time of Euro-American arrival in the region (1840s), the Kings was flowing down the south side of its delta and into Tulare Lake. Now we have constructed waterworks so that it is possible to divert part of the Kings River flow north through the San Joaquin River into San Francisco Bay. It keeps that water out of Tulare Lake, but it is essentially a loss from the point of view of Tulare Lake Basin water users.

With the construction of Pine Flat Dam in 1954, the need to divert water through this system was greatly reduced. Even so, diversions through this system have occurred in 38% of the years since the dam was completed. Likewise, the Kern naturally flowed into Buena Vista Lake and then overflowed into Tulare Lake. Now it is possible to divert part of the Kaweah, Tule, and Kern River floodwaters and send them to the Los Angeles area.

Runoff is much larger in some years than in others, resulting in a greater need to export water out of the basin. For example, because of a huge snowpack, the runoff in 1983 was the largest since record-keeping began in 1894. As a result, a record 3.1 million acre-feet was exported from the Tulare Lake Basin that year. That was almost as much as the 3.5 million acre-feet combined average annual flow of the San Joaquin and Kings Rivers.

That isn't to say that exports (or inter-basin transfers) are necessarily a bad thing; just that we need to recognize that such diversions have consequences to the groundwater aquifer. Every time that water is transferred out of the Tulare Lake Basin, there is that much less water available for use or for groundwater recharge.

One of the big lessons learned from preparing this document is that our historic droughts haven't been nearly as bad as some that occurred prior to settlement. The 1928–34 drought (part of the 1922–34 drought) was arguably the longest and most severe drought to strike the Tulare Lake Basin during historic times. However, California was subject to prolonged, severe droughts from about A.D. 900–1400. The first epic drought lasted more than two centuries before the year 1112; the second drought lasted more than 140 years before 1350. Evidence of those mega-droughts is still surprisingly visible in places like Yosemite. The takeaway message is that what we think of as a long-term drought today is mild compared to these earlier mega-droughts.

## The Effect of Floods on Tulare Lake

For various reasons, there is no longer a lake in the Tulare Lakebed, at least in most years. For an explanation of those reasons, see the section of this document: Why is there no lake in the Tulare Lakebed today?

The story of Wildlife in and around Tulare Lake is summarized in that section of this document. The lake and its associated wetlands used to provide very valuable habitat for a wide variety of wildlife. In addition, the lake and wetlands provided biological connections among the various river and stream courses in the Tulare Lake Basin. It is hard for those of us living in the 21st century to wrap our minds around what that 19th century ecosystem was like; it was truly extraordinary.

One of the lessons learned in preparing this document was that the flood cycle of the Tulare Lake Basin was critical in maintaining that ecosystem; the floods provided sufficient water storage to keep the lake going through the drought or non-flood years (see Figure 10 and the section of this document on the Role of Floods in Maintaining Tulare Lake).

Once Tulare Lake (and the other four valley lakes) had been dried up, disintegration of this remarkably complex system was sealed with the damming of the four main rivers. The functioning infrastructure of this formerly biodiverse ecoregion was so badly broken that it resulted in the loss of most of the wetland habitat and nearly all of the biological connectivity between the watersheds in the high country and the lowland floodplains. Water-dependent habitats on the adjacent land, particularly on the Kaweah Delta and other riparian corridors, were also significantly degraded during the ensuing decades.

The loss of the Tulare Lake ecosystem affected even protected areas like the national parks. For example, we speculate that the relict populations of beavers, mink, and river otters that hung on in the national parks became isolated from populations elsewhere in the parks and the basin. Their numbers gradually declined, and some of those species may now be extinct in the parks. We speculate that a similar problem occurred with many populations of birds, fish, and other animals.

Despite the best efforts of water managers, floodwaters still make it to the Tulare Lakebed on occasion, especially in heavy runoff years. Migratory waterfowl and shorebirds still return in large numbers when Tulare Lake has a major flood. Very large numbers returned when the lake had its last great reappearance in 1937–46. There was a significant but smaller appearance of waterfowl and shorebirds during the floods of 1982–83 and 1997. While exciting to see, these bird congregations are ephemeral and move on once the floodwaters recede.

Floods — with the water that they brought — created a marvelous ecosystem in the Tulare Lake Basin. Reminders of that ecosystem survive in disjointed preserves in the valley, in the foothills, and in the Sierra. The framework of the hydrologic system that powered that ecosystem still exists today. On occasion, flooding can recreate a portion of Tulare Lake. But just adding water to that lakebed is not enough to recreate the complex ecosystem that once existed. The associated habitat is highly degraded, and the ability of the ecosystem to provide connections among the various river and stream courses in the Tulare Lake Basin has largely been lost.

## Conclusion

The Tulare Lake Basin has a fascinating hydrologic history. We can and should learn from that history. Society increasingly relies on a stability of climate, but that is not supported by the last 2,000 years of history in the area covered by this study. What is reliable about our climate is its extreme and relentless variability. We should not be complacent about floods and droughts. We need to learn from our past as we plan for the future. The author hopes you enjoy and learn from this document.

# Acknowledgements

We owe a debt of gratitude to all the National Park Service (NPS) employees who have recorded their observations in the superintendent's monthly and annual reports over the years. Ward Eldredge was very helpful in going through the parks' archives to provide access to those old reports.

The U.S. Weather Bureau and its successor, the National Weather Service (NWS), have collected and preserved a wealth of observations from throughout the San Joaquin Valley. Gary Sanger at the NWS forecast office in Hanford was very helpful in providing access to that data. Gary also consulted on several parts of this document.

Tony Caprio, George Durkee, Annie Esperanza, Dave Graber, Linda Mutch, and other National Park employees provided insights on Sierra climate change and fire history. George Durkee was also very helpful in sleuthing out the routes that early explorers and fur trappers took when they came into the Tulare Lake Basin. Keith Hamm shared the flood information that he found while researching logging history.

René Ardesch, Mo Basham, Anne Birkholz, Tony Caprio, Ned Kelleher, Cal Kessner, Ed Nelson, Nate Stephenson, Kirk Stiltz, Bill Sullivan, Jerry Torres, Katrina Young, and other national park employees were generous in providing their recollections of past floods and droughts.

About 97% of Sequoia and Kings Canyon National Parks is designated wilderness. We know about the storms and floods in that part of the parks in large part because of the wilderness rangers who live and work there during the summer. Wilderness rangers Dave Alexander, George Durkee, Erika Jostad, Bob Meadows, and Cindy Wood all contributed to this report. Many others contributed by recording their observations each year. The meadow monitoring staff, especially Erik Frenzel, have become an important go-to source for trail conditions in the wilderness. Trail crews and their supervisors Tyler Johnson and David Karplus have been another valuable source of information on storms and floods in the wilderness.

Several retired national park employees have contributed from their years of knowledge. Among those were Manuel Andrade, Leroy Maloy, and Harold Werner. Steve Moffit, Sequoia National Park's former trails foreman, was particularly helpful.

Thank goodness for newspapers. Those have proved an invaluable source for information on past floods and droughts in the Tulare Lake Basin. The *Visalia Times-Delta* and its predecessors were particularly helpful. Several newspapers put together specials on particular floods. For example, John and Sarah Elliot collected stories of the 1955 flood and published those in *The Kaweah Commonwealth*. Dody and Tom Marshall at the Three Rivers Historical Museum were helpful in providing access to two old issues of the *Three Rivers Current* in the museum's archives. The *Exeter Sun* produced a special edition on the 1955 flood.

*Los Tulares*, the quarterly bulletin of the Tulare County Historical Society, was a treasure trove of information.

The U.S. Army Corps of Engineers (USACE) was very helpful in preparing this report: especially Calvin Foster, Wayne Johnson, Christy Jones, Kyle Keer, and Virginia Rynk at the Sacramento District; Phil Deffenbaugh and Valerie McKay at Lake Kaweah; and Mike LaFrentz at Pine Flat.

Others from the Sacramento District also provided valuable assistance, but requested that their names not be listed.

Joyce Fernandez in the Sacramento Office of the U.S. Bureau of Reclamation (USBR) found what appears to be the last surviving copy of a key Tulare Lake document. She spent many hours scanning that document so that others could benefit from the information in it.

Josh Courter, the Western District Divide hydrologist on the Sequoia National Forest, assisted with analysis of the debris slides on the Kern River. Jim Roche, the hydrologist at Yosemite National Park, provided insights on flooding and erosion on the Merced River. Kerry Arroues with the Hanford office of the Natural Resources Conservation Service contributed his much needed soil science expertise, especially working with the tricky Tulare Lakebed issues.

Jerry DeGraff, a geologist with the Sierra National Forest, was an invaluable resource for understanding debris flows. Robert Meyer, a forester/hydrologist, also assisted in the review of the debris flow section of this document. Robert was the surface waters specialist for California for 15 years as well as a flood specialist for the USGS.

Roland Knapp is a research biologist specializing in aquatic ecosystems of the Sierra. His expertise was very helpful in understanding the historic role of furbearers in the higher elevations of the Tulare Lake Basin.

Michael Dettinger (USGS/Scripps) is one of the country's leading atmospheric river researchers. He gave his time to consult on atmospheric rivers and the 1905–06 snowpack. Arndt Schimmelmann (Indiana University) is one of the country's leading paleo-climatologists. He gave his time to review the California mega-flood portion of this document.

Sarge Green has many years of experience working with agencies such as the California Regional Water Quality Control Board, resource conservation districts, and irrigation districts. He is currently program director of the California Water Institute at CSU Fresno. Sarge was the perfect person to review the Groundwater Overdraft section of this document.

Mark Tilchen and Valerie McKay put together the very relevant book: *Floods of the Kaweah*. Denis Kearns contributed an out-of-print book that the State Disaster Office prepared on the 1955 flood. Bill Templin (a former USGS hydrologist) was instrumental in documenting the flood and debris flow that occurred on Lewis Creek in 2008, citizen science at work. Bill also knew the history of stream gages on the South Fork Kings and was very helpful in providing access to that data.

Rob Hansen was generous in sharing his treasure trove of information about Tulare Lake and the San Joaquin Valley. Terry Ommen was incredibly helpful in providing historical accounts of floods in Tulare County. I really couldn't have pulled this document together without the support of people like Rob and Terry who have spent years collecting all the primary source material and then entrusting that to my care.

Tulare Basin Wildlife Partners was supportive of this document in many ways. Several of the people active in that organization (particularly Rob Hansen, Carole Combs, Sarah Campe, and Niki Woodard) contributed material for the document, reviewed it, and generally provided encouragement. In addition, they were the first organization to post the electronic version of this document, doing so through their new Tulare Basin Watershed Initiative (TBWI). Niki was the

14

webmaster for the TBWI. She also volunteered her considerable design skills to laying out the front and back covers of this document.

More than 65 people contributed their time to reviewing this document, far more than can be credited here. Julie Allen and Anne Birkholz were both instrumental in beating the Summary section into shape. Karen Folger reviewed the entire document; she was a great reviewer.

My sister Karen Austin was tenacious in reviewing the entire document, checking grammar, punctuation, content, citations, the whole shebang. Karen was awesome as an editor, and you couldn't ask for a finer sister. I only wish that she had lived to see this document printed.

Many people reviewed particular floods. Harold Werner, the national parks' former wildlife ecologist, reviewed the wildlife section. Jane Allen reviewed the section on the Native Americans. Sarah Campe pointed out the need to write the document (well, rewrite it) so that it could be understood by those who didn't have a national park background.

Tony Caprio was an insightful reviewer. Tony also had the skill set to fix errors detected in maps that were acquired from historic and other secondary sources. That was very handy.

Julie Allen is a delightfully detailed and constructive reviewer. She is also very well connected in the watershed planning world and was a never-ending source of encouragement during the writing of this document.

Charisse Sydoriak reviewed the entire document, looking at it from the perspective of an NPS manager. Colleen Bathe also reviewed the entire document, looking at it from the perspective of a Sequoia Natural History Association manager.

Dave Graber, the NPS regional chief scientist, reviewed various sections of this document, especially those dealing with wildlife and ecological relationships. Dave makes a great reviewer; he knows his subject matter, and he quickly finds the weakness in an argument. In addition to being a reviewer, Dave provided guidance and encouragement in the writing and publication of this document.

Bob Meadows is a walking encyclopedia of weather knowledge on the national parks. He was especially helpful in reviewing this document and straightening out long-festering data errors.

Gary Sanger reviewed the meteorological sections. Jennie Skancke reviewed the hydrology sections as did Ned Andrews and Roger Bales. The Sacramento District of the USACE reviewed the key findings.

Terry Ommen reviewed the entire document, looking at it through the lens of an experienced writer. Rob Hansen was a spectacular reviewer; he was particularly critical in reviewing the Tulare Lake sections. Nobody knows Tulare Lake and its history like Rob does.

Bill Tweed, the national parks' chief park naturalist from 1996–2006, has been a singular inspiration for this document; he built the foundation for it in so many ways. In many years, he was the force that motivated the park staff to produce the superintendent's annual report. He has researched numerous papers on weather and climate over the years (several of which are shamelessly plagiarized in this document). Bill was a reviewer and consultant on many sections of this document. And most of all, he has preached the need to be aware of the world and the weather around us.

As described above, many individuals and agencies have assisted with the preparation of this document. Including their names in this Acknowledgement section does not imply that they have reviewed this entire document or blessed it. Furthermore, any errors that remain in this document are the sole responsibility of the author.

# Background Material

## Human Perspective

By its nature, this document illustrates floods and droughts as seen through the lens of humans. Their stories reflect the human dimension, what people thought was worth recording.

Furthermore, the occurrence of floods has changed with infrastructure (dams, levee maintenance, more flood-resistant bridges, etc.). For example, as dams were constructed, flooding became less frequent downstream. When levees weren't maintained (or when a conduit became plugged), flooding could become much worse even though the hydrology didn't change.

## Peer Review Process

After long consideration, the National Park Service declined to conduct a formal peer review of this document. However, this document did undergo an extended period of informal peer review, lasting from January 2011 through October 2012. During that period, it was reviewed, in part or whole, by nearly 70 individuals and agencies. The reviewers were selected based on interest and expertise. Some of the reviewers were directly involved in the collection, analysis, or reporting of some of the information used in this document. The author served as the peer review manager and maintains the records of the review process.

## Citation of Source Material

This document was designed to tell the story of water in the Tulare Lake Basin. Our basin has experienced a lot of floods and droughts during the last 2,000 years. A rich trove of stories and data has survived to tell the story of those events. The Literature Cited section of this document contains over 1,100 source citations documenting those stories and data.

A National Park Service reviewer observed that there are "uncounted source citations" missing from this document. While the comment may be a bit melodramatic, that does reflect on the intentional design of this document. It was not meant to be a scientific treatise of the subject matter. (See the Purpose section of this document.)

If this had been a scientific treatise, it would have been necessary to strictly adhere to the rule of only stating what could be backed directly by citation. While theoretically possible, that would have been an editorial challenge. This document is dense with facts. In addition to 21 figures and 89 tables, it contains 3,000 or so paragraphs.

Many of those paragraphs were composed using information from multiple sources. Some paragraphs have four or more sources. The challenge of citing those sources each time that they were used would have been made more difficult because they are so intertwined. The same source publication can be the source for dozens of sentences and tables that are scattered throughout this document. Individually calling out those citations would have made the document significantly less readable, so that generally wasn't done.

Facts that can readily be found by searching the Internet have not always been cited. However, care has generally been taken to cite the hundreds of obscure sources that cannot be found with a simple online search.

Facts that are contained within the records management and archive systems of Sequoia and Kings Canyon National Parks have not been formally cited; these have only been identified by informal reference within the body of the document. Personal communications have also not been formally cited; these have only been identified by informal reference.

In order to avoid interfering with the flow of the document, the formal citations are contained as endnotes in a Literature Cited section at the end of the document rather than as footnotes.

## Reliability of Source Material

This document uses a mixture of hard data and personal observations to tell the story of floods and droughts in the Tulare Lake Basin. Hard data alone would be insufficient. In part, that is a reflection of the nature of the document's design. (See the Purpose section of this document.) Personal observations are necessary to provide the human dimension of the story.

The heavy reliance on personal observations also reflects that there is limited data available to tell the complete story of floods and droughts in our basin, especially as we look further back in our history. In a general sense, data can be used to give the general outline of what has happened. Then the story has to be filled in by the observations of people who were there to witness the events.

However, this brings up a problem: Personal observations, even those of sworn eye witnesses, are not as reliable as those of machines. Most of the time people probably get most of their stories right, but sometimes they get parts of their stories wrong.

So the reader is cautioned to take these stories with a grain of salt. Wherever possible, stories used in this document have been cross-checked against other stories and available data.

## Disclaimer Regarding Subject Matter Expertise

This is not meant to be a technical document or scientific treatise. (See the Purpose section of this document.) The author is neither a trained historian nor a physical scientist, and is not recognized by any entity as such. To use NPS jargon, this work cannot be ascribed to "a professional in the field of inquiry."

## Note about Completeness

This is not intended to be a complete document. By its design, it is a collection of the documentation that we have been able to find, brought together in a single place. This is meant to be a reference source for others to turn to.

There is a lot more source material out there. You could make a life's work out of this; it's hard to know when to stop tracking down loose ends. As Oscar Wilde said: books are never finished, they are merely abandoned.

But we now have a moderately complete listing of the major floods and droughts that have occurred in the Tulare Lake Basin since about 1850. We also know some of the big floods and droughts back as far as about the year A.D. 212. In general, the farther back that you look, the bigger the flood or drought has to be in order to be detectable.

## National Park Service Involvement

This document was not commissioned or authorized by the National Park Service. It was researched and written as a volunteer effort by the author, primarily on weekends and evenings.

However, Sequoia and Kings National Parks did support the preparation of this document in various significant ways. The national parks provided full access to its historical files; that proved invaluable. Following NPS national policy, the national parks also allowed incidental use of government office equipment (copiers, printers, and computers) during non-duty hours.

This document is focused on Sequoia and Kings Canyon National Parks; as shown in Figure 4, those parks occupy the headwaters of the Tulare Lake Basin. The National Park Service considered publishing this document in its technical report series. However, that proved to be infeasible. The primary obstacles were:

- This document was designed as more of a historical than a traditional technical paper. (See the Purpose section of this document.) That made it less than an ideal fit for the NPS technical report series.
- If this document were a traditional technical paper to be included in the NPS technical report series, it would have to strictly adhere to the rule of only stating what can be backed directly by citation. However, this document intentionally does not cite all of the sources that it is based on. (See the Citation of Source Material section of this document.)
- This document identifies some of the flood risks in the Tulare Lake Basin that are not mitigated by the four federal reservoirs. This is a factual condition that is based on facts and has been thoroughly reviewed. However, if the NPS were to publish this document, that might be mistaken as implying that the national parks thought that local authorities should be acting to mitigate those risks. That would have the appearance of the parks criticizing neighboring agencies, something that generally isn't done.

Therefore, the National Park Service chose not to publish or endorse this document.

## Maps of the Tulare Lake Basin

Figure 2 illustrates California's various water basins, including the Tulare Lake Basin.

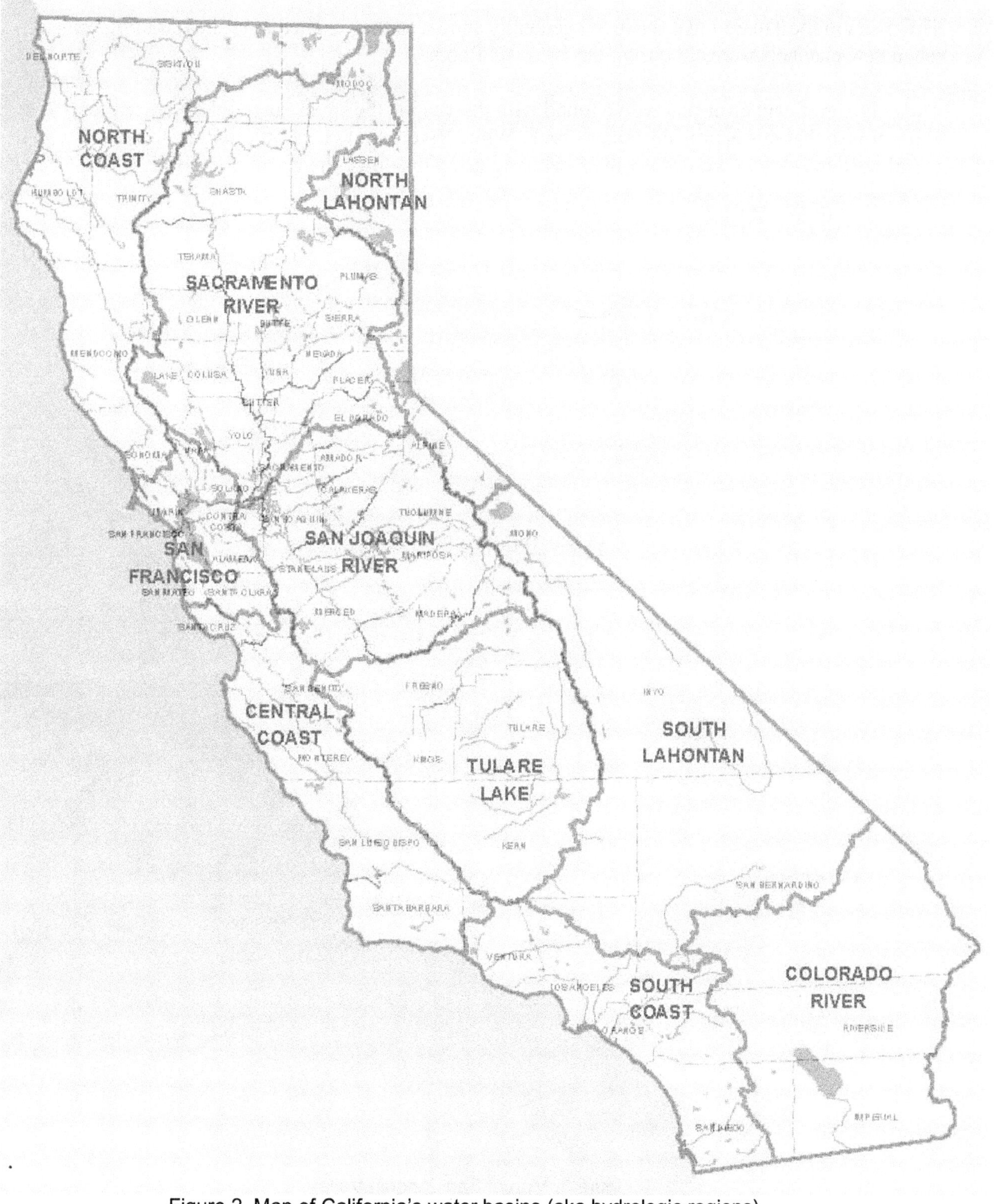

Figure 2. Map of California's water basins (aka hydrologic regions).
Source: California Department of Water Resources

**Floods and Droughts in the Tulare Lake Basin**
Background Material

Figure 3 illustrates the entire San Joaquin Valley, including both the San Joaquin River Basin and the Tulare Lake Basin.

The rivers in the Tulare Lake Basin are generally shown as they existed after the rerouting caused when new channels were cut during the 1861–62 flood.

The only significant error that we've found with this map is the stream marked "White River." Rob Hansen says that this is probably Deer Creek. White River would be a little farther south and would run almost due east-to-west rather than swinging north to approach very near Tule River like Deer Creek does.

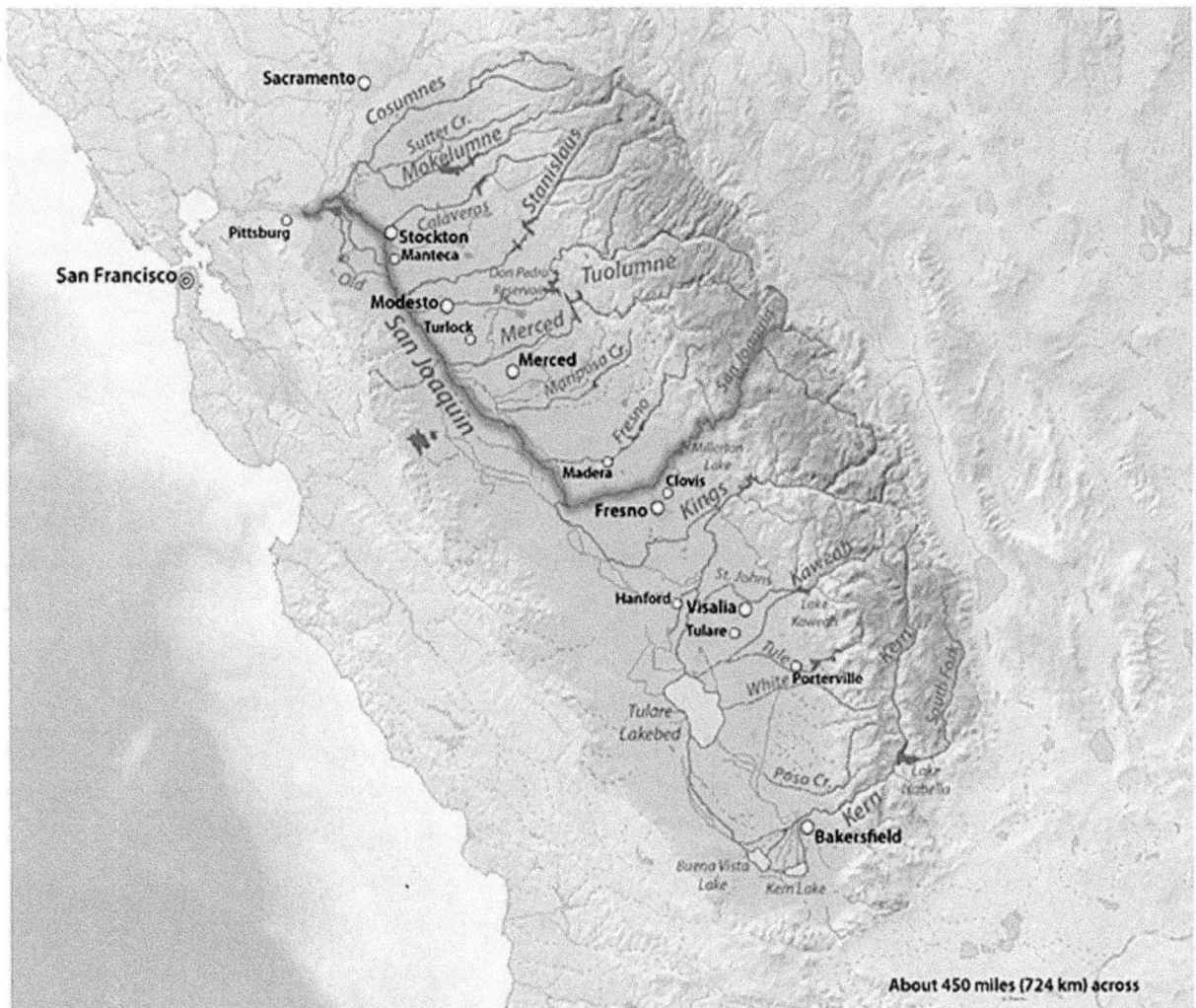

Figure 3. Map of San Joaquin Valley.
Source: Wikimedia Commons

Figure 4 is the best map that we've found of the Tulare Lake Basin. It is useful for providing an overview of the basin and its major features. However, it was prepared by others and contained some errors. The labeling errors have been corrected. The other errors could not be corrected without access to the GIS source data. In using this map, the reader should be aware of the following:

- The map doesn't represent conditions at any particular point in time. River locations have changed over time. Lakes have dried up. Reservoirs have been constructed. Cities have grown up. In general, the map shows streams, rivers, and reservoirs as they are today.
- Tulare Lake is shown more or less at its high stand in 1862 at elevation of 216 feet. However, the lake is not precisely located. It is generally too far east and too close to Hanford.
- Mill Creek is shown as a significant stream course. It is actually less than a seasonal stream.
- Jerry Slough has not carried floodwaters since 1952. Its course is not accurately mapped.

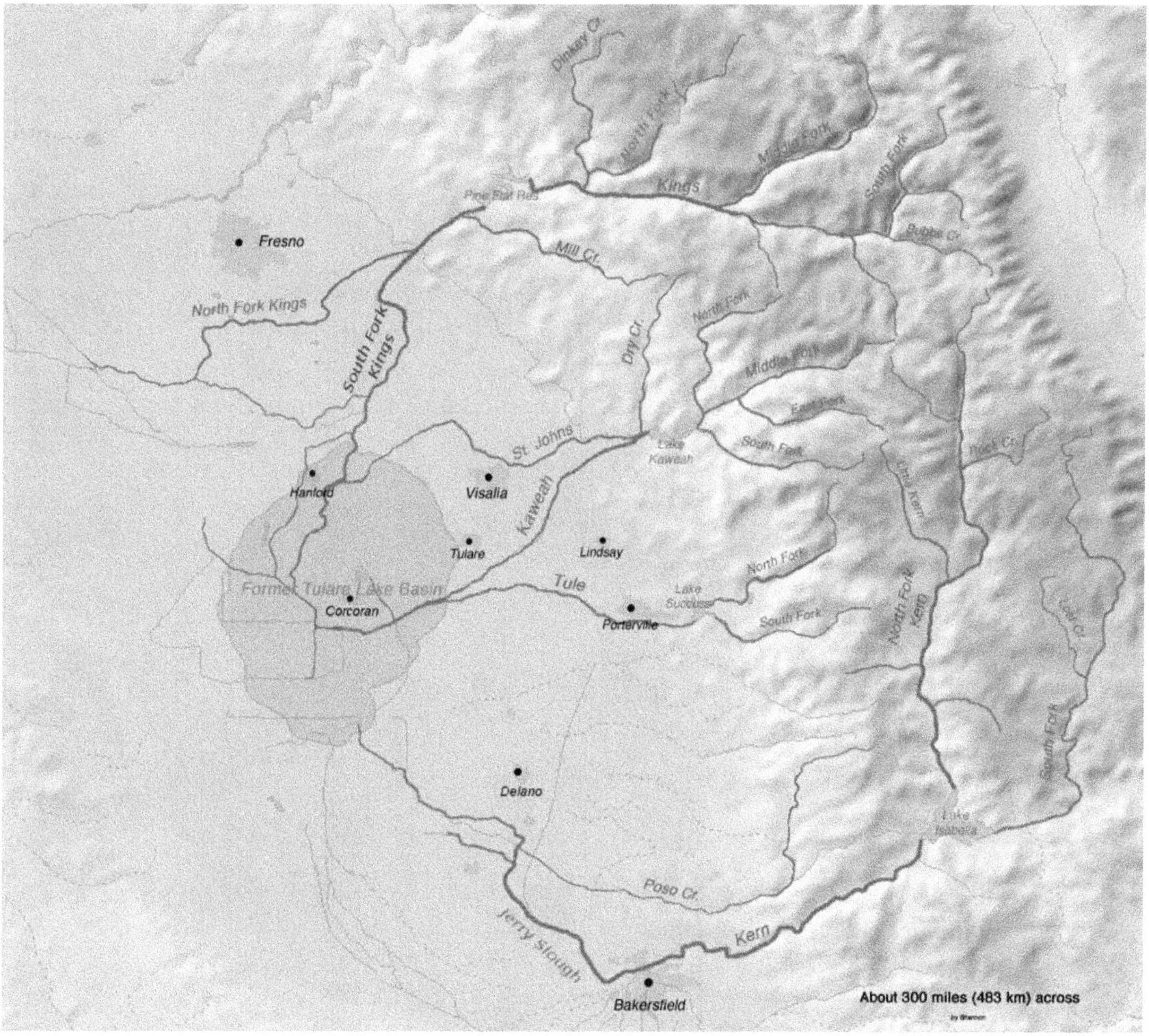

Figure 4. Map of Tulare Lake Basin.
Source: Wikimedia Commons with corrections by Tony Caprio

Figure 5 shows the natural communities of the Tulare Lake Basin as they were in about 1850.

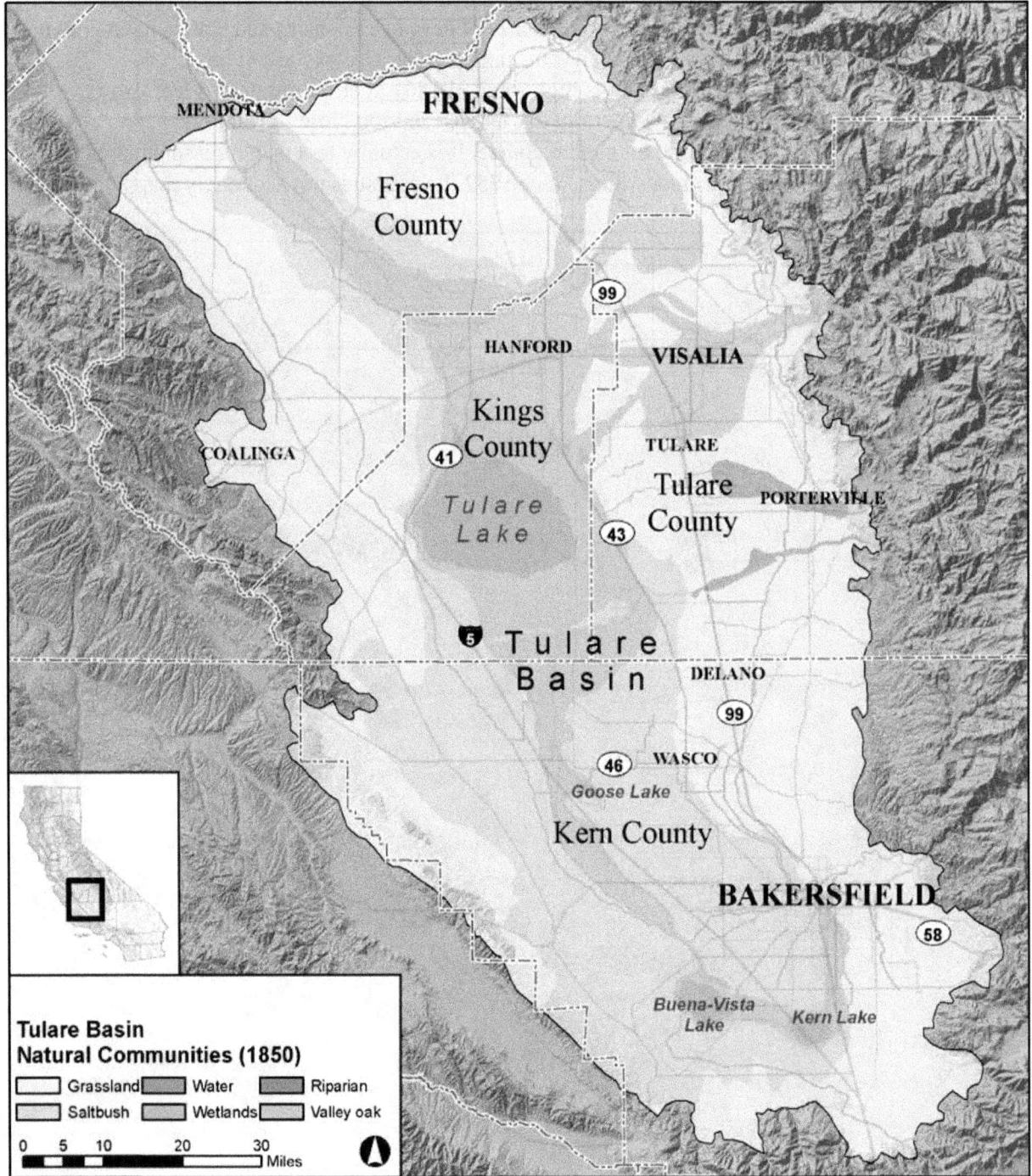

Figure 5. Map of Tulare Lake Basin natural communities.
Source: Scott Phillips, CSU Stanislaus

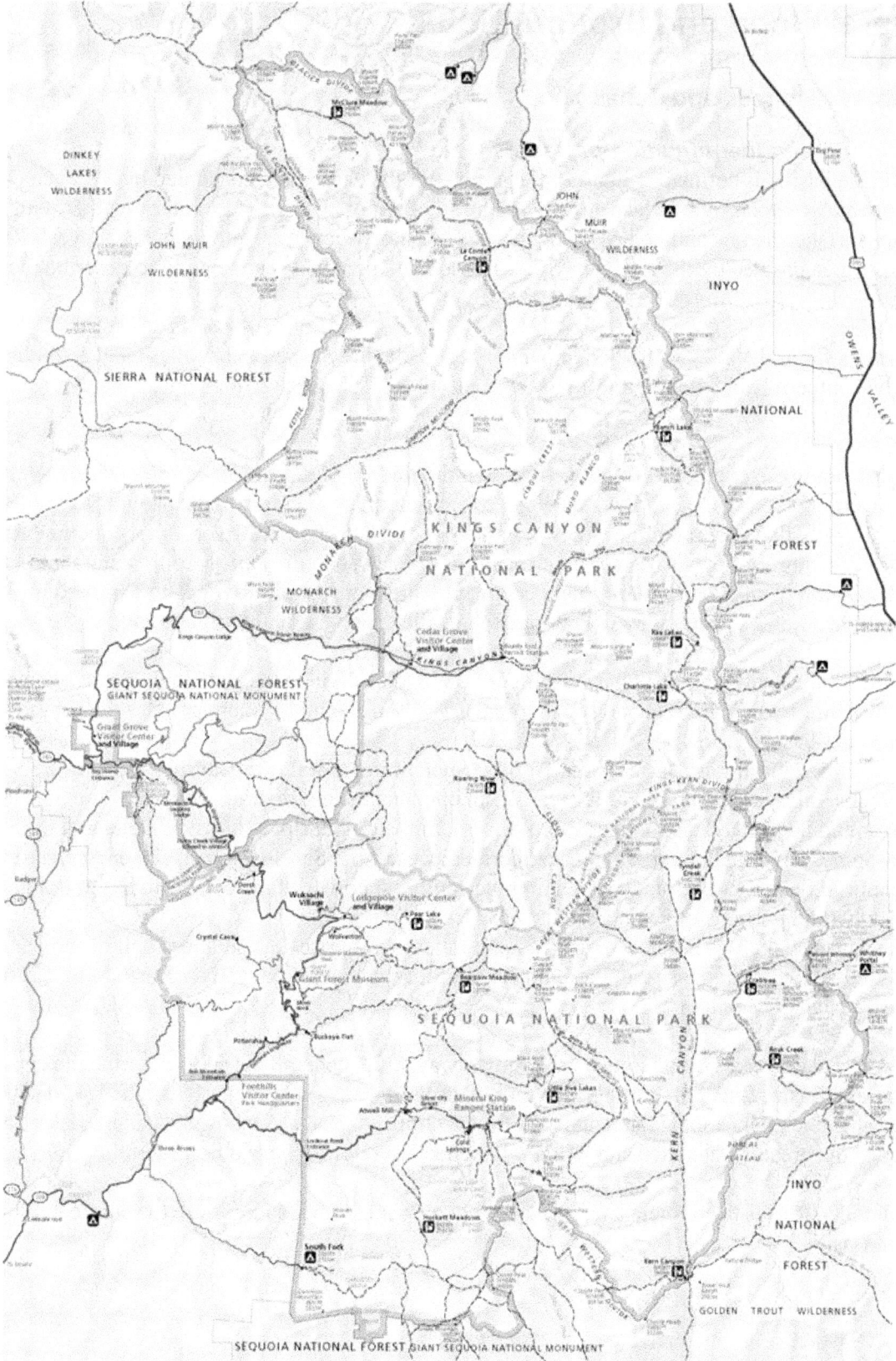

Figure 6. Map of Sequoia and Kings Canyon National Parks.

24

# General Flood and Drought Notes

## Basins, Watersheds and Deltas

### *Description and Identification of Basins*

A "drainage basin" is defined as a part of the surface of the earth that is occupied by a drainage system, which consists of a surface stream or a body of impounded surface water together with all tributary surface streams and bodies of impounded surface water. An example is the Kaweah River Basin, the area drained by the Kaweah River. The term "watershed" can be used to describe the same feature.[1]

California's Central Valley is divided into a northern half (the Sacramento Valley) and a southern half. The southern half, extending from Sacramento to the Tehachapi Mountains, is known as the San Joaquin Valley.

The San Joaquin River emerges from the Sierra in the middle of its namesake valley, and then turns north toward the San Francisco Bay (see Figure 3). Sometimes the term "San Joaquin Valley" is used to describe only that portion of the valley occupied by the San Joaquin River. In this document, the term "San Joaquin Valley" is used to describe all of the San Joaquin Valley from Sacramento to the Tehachapis, not just the northern portion occupied by the San Joaquin River. The San Joaquin Valley covers about 31,800 square miles. The floor of the valley is about 10,000 square miles.[2]

The California Department of Water Resources (DWR) has divided the state into 10 surface water basins or major hydrologic regions (see Figure 2). The San Joaquin Valley includes two of those surface water basins:

1. The San Joaquin River Basin (about 15,600 square miles)[3] drains the northern half of the San Joaquin Valley. The San Joaquin River Basin contains the entire drainage area of the San Joaquin River and its tributaries. It extends from the Sacramento–San Joaquin River Delta and the Cosumnes River in the north to the southern reaches of the San Joaquin Basin, encompassing the area from Sacramento County (including the southeast corner of the county itself) to Madera County (and portions of Fresno County).
2. The Tulare Lake Basin (about 16,200 square miles)[4] comprises the southern half of the San Joaquin Valley. It ranges from the southern limit of the San Joaquin River Basin to the crest of the Tehachapi Mountains.

The term "Tulare Lake Basin" is also used in two ways. It is generally used to describe the entire Tulare Lake Basin. But at other times, the term is used to describe the Tulare Lakebed (790 square miles). This document adopts the language of the California Water Plan, using "Tulare Lake Basin" to refer to the greater watershed and "Tulare Lakebed" to refer to the lakebed itself.[5]

Although DWR uses the nomenclature "Tulare Lake Basin," other sources use the shortened form, "Tulare Basin."

## National Park Watersheds and Rivers
Sequoia and Kings Canyon National Parks are drained by the following five rivers:
1. San Joaquin River (South Fork)
2. Kings River (Middle Fork and South Fork)
3. Kaweah River (North Fork, Marble Fork, Middle Fork, East Fork, and South Fork)
4. Tule River (North Fork)
5. Kern River (North Fork)

## Description and Identification of Deltas
When rivers flow west out of the Sierra onto the valley floor, they lose energy. When this occurs, they lose their ability to carry sediment and they often form fan-shaped deltas. In each flood, they deposit more sediment onto their delta. Or rather, that's what used to happen before the federal reservoirs were built on each of the major rivers: Kings, Kaweah, Tule, and Kern. Those reservoirs now function as sediment traps, intercepting much of the sediment load before the rivers can deliver it to their deltas.

For the bigger rivers, their deltas can be quite large, covering many square miles in area, and stretching out far across the valley floor. The Kings River Delta begins about where the town of Kingsburg is located; Visalia sits atop the Kaweah Delta.

The USACE estimates that the various channels of the Kaweah Delta (below McKay's Point) are able to absorb and distribute a flow of up to 5,500 cfs of water before flooding begins to occur. So when the discharge below Terminus Dam (including the flow from Dry Creek) exceeds that amount, then flooding in the Visalia area can be expected. Prior to construction of the dam, some sort of June flooding occurred locally every few years on average. The older homes in Visalia were built with floor levels several feet above ground level because of this routine flooding. Few houses had basements.

Water on the valley floor in the San Joaquin Valley is generally trying to get from higher elevations (near Bakersfield) to lower elevations (the San Francisco Bay). The deltas of the Kern and Kings Rivers that extend out from the foothills can be thought of as a series of ridges or impediments to this flow. These ridges act as sills or dams, creating lakes on the valley floor.

The streams that flow east from the Coast Ranges also form deltas. By an odd coincidence, the delta formed by the east-flowing Los Gatos Creek (aka Arroyo Pasajero) meets the much bigger delta formed by the west-flowing Kings River. The resulting sill has an elevation of 207 feet. Historically, that sill served to regulate the elevation of Tulare Lake in very wet years.

The Sacramento–San Joaquin River Delta is located at the confluence of the Sacramento and San Joaquin Rivers, immediately east of the upper arm of San Francisco Bay.

## Elevations
The elevations given in the literature about Tulare Lake levels must be treated carefully and do not necessarily represent what the elevations would be today with the current sea level reference datum. Some of the elevations are derived from surveys in the 1800s, and the sea level datum is usually not specified.

Sea level fluctuates from hour to hour and place to place. The sea level "datum" is the elevation that surveyors of a particular time and place use as their reference for a zero elevation. It is their vertical control point. When California was first settled, the state was on its own to establish a standard for what was the sea level reference datum. (Imagine the state engineer going to San Francisco Bay and taking the average of the tides.)

In any case, the early elevations of Tulare Lake and the surrounding area were made using the California State Engineering Department datum. By 1907, it was known that those elevations had to be reduced by 4.2 feet to get to the sea level datum as established by the U.S. Geological Survey (USGS).[6] By 1929, a standard sea level datum had been established across North America. The current sea level datum for North America (NAVD 88) was set in 1991.

S.T. Harding reconstructed Tulare Lake levels in 1949.[7] Harding determined that the elevations from the 1800s that were used by C.E. Grunsky in his graph of Tulare Lake levels[8] should be reduced by 4.2 feet to conform to the sea level datum that was being used in 1949.[9]

It's critical to know the sea level datum associated with any given elevation. Without knowing the datum that was used when measuring a particular elevation, the reader can't be sure what any given elevation really means.

An example of the confusion created when an author doesn't provide this information is that some literature continues to report Tulare Lake's high stand in 1862 as 220 feet.[10] Those authors are still using W.H. Hall's original measurements based on a sea level reference datum from the late 1800s; they just aren't saying so. If they had used a sea level datum from the early 1900s or later, they would have stated the lake's elevation as 216 feet. Sea level had been considered some 4 feet higher in Hall's day, so the elevation for Tulare Lake also appeared to be 4 feet higher.

Likewise, some authors still report the Tulare Lake sill as being at elevation 211 feet.[11] That's because they're also using the sea-level datum from the late 1800s. If they had used a sea level datum from the early 1900s or later, they would have stated the sill's elevation as 207 feet. Only in California could you have a lake with an elevation of 207 feet seemingly draining uphill over a sill with an elevation of 211 feet.

Conditions have changed significantly in the six decades since Harding's time. The most dramatic change is the land subsidence that has occurred along the western side of the Tulare Lake Basin, particularly in Kern, Kings, and Fresno Counties. It has been speculated that some areas in those counties have now subsided upwards of 50 or 60 feet.[12] See the section of this document on Groundwater Overdraft for a more detailed discussion of this topic.

However, for the sake of consistency, most major studies of Tulare Lake continue to use Harding's reconstructions, but note that they are doing this.[13, 14] This document also uses Harding's reconstructed elevations.

## Overview and Terminology

### *What Constitutes a Flood*

There's a surprising variety in what constitutes a flood. This document contains a definition of what constitutes a drought. However, it does not have an all-encompassing definition of a flood; that has proved too messy a concept to define.

USGS broadly defines a flood as an overflow or inundation that comes from a river or other body of water and causes or threatens damage.[15] That's almost like saying that a weed is any plant growing where you don't want it; you'll know it when you see it.

Some of our floods are obvious: a river overflows its banks or a downpour overwhelms a city's drainage system. At the other extreme, some of our floods have two components: hydrologic and socio-political. Society decides what their tolerance is for natural processes and where they are willing to let a river flow. Some of our floods are the result of water appearing at the wrong place at the wrong time. They're an inconvenience.

As one example, farmers wanted to drain Tulare Lake and keep it dry so that the lakebed could be used for agricultural purposes. They viewed Tulare Lake as an inconvenience, a nuisance to be prevented. Their viewpoint has prevailed. As a result, society has defined the presence of excess water in the lakebed as a flood. It's an odd situation, but it effectively represents society's values. Water managers go to great efforts to minimize this type of flood.

It's helpful to look at lakebed flooding from the point of view of the farmers there. Runoff was below average in both 1970 and 1971, bordering on drought conditions. No significant storm or flood event happened in either year. Despite this, flooding occurred in the Tulare Lakebed in both 1970 and 1971. How was that possible? By looking at Figure 11, you can see that this flooding was left over from the big 1969 flood.

Lakebed flooding is a social construct; it is counted based on the number of growing seasons that are missed. The lakebed was flooded for three growing seasons: 1969, 1970, and 1971. So this is counted as three floods from the perspective of the lakebed farmers, even though the flood event occurred only once. (Something similar happened in the lakebed in 1982–84 and 1997–99. In each of those cases, lakebed flooding continued into a non-flood year.)

From the perspective of the natural environment, events like this would not be a flood at all. Tulare Lake was a natural body of water that society is trying to prevent from reclaiming its lakebed. Whenever the lake reforms naturally, we call that a flood.

So 1971 was a year with no storm event. It experienced near-drought conditions. And yet it is counted as a flood year. This can be a bit counter-intuitive, even mind-bending. That is one of the reasons why it proved so difficult to come up with a single definition for what constitutes a flood.

## What Constitutes a Debris Flow

This document also addresses large-scale debris flows and landslide dams, both of which are associated with flood events. A debris flow is a moving mass of loose soil, rock, debris, and water that travels down a slope under the influence of gravity. It can carry material ranging in size from clay to exceptionally large boulders, and may contain a large amount of woody debris such as logs. Debris flows are typically associated with periods of heavy rainfall or rapid snowmelt and tend to worsen the effects of flooding that often accompany these events. Large-scale debris flows usually occur in small, steep stream channels and are often mistaken for floods. In fact, debris flows and flash floods often occur simultaneously in the same area.

## Water Year and Runoff Terminology

The term "water year" refers to the twelve-month period beginning October 1 in one year and ending September 30 of the following year. The water year is designated by the calendar year in which it ends. So water year 2011 would cover the 12-month period from October 1, 2010 through September 30, 2011.

By default, this document uses the term "year" to refer to calendar year. It generally makes clear when referring to water year, fiscal year, etc.

Much of the water storage in the Sierra is in the form of snow, so there is considerable interest in what the runoff will be during the April–July snowmelt period. The April–July period is a focus of agricultural interests, but is also relevant for management of montane meadows. (Snowpack on the April 1 and May 1 snow surveys is typically used as a predictor of when meadows will dry out enough for stock grazing.) The amount of snowmelt runoff varies by year and watershed, but constitutes roughly two-thirds of total annual runoff for the Tulare Lake Basin.

The term "annual runoff" is sometimes used to describe all of the runoff over the October–September water year. But at other times, the term is used to describe only the portion that occurs during the April–July snowmelt period.

## Acre-foot Water Measurement

An acre-foot of water is enough water to cover an acre of land one-foot deep. It is enough to provide a 12- to 18- month water supply for an average family in the Tulare Lake Basin.[16]

Water flowing at a steady rate of 1 acre-foot per day is equivalent to: approximately 0.504 cubic foot per second, 226 gallons per minute, 43,560 cubic feet of water per day, or about 326,000 gallons per day.

## *What Constitutes "Normal"*

Figure 7 illustrates the relative frequency of different water year types in the San Joaquin River Basin. It is based on the San Joaquin Valley Water Year Index shown in Figure 14.

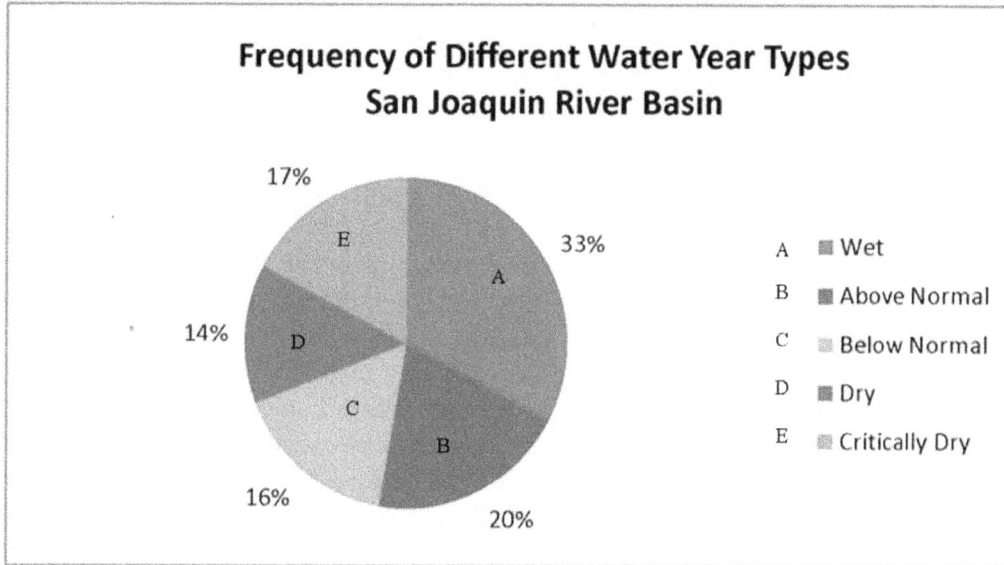

**Frequency of Different Water Year Types
San Joaquin River Basin**

| | |
|---|---|
| A | Wet |
| B | Above Normal |
| C | Below Normal |
| D | Dry |
| E | Critically Dry |

- A: 33%
- B: 20%
- C: 16%
- D: 14%
- E: 17%

Figure 7. Frequency of different water year types in the San Joaquin River Basin.

As Bill Tweed has written, the Tulare Lake Basin occupies the highly variable frontier between the wet winter climate of the Pacific Northwest and the often very dry winter climate of Southern California. We live on the unreliable southern edge of the winter storm track. Some years that storm track comes far enough south to include us, but some years it does not.

The statistical average is what weathermen tend to call "normal," but that does not make it normal in the sense that we can count on it. As shown in the above graph, about half of our years are wetter than average and half are dryer; that's just how the underlying index was set up. Other than that, the categories in this graph are somewhat arbitrary. They are standards used by the California Department of Water Resources to track and report on droughts over the past 111 water years (1901–2011). For more detail on these categories, see Table 8 and the associated text.

One of the lessons that we can learn from Figure 7 is that most of our years are far from normal. As shown in Figure 7, only 36% of our years have been in the two categories that bracket the statistical average (Above Normal and Below Normal). The rest have been either much wetter or much dryer. This is even more vividly illustrated in Figure 14 on page 113.

## Types and Duration of Floods

There are three very different types of floods that occur in the Tulare Lake Basin:

1.  Sharp, high flood peaks of short duration and comparatively small volumes, typically lasting a day or two. In this type of flood, elevated peak volumes and river heights damage developed areas along a river. These are our typical rain floods. These are the type of floods that result in damage to national park infrastructure, and the ones more likely to be recorded in park monthly and annual reports. Dams are designed to offer those who live downstream a relatively high level of protection from this type of flooding event. For an example of a typical flood of this type, see Figure 20 on page 351 which graphs the discharge of the January 1997 flood.

2.  Sustained high levels of discharge, typically lasting a month or more. These are snowmelt floods that typically occur between March or April and June. Such prolonged discharges inundate the valley floor. Dams are not designed to completely contain this type of flooding event. These floods don't necessarily cause any damage in the Sierra or even in the deltas below. The floods themselves may not be recorded in national park reports, but there may be a mention of high precipitation, late springs, or difficulty accessing the higher elevations. These floods are primarily marked by the delivery of large volumes of water to the Tulare Lakebed. Residents of Corcoran (in the Tulare Lakebed) have a very different view of such floods than do residents of Three Rivers (above Lake Kaweah).

3.  Cyclic mega floods. Floods in this category can be extraordinarily large. This category of flooding occurs approximately every 200 years and can result in a series of flooding events lasting up to 10 years. (See the section of this document that describes the California mega-floods).

## When Do Floods Occur

A look at Figure 13 will show how widely runoff varies from year to year. Wet years and dry years commonly alternate. The Tulare Lake Basin doesn't have normal conditions in the sense of a statistical average. What is reliable about our climate is its extreme and relentless variability. That is our real normal; that is the lesson of Figure 13.

Floods are amazingly commonplace in our area. A look at the Table of Contents or Figure 17 on page 139 will show just how commonplace; this document describes what we know about approximately 183 floods that have occurred during the last 2,000 years.

Floods occur at all manner of times. They occur in wet years, and they occur during multi-year droughts. They occur during the winter wet season, and they occur during the summer dry season. When they occur varies so widely because there are such a variety of causes for floods.

## Causes of Flooding

Sustained high levels of discharge (as opposed to high flood crests) result from an unusually deep snowpack. Mega-floods that occur with a quasi-periodicity of approximately 200 years have their own unique set of causes.

Other high flood crests — whether local or widespread — typically result from one of six different conditions in the Tulare Lake Basin:

1. **Heavy winter rainstorms.** This is the most common cause of major flooding events. It results from heavy, relatively warm rains during November–February. Writing in 1900, John Muir described this type of flood quite well:[17]

   > The Sierra Rivers are flooded every spring by the melting of the snow as regularly as the famous old Nile.…Strange to say, the greatest floods occur in winter, when one would suppose all the wild waters would be muffled and chained in frost and snow…But at rare intervals warm rains and warm winds invade the mountains and push back the snow line from 2000 feet to 8000, or even higher, and then come the big floods.

   When weather conditions are right, the Southern Sierra can wring amazing amounts of water out of passing storms. Precipitation events producing 10 to 20 (or many more) inches of water equivalent are possible in the region. When such superstorms are unusually warm — and particularly if they drop heavy rain on an existing snowpack and melt much of it — our main rivers can rise to peak flows well in excess of 50,000 cfs. Such extreme events have occurred infrequently in historic times. Some of the more impressive winter rain floods occurred in 1861–62, 1867, 1955–56, 1966, 1969, 1982–83, and 1997.

2. **Spring snowmelt.** These floods take place at the onset of hot weather after a wet winter has built up a larger-than-average snowpack in the mountains. Such events typically occur several times a decade, and have relatively gentle peaks compared to the winter floods. The May–June 1850 flood may have been one of the largest such floods to occur in historic times. Bigger spring snowmelts have probably occurred since that time, but a large portion of the floodwaters have been diverted out of the rivers since the late 1800s.

3. **Remnants of Pacific hurricanes.** In this document, the terms "hurricane," "cyclone," and "typhoon" are used somewhat interchangeably. Floods resulting from Pacific hurricanes are relatively rare events that occur between June and October. (The Tulare Lake Basin receives moistures from a variety of tropical storm systems, especially during the winter. However, Pacific hurricanes are a category unto themselves; something we seldom see, and never in the winter.) Because the waters off California's coast are so cool, hurricanes always degrade before they make landfall. Only one such storm has ever been recorded coming ashore in California as a tropical storm (September 25, 1939 near Long Beach.) When remnants of hurricanes do come ashore, they can be intense and deliver a huge amount of precipitation. They cover a broader area and last longer than most other summer storms do. They usually make landfall in Southern California or in Mexico, they rarely come as far north as the San Joaquin Valley. The only examples of Pacific hurricanes causing floods in the Tulare Lake Basin that we are aware of were those of 1918 (unnamed), 1932 (unnamed), 1972 (Gwen), 1976 (Kathleen), 1978 (Norman), 1982 (Olivia), 1998 (Isis), and 2002 (Huko). For a typical example, see Figure 18 and Figure 19 on page 318 which show Hurricane Olivia coming up the Pacific coast, recurving, and then coming ashore as a tropical depression. Typhoon Melor in 2009 was in a special category. The remains of Melor remained in the western Pacific, but water vapor from that typhoon moved across the ocean basin via an atmospheric river, a recently discovered phenomenon. Although never a

hurricane, the cyclic storm that struck the Buena Vista Lake Basin in February 1978 behaved in a manner similar to a hurricane. Likewise, the cyclic storm that collapsed the Interstate 5 bridges near Coalinga in March 1995 behaved in a manner similar to a hurricane. Hurricanes and cyclic storms have the ability to punch through a mountain barrier and deliver extremely large rainfalls to the rain shadow on the lee side of those mountains. For example, the 1932 hurricane delivered rain at an average rate of 6 inches an hour over the entire Cameron Creek watershed near Tehachapi, causing catastrophic flooding.

4. **High intensity non-tropical storms.** This is somewhat of a miscellaneous category: With their high intensity, storms such as these are often described as cloudbursts. They frequently cause flash floods and overwhelm drainage systems. They are often relatively localized, although they can cover a wider area when associated with a frontal system. They are often relatively brief, over within a few hours. We may think of them as summer storms, but high intensity storms can occur at any time of the year. Examples of such flood-related storm events that we are aware of were those of 1898, 1913, 1937, 1941, 1945, 1951, 1965, 1972, 1976, 1984, 1986, 1995, 1996, 1997, 2001, 2002, 2003, 2005, 2006, 2007, and 2008. There were surely many others, but it is easy to miss them in the record; they just don't get systematically recorded like other floods.

5. **Monsoonal moisture.** The North American Monsoon is associated with a high pressure ridge that moves northward during the summer months and a thermal low (a trough of low pressure which develops from intense surface heating over the Mexican Plateau and the Desert Southwest of the U.S.) The monsoon typically develops in the U.S. between July and mid-September each year. The upper-level high pressure ridge may be coincident with a surface high. The clockwise flow around the upper-level pattern brings monsoonal moisture into Central California, including the Southern Sierra. The surface high brings lower-level moisture into the Desert Southwest, including the desert areas of Southern California. Pulses of low-level moisture are transported primarily from the Gulf of California and the eastern Pacific. Upper-level moisture is also transported into the region, mainly from the Gulf of Mexico by easterly winds aloft. The low-level moisture largely impacts Arizona and Sonora. The upper-level moisture can be transported as far as California. Depending on the position of the jet stream, the high pressure shifts from being centered over northern Mexico to over the Colorado Plateau or the Great Basin. When the high pressure is centered in the northern position, it causes south to southeasterly winds aloft to move tropical moisture into southwestern California and the Sierra. When this moisture-laden air reaches the Sierra, it is forced upwards and forms thunderstorms or convective cells which can be responsible for intense periods of rainfall.[18] Examples of floods in the Tulare Lake Basin that were caused by monsoonal moisture were those of August 1983, August 1984, September 1997, August 2003, July 2008, and July 2011.

6. **Landslide dam failures.** These are localized events and therefore often go unrecorded. Some of the ones that we know of occurred in 1861–62, 1867–68, 1982–83 and 2002. When the landslide dams and the blocked streams are both small (such as in the 1982–83 and 2002 floods) there is relatively little damage. But when a large landslide dam blocks a large river, the results can be highly destructive. The largest historic dam failures that we know of all occurred during a nine-day period in December 1867 and resulted in spectacular floods. Other large dam failures that may have occurred prehistorically are the Little Kern dam, the series of slides on the west slope of Moro Rock, and the enormous slide on the north side of Dennison Peak. The local canyons of the Kaweah show geologic evidence of dozens of massive landslides. Some of those undoubtedly produced severe flooding downstream similar to those that occurred in December 1867. Sierra-wide floods are common, even the norm. If one river floods, then usually the other rivers in the vicinity flood. But floods resulting from landslide dam failures are different. Such localized flooding events generally don't have much of an impact farther downstream on the main rivers. Because of these small-scale events, places such as Cedar Grove can be expected to have

33

somewhat more large landslide dam flooding events than a downstream location such as Pine Flat would experience.

## Flash Floods

A flash flood results from heavy rainfall within a short period of time, usually less than 6 hours, causing water to rise and fall quite rapidly.[19] Typically a flash flood is associated with a thunderstorm or the remnant of a hurricane.

## El Niño/La Niña–Southern Oscillation

The El Niño/La Niña–Southern Oscillation, or ENSO, is a periodic climate pattern that occurs across the tropical Pacific Ocean at two- to seven-year intervals and lasts nine months to two years. This is the most intense short-term perturbation of the Earth's climate system. These events have a strong impact on the continents around the tropical Pacific, and some climatic influence on half of the planet.

The El Niño/La Niña–Southern Oscillation has two phases. El Niño, the warm oceanic phase, is characterized by a warming of the ocean surface from the coasts of Peru and Ecuador to the center of the equatorial Pacific Ocean. La Niña, the cold phase, is characterized by a cooling of the surface waters in the equatorial Pacific. Nineteenth century Peruvian sailors named the warm northerly current off their coast "El Niño" because it was most noticeable around Christmas.

In popular usage, the El Niño/La Niña–Southern Oscillation is sometimes called just "El Niño." That usage, rather confusingly, lumps together the warm oceanic phase, El Niño, with the cold phase, La Niña.

The causes of — and relationship between — El Niño and La Niña events are not fully understood. Sometimes these phases alternate, but often they don't. El Niño events occur roughly twice as often as La Niña events.

During the last several decades, the number of El Niño events has increased while the number of La Niña events has decreased. Studies suggest that this variation is most likely linked to global climate change.

El Niño/La Niña events are capable of causing extreme weather such as floods and droughts in many regions of the world. Bill Tweed explained their effect on the Tulare Lake Basin in one of his columns for the Visalia Times-Delta.[20] In our area, strong El Niño winters are usually wet, but moderate and weak events can be wet or dry. Moderate La Niña events are often dry, but can be wet. Strong La Niña events, as in the winter of 2010–11, can be quite wet.

Tulare Lake Basin floods that were apparently associated with El Niño events include the floods of 1876, 1906, 1918, 1931, 1952, 1958, 1978, 1982–83, 1995, 1998, and January 2010, among others. The El Niño events of 1876–77, 1905–06, 1918–19, 1982–83, and 1997–98 were particularly strong. The El Niño event that developed during the winter of 1932–33 occurred during the extended drought of 1922–34 and resulted in a remarkable period of snow in the national parks.

Just as El Niño events can cause floods, so can La Niña events. Tulare Lake Basin floods that were apparently associated with La Niña events include the floods of 1911, 1916, 1924, 1938, 1950, 1955–56, 1957, 1964, 1965, 1976, and December 2010, among others. The La Niña events of 1955–56, 1975–76, and 2010–11 were particularly strong.

Although a strong El Niño event typically causes an abundance of precipitation in the Tulare Lake Basin, it can have a very different effect in other parts of the world. For example, the El Niño events of 1876–77 and 1918–19 were associated with major droughts elsewhere in the world. These resulted in devastating famines and the deaths of millions of people.

### Atmospheric Rivers

Atmospheric rivers are relatively narrow regions in the atmosphere that are responsible for most of the horizontal transport of water vapor outside of the tropics. A strong atmospheric river can create major flooding when it makes landfall.

The nontechnical term "Pineapple Express" is popularly used to describe the meteorological phenomenon that causes moisture to be drawn from the Pacific Ocean near Hawaii and transported to the West Coast with firehose-like ferocity. The Pineapple Express is a subset of atmospheric rivers, distinguished primarily by the source of the water vapor and the strength of the southwesterly trending vapor-transport atmospheric river extending toward the West Coast. About 30% of atmospheric rivers fall into the Pineapple Express category.[21]

Atmospheric rivers are embedded within much broader atmospheric storms referred to technically as "extratropical cyclones." Extratropical cyclones are the winter-time analogue to hurricanes, but have a much different structure.

Atmospheric rivers are the business end of extratropical cyclones because where an atmospheric river hits the mountains, it can create extreme precipitation, flooding, and high winds. In terms of impacts, an atmospheric river is to the broader extratropical cyclone it is embedded within, as the hurricane eyewall is to the broader hurricane of which it is a part.

See Figure 21 for a satellite image of an atmospheric river feeding a recent flood.

Atmospheric rivers average 250–375 miles (400–600km) wide. However, they tend to move through an area in ways that result in the dumping of precipitation on an area broader than the size of the river. That is, an atmospheric river might be only 200 miles across, but as it moves through an area, its footprint is likely to get smeared out to be wider than that. While that is the way that they usually work, not every storm does that. Atmospheric rivers are known for stuttering or stalling for periods of time ranging from a few hours to almost a day as they pass over California. Those areas that are under the place where they stall get extra doses of precipitation compared to areas elsewhere. That is, a broad part of the state (say, half of it) may get dumped on by an atmospheric river, but the area beneath where it stalls gets even more.

The storms of 1861–62 arguably caused the most widespread and intense flooding that the West Coast has experienced in historic times. The atmospheric mechanisms behind those storms are unknown. However, they were likely the result of an intense atmospheric river, or a series of atmospheric rivers.[22]

Although atmospheric rivers occur in the mountains in the winter, they usually produce heavy rains rather than heavy snowfalls. That is because atmospheric rivers are usually warmer than most other storms. However, the word "usually" means that not all atmospheric rivers live up to that expectation. Some atmospheric rivers are cool and produce snow, especially at higher elevations.

The following table lists five of the largest atmospheric river floods that have impacted the Tulare Lake Basin:[23]
1. November 1861 – January, 1862
2. February 11–24, 1986
3. December 29, 1996 – January 4, 1997
4. February 1–3, 1998
5. October 13–14, 2009

## *Preparing for the Next Big Flood*

By some measures, the 1861–62 flood was the greatest of these floods. It wasn't a fluke and it is only prudent that we plan for a future flood of similar or greater magnitude.

The USGS Multi Hazards Demonstration Project (MHDP) applies science to improve the resiliency of communities in Southern California in their response to a variety of major natural hazards. The MHDP assembled experts from a number of agencies to design a large, but scientifically plausible, hypothetical storm scenario that would provide emergency responders, resource managers, and the public a realistic assessment of what is historically possible. One of the MDHP's full scenarios, called ARkStorm, addresses massive West Coast storms analogous to those that devastated California in 1861–62. Storms of this magnitude are projected to become more frequent and intense as a result of climate change.

The ARkStorm scenario is patterned after the 1861–62 historical events, but uses modeling methods and data from large storms in 1969 and 1986. The ARkStorm draws heat and moisture from the tropical Pacific, forming a series of atmospheric rivers that approach the ferocity of hurricanes and then slam into the West Coast over several weeks.

The website for the USGS's Multi Hazards Demonstration Project warns that an ARkStorm is plausible, perhaps inevitable.[24] The 1861–62 storm was not a freak event, was not the last time that California will experience such a severe storm, and was not the worst case. The geologic record shows that six mega-storms more severe than the 1861–62 have struck California in the last 1800 years (see the section of this document that describes the California mega-floods), and there is no reason to believe that similar events won't occur again. In the Tulare Lake Basin, the 1867–68 flood was bigger than the 1861–62 flood on all four of our major rivers.[25]

With the right alignment of conditions, a single intense atmospheric river hitting the Sierra east of Sacramento could bring devastation to the Central Valley. An independent panel wrote in October 2007 to the California Department of Water Resources:[26]

> *California's Central Valley faces significant flood risks. Many experts feel that the Central Valley is the next big disaster waiting to happen. This fast-growing region in the country's most populous state, the Central Valley encompasses the floodplains of two major rivers — the Sacramento and the San Joaquin — as well as additional rivers and tributaries that drain the Sierra Nevada. Expanding urban centers lie in floodplains where flooding could result in extensive loss of life and billions in damages.*

## *Measurements of Peak Flood Flows*

The peak flow in a river during a flood can be measured in a variety of ways, among them:

1. **Instantaneous flow at the peak moment of the flood.** Measurements of this type typically come from a manually read gaging station.
2. **Peak hourly flow.** This is the average flow for the peak hour of the flood. This type of measurement typically comes from an automatic gaging station.
3. **Peak daily flow.** This is the average hourly flow for the peak day of the flood. This typically comes from an automatic gaging station. It is calculated by averaging the 24 hours of the peak day (midnight to midnight). The term "peak daily flow" suggests that this is a measurement of total daily flow when that really isn't the case. To avoid that confusion, this document generally refers to this measurement using the nontechnical term "peak average daily flow."

When you see a reference saying that a flood peaked on a particular day, it is usually referring to one of the first two measurements.

Every flood is different, and there is a wide range of variation. However, in a ballpark sense, the peak hourly flow measurement on the big rivers in the Tulare Lake Basin is often about 150% greater than the peak average daily flow. As an example, the peak hourly flow for the Kern River in the 2002 flood was 26,500 cfs while the peak average daily flow (the average hour for the peak day) was 10,306 cfs.

## *Flood Rate and Flood-risk terminology*

One way of describing the size of a particular flood is by describing the risk of a flood that big (or bigger) occurring in any given year. It used to be common practice for the USACE and other risk management agencies to express the risk of floods using exceedence interval terminology (e.g., 50-year flood, 100-year flood, etc.). This is also commonly called a flood's recurrence interval or return interval.

Most of those agencies have now adopted the practice of expressing risk in terms of exceedence frequency, or flood likelihood (e.g., a 20% chance of a flood of a certain size or larger occurring in any given year).

The public still likes to think of a flood in terms of its recurrence interval. If you know the exceedence frequency for a particular flood, you can calculate its recurrence interval using the following conversion formula:

$$\left(\frac{1}{\text{Exceedence Frequency}}\right) * 100 = \text{Recurrence Interval}$$

For example, the flood that occurred on the Kings River in December 1955 had an exceedence frequency of 1%. The above formula can be used to calculate its recurrence interval:

$$\left(\frac{1}{1\%}\right) * 100 = 100$$

A flood with a recurrence interval of 100 is commonly referred to as a 100-year flood. A flood of that magnitude has a 1% chance of being equaled or exceeded in any given year.

Table 2 presents the two types of terminologies for a selection of different size floods.

Table 2. Comparison of flood-risk terminology.

| Flood Recurrence Interval | Likelihood of Flood |
|---|---|
| 10 years | 10% chance/year |
| 20 years | 5% chance/year |
| 50 years | 2% chance/year |
| 100 years | 1% chance/year |
| 200 years | 0.5% chance/year |
| 500 years | 0.2% chance/year |

Calculation of the risk of different size floods occurring is done by looking at the history of past floods for a given watershed. The results of those calculations — the predictions of the likelihood of future floods — are a series of flood frequency curves. In our area, those calculations are generally done by the Sacramento District of the USACE.

Since a flood with an recurrence interval of 100 is commonly referred to as a 100-year flood, there is a temptation to think that such a flood has a 100% chance of occurring in the next 100 years, but nothing is guaranteed in nature. There is actually only a 63.4% probability of a 100-year (or bigger) flood occurring in the next 100 years. It is the weirdness that is probability.

The probability, Pe, that a certain-size flood occurring during any period will exceed the 100-year flood threshold can be calculated using the formula $Pe - \left[1 - \frac{1}{T}\right]^n$ where T is the return period of a given storm threshold (e.g., 50-year, 100-year, etc.), and n is the number of years that you're interested in.

So for a 100-year flood, the formula would be $1 - \left[1 - \frac{1}{100}\right]^{100} = 63.4\%$. That's almost enough to make you feel like you're back in high school again.

Some caution is necessary when using flood frequency curves, since they are derived from historic data. One of the assumptions used in building flood frequency curves is that the probability distribution function is stationary, meaning that the mean, standard deviation and max/min values are not increasing or decreasing over time. If temperatures are changing and precipitation cycles are being altered, then the probability distribution is also changing. The simplest implication of this is that not all of the historical data can be considered valid as input when building the frequency curves. In an era of changing climate, the past cannot necessarily be used to predict the future.

There are two other flood-risk terms that are used by dam designers:
- probable maximum flood
- level of flood protection

The probable maximum flood is the flood that may be expected from the most severe combination of critical meteorological and hydrologic conditions that are reasonably possible in a particular drainage area. This is a very unlikely event, much less than a 500-year flood. Dams are designed to safely pass the probable maximum flood without being overtopped. This is especially critical for earthen dams like Terminus, Lake Success, and Lake Isabella. If a dam were overtopped, this could lead to failure of the dam and catastrophic consequences for the people downstream.

The other dam-related flood-risk term is "level of flood protection (aka design protection level)." This is the level of protection provided by a reservoir for people and property downstream of that reservoir. The level of flood protection is related to the storage capacity of a reservoir.

Authorized flood-control reservoirs are designed to provide a particular flood-control pool (aka flood control space or flood management reservation space). That flood-control pool is used to store high inflows from a flood event so that flows downstream of a dam do not exceed the stated channel capacity. Hydrologists manage this flood-control pool to temporarily store the rain-flood runoff which would otherwise pass by a dam.

The goal is to keep flows downstream of the dam within their stated channel capacity so that flooding conditions are avoided. The maximum size flood that a dam can control is termed a dam's "level of flood protection." The maximum flow that the channel downstream can handle without causing flooding conditions is termed the "maximum objective flow."

The term "level of flood protection" does not imply or guarantee that communities downstream of a reservoir have any such level of flood protection. A reservoir's estimated "level of flood protection" is based on the results of a probabilistic computer model. The model assumes that a certain set of conditions are present. If those conditions aren't present during a real flood, then the maximum objective flow may be exceeded, and flooding could result.

The models typically use a fairly simple set of assumptions about conditions downstream of a reservoir. To assess the level of flood protection for communities downstream of a reservoir would require a full risk and uncertainty assessment of downstream hydrologic, hydraulic, geotechnical, and economic conditions.

A reservoir's level of flood protection can change over time. An example is Terminus Dam which forms Lake Kaweah. When originally constructed in 1962, Lake Kaweah's storage capacity was estimated to be sufficient to provide a 60-year level of flood protection downstream. But as sediment accumulated in the reservoir, the level of protection had decreased by 1978 to only a 46-year level of flood protection. When fuse gates increased the flood-control pool size of the reservoir in 2004, the level of protection increased to a 70-year level of flood protection (see Table 6).[27, 28]

As described above, the flood-control pool of a reservoir is used to manage rain floods. However, the entire gross-pool capacity of the reservoir is theoretically available to control snowmelt floods.

## Landslides and Landslide Dams
Landslides sometimes dam rivers and streams. In the Sierra, dams are occasionally formed by rockslides and debris flows as well.

The USGS has urged local risk management agencies to prepare for a return of a flood as big as the 1861–62 flood. We definitely don't have plans in place for dealing with such an event.

However, the Tulare Lake Basin experienced a flood even bigger than the 1861–62 flood just six years later: the 1867–68 flood. It was a bigger flood on all four of our major rivers.[29] In addition to being a major flood, the storm that brought on that flood was a soaking rain that lasted for upwards of six weeks. Much of the Sierra foothill and montane zones consist of steep hillsides of unconsolidated debris slopes. When these hillsides are soaked to depth, landslides can be triggered.

In a short intense storm like the December 2010 storm, we get many relatively small landslides and debris flows. But in an extended event like the 1867–68 storm, we get cataclysmic landslides. In the past, some of these have formed landslide dams across our major rivers that were up to 400 feet high.

When dams such as those fail, the results downstream can be catastrophic. For example, the residents of Bakersfield woke on New Year's Day, 1868, to a 200-foot-high flood coming out of the Kern Canyon.

In the 1867–68 storm, the landslide dams on the Kaweah and Kern held the flooding rivers back long enough for the residents downstream to react and get out of the floodplain. In contrast, the dams on the San Joaquin River and Mill Flat Creek presented a less clear signal downstream, partly because those events happened at night.

The residents of Old Kernville and Weldon had about 24 hours' notice because the Kern River stopped running. They were able to evacuate their towns before the river submerged them under about 50 feet of water. The residents of Millerton, the county seat of Fresno at the time, weren't aware of what was happening. The disintegrating remnants of one or more landslide dams hit their town just before midnight on Christmas Eve, 1867, destroying it. That's why Fresno is now the county seat.

Just as USGS is urging us to prepare for a return of a storm similar to the 1861–62 flood, it might be prudent to prepare for a return of a storm similar to the 1867–68 flood, complete with landslides and landslide dams. This is especially true in high landslide hazard zones.

## Cedar Grove Flooding

### *Challenge of Modeling Flows*

The impetus for preparing this document was a project to replace the bridge over the South Fork of the Kings River in Cedar Grove. (That bridge would eventually be replaced beginning in the fall of 2010. The new bridge is scheduled to be open for use in the spring of 2013.) To design the replacement bridge, the national parks needed to know the magnitude of floods on that reach of the river. Unfortunately, there are no stream gages along the portions of the Middle and South Forks of the Kings River that are within the national parks.

The nearest long-term gaging station is far downstream at Pine Flat Dam on the mainstem of the Kings River. (Fragmentary data are available from a collection of gages on the South Fork Kings below the park. See the section of this document that describes the Stream Gages on the South Fork Kings.)

Therefore, in 2006, the Federal Highway Works Administration (FHWA) estimated the probable flood flows by modeling the hydrology of the Cedar Grove drainage basin. The results of the peak flow discharge computations are shown in Table 3.[30]

Table 3. Peak flow discharge computations for Cedar Grove.

| Flood Recurrence Interval | Likelihood of Flood | Peak Discharge |
|---|---|---|
| 50 years | 2% chance/year | 13,300 cfs |
| 100 years | 1% chance/year | 18,500 cfs |

Those peak flow discharges are calculations of how big the model projected a flood event would likely be. Bridges like the Cedar Grove Bridge are typically designed for a 50-year-flood event because that is a comparatively rare occurrence. At the same time that the model was being constructed, the national parks were working to rebuild the actual flood history for that reach of the South Fork Kings. Our goal was to ensure that the new bridge would span the 50-year flood event. We wanted to stop interfering with the natural processes of this designated Wild and Scenic River. To our surprise, we discovered that floods of this magnitude (floods that would cover an area equivalent to the 50-year-flood predicted by the model) were a relatively frequent event.

According to the Environmental Assessment for the Cedar Grove Bridge,[31]

> *The South Fork of the Kings River at Cedar Grove has experienced 50-year flow events nine times in the past 70 years (1937, 1950, 1955, 1966, 1969, 1978, 1982, 1983, and 1997).*

Obviously a flow event that occurs every eight years on average (nine times in 70 years) is significantly different from an actual 50-year flow event. (Using more complete data, we would later learn that Cedar Grove has experienced these modeled 50-year flow events at least 13 times in the past 70 years, an average of once every five years.) How to explain the huge discrepancy between the modeled flows and the observed flows? FHWA used a reliable model, so presumably the difference resulted from one or more of the three model inputs:
1. The size of the drainage basin: approximately 357 square miles.
2. A U.S. Geological Survey (USGS) regional regression equation for the Sierra that predicts the frequency of floods for given rainfall values.

3. Site-specific historic rainfall values from the National Oceanic and Atmospheric Administration's (NOAA) *Atlas 2, the Precipitation-frequency Atlas of the Western United States.*

One possible source of error was the USGS regression equation. It was developed to reflect average conditions across the Sierra; it does not necessarily reflect the actual conditions of the South Fork Kings. Another likely source of error was the NOAA precipitation data. This part of the Sierra has very few precipitation gages. That's a serious problem because some of our big floods are caused by localized storm events.

When the national parks started the bridge design process, it would have been very helpful to have had a description of the flood history of the South Fork Kings; but nothing like this document existed. The parks were always playing catch-up. The model results were completed while reconstructing the flood history was still a work in progress. That meant that the actual flood history was not on hand to validate the model results: a serious shortcoming.

### Completeness of Flood History

We have a very incomplete record of Cedar Grove floods. Many of the floods occur in the winter when no one is present in Cedar Grove to observe the event. Just as important, the national parks have no records management system in place for routinely recording flood events when they are observed. For example, we know about the 1937 flood in Cedar Grove only because the U.S. Forest Service (USFS) was designing a new bridge and recorded the high-water mark on their design drawings. We are very lucky that those drawings survived.

All that we know of the 1943 and 1945 floods at Cedar Grove is what we can infer from conditions some 50 miles downstream at Piedra. (We didn't have that information until June 2011, well after construction of the new Cedar Grove Bridge had begun.)

We initially learned about the 1950 flood only because it swept away a major bridge downstream in Reedley, and because of an NPS Washington Office memo documenting the damage in Cedar Grove and elsewhere in the national parks. We were very lucky that memo survived. We know about the 1955 flood in Cedar Grove only because Wayne Alcorn flew into that area after the flood and photographed the damage. (Helicopters weren't that common in 1955. Landing one in Cedar Grove brings to mind the opening scene of the 1972–1983 CBS M*A*S*H* television series.) We are very lucky that those photographs could be found.

All that we know of the 1966 flood at Cedar Grove is what we can infer from conditions downstream at Pine Flat Dam. (We didn't have the data to calculate the flood recurrence intervals in Table 17 until March 2011, well after construction of the new Cedar Grove Bridge had begun.)

We know about the 1969 flood on the South Fork Kings because it swept away a major trail bridge, and because we were fortunate to record Jim Harvey's recollections of the aftermath of that flood before he retired.

We know about the 1978, 1982, and 1983 floods in Cedar Grove only because Jerry Torres (Kings Canyon National Park's trails supervisor at the time) was there to observe them; we have no surviving documentation. CalTrans and Jeff Manley (a national park employee) flew into Cedar Grove after the 1997 flood, took lots of photographs to document the event, and yet none of those photographs survived. Thanks to Jerry Torres, we now know that there are some other surviving photographs of damage from that event, but we have yet to find them.

This incomplete record of floods suggests that Cedar Grove may have experienced more "50-year floods" in the last 70 years than the nine that we could document at the time that the environmental assessment was printed. For example, it now appears that the 1943, 1945, 1963, and 1980 floods should have been included in the list of modeled (as opposed to actual) 50-year floods. By compiling this document, the expectation is that the next project team won't have to start from scratch. They will at least have a general idea of the flood history of the rivers within the Tulare Lake Basin.

## Summary of Flood History

Pine Flat Dam on the Kings River was completed in 1954. The history of floods on the South Fork Kings prior to 1954 can be inferred from sources such as the California Water Plan and historical records. Some of the bigger floods from that earlier period were 1861–62, 1867, 1890, 1906, 1914, 1937, and 1950. Several of those would probably have qualified as actual (as opposed to modeled) 50-year floods. At least two (the 1861–62 and the December 1867 floods) appear to have qualified as 100-year flood events. The 100-year floodplain has been mapped in Cedar Grove, at least in general outline. The Cedar Grove Lodge appears to lie just within it, and the Cedar Grove Visitor Center appears to be just outside.

With completion of the Pine Flat Dam, much more reliable gaging data became available for the Kings River. In the absence of flow data specific to the South Fork Kings, it is not unreasonable to expect that the flood flows in Cedar Grove are roughly proportional to those downstream at Pine Flat. If that is the case, then Cedar Grove has probably experienced two actual (as opposed to modeled) 50-year or bigger floods since the dam was completed: December 23, 1955 and December 6, 1966. See Table 17 for the actual ratings (the flood recurrence intervals) of these and other floods.

## Stream Gages on the South Fork Kings

USGS operated a recording stream gage on the South Fork Kings from 1950–1957. It was located 0.3 mile downstream from Grizzly Falls, across the highway from the big rock deposit that resembles a terminal moraine. It was installed on the downstream side of a big boulder that may have been a glacial erratic. Evidence of that stream gage is still visible today. The bottom section of the 3-foot-wide corrugated metal stilling well protrudes from the streambed alongside that big boulder (photograph on file in the national parks). There's a bolt sticking out of that boulder that was either an elevation pin or part of the attachment for the catwalk to the stilling well.

Either because of good planning or incredible good luck, that gage began operation on November 16, 1950, just as the second and biggest of the 1950 storms struck. It continued to operate through September 30, 1957. The data from November 16–30, 1950 is available in a report that covered the 1950 flood.[32] The data from December 1, 1950 – September 30, 1957 is available online from the USGS.[33] This gage is known in USGS parlance as USGS 11212500 SF Kings River near Cedar Grove CA.

There has never been any electricity in this section of the canyon. Alternative power technology was rather primitive in the 1950s. This gage was probably operated by a hand-wound mechanical timer.

This is the only recording stream gage ever installed on this reach of the Kings. In addition to this stream gage, there have been two pairs of crest-stage gages, two staff gages, and a cable car on the South Fork Kings. USGS has a website that explains the various types of gages.[34]

43

Crest-stage gages provide information on the highest flow since the gage was last visited. They are similar in concept to a min/max thermometer. A pair of crest-stage gages are located in the South Fork Kings on separate boulders where the old USGS stream gage was located (multiple photographs on file in the national parks). Another pair of crest-stage gages is located on the Grizzly Creek culvert, one on each end (multiple photographs on file in the national parks).

The lower pair of crest-stage gages was operated by USGS through December 6, 1966. Presumably those gages were installed in October 1957 when the stream gage was removed. The data from the crest-stage gages are available online from USGS.[35] For the sake of continuity, the data were reported in the USGS database using the same database code and name as the old USGS stream gage USGS 11212500 (SF Kings River near Cedar Grove CA).

The Grizzly Creek station was operated from April 12, 1960 – May 20, 1973. Those data are also available online from USGS.[36] The station is known in the USGS database as USGS 11212450 (Grizzly Creek near Cedar Grove CA).

All four of these crest-stage gages are still in place and functional to varying degrees. Bill Templin has data that he has collected from them in recent years.

USGS also installed a cable car over the South Fork Kings. That cable car was located about ½ mile downstream from Grizzly Falls. The exact location of this former cable car installation is unknown.

Staff gages are located on the Lower South Fork Bridge (the lowermost bridge across the South Fork Kings within the national park). The gage numbers are painted on both ends of the center bridge pier. Floodwaters gradually erode away the paint, making the gage more difficult to read. Bill Templin last painted the downstream numbers in 2011. The upstream gage is the less accurate of the two. That gage is affected by the hydraulic jump when the flow hits the edge of the pier, making it difficult to accurately read the numbers. Although the staff gage has traditionally been located on this bridge, this is not the ideal bridge for that purpose. The problem with using this bridge for discharge measurements is that the bridge crosses the river at a severe angle, and a good discharge measurement needs to be made perpendicular to the flow line. Ideally the staff gage should be placed on a bridge that crosses the river perpendicular to the flow.

There are three bridges over the South Fork Kings in the national park. The middle bridge (the new Cedar Grove Bridge) is generally unsuitable for a stream gage because of the riprap that has been placed around the bridge piers.

The Upper South Fork Bridge (the uppermost of the three bridges) has recently been identified as the most suitable location for a stream gaging station.

## Tulare Lake and other Valley Lakes

### *General Notes on Tulare Lake*

The San Joaquin Valley was once part of a large, ocean-covered basin. Gradually the Sierra Nevada was uplifted. About 50 million years ago, parts of the current San Joaquin Valley, particularly north of present-day Coalinga, rose above sea level for the first time and the Coast Ranges were uplifted. By five million years ago, the seaways connecting the valley with the Pacific Ocean had all closed, leaving the San Joaquin Valley an isolated inland sea.

About 700,000 years ago, the continued filling of the valley by sediment from the Sierra and Coast Range mountains left Lake Clyde (aka Lake Corcoran) as the last remaining widespread ancient lake. It extended about 250 miles from present-day Bakersfield to Stockton.

The Corcoran Clay was deposited in Lake Clyde between about 600,000–800,000 years ago during the Pleistocene. The Long Valley Caldera eruption occurred about 747,000 years ago near present-day Mammoth Lakes. Tephra from that eruption (known as the Bishop Tuff) is associated with the Corcoran Clay.

The Corcoran Clay underlies 6,600 square miles of the San Joaquin Valley.[37] It extends from Stanislaus County in the north, to Kern County in the south. In the Visalia area, that clay layer extends from near Highway 99, west toward the Coast Ranges. The top of the Corcoran Clay is up to 900 feet deep, and it is up to 200 feet thick. It separates a deeper aquifer of high-quality water from a shallower aquifer that in some places is lower quality.

At least 25,000 feet of sediment have accumulated in the San Joaquin Valley since the Sierra started uplifting. By the end of the Pleistocene, 10,000 years ago, the valley was completely filled, except for Tulare, Kern, and Buena Vista Lakes, which remained in the depressions.

At the time of Euro-American settlement, there were five generally recognized valley lakes. In order, stretching upstream from near present-day Lemoore to Bakersfield, those lakes were:
1. Summit Lake
2. Tulare Lake
3. Goose Lake
4. Buena Vista Lake
5. Kern Lake

These lakes were the anchors of a wetland complex of over 400,000 acres (see Figure 5). That complex connected with the wetlands that fringed the San Joaquin River, making a continuous wetland all the way to the Sacramento–San Joaquin River Delta.

Those lakes supported an extensive fringing tule marsh. The tules (also known as bulrushes) grew in very dense stands. The plants were up to 20 feet tall and the stems were 2–3 inches in diameter.[38] The tule-bed around Tulare Lake was typically a hundred yards or more wide. However, at the south end of the lake, the tule-bed extended outward from the shore for about 15 miles.

Floating islands of tules, many large enough to support the weight of several people, are reported to have drifted windblown across Tulare Lake's surface. Where the rivers and streams entered the lake, dense thickets of willows and buttonbush (*Cephalanthus occidentalis*) crowded the shore.[39]

John W. Audubon (one of John James Audubon's sons) described his encounter with the tules when he traveled through the San Joaquin Valley in 1848.

> *Following down the San Joaquin southwest and west, we came to the river of the lakes, and stood off northwest (its general course) for nearly two days, but were so impeded in our progress by the bull-rushes that we turned aside to a clump of trees, where we expected to find water and grass; but not succeeding, returned to the river, about eight miles, and with great difficulty reached the edge of it for water at dusk cold, tired, and regretting our lost time. We resolved, nevertheless, to steer off from the rushes next day. This is the locality from which, I suppose, the valley takes its name, "tulare" meaning "rush," this plant taking here the place of all others.*
>
> *There is no trail but that of wild horses and elk, all terminating at some water-hole, not a sign of civilization, not the track of a white man to be seen, and sometimes the loneliness and solitude seem unending.*

Another traveler passing through the San Joaquin Valley during that period observed:[40]

> *The San Joaquin? What, that's the end of the World.*

Tulare Lake fluctuated in size depending largely on the amount of runoff coming from the Sierra. In very wet years, it could grow to at least 790 square miles. That's over four times larger than Lake Tahoe (192 square miles), easily qualifying it as the largest freshwater lake west of the Mississippi. One author described what it was like to be on Tulare Lake in the mid-1870s:[41]

> *At this time Tulare Lake was beautiful to look at. It was nearly always as smooth as glass and the water as clear as crystal. One could take a small boat and row out far from shore where the sensation would be that of swinging in midair. During any hour of the day the outlook was a pleasing one. Far to the east the outlines of the Sierra Nevada Mountains could be seen rising above a pale blue mist that hung over the vast stretch of lowland. To the south the Tehachapi Range loomed up, and on the west the Coast Range looked as close as if you could put your hand on it. But how deceiving it all was, for the nearest mountain was at least 25 miles away.*

Although classified as a freshwater lake, Tulare Lake tended to be more brackish and alkaline along its margins and when it became very shallow, particularly so in dry years. Orlando Barton described pioneers driving their wagons ½–¾ mile into the lake in order to fill their water tanks with fresh, cold water from the inflow of the rivers.[42] Sounds like the old-time equivalent of a visit from the Culligan Man.

Various sources purport that Tulare Lake was at one time or another upwards of 80 or 100 miles long.[43] Those numbers are improbably large and belong in the realm of legend. John C. Fremont reported that the lake was about 60 miles long when he explored it in the winter of 1845–1846. S.T. Harding concluded that this 60-mile figure probably included flooded areas outside of the actual lakebed, such as the Fresno Slough area.

George H. Goddard published a map of the State of California in 1857 that showed Tulare Lake as being about 40 miles long. The Goddard map was produced by order of the Surveyor General of California and may be one of the more reliable source documents from that time period. According to the 1892 Thompson Historic Atlas Map, the lake had a length of 44 miles during the early years of American occupation.[44] During the time of highest water, the lake's maximum width was about 22 miles. The shoreline was quite irregular.

See Figure 4 for an overview map of Tulare Lake and the entire basin. At its maximum size, the eastern shore of Tulare Lake was within five miles of present-day Highway 99 near Pixley and the south shore was near the state historical monument at Allensworth. The western boundary was at present-day Kettleman City. Lemoore marks the approximate northern boundary of the lake.

The outline of the lake as shown in Figure 4 is only a general approximation at best. Frank Latta researched and mapped Tulare Lake in much more precise detail (see Figure 8).[45]

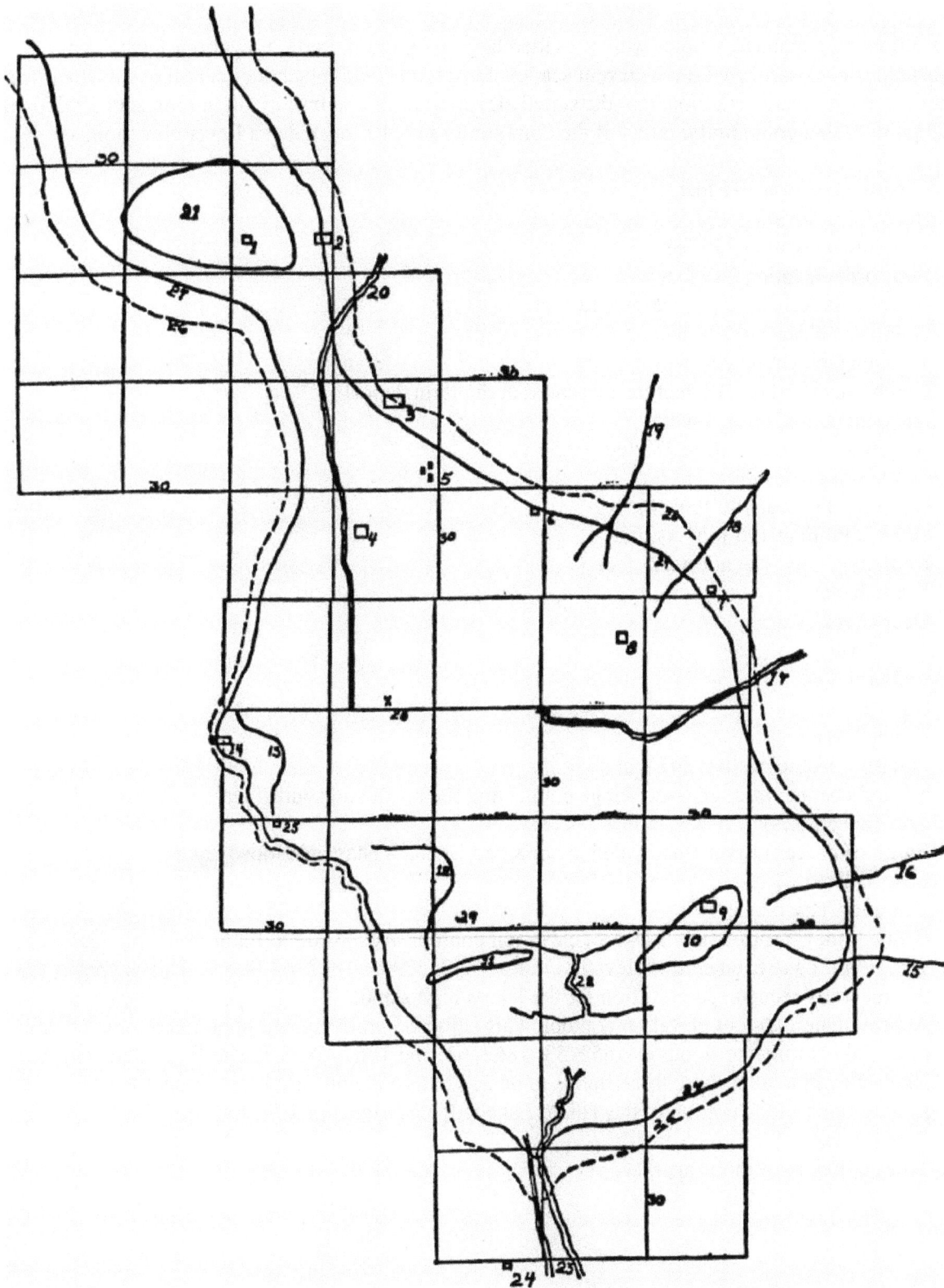

Figure 8. Historic map of Tulare Lake.
Source: Frank Latta

## Legend for Figure 8

The following explanatory notes were provided largely by Frank Latta. Elevations have been adjusted for consistency with the current sea level reference datum (see the section of this document that describes Elevations). Note that the stated elevation for a community often represent a relatively high point in that community; much of the community may be located at a lower elevation.

1. Lanare, elevation 207 feet
2. Riverdale, elevation 223 feet
3. Lemoore, elevation 230 feet
4. Stratford, elevation 203 feet
5. Santa Rosa Rancheria, elevation 218 feet
6. Guernsey, elevation 223 feet
7. Waukena, elevation 226 feet
8. Corcoran, elevation 206 feet (as measured at the train depot)
9. Alpaugh, elevation 213 feet
10. Atwell's Island, elevation 212 feet
11. Skull Island, elevation 212 feet
12. Gordon's Point at 190 foot contour
13. Orton's Point at 196 foot contour
14. Kettleman City, elevation 253 feet
15. White River
16. Deer Creek
17. Tule River
18. Cameron Creek
19. Cross Creek
20. Kings River
21. Summit Lake, elevation 207 feet
22. Natural channel through Sand Ridge connecting the north and south Tulare Lakes (aka Taché and Ton Taché lakes)
23. Entrance of Kern River floodwaters from Kern, Buena Vista, and Goose Lakes
24. Lost Hills
25. Cox and Clark adobe
26. (Dotted line) Approximate 221-foot elevation contour. According to Latta, some pioneers said that Tulare Lake reached this elevation in the 1852–53 and 1862–63 floods. This is greater than the generally recognized elevation for the lake's high stand.
27. Approximate 216-foot elevation contour. This is the generally agreed upon maximum elevation of the lake during the floods of 1852–53, 1862–63, and 1867–68. S.T. Harding calculated the lake's elevation after these three floods to be 215.5, 216.0, and 215.4 feet respectively.[46]
28. Lowest point in the bottom of the Tulare Lakebed, elevation 179.1 feet.
29. Terrapin Bay
30. Township lines, 6 miles apart

The annual elevation of the water in Tulare Lake is available for each year since 1850. It is as if a permanent gaging station had been established in the deepest part of the lake, two years before Nathaniel Vise and the first settlers came to the Four Creeks Country that would later be known as the Kaweah Delta. There clearly was no such gaging station back then. And yet we are blessed with this wonderful dataset. The two people most instrumental in its creation were:

- C.E. Grunsky, a civil engineer who first examined the Tulare Lake area in the 1870s
- S.T. Harding, a professor of irrigation at the University of California at Berkeley for 35 years and a long-time consultant to the Tulare Lake Basin Water Storage District

The very involved story of how Grunsky and Harding pieced together the Tulare Lake elevation dataset is told elsewhere.[47] But to greatly oversimplify their work, they talked to all the watermasters and collected all the gaging data. They dedicated many years to studying how the lake worked and built an elaborate model of the lake. From this they were able to fill in the gaps and calculate the lake elevations for those years when the gaging data weren't available.

The highest lake stage was attained three times during the historic period:

- After the 1852–53 flood raised the lake to elevation 215.5 feet
- After the 1861–62 flood raised the lake to 216 feet
- After the 1867–68 flood raised the lake to 215.4 feet

The above elevation figures are based on research done by S.T. Harding.[48]

The lowest elevation of the lakebed is 179.1 feet. C.E. Grunsky estimated that the bottom of the lakebed — the area that had roughly this elevation — covered about 100 square miles.

Deserting Spanish soldiers from the mission at San Diego are believed to be the first Europeans to enter the San Joaquin Valley. Pedro Fages came in search of them in the fall of 1772 and was the first person to make a written report of the San Joaquin Valley. (In his report, Fages referred to a Native American village located on the valley's southernmost lake, calling that village "Buena Vista," hence the name that we now use for that lake.)

The Yokuts were the dominant Native American group on the San Joaquin Valley floor and in the adjacent foothills; their population throughout the region at the time of European contact was at least as high as 40,000 and probably much higher.[49, 50] At least 19,000 Yokuts lived in the Tulare Lake Basin or visited it seasonally.[51] In the Kaweah River Basin, the Yokuts lived as far upstream as the present-day town of Three Rivers (junction with South Fork Kaweah).

One source said that the Native American population around Tulare Lake was the densest in the Americas north of Mexico's Lake Texcoco; but that seems like a bit of an exaggeration, bordering on urban legend. What can be said is that population densities in the Tulare Lake Basin were high compared to those of non-farming aborigines in other parts of North America. The highest population densities by far — six to seven people per square mile — occurred in the lush stream delta and delta foothill areas along the Kings and Kaweah Rivers.[52] The total population along the lower Kaweah was approximately 3,800.[53]

The first European to reach Tulare Lake was Father Juan Martin who visited the rancheria of Bubal (which was then located on Atwell's Island) in 1804.[54, 55] Bubal was one of the largest Yokuts villages, with a population of about 1,300. Martin estimated that there were at least 4,000 Native Americans living around the lake.

The Native Americans who lived around the valley lakes had a very different lifestyle than those who lived in the foothills. Frank Latta gave a vivid description of what life was like for those who lived in the vicinity of Goose Lake:[56]

> *The Tuhoumne Yokuts were on Kern River...and on Buena Vista, Jerry, Goose Lake, and Bull Sloughs, from the eastern portion of the Elk Hills past Goose Lake and Adobe Holes toward Tulare Lake. Except for an occasional antelope surround, or a ground squirrel smoke-out on the West Side, theirs was strictly a goose, duck, mudhen, swan, blue heron, egret, pelican, lake, slough, swamp-and-overflow culture; water and mosquitos, willows and mosquitos, tules and mosquitos everywhere; tule boats, tule bags, tule skip-rings, and other tule equipment — and mosquitos; tule houses, tule sunshades, tule windbreaks, piled up tules for sails on tule boats; tule clothing — caps, capes, hoods, parkas and skirts; tule mattresses, tule mats, tule blankets, pounded tule-fibre disposable diapers for babies, tule baby cradles, tule fuel, tule blinds for hunting, tule-seed mush, tule-root bread, tule baskets, tule shrouds, tule rope, tule string, tule elk, beaver, sea and freshwater otter, tules, tules, tules — and mosquitos; seal, raccoon; waterfowl and fish in myriads; more tules, tules, tules — and mosquitos.*

Malaria was unintentionally introduced into the San Joaquin Valley from Oregon in early 1833 by a party of beaver trappers from the Hudson's Bay Company. More than 20,000 Native Americans died from the disease that spring. These included Yokuts, Chumash, Miwok, and others.[57]

By 1853, the population of Yokuts in the Tulare Lake Basin had been reduced to no more than 1,100. The survivors were located primarily in the foothills, on the eastern shore of Tulare Lake, and among the timber on the upper Kaweah Delta. In 1876, the Tule River Indian Reservation was created east of Porterville where 1,200 people were taken from various aboriginal groups. A few Yokuts managed to stay outside the reservation, but even that small aboriginal population proved incompatible with many of the American settlers, who outnumbered them by three to one as early as 1860, becoming the new dominant culture in the basin. Most of the remaining Yokuts were eventually rounded up and taken to the Santa Rosa Rancheria near Lemoore.[58, 59]

In 1806, Father Pedro Munoz and Second Lieutenant Don Gabriel Moraga visited San Joaquin Valley as far south as the Kings River. They noted salmon and beaver. Moraga changed the name from *Valle de los Tulares* to the San Joaquin Valley.

Jedediah Smith visited Tulare Lake in 1827 and quickly discovered the need to stay way back from the tule swamps in order to travel through the San Joaquin Valley.

The Kings, Kaweah, and Tule Rivers historically flowed into Tulare Lake, while the Kern River flowed into Kern and Buena Vista Lakes (which occasionally overflowed to Tulare Lake). These four rivers formed expansive, low-gradient, fan-shaped deltas near the lakes that were covered by vegetation. The fans formed by the Kings and Kern Rivers extended far out into the valley. On the western side of the Tulare Lake Basin, coalesced fans originating from the Coast Ranges are comparatively short and steep.

Surface waters were periodically exchanged between the San Joaquin River Basin and Tulare Lake Basin through a complex of slough channels. Some of the channels branching off the mainstem of the San Joaquin River near Firebaugh extended southward, and eventually formed a deep slough channel about 40 miles long and 250 feet wide. This feature (Fresno Slough, aka Fresno Swamp) eventually branched into smaller channels 8 to 10 miles from the San Joaquin River, which became intricate and ramified as they entered Tulare Lake, completing the surface connection. A large bar at the mouth of the slough (on the Tulare Lake side) prevented water exchange between Tulare Lake and the San Joaquin River except during periods of high flows.[60] Fresno Slough merges with the San Joaquin River at Mendota Pool near the present-day city of Mendota.

Flow in the Fresno Slough system was generally from south to north, bringing in seasonally high water from a Kings River distributary, groundwater, and the occasional overflow from Tulare Lake. (A distributary is a branch of a river that flows away from the mainstem. Distributaries are common on deltas. The Kings, Kaweah, and Kern Rivers all have distributaries.)

Eyewitness reports variously describe flows in the Fresno Slough system at different times as both south from the San Joaquin toward Tulare Lake, as well as north from Tulare Lake into the San Joaquin. C.E. Grunsky believed that Lieutenant George Derby had crossed the delta of the Kings River in May 1850. Based on Derby's account, Grunsky concluded that the water in the Fresno Slough was flowing from the Kings River Delta north toward the San Joaquin River and that part of the Kings River was flowing south to Tulare Lake.[61]

The Kings River is the largest of the four rivers in the Tulare Lake Basin; it was historically the largest source of supply for Tulare Lake. Prior to 1862, all the water in the Kings River flowed into Tulare Lake via what is now called the South Fork Kings, along the south side of the delta. Since then, a system has developed that routes a portion of the Kings River floodwaters along the north side of the delta. That is a heavily engineered system, but it was made possible by huge new sloughs eroded during the 1861–62 and 1867–68 floods. See the section of this document on Pine Flat Dam for a discussion of the system that has been developed to manage Kings River waters on its delta.

Tulare Lake was divided into two parts by a sand ridge about 12 feet high which extended across the lake. The ridge extends from present-day Alpaugh to Kettleman City. It consists of Atwell's Island, Skull Island and Dudley Ridge, all more or less connected.[62] The ridge varies in width, but tends to be about 100 yards wide.[63]

During much of the year, the prevailing winds across Tulare Lake are from the north and northwest. In the distant geologic past, that sand ridge was formed when those winds piled up a ridge of sandy material along the southeast side of the lakebed during a period when the lake was not present.[64]

A slough connected the two parts of Tulare Lake, passing through Sand Ridge. During very high-flow years, this slough served (and still serves on occasion) as part of the extension of Kern River. Kern River floodwaters last reached Sand Ridge in 1983.

At moderately high-water levels, Sand Ridge was mostly submerged with parts of it forming islands. The biggest island was about two miles wide by nine miles long and was known variously to early settlers as Root Island, Hog Island, or Atwell's Island. Today that island is the site of Alpaugh.[65]

At relatively lower water levels, Sand Ridge divided Tulare Lake into two separate lakes. To the Yokuts Indians who lived there, the southern lake was known as Ton Taché, in contrast to the northern lake which was known as Taché. As the water level lowered farther, the southern lake would become little more than a marsh, having drained completely into the northern lake. The elevation of the lowest point in the Ton Taché lakebed is 204 feet.[66]

Today, the southern lake, Ton Taché, might still be considered to exist but in a highly modified form. Tulare Lake water storage districts and irrigators know it as the South Flood Area and use it to store floodwaters. It includes all three of the Hacienda Reservoirs that are south of Sand Ridge as well as the South Wilbur Flood Area which is north of Sand Ridge.

Ton Taché was more than just the overflow for the northern part of Tulare Lake. In addition to being fed by floodwaters from the Kern River, Ton Taché also received the runoff from Deer Creek, White River, and Poso Creek.[67, 68]

Tulare Lake sits in a natural depression. But it is also dammed on the north by the meeting of the Kings River Delta and the much smaller Los Gatos Creek (aka Arroyo Pasajero) delta. In this document, the broad feature formed by the meeting of these two deltas is referred to as the Tulare Lake sill or the delta sill. Other documents have described this feature by a variety of terms, including a ridge.[69] The Kings River Delta is of glacial origin and has been described as an alluvial fan-dam.[70]

As explained earlier, all the water in the Kings River used to flow into Tulare Lake via what is now called the South Fork Kings, along the south side of the delta. Some Kings River water still flows into the lakebed during larger flood events.

On those occasions when the lake used to fill, the lake would overflow the delta sill and flow north toward the Fresno Slough and thence into the San Joaquin River.

The two-way channel where Tulare Lake spills north over the delta sill (and where the Kings River flows south into Tulare Lake) is about 15 feet across at the bottom and 60 feet across at the top.[71] The elevation of this point on the delta sill is 207 feet. When Tulare Lake was full to that elevation, it was about 28 feet deep at its deepest point (elevation 207 - 179 feet).

If only it were so simple, but it's not. The Tulare Lake sill was broad and densely vegetated with tules. This greatly reduced the rate at which water flowed out of the lake. C.H. Lee found that although some outflow started when the lake reached a depth of 28 feet (elevation 207 feet), significant outflow didn't really start until the lake reached a depth of 31 feet (elevation 210 feet).

There is one further complication. Left to their own devices, rivers like the Kings, Kaweah, and Kern have moved back and forth across their alluvial fans. At the time of Euro-American arrival in the region (1840s), the Kings was flowing down the south side of its delta and into Tulare Lake.

In earlier geologic time, that river would have been flowing across other parts of its fan and contributing much less water to the lake. For example, the Kings River appears to have been flowing northward from about 21,000–15,000 B.P. (Before Present). This was during the recessional phase of the Tioga glaciation in the Late Pleistocene, when lakes east of the Sierra were reaching their post-glacial high stands. The Buena Vista Basin has deposits that record the runoff from that time period, but the Tulare Lakebed has no comparable record. So apparently the Kings River was flowing north across its fan during that period, contributing no water to Tulare Lake.[72]

As Euro-Americans have settled and developed these alluvial fans, we have fought to constrain the rivers and prevent their migration.

At the time of settlement, most of the water in Tulare Lake came from the Kings and Kaweah rivers. The Tule River flowed into the lake, but provided less volume than the Kaweah. In wet years, Kern River water entered the lake from the south. In high runoff years (*perhaps* every 5–10 years on average), Tulare Lake overflowed the delta sill and connected through the Fresno Slough to the San Joaquin River. From there, the water flowed on to San Francisco Bay.

Summit Lake is a tiny lake set in the fan-shaped delta of the Kings River. It sits in the throat of the channel at the north end of Tulare Lake. Whenever there was outflow from Tulare Lake, those waters flowed north through Summit Lake. Likewise, whenever Kings River waters flow into Tulare Lake, that water flows south through Summit Lake.

Summit Lake is located west of present-day Lemoore. This overflow point in Tulare Lake is immediately west of the intersection of Eight Ave and the south end of 26¼ Ave. In recent years, Summit Lake has been reduced to a circular alkaline flat. Imagine a highly reflective white disc. Apparently it is used as a landmark by pilots landing at the nearby Lemoore Naval Air Station (NAS).

When Tulare Lake overtopped the delta sill, water would flow northerly in a well-defined channel toward the Fresno Slough and thence into the San Joaquin River. At the lake's highest stage, about six feet of water flowed in a broad expanse northerly over the delta sill.

The sill has an elevation of 207 feet. At its highest stage, Tulare Lake had an elevation of 216 feet. For comparison, the highest point in the city of Corcoran, which was built within the lakebed, is the train depot: elevation 206 feet. Emergency levees have to be constructed to protect Corcoran when lake levels approach about 190 feet. Stratford, with a nominal elevation of 203 feet, has a similar problem with lake flooding.

In 1857, the California State Legislature granted a private company the right to construct a canal from the San Joaquin River to Tulare Lake, and then on to Buena Vista and Kern Lakes. That canal would have drained those three lakes and would have carried boats of up to 80 tons. The company was also given the exclusive right to reclaim all the swamp and overflow lands in the huge area represented by the lakebed of those three lakes and all the surrounding wetlands. That was one of the grandest opportunities ever given to enterprise in California, second only to the grants for the construction of the Pacific railroads. Tulare County bitterly fought back as soon as it learned of the act, but the State Supreme Court ruled that the Legislature could not revoke the franchise that it had granted. The company struggled to build the canal but never succeeded in that grand enterprise.[73]

Tulare Lake lies in the rain shadow of the Coast Ranges and is normally protected from large Pacific storms. The average annual rainfall in Tulare Lakebed is six inches.[74] Water comes into the lakebed from the various tributary rivers and creeks. It leaves via a variety of ways:
- Natural outflow. When the elevation of the lake is higher than the elevation of the lowest point on the Tulare Lake sill (elevation 207 feet), the water begins to flow out of the lake. That last occurred in 1878. Since 1878, the Tulare Lake Basin has functioned essentially as a closed basin, an inland sink without a regular outlet to the ocean.

- Evaporation. S.T. Harding calculated gross evaporation from Tulare Lake at 4.6 feet per year. Subsequent measurements showed that actual evaporation is somewhat higher: 5.2 feet per year. That is the primary way that water leaves the lake.[75]
- Absorption into the ground. That is pretty minor. For example, during the 11 years between 1906–16, Harding calculated that only 4% of the water flowing into the lake was absorbed into the ground.
- Used for irrigation within the lakebed. This varies but is relatively minor in a big runoff year.
- Pumped out of the lakebed. This is generally minor.

There is much misinformation and outdated information out there about Tulare Lake. However, there are at least three good publications that are based on exhaustive searches of the literature, both published and unpublished, including gray literature:

- The Bay Institute. 1998. *From the Sierra to the Sea: The Ecological History of the San Francisco Bay-Delta Watershed.*[76]
- United States Bureau of Reclamation. 1970. *A Summary of Hydrologic Data for the Test Case on Acreage Limitation in Tulare Lake.*[77]
- ECORP Consulting Inc. 2007. *Tulare Lake Basin Hydrology and Hydrography: A Summary of the Movement of Water and Aquatic Species.* Prepared for U.S. Environmental Protection Agency.[78]

## *General Notes on Kern, Buena Vista, and Goose Lakes*

The three historic lakes in the south end of the Tulare Lake Basin are:

- Kern Lake (south of Bakersfield)
- Buena Vista Lake (just east of Taft)
- Goose Lake (southeast of the junction of Highway 46 and Interstate 5)

One source said that the five lakes (Summit, Tulare, Buena Vista, Kern, and Goose) were commonly joined in very wet years, but that is misleading. It would be more accurate to say that those lakes were set within a huge wetland complex. In very wet years, it would have been a challenge to go cross-country by foot between any two of those lakes; dry ground was probably quite scarce. However, it would have been easy to go among the lakes by boat if you had known how to navigate through the maze of interconnecting and frequently changing sloughs and waterways.

Before the 1861–62 flood, the Kern River channel ran where the Kern Island Canal now runs in Bakersfield: by the Beale Library (between Chester and Union Ave) on its way to Kern Lake. That flood shifted the river to the west. The new channel began at Gordon's Ferry (just north of present-day Bakersfield College) and passed through what is now Old River and into the Las Palomas slough system between Kern Lake and Buena Vista Lake on its way to Tulare Lake (see Figure 9).[79, 80] Not only did that new channel bypass Kern Lake, but one source said that it also bypassed Buena Vista Lake, meaning that those lakes would only get water during years with very high runoff. In any case, the river would shift even farther northwest in the 1867–68 flood.

The Sinks of the Tejón was the first Butterfield Overland Mail stage stop north of Fort Tejón. It was located at the intersection of present-day David and Wheeler Ridge Roads, roughly 10 miles northeast of where Interstate 5 and Highway 99 diverge.

The winter of 1861–62 was a high water year on all the rivers in the Tulare Lake Basin. When the Kern River came out of the canyon that winter, it created one vast sea of water from Gordon's Ferry to the Sinks of the Tejón. Kern Lake was located in the southeast corner of that huge sheet of water.[81]

Estimates of the size of Kern Lake at full pool vary a good bit, but it was somewhere between 9,000–13,000 acres. It was roughly triangular in shape, some 9 miles long (measured along its northern edge) and 4 miles wide in the middle. It would be possible to determine its exact size rather precisely using the Natural Resources Conservation Services (NRCS) soils map for southwest Kern County. Tejón Creek flowed into the southeast end of Kern Lake in a channel two feet deep and ten feet wide.[82] Tejon Ranch Headquarters were located some miles south of that point.

Although it is convenient to talk about the Kern River as if it were a regular river channel flowing into Kern, Buena Vista, and Tulare Lakes, that's not exactly how it worked. The mainstem channel of the Kern River effectively terminated not too far after emerging from its canyon near Bakersfield. After that, the waters of the Kern passed through an ever-changing network of sloughs, some larger than others, across its very sandy alluvial fan.

For example, the soils map shows that the Kern flowed south to Kern Lake via a number of finger sloughs. The connection from Kern Lake west to Buena Vista Lake was via the 7-mile-long Connecting Slough. It was much the same throughout the area north to Tulare Lake; this was one vast wetland. The oft-cited comparison to the Everglades isn't all that farfetched.

Prior to 1861, the Kern River flowed directly through Kern Lake in high-water years. The 1861–62 flood rerouted the Kern River to the west, through the Las Palomas slough system, bypassing Kern Lake. In subsequent years, Kern Lake received waters from the Las Palomas slough system.

The 1867–68 flood moved the river channel even farther northwest to its present location, ending in the Buena Vista Slough (aka Kern River Flood Channel), a few miles north of Buena Vista Lake and entered that lake from the northwest as shown in Figure 9. This reduced the flood flows into Kern Lake still farther.

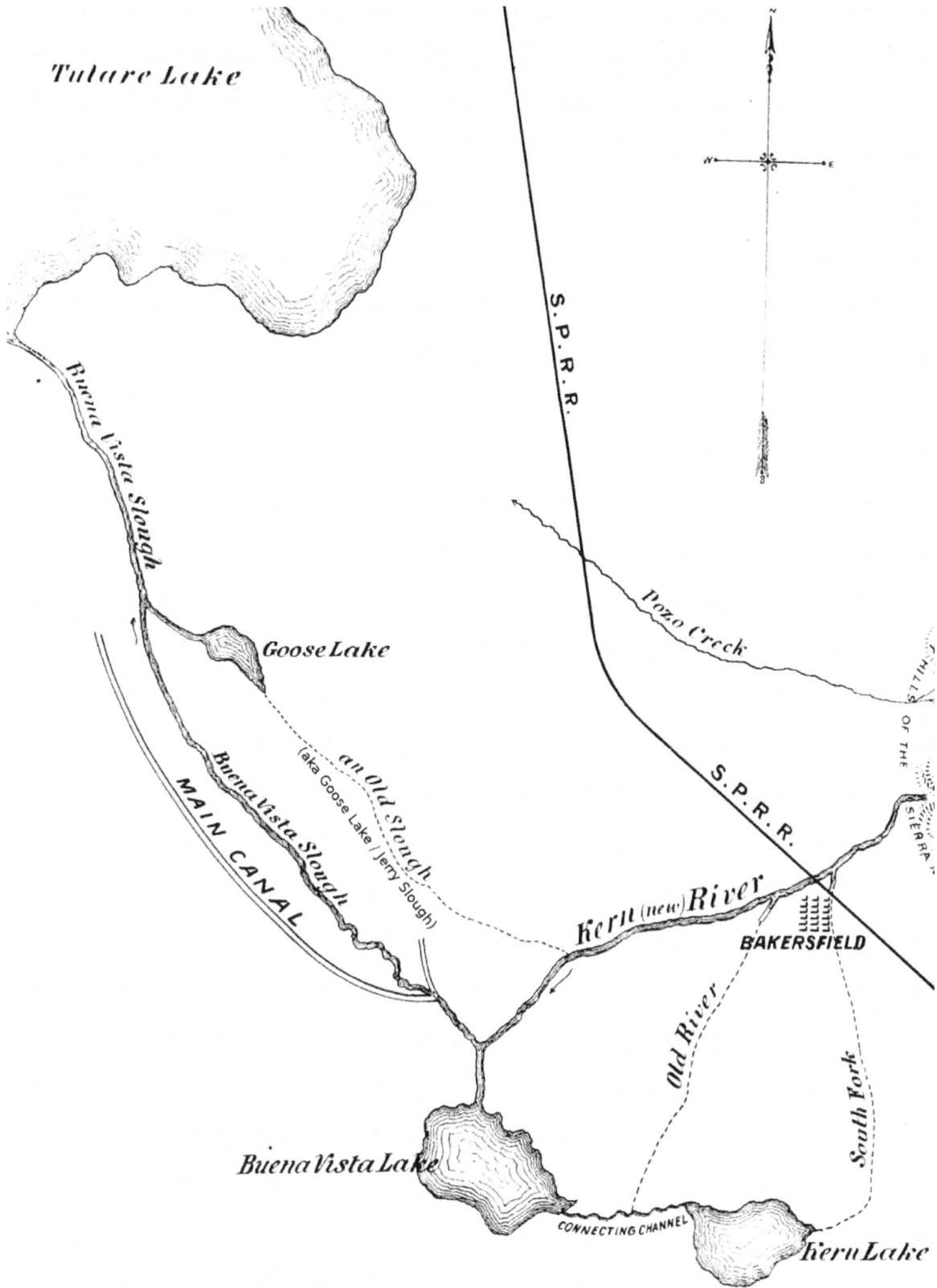

Figure 9. 1880 map of Goose, Buena Vista, and Kern Lakes.[83]
Source: Charles Lux with modifications by Tony Caprio

The Kern Lakebed is located about 10 miles northwest of where Interstate 5 and Highway 99 diverge. The interstate passes through the western portion of the lakebed. The lakebed is immediately northeast of the small community of Lakeview, but that community no longer has a lake view.

The Kern Lakebed is largely or entirely owned by the J.G. Boswell Co. Virtually all of the acreage has been converted to cropland except for an 83-acre tract on the south edge of the lakebed. For many years, it contained the last remnant of Kern Lake. (That pond was known locally as Gator Pond, named for a small alligator that supposedly lived there in the 1930s.) From 1984–1995, the J.G. Boswell Co. partnered with The Nature Conservancy, allowing them to manage this valuable tract as the Kern Lake Preserve. Recent accounts are sketchy, but the tract has apparently since been dewatered and it no longer appears to be a functioning wetland.

Buena Vista Lake is immediately west of Kern Lake. Estimates of the size of Buena Vista Lake at full pool vary a good bit, but it was somewhere between 25,000–50,000 acres, some 8 miles wide by 12 miles long.[84] Both Buena Vista and Kern Lakes were 290 feet in elevation at full pool.

In 1973, two manmade lakes (998-acre Lake Webb, and 86-acre Lake Evans) were created within the Buena Vista Lakebed for irrigation and recreation purposes. Kern County Parks and Recreation manages those two lakes as the Buena Vista Aquatic Recreation Area. The rest of the lake bottom is farmed, part of the J.G. Boswell Co. cropland.

Waterworks have been constructed to allow the choice of whether to impound Kern River floodflows in Buena Vista Lakebed or pass them through to the Tulare Lakebed. In the 1952 flood, 232,000 acre-feet of Kern River floodwaters were stored in the Buena Vista Lakebed. The water storage district has apparently never chosen to use that lakebed for such purposes in subsequent floods, particularly in the big 1969 flood. That decision would eventually be ruled on in a landmark decision by the U.S. Supreme Court. For a more complete description of this event, see the section of this document that describes the 1969 flood.

The Kern River is traditionally considered to terminate in Buena Vista Lake, but that is somewhat misleading. In low water years, the waters of the Kern never reached the lake. However, in high-water years, the lake would overflow or spill.

When Buena Vista Lake overflowed, water then flowed northwest through Buena Vista Slough to Tulare Lake (see Figure 9). About midway to Tulare Lake, Buena Vista Slough passed through what Miller and Lux called the Buttonwillow Swamp along the west side of Buttonwillow Ridge. At the northern toe of Buttonwillow Ridge, Jerry Slough (aka Bull Slough or the northern extension of Goose Lake Slough) joined with Buena Vista Slough and these waters then passed through Sand Ridge to enter the main body of Tulare Lake. During high-water periods, there was a smaller portion of Tulare Lake on the south side of Sand Ridge, known to the Native Americans as Ton Taché.

But that was only one of two ways that Kern floodwaters found their way into Tulare Lake. The other way was via a natural overflow flood channel, a distributary that began much farther upstream (see Figure 9). In high-water years, the Kern would overflow its banks just west of Bakersfield, a few hundred yards east of the present-day Stockdale Bridge, at a place where the historic wooden Bellevue Weir was built across the river. That weir is now gone, but there is a rock spillway across the river at the same location, which slightly raises the level of the river upstream. See the 1950 flood for an account of private interests working to contain flooding in this flood channel.

This flood channel (Goose Lake Slough) flowed 14 miles west through what is today the Rosedale neighborhood until, just before it joined the Buttonwillow Swamp, it was deflected by the east end of a gentle rise known as Buttonwillow Ridge. Here the slough channel (now called Jerry Slough) veered 14 miles northwest to Goose Lake. Jerry Slough formed a network of forking and rejoining channels; it was a challenge to navigate.

Goose Lake is the widest point on what has been termed the Jerry Slough Delta. That landform is located in the northwest to southeast-trending trough of low-lying land between Buttonwillow Ridge (southwest of Goose Lake) and Semitropic Ridge (northeast of Goose Lake).

Goose Lake is a slight depression, averaging a little more than a mile in diameter. At full pool, Goose Lake and Jerry Slough formed a single body of water, elevation 250 feet, up to 20 miles long, and 1–4 miles wide.

Today, the Goose Lake Bottom is divided into quadrants by various canals and channels. The southwest quadrant of the lake bottom is a nearly permanent body of shallow water surrounded by emergent marsh vegetation. This "surge pond" is the one quadrant of the lake bottom that most often contains water. Were Goose Lake to fill to a surface elevation of 235 feet, Goose Lake would measure approximately 3 miles long by 1–1½ miles wide and would cover approximately 1,600 acres. When duck club lands in the Goose Lake Bottom are flooded to an elevation of 237 feet, water covers approximately 922 acres of wetlands inundated to a depth of 6–8 inches. This is in addition to the 1,600 acres in the surge pond.[85]

Today, Interstate 5 runs along the crest of Buttonwillow Ridge. The ridge begins near Stockdale Highway and peters out a few miles south of Highway 46, just a short distance downstream from Goose Lake. The ridge separates Buttonwillow Swamp from Jerry Slough.

Where Buttonwillow Ridge peters out, the waters of Buttonwillow Swamp and Jerry Slough come back together. The slough channel north of Goose Lake and Buttonwillow Ridge is known as Bull Slough.

The actual location of Bull Slough is somewhat uncertain; this is a nomenclature issue. In 1880, Buena Vista Slough was the main waterway and Bull Slough was apparently only the northern portion of the short slough that connected Goose Lake and Buena Vista Slough.[86] Possibly the name Bull Slough later came to apply to a larger area extending farther to the north. In any case, Goose Lake Slough, Jerry Slough, and Bull Slough have not carried Kern River floodwaters since 1983.

In the 1800s, Henry Miller (of Miller & Lux fame) set about draining Buttonwillow Swamp under the provisions of the Swamp and Overflow Act. He accomplished this by building a huge levee along the west side of the swamp, which created a manmade channel (known today as the Lokern Flood Channel) running along the toe of the Elk Hills, McKittrick and Belridge alluviums between Buena Vista Lake and Highway 46. That confined the Kern River to a narrow channel, thus drying up a lot of the wetland. However, in high-water years, the Kern would still overflow into the Goose Lake system and enter Miller's empire through the back door.

To solve that problem, Miller built another huge levee across the narrowest spot in the Goose Lake system, which happened to be at the west edge (downstream side) of the Goose Lake depression. That levee connected Buttonwillow Ridge to the Semitropic Ridge, right at the toe of Semitropic Ridge, running northwest along the toe, all the way to the present-day Kern National Wildlife

60

Refuge, thereby holding this "back door" water against the Semitropic Ridge. These massive earthmoving projects effectively reclaimed Buttonwillow Swamp so that it was no longer a wetland.

The Goose Lake system was left unaltered, except that when water does come down the Kern, it is held slightly deeper than it would otherwise be in Goose Lake. The last significant floodwaters to come down Jerry Slough were in the fall of 1951 and the spring of 1952. Goose Lake has been filled three or four times since 1952, when there was a need in some high-water years to divert water to anyplace they could find a spot that would take it.

During high-flow events, Kern River water continues flowing to the north, passing near the west side of present-day Kern National Wildlife Refuge, crossing through Sand Ridge between Alpaugh and Dudley Ridge, and enters the south end of Tulare Lake.

## *General Notes on Bravo Lake*

In addition to the five generally recognized valley lakes, there was reported to have been a sixth natural lake in the valley: Bravo Lake (aka Wood Lake). This lake was located near the upper end of the Kaweah Delta in the Kaweah River Swamp. It was immediately north of the St. Johns River, about two miles west of McKay's Point (Township 17 S, Range 27 E, Section 31). Today Bravo Lake is in the southeast edge of the city of Woodlake.

Relatively little is known about Bravo Lake compared to the five better-known valley lakes. Assuming that reports are correct, the lake was originally a natural feature, but was converted into a reservoir in the 1870s. We have found no written description of what the natural lake looked like or how it was formed. However, we can make reasonable inferences based on its location on the Kaweah Delta and old maps. There really is very little hard documentation to go on, so this should all be read cautiously.

The earliest reference to the lake that we have found was as a landmark for the small community of Stringtown that was caught up in the 1867–68 flood. Stringtown was a settlement of five families described as living in a line south of present-day Woodlake along the Kaweah River, east of Bravo Lake. (This reference to Stringtown's location is somewhat unclear. Bravo Lake was located adjacent to the St. Johns River.) Although Stringtown may not have been much of a town, this suggests that Bravo Lake may have existed prior to the 1867–68 flood.

The lake does not appear to have occupied a natural depression or to have been on a stream course. Judging from available data, the contours under Bravo Lake appear to slope toward the southwest, generally toward the St. Johns River. The lake appears to have been a floodplain feature associated with the St. Johns River. The south edge of the lake generally paralleled the north bank of the St. Johns River. That suggests that the lake may have been formed by the natural levee created on the river's north bank during floods.

Levees are commonly thought of as man-made, but they can also be natural. When a river floods over its banks, the water spreads out, slows down, and deposits its load of sediment. The coarsest sediment is dropped first as the river no longer has the energy to carry it. This coarse material forms a natural embankment (levee) along the edge of the river channel.

It is tempting to think that Bravo Lake was created primarily during the great floods of 1861–62 and/or 1867–68. Those were the two floods that created the St. Johns River. Both of those floods deposited large amounts of silt and debris on the Kaweah Delta.

Ed Reynolds, an early Tulare County pioneer, recalled how Bravo Lake got its name. According to Reynolds, a fight took place near the lake in 1870 between Tom Fowler and "Swamp John" Asbill. Each man had a rooting section. Many of Fowler's supporters yelled "Bravo, Bravo", and that's how the lake apparently got its name. This suggests that the lake may have had no generally accepted name prior to 1870.

In 1872 the Wutchumna Ditch Company organized and commenced the construction of an irrigation system which eventually consisted of about forty miles of main and branch ditches. The water was taken from the Kaweah just above McKay's Point. Bravo Lake, situated near the intake of the canal, was used as a storage reservoir for floodwaters so that a supply was maintained throughout the year.[87] Water from Wutchumna Ditch entered on the east side of Bravo Lake and exited on the west. The ditch system was largely completed by 1877.

62

Joe Childress was the manager of the Wutchumna Water Company for many years. While he has heard that a natural lake once existed where the reservoir is today, he has never seen any records to substantiate that. The company has no surviving engineering records to document how the reservoir was constructed in the 1870s. Joe speculated that the original (natural) Bravo Lake may have been a shallow ponding basin on the edge of the St. Johns. The ditch company may then have built a low levee around that to raise the water level and convert it into a reservoir.

The Tulare County Times reported that Bravo Lake was full of water in July of 1886.[88] Presumably that was a reference to the use of the lake as a reservoir for Kaweah River water delivered via the Wutchumna Ditch. The 1892 Thompson Historic Atlas Map labeled the lake as "Bravo Lake (Reservoir)" and showed it as connected to Wutchumna Ditch on both the east and west sides (map on file in the national parks).[89] A photograph of the lake taken in about 1900 shows it to have a relatively natural shoreline with a scattering of small islands (photograph on file in the national parks).

The original (natural) Bravo Lake had no creek or river flowing into it. The nearest cross drainage of any size is Antelope Creek which is about half a mile west of the lake. The lake would have received water only when the St. Johns overflowed its banks. Perhaps the Kaweah Delta contained other lakes of this type, but Bravo Lake is the only one of this type that we know of.

Our understanding of the hydrologic operation of the natural Bravo Lake is based largely on supposition. We have found no records about when the St. Johns overflowed in the vicinity of Bravo Lake in the 19th century. The floods of 1861–62 and 1867–68 almost surely inundated this area, quite possibly creating the lake. It is also possible that the St. Johns River overflowed its banks in the floods of 1872, 1874, 1875, 1876, and 1877, putting water into Bravo Lake. In the 1877 flood, the levee on the south bank the St. Johns failed farther downstream, causing flooding in Visalia.

Once Bravo Lake began functioning as a reservoir, it presumably got most of its water from the Wutchumna Ditch. As man-made levees were raised around the reservoir, it would have been completely cut off from the St. Johns' floodwaters.

The community of Woodlake was founded in 1912 by Gilbert F. Stevenson, a wealthy land developer from Southern California. He planned for the town to be a resort community with the lake the center of a "Mecca of prosperity." Stevenson constructed a large levee around the lake. That is the levee that we see today.

Stevenson had big plans for the lake. The lake level was to be raised substantially. Several islands were to be constructed within the lake which would be serviced by pleasure craft. On these islands would be elaborate dance pavilions, bathhouses, and restaurants. A narrow gauge railroad would run around the levee for scenic excursions (one source said that the train was intended to connect the islands).[90]

Stevenson also planned to develop several parks outside the levee, featuring shade trees, fountains, walkways, baseball diamonds, etc.[91] The Tulare County Times reported in 1913 that "Steps are now being taken to make the lake, formerly Bravo and now Woodlake, the most popular picnic ground in the county."[92] Stevenson's numerous financial commitments resulted in his downfall during the Great Depression; his plans for the resort and the lake were never realized. The Wutchumna Water Company took over control and operation of Bravo Lake and its irrigation system.

The USGS 1928 Lemon Cove quad sheet labeled the lake as Wood Lake, maintaining the name that Stevenson had apparently applied to it. However, the 1952 Woodlake quad had switched to using Bravo Lake for the lake's name.

### *Wildlife in and around Tulare Lake*

Thanks to the writings of James Carson and others, we have a pretty good idea of the wildlife that lived in and around Tulare Lake in the middle of the 19th century.[93] For example, Carson wrote that beavers, mink, and otters were all present.

Rob Hansen cautions that these early accounts need to be viewed critically. For example, Carson wrote that muskrats were also present. However, muskrats were almost certainly not present in the Tulare Lake system until 1943. So it's worth examining what evidence we have that beavers, mink, and river otters were actually present in the Tulare Lake system in the 19th century.

Some have questioned whether beavers were really present in the San Joaquin Valley, but Carson wasn't the first or the only one to report them. Father Pedro Munoz and Second Lieutenant Don Gabriel Moraga observed beavers when they explored the central portion of the San Joaquin Valley in 1806. Felipe Santiago Garcia reported seeing many beavers in all the rivers of the Tulare Lake Basin.[94]

The Spanish were aware of the beavers in the San Joaquin Valley, but they generally didn't exploit the valley's many resources. All of California, including the San Joaquin Valley, was Spanish territory. The Spanish settlements were largely on the coast, so that left the beavers in the San Joaquin Valley as fair game for the English and American trappers to exploit. The history of the early beaver trappers in California and the Southwest has been well researched by Dr. Robert Cleland and others.[95, 96]

The first and most famous of the trappers was Jedediah Strong Smith.[97] When Smith and his band of trappers arrived in California, they came by a route far south of the Sierra. After crossing the Great Basin from the Great Salt Lake, they struggled up the Mojave River, finally reaching the haven of Mission San Gabriel (southeast of present-day Pasadena) at the end of November 1826.

They stayed there for the next two months, learning much about California. Harrison G. Rogers, (Smith's clerk) wrote on December 26, 1826, that there were supposed to be plenty of beavers at both Buena Vista and Tulare Lakes.[98] When Mexican Governor Echeandia learned of Smith's presence, he ordered Smith to immediately leave by the way that he had come.

Smith had no intention of abandoning his search for new beaver country. So, although he departed San Gabriel as ordered on January 18; instead of going back by the Mojave, he headed northward over the Tehachapis and entered the San Joaquin Valley.[99]

There has been speculation about which lakes and streams Smith would have trapped as his expedition wandered north through the Tulare Lake Basin.[100, 101] It does seem reasonable that he would have checked to see if Buena Vista and Tulare Lakes really did have the large number of beavers that he had been told.

On the other hand, the map of his expedition shows Smith's route touching Tulare Lake at the south and gradually bearing away from it as he went northward. That suggests that he might not have hunted beavers on the Kern, Tule, or Kaweah Rivers or on the other small streams that flow out of the Sierra before he reached the Kings. Smith wrote a letter to General William Clark, Superintendent of Indian Affairs, saying that he began his spring hunt there. By the end of April, Smith had trapped his way north through the San Joaquin to the American River and his horses were packing great bundles of beavers.

On May 20, 1827, Smith and two of his group began an attempt to cross the Sierra, following generally the route of present-day Highway 4. They reached the crest near Ebbetts Pass eight days later, becoming the first white people to cross a Sierra pass.[102]

Smith returned to the San Joaquin by the Mojave route in 1828, rejoining his party.[103] After some difficulties with the Spanish authorities, they were allowed to travel to San Francisco. There they sold 1,568 pounds of beaver pelts and 10 otter skins for $3,940. Smith's party spent the next few months trapping beavers in the lower tributaries of the San Joaquin.[104]

Peter Skene Ogden led a 60-man trapping expedition for the Hudson's Bay Company in 1829–30; he was one of the first English trappers to explore California. His expedition trapped down the Colorado River nearly to the Gulf of California, then worked their way trapping up the San Joaquin Valley.[105]

After Ogden came into the San Joaquin Valley, his trapping route took him along what he called the South Branch of the Bonaventura. The Bonaventura is the stream now known as the San Joaquin River, so his route up the "South Branch" presumably took him along the Kern River through Kern, Buena Vista, and Tulare Lakes.

Ogden is the only English trapper to venture so far south. After Ogden's expedition, Hudson's Bay Company trappers, working from their post at Fort Vancouver, generally trapped only as far south as the San Joaquin River. Most or all the trappers in the Tulare Lake Basin after Ogden were Americans. According to Donald Tappe, trappers working for the Hudson's Bay Company took beaver furs as far south as Buena Vista Lake, but they usually considered it unprofitable to work farther south than the shores of Tulare Lake.[106]

Ogden's English trappers weren't the only trapping party to enter the San Joaquin Valley in 1830. Also in that year, an American trapping party led by Ewing Young which included Kit Carson traveled from Taos, New Mexico to Mission San Gabriel. After crossing into the San Joaquin Valley, Young's party initially trapped beavers along the Kern River.[107] Following Ogden's trail, they caught up with his group in a few days.

The two groups trapped the San Joaquin Valley together for 10 days until they came to the Sacramento–San Joaquin River Delta.[108] Although Ogden had collected a thousand skins by then, he generally found that the California waters were the poorest in furs of any area that he had yet explored.

In 1833, Zenas Leonard recorded beavers living in the San Joaquin River and being traded out of Monterey.[109] Leonard did not say where those beavers were trapped or by whom.

Stephen Hall Meek recorded that his fur trapping party pitched their camp for the winter on the shore of Tulare Lake in December 1833.[110]

Clearly beavers occupied the lower elevations of the San Joaquin River and Tulare Lake Basins in the early 19th century. If beavers behaved in those basins as they did in the Rockies and elsewhere, they would go upriver as far as they found suitable habitat. That habitat existed in the form of cottonwood and aspen; the only question is whether beavers made use of it.

The various subspecies of American beaver that live in the Rocky Mountains have adapted to occupy virtually all areas of suitable habitat, even in the alpine. For example, the subspecies that lives in Colorado (*Castor Canadensis concisor*) lives throughout that state in suitable habitat, although it is most abundant in the subalpine zone. Many of the high alpine ponds and meadows that exist in Colorado today are the work of generations upon generations of beavers.[111]

The subspecies of beaver that lives in our state is the California Golden beaver (*Castor Canadensis subauratus*). Some early 20th century naturalists (Joseph Grinnell, Donald Tappe, etc.) questioned whether our subspecies lived above the 1,000 feet elevation in the Sierra.[112, 113] In effect, they were saying that they had no evidence of beavers living above that elevation. They certainly didn't present any biological argument that our subspecies was incapable of colonizing suitable habitat above the 1,000-foot elevation limit. Information has since become available documenting that our subspecies did dwell well above that elevation, making use of the habitat that they found.

Much of that evidence comes from areas in the Central Sierra such as the upper reaches of the Carson River. For example, Donald Tappe recorded an eyewitness who said that beavers were plentiful on the upper part of the Carson River and its tributaries in Alpine County until 1892 when they fell victim to heavy trapping.[114] However, accounts also survive of beavers at relatively high elevations in the Tulare Lake Basin including the area that is now Sequoia and Kings Canyon National Parks.

A 14-man party led by Ewing Young which included Colonel Jonathan J. Warner trapped the Kings River in the fall of 1832 "up to and some distance into the mountains and then passed on to the San Joaquin River, trapped that river down to canoe navigation in the foothills..."[115]

Young and Colonel Warner are believed to have trapped the North Fork of the Kings River up to about present-day Courtright Reservoir (elevation 8,170 feet). They then crossed the divide to the South Fork of the San Joaquin River. A logical route would have been via Hell for Sure Pass, taking them into present-day Kings Canyon National Park.[116, 117]

Aspen Meadow (elevation 8,206 feet) and Blaney Meadows on the upper reaches of the San Joaquin have what would seem to be quite suitable beaver habitat. As Young's party progressed down the San Joaquin, they came upon the trail of another trapping group. When they caught up with that group, it turned out to be a Hudson's Bay Company party led by Michel Laframboise.[118]

Earle Williams similarly interpreted accounts of Colonel Warner's expedition, stating that "Warner had been trapping fur-bearing animals at the headwaters of the Kings River about the same time that the Walker party was descending the Merced River."[119]

Donald Tappe recorded an eyewitness account from a retired game warden in 1940, who stated that beavers were "apparently not uncommon on the upper part of the Kings River" until 1882–1883.[120]

On the Kern River, Roy De Voe, a native of the lower Kern Canyon, recalled that he had seen "very old beaver sign" at Lower Funston Meadow (elevation 6,480 feet) in 1946. De Voe also reported that his friend Kenny Keelor trapped the Kern River for beavers around 1900, making his camp at the mouth of Rattlesnake Creek (elevation 6,585 feet) until they were largely trapped out by about 1910–1914.[121]

The presence of Beaver Canyon Creek, tributary to the lower Kern River just east of Delonegha Hot Springs, is also consistent with the Kern River Basin having historically supported a population of native beavers.

Officially, mink records extend only as far south as Fresno County. We have only James Carson's account that mink were in the Tulare Lake area in the 19th century. Because all the wetlands were interconnected, it seems logical that beavers, mink and river otters were able to move about from one end of the Tulare Lake Basin to the other.

There are a number of well-documented mink sightings from Kaweah Oaks Preserve in the 1980s and possibly later. These include numerous photographs taken from a blind set up on the edge of Deep Creek, a mink that entered that blind to take food from the photographer, and a road-kill mink on Highway 198 next to the preserve. Rob Hansen recalled an anecdotal reference from another observer who saw a mink on the 3¼ mile stretch of the Kaweah River channel between McKay's Point and Highway 245 (Road 212).

Kirk Stiltz recalled observing mink twice in the Kaweah River. The first time was in about 1979 just below the Slicky swimming hole in Three Rivers. The other time was in about 1986 just above where Highway 245 (Road 212) crosses the river south of Woodlake.

There are four records of mink in Sequoia and Kings Canyon National Parks. All of these were by reasonably reliable observers, but none were backed up with photographs or specimens. Therefore, the national parks consider the current status of mink in the park to be unconfirmed. The last reported mink observation in the parks was in 1994 near Bearpaw Meadow (elevation about 7,700 feet).

The presence of river otters in the Tulare Lake Basin ecosystem has also been questioned, but there have been observations in addition to those of James Carson.

Frank Latta recorded that the Yokuts who lived in the vicinity of Goose Lake hunted both beavers and otters.[122]

In addition, a group of sailors arrived at the south end of the Tulare Lake Basin in 1853 and began trapping beavers and otters from Kern and Buena Vista Lakes (see the section of this document on the 1852–53 floods.)

Joseph Dixon interviewed J.W.B. Rice who lived on the Kaweah River, four miles northeast of Lemon Cove. This would have been in the vicinity of present-day Terminus Dam. Rice reported having trapped three river otters in the Kaweah near his place and knew of others being taken on that stream.[123]

There are four reported observations of river otters in the national parks:[124]
1. William Colby and Poly Kanawyer reported river otter in the Kings Canyon at about 5,000 feet elevation in about 1910. That would have been on the Middle Fork Kings River, about four miles upstream from Tehipite Valley.
2. Poly Kanawyer also reported river otter in Simpson Meadow (on the Middle Fork Kings River) in about 1910.

3. Ray Walls, an electrical foreman, reported seeing a river otter on the South Fork Kaweah River near Ladybug Camp in March 1941.
4. CCC educational adviser McDonald of the Maxon Ranch CCC Camp also reported seeing a river otter near Ladybug Camp one month later in April 1941. Presumably this was the same animal that Walls had seen.

The fur-bearing population in the San Joaquin Valley was rapidly hunted to near extinction, so no fur post was ever established here. James "Grizzly" Adams noted that a few beavers survived in secluded spots up to the late 1850s.[125] Walter Fry reported beavers at Ash Mountain (on the Kaweah River) in the national parks in 1920.

The California Department of Fish and Game (CDFG) reintroduced beavers to the upper tributaries of the South and North Forks of the Kern River between 1949–52.[126]

According to James Carson, deer (both "red and black-tailed") were present in large numbers in the general vicinity of Tulare Lake in the middle of the 19th century.[127] Apparently "red deer" was a reference to California mule deer. What we think of as red deer today are native to Europe. The California mule deer that lived in the mountains moved down to the area around Tulare Lake in the winter.[128]

The reference to black-tailed deer is a more interesting question. Today Columbian black-tailed deer are found largely in Northern California. They range south through the Monterey Bay and Big Sur areas to Ragged Point where they are replaced by California mule deer. That is some 150 miles west of Tulare Lake. There is no official record of Columbian black-tailed deer having ever lived in the San Joaquin Valley. However, perhaps Carson was right that a population of Columbian black-tailed deer was living in the Tulare Lake area in the 1850s. We really don't know what was present in that ecosystem before it was radically altered.

Coyotes, elk, and pronghorn were also present in very large numbers in the general vicinity of Tulare Lake in the middle of the 19th century.[129]

The following excerpt from a 1904 article in the Bakersfield Daily Californian entitled "The Phantom Antelope" provides a sense of the historical extent of the grassland-wetland ecosystem along Cross Creek just north of Visalia and offers some notion of the wildlife populations encountered by early settlers in this part of the San Joaquin Valley.[130]

*In the early (eighteen) fifties the plains between Kings River on the north and the Four Creeks timber (Kaweah River in vicinity of Visalia) on the south were the ranging ground of vast herds of antelope. A stream of water known as the Elbow swamp, caused by the spreading out of a branch of the Kaweah River that waters the Visalia country, split the plain between Kings River and the Four Creek timber into about two equal portions, debauching into Tulare Lake about midway between the two above named points. A level plain of about ten or twelve miles in width lay on each side north and south of Cross Creek and from the point where it left the Elbow swamp it was about eighteen or twenty miles in length to the point where it emptied into Tulare Lake. A narrow fringe of willows grew along its banks, with occasional bare breaks, and all along its banks on either side it was a favorite watering place for the herds of antelope that ranged upon the plain, which was covered at the time I speak of, with a luxuriant growth of grass extending all the way from the Sierra Nevada low hills, a distance of almost thirty to forty miles to Tulare Lake. Along the banks of this creek was an ideal hunting ground and it was the principal source from which the larders of the settlers at and around the vicinity of Visalia were supplied with flesh; and extremely palatable and juicy flesh it was, for nothing in the shape of meat can excel for flavor and excellence the roasted ribs of a fat antelope or a steak from one of his hind quarters. They were usually at their best in June and a prime buck at that stage carried globes of tallow on his kidneys that would rival the fattest of our south-down sheep at their best.*

*It was customary for a hunting party to encamp in one of the depressions easily found along the creek banks, and then spread out a mile or more apart and await the advent of a band of antelope at an accustomed watering place three or four men could secure a large supply of game in a day's hunt. I have been one of a party of five that killed ninety antelope in one day in this manner.*

Pronghorn and elk populations were quickly decimated by market hunters in the days before game laws. Pronghorn and elk were essentially eliminated by 1870.

Black and grizzly bears were also present. (The last grizzly bear in the state was killed in 1922 at Horse Corral Meadow in what is today Giant Sequoia National Monument.)

Carson and others reported gray wolves around Tulare Lake. Some have suggested that Carson was mistaking coyotes for gray wolves. It isn't universally accepted that gray wolves were even native to California. However, a review of early pioneer diaries shows that gray wolves were most likely present throughout California, including in the Central Valley.[131] Joseph Grinnell concluded that "unquestionably wolves ranged regularly over the northern one-fourth of the State and south along the Sierra Nevada to Inyo County at least…"[132]

Gray wolves had a significant presence in the San Joaquin Valley based on accounts left by early visitors. John C. Fremont noted in 1844 that he saw "wolves frequently during the day — prowling about for the young antelope, which cannot run very fast."[133]

John W. Audubon reported that gray wolves were very numerous when he traveled through the San Joaquin Valley in 1848. He said that "their long, lonely howl at night … tell the melancholy truth all too plainly, of the long, long distance from home and friends." The wolves were so bold at night that Audubon had "several pieces of meat and a fine goose stolen from over (his) tent door." He assumed that the wolves preyed on the elk that were abundant in the area.[134]

Hale Tharp was the first Euro-American settler in the Three Rivers area. He recalled that wolves were very plentiful in that area when he arrived in 1856. He saw six wolves when he took a trip to Log Meadow in what is now Sequoia National Park in the spring of 1861.[135]

Resident populations of gray wolves are generally thought to have been extirpated from California sometime in the 1800s. Gray wolves seen or trapped in the state in the late 1800s and early 1900s are generally presumed to have been wanderers from Oregon and Nevada.[136]

However, wolves apparently held on in parts of the Southern Sierra into the early 20th century. In describing the principal animals of Sequoia National Park, the superintendent's annual report for 1900 listed both coyote and "black wolf."

On September 25, 1908, Charlie Howard killed a wolf at Wolverton in Sequoia National Park. At the time, Howard was slaughtering beef for a troop of soldiers, and the wolf came up within 50 yards of his camp in broad daylight and was eating some of the beef offal. Someone, almost certainly Walter Fry, inspected the carcass. (The event happened in his ranger district, and he was a very curious naturalist.) Fry later reported that the wolf was a large male in fairly good condition, but quite old, as evidenced by badly worn teeth.[137] Guy Hopping, a long-time national park ranger and former superintendent of General Grant National Park, reported seeing and hearing a wolf in the Roaring River country of what is now Kings Canyon National Park in the summer of 1912. He described the howl as deep, like that of a big old hound.[138] Those are the last two reliable records of gray wolves in the Tulare Lake Basin.

Walter Fry served Sequoia National Park for 25 years as its chief ranger, superintendent, judge, and naturalist.[139] He concluded from his research that the gray wolf was native to the park.[140]

The last known specimen of a native gray wolf to be collected in California was killed by a government trapper in Lassen County in 1924.

The few wolves seen in California since then are presumed to have been captive-bred wolves released by humans. One of the few examples of such a record is a wolf shot and killed while raiding a chicken coop near Woodlake in 1962.[141] That wolf turned out to be an animal of Asian descent that was apparently an escaped pet.

On Dec. 28, 2011, a male gray wolf with a GPS collar crossed the state line from Oregon into Lassen County, becoming the first gray wolf known to live wild in California since 1924.[142]

Mountain lions, bobcats, and gray foxes were all present around Tulare Lake. Carson reported that ocelots were also present. Whether that is a reliable account or not is uncertain. Much has changed since the 1850s. Although the ocelot's present-day range comes close to California's border in Mexico, there have been no confirmed sightings in the state during recent times. The ocelot's historic range probably included at least Southern California. Harold Werner, the national parks' former wildlife ecologist, sees no reason that it couldn't have extended into the Tulare Lake Basin. So it's tempting to think that Carson was correct and a few ocelots were living in the area around Tulare Lake. Unfortunately, there was nothing like a thorough inventory done of that ecosystem before it was radically altered.

Felipe Santiago Garcia recorded in 1807 that wild horses and cattle were present in large numbers.[143] Those feral animals were descended from stock escaped and stolen from Spanish settlements along the coast. Some accounts from early settlers said that they didn't bother to hunt wild game because of the abundance of wild cattle.

Carson also reported that red foxes were present, but that seems suspect. The red fox is a relative newcomer to the Tulare Lake ecosystem. It was introduced into the lowlands of California beginning in the late 1800s for fur farming and fox hunting.[144] Today the Sierra Nevada red fox lives at relatively high elevation (e.g., the Sonora Pass region). Based on that, it is assumed that this species would not have been present in the Tulare Lake ecosystem before these non-native red foxes were released. Hopefully this is a correct assumption.

Carson reported that large numbers of gulls and band-tailed pigeons lived around the lake. He didn't say whether those populations were transient or resident. In more recent times, gulls have not nested at the lake. Similarly, band-tailed pigeons today are primarily a bird of forested parts of the Sierra and foothills where blue oaks and California bay grow. At the time when Carson was writing, the Kaweah Delta was a forested area with an abundance of valley oaks. With our 21st century perspective, it's hard for us to imagine just how different the ecosystem was back then. It may well be that large flocks of band-tailed pigeons resided on the delta or at least visited the area seasonally.

The Tulare Lake ecosystem was a significant stop for hundreds of thousands of ducks, geese, sandhill cranes, swans, curlews, snipe, and other birds migrating along the Pacific Flyway. Colonel Andrew Grayson described the density of waterfowl in the fall of 1853:

> *On October 31 our surveying operations brought us to the main Kern River. Here we found any quantity of elk and waterfowl, and such a place for hunters I never saw! The mallard duck abounded, but of every description of waterfowl my pen could scarcely describe the numbers or the excitement they would create in the breast of a sportsman. Your ears are confused with the many sounds — the quacking of the mallard, the soft and delicate whistles of the baldpate or teal, the underground-like notes of the rail or marsh hen, the flute-like notes of the wild geese and brant, the wild rantings of the heron, not to forget the bugle-like notes of the whooping crane and swan and a thousand other birds mixing their songs together — creates that indescribable sensation of pleasure that can only be felt by one fond of nature in its wildest and most beautiful form.[145]*

There are no valid records of whooping cranes in California. So in the above account, Grayson was almost certainly referring to the call of sandhill cranes.

The following account describes what conditions were like in the mid-1870s:[146]

> *The surface of the water was nearly always covered with some sort of water fowl. Geese, ducks, pelicans, snipes, mudhens, cranes and other birds were there by the millions at their own season of the year. Men went out with their blunderbusses and killed ducks by the thousands and they never seemed to grow less. The waters were filled with fish that could be caught with bare hook. During the hunting season parties left all parts of the state to go shooting on Tulare Lake. Arks were built and keepers engaged to take care of them. Hundreds of sportsmen made their annual pilgrimage to this big body of water.*

The lake and surrounding wetlands were a major waterfowl hunting area until at least the mid-1880s. (For example, a wagonload of swans was sold on Main Street of Visalia on the morning of January 7, 1886. The birds, weighing 18–20 pounds each, brought $1 each.)

Sequoia National Park was created in 1890. Walter Fry recorded the changes that he observed in bird life between 1906 and 1931 and then summarized these in a report.[147] Because of Sequoia National Park's location, it was never a significant breeding ground for waterfowl. However, Fry said that it served as a "splendid refuge" for waterfowl and shorebirds during the winter months. Until about 1930, it was a common sight in the park to see many ducks, geese, swans, and other such fowl during the autumn, winter, and spring, some of which remained in the park throughout the year. But by about 1930, most of those species were seldom seen, and when they were seen, they were few in number. Fry reported that the rapid decrease in water- and shore-birds started about 1909 and continued through the time of his report (1931).

Fry attributed the decline of these birds to four principal causes:
1. the settlement and drying up of their breeding grounds
2. the length of the open hunting season and bag limit
3. the increase in the number of hunters
4. disease

Fry provided a table in his report comparing the number of birds present by family in 1906 and 1931. That table is reproduced in part in Table 4:

Table 4. Number of species per family for selected bird families.

| Family | 1906 | 1931 |
|---|---|---|
| Grebes | 1 | 1 |
| Loons | 1 | 1 |
| Cormorants | 1 | 1 |
| Ducks, geese, swans | 17 | 2 |
| Herons, egrets, bitterns | 4 | 2 |
| Rails and coots | 2 | 1 |
| Stilts | 1 | 0 |
| Snipes and sandpipers | 2 | 1 |
| Plovers | 2 | 1 |
| Vultures | 2 | 1 |
| Hawks and eagles | 12 | 10 |
| Cuckoos | 2 | 1 |
| Flycatchers | 7 | 6 |
| Crows, jays, magpies | 6 | 5 |
| Blackbirds, orioles, meadowlarks* | 5 | 3 |
| Shrikes | 1 | 0 |
| Wood warblers | 11 | 10 |
| Total of above families | 77 | 46 |

* 3 species were lost from this family during this period, but the western meadowlark was added.

This represents a loss of 30 species in 16 families during this 25-year period. One new species was added: the western meadowlark. Fry observed that the losses were all among the water birds and migratory species. The period that Fry documented (1906–31) was generally the period during which the Tulare Lake ecosystem was being lost and the five valley lakes were drying up.

We understand the magnitude of the loss of Sequoia National Park bird life during this time, but we are unable to describe it in detail. We have a very incomplete list of which species were lost. In general we can only identify the loss at the family level. The tundra swan is the one notable exception: Fry recorded that it used to roost near Potwisha where the Marble and Middle Forks of the Kaweah River join.

We also lack a census of the number of birds by species. It is apparent from Fry's description that waterfowl and shorebirds used to visit the park regularly during the early decades of the park's existence. Since then, the park has remained largely undeveloped, but adjacent lands have changed radically. It is a reminder that our national parks do not exist as islands.

American white pelicans were migrants to Tulare Lake and also bred there periodically until at least 1942. Waterfowl returned in spectacular numbers when Tulare Lake had its last great reappearance in 1937–46, and there was abundant breeding by waterfowl, colonial water birds (grebes, cormorants, herons, egrets, and ibises), stilts, avocets, and terns in the South Flood Area during the floods of

1982–83 and 1997. While exciting to see, these bird congregations were ephemeral and moved on once the floodwaters receded.

During wet years, Tulare Lake was the terminus of the Western Hemisphere's southernmost (chinook) salmon run. The Kings River supported both spring and fall runs of salmon. On November 2, 1819, Spanish Lieutenant José María Estudillo observed Tachi tribesmen catching salmon and other fish in the Kings River by means of hand nets:

> *This they did before my very eyes, with great agility, diving quickly and staying under the water so long that I prayed.*

Others left accounts of Native Americans spearing fish in Tulare Lake and elsewhere in the Central Valley. The Native Americans dried and smoked large quantities of fish, prizing salmon above all other.

Yokuts tribesmen built tule balsa (rafts) and fished in Tulare Lake. These craft were reminiscent of the reed boats used on Lake Titicaca in Bolivia. The tule balsas could stay out on the lake for days, held up to a dozen people, and often had mud fireplaces for cooking.[148, 149]

In addition to the occasional spring salmon run, Tulare Lake had small populations of white sturgeon and steelhead and a very large population of "lake trout." The lake trout lived year-round in Tulare Lake and in the larger tributary rivers. A very wet winter such as 1860–61 would allow them to come up the smaller streams, rising as high as Antelope Valley on the Kaweah Delta (vicinity of present-day Elderwood). The lake trout was a fine, white-fleshed fish that grew to 30 pounds and was appreciated for its taste. Despite its name, it wasn't a salmonid; it was most likely the Sacramento pikeminnow (aka Sacramento squawfish or pike). Sacramento pikeminnow are still present in the larger rivers of the Tulare Lake Basin.

With the coming of American settlers, Tulare Lake became an important commercial fishery, shipping tons of fish to San Francisco each year. The fishery included lake trout, chinook salmon, Sacramento perch, and white catfish (after the introduction of that fish in 1873). Freshwater mussels (aka lake clams) were abundant. In addition to white catfish, a number of other exotic fishes were introduced into Tulare Lake and other lakes within the basin.

Tulare Lake was also known for its population of western pond turtles (locally called terrapin). Harold Werner, the national parks' former wildlife ecologist, recalled reading that the turtles were once so abundant that a roar was created when sunning turtles were disturbed and took flight into the water.

Those turtles were the source of a regional favorite. They were caught in seines and shipped live in sacks to San Francisco. There they were relished in terrapin soup and other delicacies.

The last category of animal that might have been present in the Tulare Lake ecosystem was marine mammals. This is not as farfetched a proposition as it might sound. During high-water periods, Tulare Lake was connected to San Francisco Bay by the San Joaquin and Kings Rivers. That was how chinook salmon entered the lake, so it is plausible that marine mammals could have used the same route.

73

If they did do this, they would have encountered apparently suitable habitat. Tulare Lake was brackish, rather like an estuary. There was a food supply including an abundance of freshwater mussels and a wide variety of fish.

Frank Latta reported that the Yokuts who lived around the lakes harvested both seals and sea otters. He also observed that Spanish expeditions reported seals and sea otters 150 leagues (375 miles) upstream from San Francisco Bay.[150] That measurement was presumably made along the Sacramento River. The same distance along the length of the San Joaquin would encompass the Tulare Lake ecosystem.

Marine mammals, at least sea lions, are still occasionally observed coming up the San Joaquin River. In February 2004, a male California sea lion came up the river and canals as far as Henry Miller Road north of Los Banos. He just kept going upstream until he ran out of water. At that point, he was about 65 miles from San Francisco Bay and only 100 miles from Tulare Lake. When a California Highway Patrol car arrived, the animal lumbered over, jumped up on the trunk and lay down.[151, 152, 153]

In its pristine state, the Tulare Lake Basin was like a wheel of water, with Tulare Lake as the hub and all the Sierra streams as spokes in the wheel. Once Tulare Lake (and the other four valley lakes) had been dried up, disintegration of this remarkably complex system was sealed with the damming of the four main rivers. The functioning infrastructure of this formerly biodiverse ecoregion was so badly broken that it resulted in the loss of most of the wetland habitat and nearly all of the biological connectivity between the watersheds in the high country and the lowland floodplains. Water-dependent habitats on the adjacent land, particularly on the Kaweah Delta and other riparian corridors, were also significantly degraded during the ensuing decades.

The loss of the Tulare Lake ecosystem affected even protected areas like the national parks. For example, we speculate that the relict populations of beavers, mink, and river otters that hung on in the park became isolated from populations elsewhere in the parks and the basin. Their numbers gradually declined, and some of those species may now be extinct in the parks. We also speculate that a similar problem occurred with many populations of fish, birds, and other animals.

### *Why is there no lake in the Tulare Lakebed today?*

One major cause is that we're using a lot more water, primarily for agricultural purposes. We're taking that water from our rivers before the water reaches the historic lakebed. In a big picture sense, there is not enough water to sustain both Tulare Lake and the needs of people. Society has found what it considers to be a better use for the water: serving the needs of people, rather than serving the needs of a natural resource.

Table 5 details both total runoff and inflow to Tulare Lake for the 19 largest runoff years that we are aware of. The total runoff shown in the center column is based on the data behind Figure 13. The inflow shown in the right-hand column is based on data covered under individual floods.

Diversion of river water for irrigation began on the Kings and Kaweah River Deltas in about the 1870s. By the end of the 19th century, there was a large network of canals diverting water out of the rivers in order to support extensive irrigated agriculture. By 1900, the Kings River — historically Tulare Lake's most important source of water — was irrigating more land than any other stream in the world except the Nile and Indus Rivers, over a million acres.[154]

Table 5 illustrates how dramatically inflows to Tulare Lake decreased as a result of those diversions. The large runoffs that used to sustain Tulare Lake continued to come. But after the turn of the 19th century, farmers were very successful in diverting most of those waters from the lakebed onto their irrigated lands. When the four federal reservoirs began operation during the 1954–61 period, they had an important, but relatively less noticeable effect on the amount of inflow to Tulare Lake.

Table 5. Flow measurements for the 19 largest runoff years: 1850–2011.

| Water Year | Total runoff of 4 major rivers[1] (acre-feet) | Total inflow to Tulare Lake[2] (acre-feet) |
|---|---|---|
| 1853 | [3] | 5,096,000 |
| 1862 | [3] | 6,290,000 |
| 1868 | [3] | 5,360,000 |
| 1906 | 7,195,240 | 1,530,000 |
| 1909 | 5,689,840 | 1,175,000 |
| 1916 | 6,512,710 | 1,041,700 |
| 1938 | 5,773,470 | 126,000 |
| 1952 | 5,375,050 | 583,000 |
| 1967 | 6,253,344 | 94,300 |
| 1969 | 8,379,585 | 1,155,000 |
| 1978 | 6,078,925 | [4] |
| 1980 | 5,821,879 | [4] |
| 1982 | 5,201,438 | [4] |
| 1983 | 8,746,222 | 1,069,000 |
| 1986 | 5,692,766 | [4] |
| 1995 | 5,814,847 | [4] |
| 1997 | 4,931,557 | [4] |
| 1998 | 4,883,910 | [4] |
| 2011 | 5,910,342 | [4] |

[1]This is the total runoff of the Kings, Kaweah, Tule and Kern Rivers.
[2]See Table 9 for examples of exports that have been made in recent decades to keep floodwaters out of the Tulare Lakebed.
[3] No runoff data is available for these rivers prior to 1894.
[4]There were some relatively small inflows to Tulare Lake in each of these years, but no measurements are available.

But there's another reason that the lake isn't in the lakebed, at least in most years. That also involves society's values: the value that it places on the lakebed itself. Society used to place a high value on Tulare Lake as a resource for all the food that the lake and its associated wetlands produced. However, today the resource that society values is the irrigated agricultural crops, improvements, and the towns that occupy the lakebed. Therefore, society marshals its resources to try to prevent the lake from returning to its lakebed. And when the lake does return in varying degrees, people strive to minimize the damage that it causes. Some people view Tulare Lake as an inconvenience, a nuisance to be prevented. Society has defined the presence of excess water in the lakebed as a flood.

A variety of steps have been taken in recent decades to keep water out of the Tulare Lakebed:

- Encouraging users below the four federal reservoirs to take all the water that they can productively use or store in retention basins so that the reservoirs can be drawn down in anticipation of an oncoming flood.
- Coordinating the operation of the federal reservoirs to keep floodwaters out of the lakebed. This has included engaging the assistance of private interests and of PG&E in this complex effort.[155]
- Installing temporary sandbag (or sack concrete) barriers on the spillway of the federal dams, thereby allowing the reservoirs to operate as much as 5½ feet above full pool level; thus keeping this water from flowing down to the Tulare Lakebed. See Figure 4 for a map showing how the federal reservoirs sit upstream of the lakebed. At Terminus Dam, the need for such temporary measures was eliminated when fuse gates were installed in 2004.
- Blocking the Kern River at Sand Ridge, causing a huge holding pond to form at the south end of Tulare Lake. This holding area has since been further developed and is now known as the South Flood Area.
- Diverting Kings River floodwaters to the San Joaquin River in order to minimize flooding in the Tulare Lakebed. This is done using the North Fork / Fresno Slough / James Bypass channel. The Fresno Slough Bypass (now known as the James Bypass) began operation in 1872. The capacity of the associated system has since been increased several times. Prior to about 1872, all of the Kings River water flowed into Tulare Lake. See the section of this document on Pine Flat Dam for a more detailed description of the James Bypass. Water that is sent through this system winds up in San Francisco Bay; it is essentially a loss from the point of view of Tulare Lake Basin water users. With the construction of Pine Flat Dam in 1954, the need to divert water through this system was greatly reduced. Even so, diversions through this system have occurred in 38% of the years since the dam was completed.[156]
- Diverting Kern River floodwaters into the California Aqueduct rather than into Buena Vista and/or Tulare Lakes. This is done using the Kern River Intertie and Cross Valley Canal. Once the water enters the California Aqueduct, it is pumped over the Tehachapi Mountains and sent to the Los Angeles area. The Kern River Intertie was completed in 1977.[157] Prior to that, a big flood on the Kern would first fill Buena Vista and/or Goose Lakes, and then spill into Tulare Lake. The term "beneficial use" refers to a reasonable quantity of water applied to a non-wasteful use. Transferring Kern River floodwaters over the Tehachapis to the Los Angeles area has now been determined to be a beneficial use.[158]
- Transferring water from the Kings, St Johns, and Tule Rivers into the Friant-Kern Canal in order to minimize flooding in the Tulare Lakebed. This is done by using pumps at the point where each of those rivers cross the canal. Once the river water enters the canal, it flows by gravity to the canal's terminus near Bakersfield. There it is emptied into the Kern River. The water is then routed to the Los Angeles area using the Kern River Intertie and Cross Valley Canal as described above. A combined total of over 472,000 acre-feet of floodwaters was pumped into the canal

during the years 1978, 1980, 1982, 1983, 1986, 1995, 1997, 1998 and 2006. (The total amount may have been a good bit more than this; records are incomplete.) Transfers may have been made in later years as well. Including all sources (four rivers), exports to the LA area have occurred in 30% of the years since the Kern River Intertie began operation in 1977.[159]

Despite all of the above efforts, floodwaters still make it to the Tulare Lakebed on occasion, especially in heavy runoff years (see Figure 11). To assist in the reclamation of the lakebed, over 20 reclamation districts were formed under California general reclamation district laws between about 1896 and 1925. The reclamation districts have built levee systems which divide the lakebed into cells or sumps. As floodwaters come into the lakebed, the sumps are filled, more or less in order. The first four cells (the South Wilbur Flood Area and the three Hacienda Reservoirs) are devoted to holding floodwater; they are never planted in crops. This use of lakebed levees minimizes the damage and allows the remaining portions of the lakebed to be used for agricultural purposes. In huge runoff years, emergency levees still have to be constructed within the lakebed to protect the towns of Corcoran and Stratford.

## *Role of Floods in Maintaining Tulare Lake*

Our first-hand knowledge of Tulare Lake dates back over 150 years to the middle of the 19th century. Floods would abruptly raise the level of the lake after which it would gradually shrink during the drought or non-flood years that followed (see Figure 10).

As described earlier, Tulare Lake and the other four valley lakes were not landlocked, an inland sink, the way that we think of them today. They were the anchors of a wetland complex of over 400,000 acres (see Figure 5). That complex connected with the wetlands that fringed the San Joaquin River, making a continuous wetland all the way to the Sacramento–San Joaquin River Delta.

The flood cycle of the Tulare Lake Basin was critical in maintaining that ecosystem; the floods provided sufficient water storage to keep the lake going through the drought or non-flood years. Once Tulare Lake and the other four lakes had been dried up, the last remnants of the ecosystem totally disintegrated with the damming of the four main rivers. The functioning infrastructure of this formerly biodiverse ecoregion was so badly broken that it resulted in the loss of most of the wetland habitat and nearly all of the biological connectivity between the watersheds in the high country and the lowland floodplains.

Floods — with the water that they brought — created a marvelous ecosystem in the Tulare Lake Basin. Reminders of that ecosystem survive in disjointed preserves in the valley, in the foothills, and in the Sierra. The framework of the hydrologic system that powered that ecosystem still exists today. On occasion, flooding can recreate a portion of Tulare Lake. The last significant reappearances of the lake were brought on by the floods of 1982–83 and 1997.

But just adding water to the Tulare Lakebed is not enough to recreate the complex ecosystem that once existed. The associated habitat is highly degraded, and the ability of the ecosystem to provide connections among the various river and stream courses in the Tulare Lake Basin has largely been lost.

### *Chronology of Tulare Lake*

A popular perception is that Tulare Lake was relatively stable before agricultural diversions began. Perhaps reflecting this mythology, one source said that when the Spanish first visited Tulare Lake in 1772, it was about 50 miles long and 35 miles wide. No source was given for this measurement, so it should probably be attributed to legend or wishful thinking. What information we do have on lake levels prior to 1844 indicates that it was not constant, but varied as a function of runoff and perhaps other climatic factors.

Annie Mitchell wrote that Native Americans said that Tulare Lake went dry about 1825.[160]

John C. Fremont (along with his scout, Kit Carson) led two government expeditions through the San Joaquin Valley. Carson was a good choice because he had traveled from Taos to the San Joaquin Valley in 1830 on a trapping expedition. On his first expedition in 1844, Fremont explored the east base of the Sierra as far south as present-day Bridgeport. Short on supplies, Fremont then decided to make the first ever mid-winter crossing of the Sierra.

In late January, the party turned west and started pushing their way up the East Fork of the Carson River. By February 6, conditions were appalling: they were lost, out of food, and the stock was in poor shape. Quoting from Fremont's diary for that day:

> *Two Indians joined our party here, and one of them, an old man, immediately began to harangue us, saying that ourselves and animals would perish in the snow; and that if we would go back he would show us another and a better way across the mountain. He spoke in a very loud voice, and there was a singular repetition of phrases and arrangement of words, which rendered his speech striking, and not unmusical.*

> *We had now begun to understand some words, and, with the aid of signs, easily comprehended the old man's simple ideas. "Rock upon rock—rock upon rock—snow upon snow—snow upon snow" said he; "even if you get over the snow, you will not be able to get down from the mountains." He made us the sign of precipices, and showed us how the feet of the horses would slip, and throw them off from the narrow trails which led along their sides. Our Chinook, who comprehended even more readily than ourselves and believed our situation hopeless, covered his head with his blanket, and began to weep and lament. "I wanted to see the whites," said he; "I came away from my own people to see the whites, and I wouldn't care to die among them; but here"—and he looked around into the cold night and gloomy forest, and, drawing his blanket over his head, began again to lament.*

> *Seated around the tree, the fire illuminating the rocks and the tall bolls of the pines round about, and the old Indian haranguing, we presented a group of very serious faces....*

Fremont decided to continue on up the Carson River, and the party crossed what would later be known as Carson Pass on February 14. Traveling through deep snow and blizzard conditions, the expedition reached Sutter's Fort (in present-day Sacramento) on March 6, 1844. Only 33 of their 67 horses and mules survived the passage, most of the rest had been eaten. Continuing south, Fremont reached the Kings River on April 8, and noted that most of the flow of that river was going into Tulare Lake.

He also observed that Tulare Lake was overflowing into the San Joaquin River, indicating that the lake was sufficiently high to overtop the delta sill that dams it. This is the earliest reliable measurement of the size of Tulare Lake.[161]

In 1844, the southern lake (Ton Taché) was nearly as extensive as the northern lake (Taché). A slough connected the two lakes, passing through Sand Ridge. During very high-flow years, this slough served (and still serves) as part of the extension of the Kern River.

Fremont returned to the San Joaquin Valley on his next expedition in the winter of 1845–1846. This time, he split his party and led one portion west up the Truckee River and crossed the Sierra at Donner Pass, arriving at Sutter's Fort on December 8, 1845. From there, he headed south to rendezvous with the rest of his expedition at the river that he knew as the Lake Fork of the Tulare River. We know that river today as the Kings River. Fremont's group arrived at the rendezvous first.

When he didn't find the rest of his group at the rendezvous spot, Fremont led his portion of the expedition (16 men on horseback herding a number of cattle) up the Kings River in search of them. What began as a search would develop into a mid-winter exploration of the Sierra.

After a one-day rest, they began their trek east on December 24, driving their cattle with them. They rode through the oak woodlands along what was likely the mainstem and North Fork of the Kings River for several days until they started to climb higher.

They worked their way up through the oak and conifer forests and some "extremely large" trees. The historian Francis Farquhar interpreted those trees to be the McKinley Sequoia Grove, seven miles west of present-day Wishon Reservoir in the North Fork Kings River Basin. The party eventually reached the 11,000 foot elevation, coming out onto a bare granite ridge that divided the North Fork of the Kings and the South Fork of the San Joaquin River. One of the few places on the divide where they could have driven cattle to this elevation was in the vicinity of Hell for Sure Pass at the west boundary of present-day Kings Canyon National Park. It was a beautiful day to be in the Sierra, the weather comfortably warm.

The next day, December 31, Fremont's party headed back down to the San Joaquin Valley. They had almost waited too late to begin their return. The weather quickly turned bad as a big snowstorm blew in. Fremont's decision to check out the Sierra had been an incredibly rash act, and the storm almost cut off their escape from the mountains.

*The old year went out and the new year came in, rough as the country.*

They soon had to abandon their cattle and had difficulty getting themselves out from the snow. But within a few days they returned to the valley floor, completing their grand scouting adventure.[162] On January 4, 1846 they returned to the Kings River, camping at the east end of present-day Pine Flat Reservoir.

After returning to the valley, the expedition engaged in more mundane explorations. Among other accomplishments, they mapped Tulare Lake as being about 60 miles long. If that accurately represents the amount of water in the winter of 1845–1846, it would suggest that the lake was close to being full. However, S.T. Harding concluded that Fremont's length measurement probably included flooded areas outside of the actual lakebed, such as the Fresno Slough area.

Lieutenant George H. Derby of the U.S. Army's Topographical Engineers visited the Tulare Lake area in May 1850.[163] By then, the southern lake (Ton Taché) was essentially dry, having been drained by the slough that passed through Sand Ridge. The remains of that southern lake formed a tule swamp 10 miles wide and 15 miles from north to south.

Derby reported that the gradual receding of the water was distinctly marked by a ridge of decayed tules upon its shore, and that he had been informed, and had no reason to disbelieve, that 10 years previous it had been nearly as extensive a sheet of water as the northern lake. That seems plausible. Fremont's measurement of the lake suggests that it was full or nearly so when he visited it in 1844, six years before Derby's visit.

One source said that Tulare Lake measured 570 square miles in 1849, but the reliability of that measurement is questionable. There were no known surveyors in the area at the time. *Possibly* this is a reference to the lake's size in 1850 when Derby measured it.

On May 9, 1850, Derby came to the Kern River, which was discharging into Buena Vista Lake by two separate mouths. At that time, Buena Vista Lake was 10 miles long and 4–6 miles wide.

After leaving the Kern, Derby crossed the Tule, Kaweah, and Kings Rivers, heading north. He then turned west to explore the huge wetland complex between Tulare Lake and the San Joaquin River (i.e., the Fresno Slough system). 1850 was a very heavy runoff year, the heaviest in the memory of the Native Americans who lived on the Kings River. Derby discovered that the entire flow of the Kings and a significant portion of the San Joaquin were flowing toward Tulare Lake. As the peak of the flood was approaching, the lake was still not overflowing the delta sill. Tulare Lake had been about one foot above the delta sill prior to the flood (elevation 208 – 207 feet). But because the sill was densely vegetated with tules, significant outflow didn't really start until the lake reached an elevation of 210 feet. The flood would eventually raise the lake to a maximum elevation of 211.5 feet, sending water flowing back toward the Sacramento–San Joaquin River Delta. (Derby was lucky. The waters were rising rapidly while he was in the area and his party barely escaped entrapment.)

In 1851 and 1852, the lake remained almost brim-full, nearly to the level of the delta sill that serves to regulate its maximum height. However, the flood of 1852–53 raised Tulare Lake by 11.5 feet. At this point, the lake had a depth of about 37 feet at its deepest point and a maximum elevation of 215.5 feet. Tulare Lake would reach this size only twice more: in 1862 and 1867. Over the next eight years (1853–61), the lake dropped 16 feet. This became the pattern for the lake over subsequent decades. Floods would abruptly raise the level of the lake after which it would gradually shrink.

While Tulare Lake was at this high stage during the 1852–53 flood, some sailors jumped ship in San Francisco and stole a whaleboat. They hoisted the sail and headed inland. Taking advantage of the prevailing winds, they sailed south up the San Joaquin River, through the Fresno Slough, and entered Tulare Lake. This is the first of five documented trips between that lake and San Francisco Bay to occur in historic times. (The other four were in 1868, 1938, 1969, and 1982–83.)

John M. Barker lived on a cattle ranch on the Kings River near Tulare Lake. One morning in the winter of 1857, he and a neighbor started out on horseback to search for some horses that had strayed. They skirted the shores of Tulare Lake between Cross Creek and the Kings River (apparently just west of present-day Corcoran). For a couple of miles from the shore, the waters in the shallows were covered with burnt tules and other refuse matter unfit for use by man or beast.

They knew that their horses would not drink from the lake (presumably because it was brackish and alkaline), but there were sloughs and water in depressions outside of the lake, where the water was clear and fit for use. They headed to one of those waterholes in order to look for tracks of their missing stock. As several of them were shod, they knew if they found shod tracks that they would be on the right trail.

Barker dismounted and walked to the edge of the water. Just as he reached it, a massive earthquake struck. The lake commenced to roar like the ocean in a storm, and the cowboys rode as fast as they could to get away from there. They returned the next day and found that the lake had run up on the land for about three miles. Fish were stranded in every direction and could have been gathered by the wagon-load.[164]

Apparently the earthquake that Barker experienced was the Southern California Earthquake of 1857 (aka the Fort Tejon Earthquake). At magnitude 7.9, this was the most powerful earthquake to hit Southern California in historic times.

Tulare Lake had been at a very high stage after the 1852–53 flood, the second highest ever recorded. After 1853, there was a gradual shrinkage of the lake until the fall of 1861. Over those eight years, the lake dropped about 13 feet in elevation.

There were multiple causes of this:
- The maximum elevation in 1853 (215.5 feet) was higher than the elevation of the Tulare Lake sill (207 feet). The water above this elevation simply flowed out of the lake and connected through the Fresno Slough to the San Joaquin River, and from there it flowed on to San Francisco Bay.
- Normal evaporation in our hot valley summers (averaging 5.2 feet per year).
- Eight years with only low or average runoff and no floods.
- Two years of drought (1856 and 1857).
- Diversion for irrigation was just getting underway (negligible).

Gordon's Ferry (aka Gale's Ferry) was located just north of present-day Bakersfield College. The Sinks of the Tejón was the first Butterfield Overland Mail stop north of Fort Tejon. It was located at the intersection of present-day David and Wheeler Ridge Roads, roughly 10 miles northeast of where Interstate 5 and Highway 99 diverge. When the Kern River came out of its canyon in the winter of 1861–62, it created one vast sea of water from Gordon's Ferry to the Sinks of the Tejón. Kern Lake was located in the southeast corner of that huge sheet of water. Buena Vista Lake backed up to within 12 miles of Fort Tejon.[165]

In the summer of 1861, Tulare Lake reached a low of 200.3 feet. The 1861–62 flood raised the lake by 15.7 feet to elevation 216, the highest that the lake has been during historic times. At elevations above 207 feet, the lake over-topped the lowest point on the Tulare Lake sill. At the lake's highest stage, about 9 feet of water flowed in a broad expanse northerly over this sill (elevation 216 - 207 feet). From there, the water flowed into the Fresno Slough and the San Joaquin River. At the height of this flood, the lake was about 37 feet deep at the deepest point (elevation 216 - 179 feet). The surface area increased from about 350 square miles in 1861 to about 790 square miles in July 1862.

S.T. Harding estimated that 6,290,000 acre-feet of water flowed into Tulare Lake in the single season 1861–62. For comparison, that is 3.9 greater than the combined current capacity of all four of the federal reservoirs in the Tulare Lake Basin.[166]

Before the 1861–62 flood, the Kern River channel ran where the Kern Island Canal now runs in Bakersfield: by the Beale Library (between Chester and Union Ave) on its way to Kern Lake. The flood shifted the river to the west. The new channel began at Gordon's Ferry (just north of present-day Bakersfield College) and passed through what is now Old River and into the Las Palomas slough system between Kern Lake and Buena Vista Lake on its way to Tulare Lake. Not only did that new channel bypass Kern Lake, but one source said that it also bypassed Buena Vista Lake, meaning that those lakes would only get water during years with very high runoff. In any case, the river would shift even farther northwest in the 1867–68 flood.[167, 168]

Bill Tweed calculated that when all the tributaries of the San Joaquin River were swollen with snowmelt, the total flow of that river as it approached the Sacramento–San Joaquin River Delta could exceed 100,000 cfs. At such times, the river spread out for miles across the flat valley floor. That would presumably describe the condition that existed in the flood of 1861–62.

Tulare Lake gradually declined in elevation after the 1861–62 flood. In the summer of 1867, the lake level was elevation 200.7 feet. However, the 1867–68 flood raised it by 14.7 feet, bringing it back to a maximum elevation of 215.4 feet. At the height of the flood, Tulare Lake was almost 37 feet deep at the deepest point. The lake has not been this deep since (see Figure 10).

In 1868, Richard Smith loaded a 16-foot scow with a one-ton cargo of honey and made the 170 mile journey from Tulare Lake to San Francisco Bay.[169] That remains the only recorded commercial trip ever made from the lake to the bay. (There were four *non-commercial* trips: 1852, 1938, 1969, and 1982–83.) Apparently Smith was able to make the return trip back through the tules to Tulare Lake.

In 1872, the Fresno Slough Bypass (now known as the James Bypass) began operation. The Kings River Handbook says that the James Bypass was "developed" in 1912–14.[170] This may mean that the bypass was further improved at that point. In any case, this channel works with the North Fork to route a portion of the Kings River floodwaters to the San Joaquin River. Prior to about 1872, all of the Kings flowed into Tulare Lake. See the section of this document on Pine Flat Dam for a more complete description of the North Fork / James Bypass.

White catfish were introduced to Tulare Lake in 1873. This is just one of several exotic fish that would be introduced to the lake. See the section of this document on Wildlife in Tulare Lake, for a discussion of the lake fishery.

The landlocked form of Atlantic salmon (*Salmo sebago*, *S. salar sebago*, or *S. salar ouananiche*) was introduced to Tulare Lake in about 1878.

From 1854 to 1872 the lake changed very little in area. Almost due west from Bakersfield there was a shrinking, but otherwise its area remained about the same. It was about these years that irrigation started in the valleys around Visalia and Bakersfield and the shrinking became very rapid. The rivers were tapped in several places and the water that would have gone into Tulare Lake was spread out over the dry pastures and cotton fields. The shrinking was most marked from 1872 to 1875. The southern end of the lake contracted and became somewhat in the form of a creek. It narrowed until it was not more than a mile wide and had drawn up from the southern end at least 15 miles.[171]

The 1878 flood filled Tulare Lake to elevation 207.5 feet, causing it to spill over the delta sill and into Fresno Slough and the San Joaquin River for the last time. That was the last natural overflow of the lake; Tulare Lake has never filled again. Since 1878, the Tulare Lake Basin has functioned as essentially as a closed basin, an inland sink without a regular outlet to the ocean.

A number of sailboats and at least two steamboats plied the lake in the 1870s and 1880s. The *Mose Andross* was a 50-foot long, side-wheel steamboat that A.J. Atwell built and operated from 1875 until 1879. This is the same Atwell who owned the lumber mill that began operation at Atwell Grove in 1879. The *Mose Andross* was built primarily to service Atwell's farming interests at Atwell's Island (site of present-day Alpaugh), but it also served as general transport during the years that it operated.[172]

The *Mose Andross* was flat-bottom, so it could pull in almost anywhere. However, there were six regular landings that it serviced in addition to Atwell's Island (that was the original spelling).[173] Those landings were located around the lake at the following points (see Figure 8):
- Cox and Clark ranch (an adobe, just south of present-day Kettleman City)
- Gordon's Point (6 miles north of Kettleman City)
- Dan Rhoades ranch (an adobe, south of Orton Point, near present-day Lemoore)
- Buzzards Roost Landing (immediately south of present-day Waukena)
- Near the Artesia Schoolhouse (at the mouth of Cross Creek, south of present-day Waukena)
- Creighton Ranch (at the mouth of the Tule River)

The *Mose Andross* was used as much for pleasure trips as for freighting. The following announcement appeared in the *Tulare Times* in 1875:[174]

> *EXCURSION ON THE LAKE: There is to be a May Day excursion on Tulare Lake and a dance on a barge in the evening. We acknowledge receipt of complimentary tickets from Captain Atwell, owner of the lake steamer to be used on this occasion.*

Gustav Eisen, the Swedish natural scientist, visited Tulare Lake in 1878, and recounted his adventure 20 years later:[175]

> *In 1878 I crossed Tulare Lake on a steamboat. This was a regular packet that ran between Hanford and a small town on the west side of the lake. (Presumably he was referring to Kettleman City.) The distance across was about thirty miles. There were one or two other steamboats running on the lake at the time. Sailboats were numerous and altogether Tulare Lake was of considerable use to the commerce of the region.*

> *On the occasion that I made my trip across the lake we were all treated to a surprise. When we were about twelve miles from Hanford, and almost out of sight of land, the boat ran over the ruins of an old ranch. We could look down through ten or twelve feet of clear water and see the fence posts of an old pig sty. There was also the foundation of a house and several metal utensils scattered about. Nobody on the boat knew whether the ranch had been on an island that had sunk from sight or whether it had been on the mainland during some previous dry year. It was a mystery.*

That ranch presumably became established during the low-water years of either 1857–61 or 1865–67. Each of those dry periods ended with a dramatic flood, causing the lake to rapidly rise by 10 feet or more. The owners of the ranch Eisen saw must have been stunned when that occurred. The lakebed floods can arrive with surprising swiftness and virtually no warning. See the section of this document that describes the 1867–68 flood for an account of several hog camps that were caught up in that flood. The onset of flooding in the lakebed swept in abruptly on Christmas Eve, 1867, catching people by surprise. They had to beat a hurried retreat, the rising waters on their heels the whole way.

The schooner *Water Witch* (formerly the *Alcatraz*) was brought from San Francisco to Tulare Lake by "Eating" Smith in 1878. (Smith earned his nickname for his big appetite.) The ownership of the *Water Witch* changed hands twice after its arrival at the lake. First Smith traded it to the McCoy brothers for some cows. The McCoys used the *Water Witch* for two seasons of harvesting turtles, sending as many as 300 dozen to San Francisco in one season. They then sold it to Captain Thomas J. Conley in 1880. Conley was described as living "near the notorious Work's adobe" near the South Fork of the Kaweah.[176]

Hopkins (Hop) Work had arrived in Three Rivers in 1859. His family was one of the three original pioneer families to settle in the Kaweah canyons. He built an adobe cabin on the east side of the South Fork, apparently at the junction of that river with the mainstem of the Kaweah. We don't know why that cabin later gained a reputation of being notorious.

In 1880, at the time of the Creighton Survey, Tulare Lake had dropped to an elevation of 200 feet; its surface covering only 445 square miles.

In 1881, Thomas Conley patented 80 acres of land on the west side of the South Fork, near where the Shoshone Inn is today. That would have been right across the South Fork from Work's adobe, apparently a well-known landmark of the day. Presumably Conley Creek is named for Thomas Conley or his family.

In 1881, Captain James W.A. Wright of Hanford met Conley in Slapjack Canyon on the road to Mineral King. Conley had just cut new masts for his ship. The following May, Conley and Wright would embark on a six-day excursion to map Tulare Lake in detail.

Mussel Slough is located along the western edge of present-day Hanford. In 1881–82, Tulare Lake reached what was then considered an unusually low stage, about elevation 192 feet. At that point, the lake margin laid bare an area near the mouth of Mussel Slough which was covered with the broken stumps of long submerged trees (illustration on file in the national parks). C.E. Grunsky made a number of visits to that area while doing a study of the water resources of the San Joaquin Valley during the years 1881–88.

At that location, he could see the location of an old channel entering the lake from the northeast. He deduced that this was the former channel of Mussel Slough during a protracted period in which the lake was at or below its then low stage. Some of the stumps had a diameter of about four feet. Their dimensions and position indicated that they were the remnants of a grove of willows which had reached mature growth along the bank of the watercourse and the margin of the lake.

Grunsky concluded that low lake stages with conditions favorable to the growth of those willows must have been continuous for a period of some 40–50 years or perhaps much longer. Therefore, at sometime in the past before the arrival of the white man, the lake had been at or below that elevation for a long period of time. After the lake rose to a stage high enough to drown the willows, it remained at or above that stage for 50–100 or more years, keeping the stumps submerged until their discovery. The long period of persistently light or moderate rainfall favoring the growth of the willows was followed by a long period in which the frequency of fairly wet winters kept the lake at fairly high stages, culminating with the very high waters of 1853, 1862, and 1868.

Grunsky presented his findings to a meeting of the American Meteorological Society in June 1930.[177] At the time of his talk, the San Joaquin Valley was undergoing a severe drought. Tulare Lake had been repeatedly dry during the preceding 30 years. The prevailing assumption was that the lakebed would remain generally dry forever more. In his talk, Grunsky made the point that we needed to take a longer term view toward climate change. Tulare Lake had been dry for an extended period in the not too distant past, and he expected that it might come back to life in the foreseeable future.

In 1878, Tulare Lake had filled to elevation 207.5 feet and spilled over the delta sill. However, in 1883, just five years later, the lake had decreased to just nine feet deep (elevation 188 - 179 feet). While that was remarked upon as the lowest elevation in memory, the lake continued to dwindle in size over the next 16 years.

The lake shrank from the south to the north. In 1882 the southern border of the lake left Kern County altogether.[178]

In May of that year, Captains James W.A. Wright (a former Civil War officer) and T.J. Conley (a man with sea-faring experience) made a six-day excursion across and around the lake in the *Water Witch*, making careful measurements and soundings of the lake. They found the greatest depth to be 22 feet, in a comparatively narrow depression like a river channel, from the mouth of Kings River on the north to Terrapin Bay on the south (see Figure 8). Other than this channel, the rest of the lake had a maximum depth of eight feet or less. The area of the lake was then 417 square miles.[179, 180]

Wright left a detailed and highly readable account of their six-day mapping adventure.[181] The *Water Witch* sailed the lake for a few more months until capsizing in a severe storm about three miles southeast of the mouth of the Kings River in the winter of 1882. That location was just south of present-day Stratford.[182, 183]

The rapid drying up of Tulare Lake was written up in the *New York Times* in 1884.[184] A few years earlier, the lake had been 33 miles long by 21 miles wide. By 1884, it had shrunk to 15 miles long with an average width of 8 miles.

As Tulare Lake shrank, the fishing technology changed. By 1887, large scale, land-based seining had become possible. Seines were taken out into the lake about a mile and drawn to shore by horse-powered windlass. There were at least five seines running, making two hauls each day with up to 2,200 pounds per haul. Each haul included 500–1200 pounds of perch in addition to catfish and lake trout. The lake trout were plentiful and ranged in size from 2–20 pounds, occasionally as large as 30 pounds.[185, 186]

As Tulare Lake shrank, it changed shape and configuration. The lake had been described as having the shape of an oyster when it was at full pool. By 1888, the remnants of the lake had become almost circular in shape.[187, 188]

As the lakes shrank, the alkalinity rose. The ecosystem started to go into a tailspin. 1888 seems to have been the pivotal year. The fishing (or seining) was apparently terrific that year as the ecosystem crashed. Over 133,600 pounds of fish from Tulare Lake were shipped to San Francisco in one ten-week period in the fall of 1888.[189] By the end of that year, the catfish, lake trout, pond turtles, mussels, and clams had reportedly died out of all three lakes (Tulare, Kern, and Buena Vista) due to the increasing alkalinity.

The *New York Times* reported in the summer of 1889 that Tulare Lake had dropped to less than three feet deep at its deepest part.[190] Elevation that year was 183.5 feet.

An editorial in the *Visalia Weekly Delta* in 1889 captured the current thinking of the day:[191]

> *Tulare Lake, from present appearances, will soon have to be erased from the maps of the state of California. As a geographical fact it exists today as a "lake" by courtesy only — for it is not a lake.*

By the time that Kings County was formed in 1893, the lake had shrunk to about 220 square miles.

The *New York Times* reported in 1898 that the lake had dried up completely.[192] This was the first time that had happened in historic times.

The drying up of the lake was best described in an August 1898 issue of *The San Francisco Call*. That newspaper article was essentially an obituary for the lake.[193]

The lake had gone from full-pool to bone-dry in just 20 years (1878–98). There were apparently five causes for the lake drying out during this period:

1. Diversion of river water for irrigation; this began on the Kings and Kaweah River Deltas in about the 1870s. The effectiveness of these diversions is dramatically illustrated in Table 5 on page 75.
2. The Fresno Slough Bypass (now known as the James Bypass) began operation in 1872. Prior to about 1872, all of the Kings River water flowed into Tulare Lake.
3. The general absence of winters with heavy precipitation. The 20-year period from 1878–98 had very few winters that were excessively rainy. There were three water years with moderately heavy runoff (1884, 1886, and 1890) as measured by inflows to the Tulare Lakebed. But those weren't nearly enough to make up for the deficit in the other years. This was in contrast with the 1860s which had storm systems that put the greater portion of the San Joaquin and Sacramento Valleys under water several times.[194]
4. Only one major flood during this 20-year period: 1890.
5. At least six years of fairly serious drought during this 20-year period: 1879, 1882, 1887, 1888, 1897 and 1898.

Gustav Eisen was a Swedish natural scientist and a member of the California Academy of Sciences. He provided a short article for the *San Francisco Call* issue on Tulare Lake giving his observations about the reasons for the lake drying up. Eisen didn't have any data or claim subject matter expertise. However, he thought that the principal cause of the drying- up of the lake was the use of the waters of the tributary rivers for irrigation purposes. In addition, he recognized that the general absence of excessively rainy winters since 1874 had a good deal to do with it.[195]

As the lake shrank, agriculture moved in. By May 1895, there was 50,000 acres of grain growing in what had been the Tulare Lakebed. The lake was essentially not to be seen.

The Tulare Lakebed dried completely for the first time in August 1898 and remained dry through 1900. The 1901 flood brought the lake back to life, if only modestly. After that flood, the lake was about six feet deep at the deepest point (elevation 185.5 - 179 feet).

Tulare Lake was virtually dry when Hobart Whitley visited it in 1905. However, the high runoff of the 1906 flood brought the lake back. That flood left the lake about 12 feet deep at the deepest point, submerging 300 square miles. (This compares to 790 square miles at its maximum in 1862 and 1868.) As a relative measure of the volume of the runoff, that was the biggest increase in the lake's depth since the 1890 flood. Combined runoff of the four rivers in the Tulare Lake Basin during water year 1906 was 7,360,000 acre-feet, the second largest runoff of record. The total floodwater entering the lake in that year was about 1,530,000 acre-feet. This inflow exceeds that of any year since that time.[196]

Many levees were constructed in the Tulare Lakebed between 1903–1905 when lake levels were low. Unfortunately, those levees were light and poorly constructed. As a result, they failed when the high flows of 1906 entered the lake. The failure of those levees resulted in large financial losses, as almost 175,000 acres of wheat and barley had been planted that year. Most of that land was flooded before the crops could be harvested.

The lake continued to rise with the floods of 1907 and 1909, and then gradually receded for the next seven years.

By September 1914, the lake had dropped to an elevation of 180.0 feet, less than one foot deep. Avian botulism became a problem in the lake that year. Corcoran was incorporated in 1914, the year that the 1912–13 drought ended.

In 1907, a massive levee had been built around four sides of Tulare Lake, attempting to constrain it to a fraction of its full natural size. Ripley's *Believe It or Not* featured the "Square Lake" in its syndicated cartoon. The lake was now harnessed, the lakebed declared safe for growing orchards. However, the 1916 flood brought the lake back to life and put an end to those hopes, at least temporarily.

The lakebed again went completely dry on April 30, 1919.

Floodwaters from the Kings and Kaweah arrived in the lakebed in May 1922. A total of 23,680 acres of lakebed cropland was inundated that year. A little more water (from both the Kings and the Kaweah) was added from heavy rains during the winter of 1923–24, but the lakebed was completely dry again early in 1924. Thanks in part to the extended droughts of 1922–34, it would be 13 years before the lake would come back to life.

The lake did get some inflows from both the Kings (1923, 1927, 1932, and 1935) and the Kaweah (1923, 1932, and 1936) during the drought years. However, most of the quantities during those years were small, and S.T. Harding believed that much of that water was quickly absorbed by the soil or used directly for irrigation of crops growing in the lakebed. Tulare Lake would not reappear as a large lake until 1937.

The 1937 flood was a major flood, bringing an end to the drought years. The Kern River sent floodwaters into Tulare Lake for the first time since 1916. Tulare Lake reappeared on February 7, 1937 for the first time since 1924. The lake rose to an elevation of 191.9 feet and would stay at roughly that elevation for nine years. American white pelicans, waterfowl, and shorebirds reappeared almost instantly and in incredible numbers.

Ward B. Minturn was a prominent Fresno businessman and one of the most prominent field ornithologists in the San Joaquin Valley. He made at least 43 visits to Tulare Lake between 1937–54. Thanks to Rob Hansen's research, we have a wonderful collection of Minturn's field notes. An entry from his field notes of October 16, 1937 gives an idea of how quickly the bird life of Tulare Lake responded when the water returned:

> *After being dry for several years, Tulare Lake is back. Main body is confined by levees in an area 6 miles by 8 miles. Quite a sea! Today I saw there one of the greatest bird sights of my experience. Great masses of white pelicans so thick on the levees that those in the center could not spread their wings to rise until those on the edges had taken flight! Pelicans flying, pelicans swimming, pelicans feeding so thick in nearby fields as to look at a distance like great snowbanks. Truly a sight I shall never forget. How many? I wish I knew. Possibly 40,000 to 50,000. I did not know there were so many left in the U.S.A.*

(To some extent, this resilience still exists. Just add water and almost instantly an enormous number of waterfowl will appear. Prime examples are the floods of 1982–83 and the flood of 1997.)

In February and March, 1938, heavy storms flooded the San Joaquin Valley. When the elevation of Tulare Lake reached 192 feet, one of the main levees in the lakebed broke and the lake spilled over 49 square miles of land. The lake continued rising, eventually cresting at 195 feet. By June, 135,600 acres of the lakebed was underwater. That was the maximum acreage flooded since the 1906 flood. Tulare Lake has not been this big since.[197]

While the high lake levels of 1938 were a disaster for the lakebed farmers, others saw opportunity. Near the height of the flood, Frank Latta and three boys took a 15-foot homemade motor boat from Bakersfield to San Francisco.[198, 199]

When World War II started, the Navy needed a place in the southern San Joaquin Valley where seaplanes could land in an emergency. The realization of this need occurred after 11 seaplanes from Hawaii arrived over Alameda Naval Air Station (NAS) from Hawaii only to find the entire region fogged in. Diverted south to Los Angeles, the pilots, low on fuel, were surprised to find themselves flying over a body of water that they knew nothing about — Tulare Lake. They landed without incident.

In January 1942, the Navy leased 3,000 acres of the Tulare Lakebed for an emergency seaplane landing base (Tulare Lake Outlying Field) and other purposes. In addition to a radio building and lighted buoys for navigation, the facility consisted of about two tents and six sailors. It was located about 10 miles south of Stratford. The only activity after that first emergency landing was a monthly seaplane visit from the commanding officer stationed at Alameda and a few low-level practice torpedo runs. In at least a conceptual sense, Tulare Lake Outlying Field might be thought of as the precursor of Naval Air Station Lemoore. Regrettably, NAS Lemoore does not have facilities for seaplane landings.

In 1943, enough runoff made it to the valley to raise the level of Tulare Lake to near the top of the lakebed levees. Wave action caused levee breaks and the flooding of 28,000 acres. Those levee breaks increased the size of Tulare Lake from 46,000 acres to 74,000 acres. By summer, 100,000 acres would be flooded.

1945 was a big flood year in the Tulare Lake Basin. It was also the first year of use of the new works, built by the USACE, to keep the Kings River out of the Tulare Lakebed. They did not work quite as designed. A break in that bypass occurred on February 3, 1945, about 20 miles south of Hanford at the height of the flood. Some ranchers were driven from their homes on the east side of the bypass and considerable grain was flooded on the west side.[200]

The J.G. Boswell Co. bought the Cousins Ranch in 1946. At that time, the ranch was still under water as a result of the 1938 flood.[201]

Thanks in large part to the extended drought of 1943–51, Tulare Lake again went completely dry on July 17, 1946.

The lake reappeared briefly four years later:
- November 19, 1950 – March 10, 1951 (maximum elevation 184.8 feet)

The winter of 1951–52 brought near-record snows to the Southern Sierra. The addition of rain to this snowpack caused Tulare Lake to reappear on January 19, 1952. That was the biggest episode of lake flooding (both by height and duration) between the early 1940s and 1969. See the section of this document that describes the 1952 flood for the considerable efforts that were made to store the floodwaters of the Kings and Kern Rivers before they reached the Tulare Lakebed. Despite these efforts, the 1952 flood still raised Tulare Lake by 15.5 feet to a maximum elevation of 194.6 feet, flooding 72,700 acres. The lake has never been this high since, although the 1969 flood would come close.

Tulare Lake again went dry late in 1953. It would reappear briefly on three occasions over the next 16 years
- December 23, 1955 – April 21, 1956 (maximum elevation 187.4 feet)
- March 31, 1958 – August 15, 1958 (maximum elevation 187.9 feet)
- December 6, 1966 – August 9 1967 (maximum elevation 183.1 feet)

Tulare Lake next reappeared on January 20, 1969. By the end of March, 125 square miles (80,000 acres of farmland) had been inundated. The total estimated lakebed inflow in 1969 was about 1.155 million acre-feet. This is the second biggest flood since the federal reservoirs were completed (both by volume and by area flooded); only the 1983 flood was bigger.

In 1969, 960,000 acre-feet were impounded in the Tulare Lakebed, inundating 88,700 acres, significantly more than was flooded in 1952. The J.G. Boswell Co. had more land flooded in the Tulare Lakebed than any other landowner (almost 50,000 of the total 88,700 acres).

All of the flow from the Kings was diverted into the San Joaquin River in 1969, at least during the January and February floods. The Kaweah and Tule Rivers both contributed significant flows to the lake. Even ephemeral Deer Creek (which fed the historic Ton Taché Lakebed near present-day Alpaugh) was flowing into the lake just south of Sand Ridge that spring. However, the majority of the inflows to the Tulare Lakebed came from the Kern River.[202]

Historically, the Kern would fill Buena Vista Lake before spilling over into Tulare Lake. However, in 1969, a giant dike protected two-thirds of Buena Vista Lake from being filled. When the other third of the lake filled, the Kern then spilled or passed through to Tulare Lake. The decision to keep the remainder of Buena Vista Lake dry was not appreciated by those downstream in the Tulare Lakebed.

In 1952, the Tulare Lake Basin Water Storage District had stored Kern River floodwaters in Buena Vista Lake. That was presumably possible because the J.G. Boswell Co., which had a long-term agricultural lease for the Buena Vista Lakebed, was willing to have its land flooded. In any case, no such water storage was allowed in 1969. That created hard feelings among some who were being impacted by the flooding that was occurring in the Tulare Lakebed in 1969. Emotions ran high as did financial losses.

The decision to pass through the Kern River floodwaters was challenged in court. But in the meantime, the floodwaters continued to come.[203] As a result, about 222,000 acre-feet of Kern River water flowed into the Tulare Lakebed. (The majority of the Buena Vista Lakebed remained dry, safe behind its giant levee.) For a more complete description of that event, see the section of this document that describes the 1969 flood.

On May 8, 1969, the USACE received approval for a half-million-dollar project to throw up levees to connect the separated segments of Sand Ridge, south of the current Tulare Lake, creating a gigantic holding pond capable of containing 100,000 acre-feet of Kern River floodwater.[204] Along with the South Wilbur Flood Area (located north of Sand Ridge), that is the area known today by Tulare Lake water storage districts and irrigators as the South Flood Area.

On June 24, 1969, Tulare Lake reached its highest modern level at 192.5 feet. (The last time that the lake had been this high was in 1952.) An emergency levee was hurriedly built just west of the Corcoran Airport. Tulare Lake was deep enough to cause significant erosion to that levee.

When the lake came back, it brought an abundance of crayfish. Mo Basham's family lived in Corcoran at the time, and her father recognized that this was a natural resource not to be wasted. For the next two years, he organized the neighborhood kids to go on crawdad hunts along the edge of the lake. The kids would bring back hundreds at a time, and their families would eat them just like lobster. To read Mo's description of these hunts, see the section of this document that describes the 1969 flood.

In 1969, two fathers and their sons took advantage of the high water to boat from Bakersfield through Buena Vista Lake and Tulare Lake to San Francisco Bay. For a few more details on this adventure, see the section of this document that describes the 1969 flood. This was the fourth documented trip between the lake and the bay. (The other trips were in 1852, 1868, 1938, and 1982–83.)

Minor flooding occurred in the Tulare Lakebed in 1970, 1971, 1973, 1978, 1980 and 1982.

This was followed by the major flood of 1983. In that year, the lake rose to 191.44 feet and flooded a slightly larger area than in 1969. Bill Tweed recalled that Tulare Lake was so big in the summer of 1983 that you could see it from the High Sierra, shining through the valley haze. To see it was like seeing a ghost, a relic of another time.

In 1982–83, two young naturalists from Bakersfield took advantage of the high water to kayak from Tulare Lake to San Francisco Bay. They had to do a good bit of portaging, but they made it. Dave Graber, NPS regional chief scientist, recalled that their trip was written up in the *Fresno Bee*. This was the fifth documented trip between the lake and the bay. (The other four were in 1852, 1868, 1938, and 1969.)

After two years of flooding (1982 and 1983), cotton growers decided to drain their lands, and also save the lake towns of Corcoran, Stratford, and Alpaugh in the process. They proposed to pump the excess water over the top of the Tulare Lake sill. The water would then flow into the North Fork of the Kings River, and from there to the San Joaquin River and the Sacramento–San Joaquin River Delta.

The Tulare Lake Irrigation District applied for a permit to pump the excess water over the top of the Tulare Lake sill. It appears that there was considerable opposition to granting this permit. Under an emergency proclamation issued by the USACE during the spring of 1983, reclamation districts and land companies remade the channel along some 29 miles of the Kings River to dewater the lake and drain the water north into the Sacramento–San Joaquin River Delta region.

A series of pumps were installed with a total lift of 43 feet. The project was designed to remove approximately 2,000 acre-feet of water per day from the lakebed. Pumping began on October 7, 1983 and continued intermittently until the program was terminated on January 19, 1984. Only about 90,000 acre-feet was pumped northward over the Tulare Lake sill under that program. Pumping was stopped earlier than scheduled due to concern that white bass might be transferred from the Tulare Lakebed into the San Joaquin River. The lakebed would not be fully drained until water year 1985.[205]

Exotic white bass had been illegally introduced into Lake Kaweah by fisherman during the 1970s. Large numbers escaped into waters downstream of Lake Kaweah during the record 1982–83 flood runoff. A large population became established in the flooded Tulare Lakebed and connecting waterways. Lake Kaweah and downstream waters of the Tulare Lakebed were treated with rotenone in the fall of 1987. This was one of the largest such chemical treatments ever carried out in the U.S., and certainly California's largest. The cost of the project was about $9.7 million.[206] Apparently, a complete kill of white bass was achieved; they were completely eliminated from Lake Kaweah and from throughout the rest of the Tulare Lake Basin.[207]

Minor flooding occurred in the Tulare Lakebed in 1985 and 1986.

Tulare Lake had appeared in 1937–46 and 1951–53. However, since the damming of the Kings (1954) and the Kaweah (1962), only very wet years have seen water return to the lakebed in significant amounts. The lake occasionally reappears during unusually wet years, as it did in 1969, 1983 and 1998.

In order to minimize flooding in the Tulare Lakebed, a combined total of over 472,000 acre-feet of floodwaters was pumped into the Friant-Kern Canal during the years 1978, 1980, 1982, 1983, 1986, 1995, 1997, 1998 and 2006. Once those floodwaters reached the Bakersfield area, they were emptied into the Kern River and then routed via the Kern River Intertie (a structure that was completed in 1977) into the southbound California Aqueduct. That water then made its way to the Los Angeles area rather than being available for use or groundwater recharge within the Tulare Lake Basin.

Additional amounts of water from the Kings and Kern Rivers have been diverted out of the basin in many years. See the section of this document that describes Groundwater Overdraft, for a discussion of those diversions.

Parts of the Tulare Lakebed still become periodically inundated during major flood events (see Figure 11). However, only remnants of the historic wetland area in the lakebed remain, confined primarily to privately owned waterfowl hunting clubs, former agricultural ground that has been enrolled in wetland reserve programs, and the Pixley and Kern National Wildlife Refuges (although neither one of those refuges is in the actual lakebed).

Figure 10 illustrates how the elevation of Tulare Lake has varied for the 120 years for which we have data: 1850–1969. Data since 1969 are only available from the J.G. Boswell Co., and that has proved impossible to obtain.

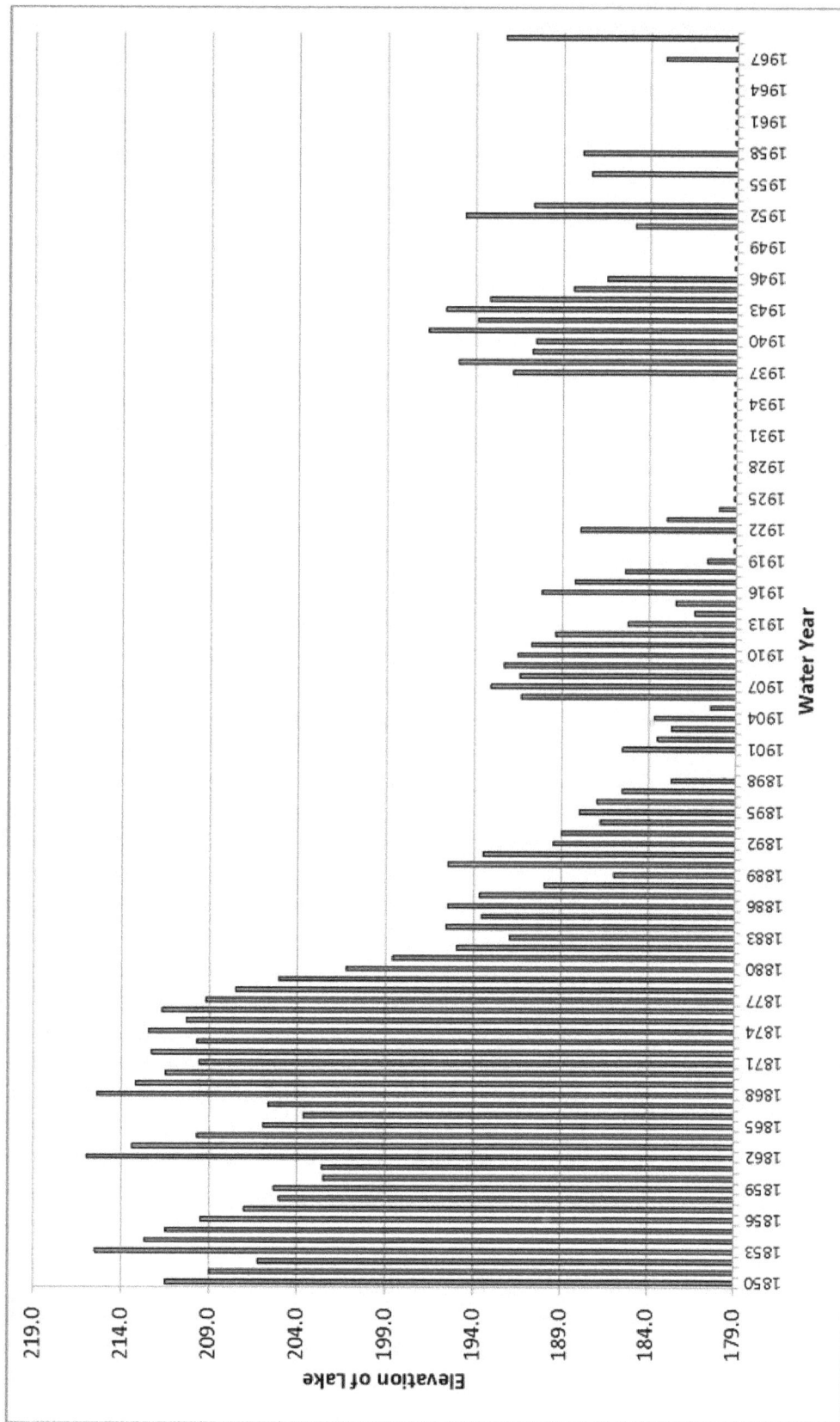

Figure 10. Elevation of water in the Tulare Lakebed for 120 years: 1850–1969.
Source: Data from USBR which obtained it from USACE which compiled it from a variety of sources.[208]

94

# CONDITIONS IN THE TULARE LAKE AREA
# SINCE COMPLETION OF PINE FLAT DAM

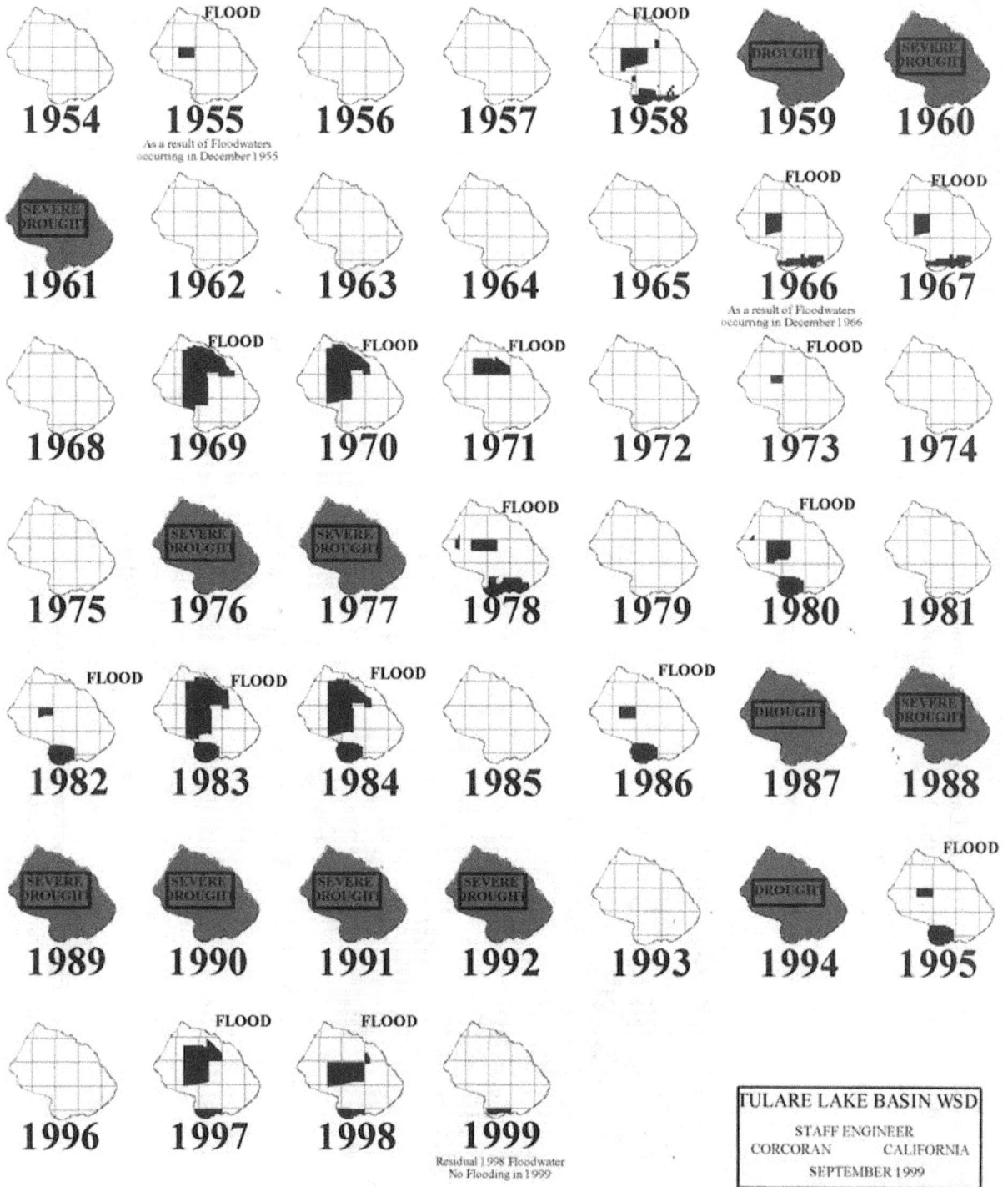

Figure 11. Portion of the Tulare Lakebed flooded each year 1954–99.
Source: Tulare Lake Basin Water Storage District

## Federal Dams and Reservoirs

### *Friant Dam*
This 319-foot-high concrete dam forms Millerton Lake on the San Joaquin River. The watershed drainage area is 1,675 square miles. The dam was built by the U.S. Bureau of Reclamation and completed in 1942; it is part of the Central Valley Project. Storage capacity is 520,528 acre-feet.

### *Pine Flat Dam*
This 429-foot-high[209] concrete dam is on the Kings River a few miles above Piedra, about 28 miles northeast of Fresno. The watershed drainage area of the Kings as measured from above the dam is 1,545 square miles.[210] The drainage area as measured from above the USGS gage at Piedra (USGS 11222000 Kings R A Piedra CA) is 1,693 square miles.

Regulation of discharges from Pine Flat Reservoir began December 4, 1951, although the dam was not completed until 1954. Full-scale operation of the reservoir began in February 1954.

Pine Flat Dam was built with a gross-pool capacity of 1,000,000 acre-feet.[211] That amount of gross storage can hold 59% of the 118-year average runoff (1894–2011) for the Kings River. That gross storage percentage is important primarily from the standpoint of irrigation, not flood control.

Pine Flat Reservoir has a gross-pool capacity of 1,000,000 acre-feet, of which 475,000 acre-feet is reserved for a flood-control pool.[212] That leaves 525,000 acre-feet available for a conservation pool in the winter.

The dam is operated to reduce floodflows to a downstream objective release of 4,750 cfs below Crescent Weir. Dam operators also try to minimize floodflows into the Tulare Lakebed.[213]

Although the maximum objective flow of Pine Flat Dam is 4,750 cfs below Crescent Weir, that location is many miles downstream of the dam, several miles west of Highway 99 (see Figure 12). There are many creeks that enter the Kings River prior to that point. There are also many diversion canals upstream of Crescent Weir to divert water out of the river. Wayne Johnson is chief of the Water Management Section in the Sacramento District of the USACE. He recalled that in the past, releases of 10,000 cfs from the dam have occasionally resulted in flows less than 4,750 cfs at Crescent Weir.

Pine Flat Reservoir is more formally (but perhaps less commonly) known as Boone Lake. At full pool, the reservoir covers about 5,956 acres and extends 20 miles back from the dam[214] with 67 miles of shoreline. Its gross pool elevation is 951.5 feet.[215]

Under the Kings River Fisheries Management Program, Kings River Water Association member units have agreed to maintain a minimum reservoir storage of not less than 100,000 acre-feet. The purpose is to maintain a pool of cool water for use in the reservoir and downstream fisheries under many, although possibly not all, critically dry conditions.

The Kings River divides in its lower reaches, about 60 miles below the dam (see Figure 12). The distributary point is located north of Lemoore, about 1½ miles above Highway 41. (A distributary is a branch of a river that flows away from the mainstem. They are common on deltas. The Kings,

Kaweah, and Kern Rivers all have distributaries.) Two diversion structures at this distributary point, Army Weir and Island Weir, control where flows are sent.[216]

The southerly channel flows southeasterly and southerly into the Tulare Lakebed. It is known by a variety of names in different stretches. It begins as the Clark's Fork. Farther along, it empties into the South Fork of the Kings which turns south toward Stratford and terminates in the Tulare Lakebed. However, since that lakebed is normally dry, the river has been extended another 10 miles in the South Fork Canal, which intersects the Tule River Canal at a point 12 miles west of Corcoran. It's an ignominious end for a fine river. The USACE lists the capacity of the South Fork Kings as 3,200 cfs.

The northerly channel is also known by a variety of names in different stretches. It begins as the North Fork of the Kings. (Not to be confused with the river of the same name that originates in the High Sierra.) At the Crescent Weir, the North Fork empties into the Fresno Slough which later empties into the 12-mile-long James Bypass channel. The James Bypass merges with the San Joaquin River at Mendota Pool near the city of Mendota.

The Kings River Delta begins about where the town of Kingsburg is located. Prior to 1862, all the water of the Kings River used to flow into Tulare Lake via what is now called the South Fork system, along the south side of the delta. Since then, a system has developed that routes a portion of the Kings River floodwaters along the north side of the delta.

The 1861–62 flood on the Kings River began the formation of Cole Slough, cutting the head of that slough. The slough was named for William T. Cole, who dug the irrigation ditch that the Kings River enlarged to form the slough. The 1867–68 flood completed the formation of Cole Slough. From Cole Slough, the floodwaters flowed on through Murphy Slough, possibly also formed in the 1861–62 and/or 1867–68 floods. These two floods created the conditions necessary to start moving a significant portion of the flow of the Kings River from the south side of its delta to the north.

The Zalda Canal was constructed in 1872. That canal was enlarged in the 1878 flood. This reach is now known as the North Fork of the Kings or the Kings River North Channel. The USACE increased the capacity of this channel from 3,500 cfs to 5,500 cfs just prior to the onset of the January1969 flood. The North Fork Channel now has a rated capacity of 6,300 cfs.

The James Bypass is a manmade channel. The Fresno Slough Bypass (now known as the James Bypass) began operation in 1872. The bypass reached its present configuration during the years 1912–14. It was built to bypass the meanders of a portion of the Fresno Slough and is sometimes still referred to as the Fresno Slough Bypass. The USACE lists the capacity of the Fresno Slough and James Bypass as 4,750 cfs. However, flows up to 6,000 cfs have passed through this reach.[217] A USGS stream gage (USGS 11253500 James Bypass (Fresno Slough) NR San Joaquin CA) (a water-stage recorder) was maintained on this stretch from October 1, 1947 – September 30, 2009.

It is hard to say at what point the Kings River began to flow across the north side of its delta. Some flow may have begun after the 1861–62 flood. There was almost certainly flow after the 1867–68 flood completed the formation of Cole Slough. Flow would have increased much more in 1872 with construction of the Fresno Slough Bypass (now known as the James Bypass) and the Zalda Canal. In this document, 1872 is used as the date of the first significant flow along the north side.

The small original Fresno Slough channel (the part that has been bypassed) meanders for nearly 15 miles, from southeast of San Joaquin to north of Tranquillity. It is no longer a part of Kings River operations. A large portion of the old slough is now managed as a wetlands area by the California Department of Fish and Game. (This agency will be known as the Department of Fish and Wildlife after January 2013.)

Total channel capacity in the lower reaches of the Kings is 7,950 cfs (3,200 cfs for the South Fork system and 4,750 cfs for the North Fork system). For comparison, the flow of record on the Kings River is 112,000 cfs, set on January 3, 1997.

Under typical flood operations, the first 4,750 cfs of flood release water from Pine Flat is directed through the North Fork / Fresno Slough / James Bypass channel to the San Joaquin River and, ultimately, San Francisco Bay. When the capacity of that channel has been reached, floodwater is sent into the Kings River South system up to its published channel capacity of 3,200 cfs. Flow in excess of 7,950 cfs is supposed to be divided equally between the two channels. In practice, the stage of the San Joaquin River during large floods may affect how water is divided between the two channels.[218]

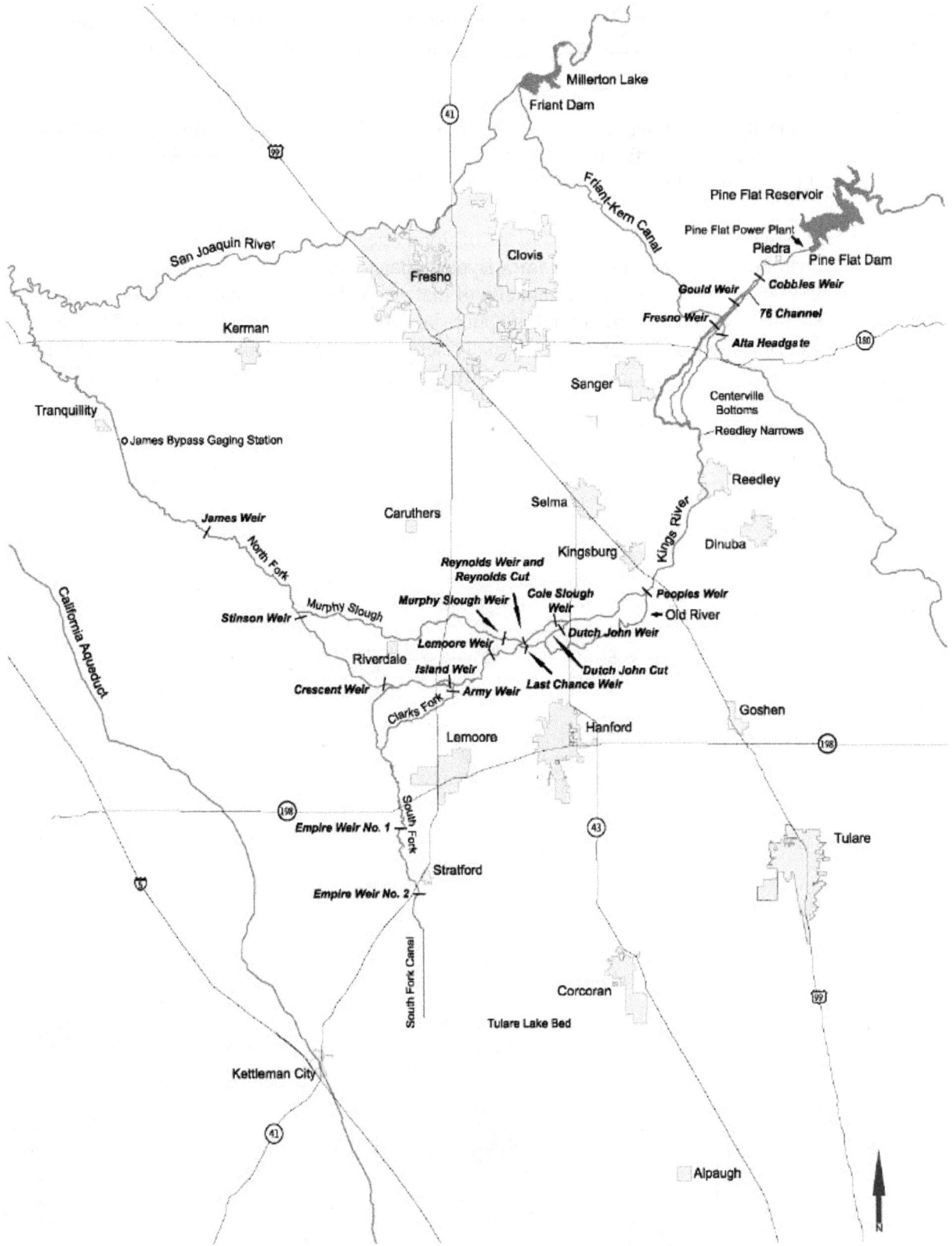

Figure 12. Map of Lower Kings River features.
Source: Kings River Handbook

99

### *Terminus Dam*

This 250-foot-high[219] earthen dam forms Lake Kaweah on the Kaweah River. The watershed drainage area of the Kaweah as measured from above the dam is about 561 square miles.[220, 221, 222] The Dry Creek watershed drainage area immediately below Terminus Dam is about 80 square miles.[223] The watershed of the Kaweah as measured from above McKay's Point is about 647 square miles.[224, 225] The entire watershed drainage area of the Kaweah as measured from above the Tulare Lakebed is about 952 square miles.[226]

One USACE report said that the dam was closed on November 1, 1961.[227] However, most sources say that the dam was completed and storage began in February 1962.[228]

Terminus Dam was built with a gross-pool capacity of 149,600 acre-feet (often rounded to 150,000). We haven't been able to find a record of how much of this was reserved for the flood-control pool, but it may have been 148,600 acre-feet.

The dam is operated to reduce floodflows so that the Kaweah doesn't exceed its rated channel capacity (5,500 cfs) three miles downstream at McKay's Point. Since Dry Creek enters the Kaweah below Terminus Dam, releases have to take the flow of that creek into consideration. The floodflow as measured at McKay's Point is considered the maximum objective flow for the dam. Dam operators also try to minimize floodflows into the Tulare Lakebed.[229]

When originally constructed, the lake's level at full pool (its gross pool elevation) was 694 feet. That provided a storage capacity of 150,000 acre-feet, which was estimated to be sufficient to provide a 60-year level of flood protection downstream.[230, 231]

A reservoir's gross-pool capacity is defined by its area-capacity curve. That curve is a graph showing the relation between the surface area of the water in the reservoir and the corresponding volume.

A reservoir's rated gross-pool capacity changes through time. The statement of a reservoir's gross-pool capacity is based on three factors:
- The volume of the reservoir as of the date of the most recent area-capacity curve.
- The assumed long-term annual rate at which sediment is accumulating in the reservoir.
- The year that the reservoir's capacity is being projected to.

A reservoir's area-capacity curve may be revised periodically for various reasons. This revision can be triggered by new information about the rate of sediment accumulation or by a change in the elevation of the pool level.

The Kaweah River, upstream of Lake Kaweah, has an unusually high potential to erode and carry sediment. That is due to its gradient and the type of soil that it is eroding.

The Kaweah is the only river in the U.S. that drops 10,000 feet in less than 100 miles. Measured over its entire length, the Kaweah is the steepest river in the U.S. It drops 10,826 feet in 76.5 miles, a gradient of 142 feet per mile. Individual reaches of rivers can be much steeper. The Marble Fork of the Kaweah has a gradient of 559 feet per mile, dropping 8,549 feet in just 15.3 miles.

From a geomorphic standpoint, a key metric is the drop from the headwaters to the range front. By that metric, the Kaweah is also the steepest river in the U.S. When measured from its headwaters to the range front at Terminus Dam, the Kaweah drops 10,505 feet in 37.5 miles, a gradient of 280 feet per mile.[232]

Lower elevation soils are more erodible than higher elevation soils. Compared to rivers such as the Merced, the various tributaries of the Kaweah have many more miles in which to erode these lower elevation soils. In addition, the geology of the lower elevation of the Kaweah River Basin consists of sedimentary or metamorphic rock that is far more erodible than the granite of basins such as that of the Merced. This combination of high-energy streams and erodible soils give the Kaweah the ability to carry a high sediment load relative to other Sierra streams.[233]

The Kaweah's sediment load used to be delivered to the Kaweah Delta. See the section of this document on Description and Identification of Deltas. But once Terminus Dam was closed in 1962, Lake Kaweah became a sediment trap. There has never been a sediment gage on the Kaweah, so the USACE had to estimate what the sediment load was when they were designing Terminus Dam. One source said that the initial design was based on an estimated long-term sediment yield for the watershed upstream of Lake Kaweah of 150 acre-feet a year. We haven't been able to find any documentation to clarify this.

Once the dam was in place, the sediment load could be measured through periodic surveys of the lakebed during the summer when the reservoir was drawn down. Detailed surveys were made in 1961, 1967, 1977, and 1988. Additional samples of bank and bed materials were analyzed and a field reconnaissance was done in 1988 and 1989.[234, 235]

The reports on the four reservoir surveys (1961, 1967, 1977, and 1988) have apparently all been lost. There have not been any general surveys of the reservoir bed since 1988.

The 1967 and subsequent surveys showed that the quantity of deposition in the reservoir bed was quite high. USACE studies attributed that largely to the 1966 flood, a storm that had a recurrence interval in excess of 100 years.[236]

In the lake's first 17 years, it received an average of 474 acre-feet a year of sediment.[237] As noted above, that was the largely the result of an unusually large flood event in 1966. But as noted elsewhere, the Kaweah has been unusually quiet of late. As illustrated in Table 17, the Kaweah hasn't seen any 20-year or larger floods in over 40 years.

The USACE revised Lake Kaweah's area capacity curve significantly downward in 1978 as a result of the 1977 survey. They documented this in a 1978 report, but apparently all copies of that report have been lost. Fortunately, Allen Wilson of the Kaweah Delta Water Conservation District (KDWCD) has a clear recall of some of that report's main findings. There are also allusions to that 1978 report (and to the 1977 survey) in various USACE reports. Apparently Lake Kaweah's area capacity curve was not revised again until after the fuse gates were installed in 2004.

The new area capacity curve set the lake's gross-pool capacity at 143,200 acre-feet (generally rounded to 143,000 acre-feet). Of this total, 142,000 acre-feet was reserved for the flood-control pool.[238] That was sufficient to provide only a 46-year level of protection.[239, 240]

The USACE studied the situation in 1989 and concluded that sediment yield for the watershed upstream of the reservoir should be approximately 100 acre-feet per year.[241] All copies of that report have apparently been lost.

Lake Kaweah's flood-control pool is small compared with the drainage area tributary to the lake. Because of this, the lake provides a relatively low level of protection from rain floods (see Table 6). This is also illustrated by the 1997 flood. That had a recurrence interval of only 14 years for the Kaweah (see Table 17). Even so, Lake Kaweah filled and emptied twice during that flood.[242]

Fuse gates were installed in the spillway in 2004. This raised the lake level at full pool 21 feet (from elevation 694 to 715 feet). When the Lake Kaweah Enlargement Project was originally designed in the 1990s, it was estimated that this would increase the lake's storage capacity to an estimated 183,300 acre-feet (generally rounded to 183,000). As the project neared completion, this estimate was revised to 185,630 acre-feet (generally rounded to 185,600).[243] There have been no new measurements of sedimentation rates since 1988.

As illustrated in Table 6, the level of flood protection for downstream communities has varied since the dam was built. When originally constructed in 1962, Lake Kaweah's storage capacity was estimated to be sufficient to provide a 60-year level of flood protection downstream. That is, it could catch that level of rain flood while keeping flows downstream of the dam within stated channel capacity / maximum objective flow. But protection decreased significantly after 1966 due to an accumulation of sedimentation. When fuse gates increased the flood-control pool size of the reservoir in 2004, the level of protection increased to a 70-year level of flood protection. [244, 245]

Table 6. Change in level of flood protection provided by Terminus Dam.

| Year | Elevation of lake at full pool | Gross-pool capacity (acre-feet) | Level of flood protection |
|---|---|---|---|
| 1962 | 694 | 150,000 | 60-year |
| 1978 | 694 | 143,200 | 46-year |
| 2004 | 715 | 185,600 | 70-year |

Wayne Johnson is chief of the Water Management Section in the Sacramento District of the USACE. He said that in calculating the 70-year level of flood protection, the USACE estimated that the actual storage in the reservoir was probably less than 185,600 acre-feet due to sediment that has occurred within the lake area since the sediment survey that was completed in 1977. (Records from the sediment survey completed in 1988 have apparently been lost.)

If Lake Kaweah's current gross storage capacity were 185,600 acre-feet, that could hold 43% of the 118-year average runoff (1894–2011) for the Kaweah River. That gross storage percentage is important primarily from the standpoint of irrigation, not flood control.

At full pool, Lake Kaweah covers about 2,154 acres and extends 6 miles back from the dam. Its gross pool elevation is 715.0 feet.[246]

Lake Kaweah has a gross-pool capacity of 185,600 acre-feet, of which 184,600 acre-feet is reserved for a flood-control pool.[247]

Prior to the Lake Kaweah Enlargement Project, Lake Kaweah had to be kept practically dry each winter. The winter conservation pool was only 1,000 acre feet (143,000-142,000 acre-feet).[248] The rest of the reservoir was kept dry for the flood-control pool of 142,000 acre-feet.

The Lake Kaweah Enlargement Project Water Control Plan established a more flexible plan for winter operations. It established a conditional winter rain flood storage pool of 12,000 acre-feet.[249] Wayne Johnson said that this value may be reduced based on a rain flood variable (aka rain parameter). This parameter is based on the precipitation that has occurred in the Kaweah River Basin above the dam. The wetter the basin is, the greater the flood pool requirement is, and so the lower the water conservation pool must be. In other words, if the basin is wet, the lake must be lower. If the basin is dry, the lake can be higher, up to 12,000 acre-feet. This conditional winter rain flood storage pool has essentially no effect on the level of flood protection.

The flow of record on the Kaweah is 105,000 cfs, set on December 6, 1966. The dam with fuse gates is designed to withstand a flood of 300,000 cfs. The first fuse gate is designed to tip at 190,000 cfs.

By one account, Lake Kaweah had seen 10 different floods as of 2001, starting in 1966 with the last coming in the 1998 season. Presumably those were 1966, 1969, 1978, 1980, 1982, 1983, 1986, 1995, 1997, and 1998.

The Kaweah River divides in its lower reaches. Major floods used to periodically result in the relocation of the distributary point where those channels divide. However, since the December 1867–68 flood, the channels have divided at McKay's Point, about a mile northwest of present-day Lemon Cove and three miles below Terminus Dam. A variety of structures have been built at this location since 1870 in an attempt to control the Kaweah and split the flow between the two channels. For a more complete description of McKay's Point and the diversion structures that have been built there, see the section of this document that describes the 1867–68 flood.

The southerly channel, known as the Lower Kaweah River, flows to the southeast. That channel ends east of Visalia, just north of the Ivanhoe turnoff (Road 156/158) on Highway 198. From there, it feeds distributaries on the south side of the Kaweah Delta, principally Mill Creek, Packwood Creek, and Cameron Creek.

The northerly channel, known as the St. Johns River, flows along the north side of Visalia. It merges with Cross Creek northwest of the city, a few miles upstream of Highway 99. The Shipp Cut was made in 1854, a small drain ditch from the Kaweah River Swamp near Rocky Ford (north of present-day Kaweah Oaks Preserve) west to Canoe Creek. The 1861–62 flood cut a new channel along the northern border of the swamp. Shipp Cut and a section of Canoe Creek were enlarged by the floodwaters and became a part of this new channel, and finally a connection was established with the Cross Creek channel, creating what we now know as the St. Johns River. The 1867–68 flood further enlarged the St. Johns and eroded a new head for that river about a mile farther upstream, farther into the swamp.

The USACE considers the total channel capacity in the lower reaches of the Kaweah / St. Johns (below McKay's Point) to be 5,500 cfs. Above this, flooding occurs in Visalia and other delta towns. This compares to the flow of record on the Kaweah River of 105,000 cfs that occurred during the December 1966 flood.

The weir at McKay's Point is used to send the majority of the Kaweah River floodwaters around the north side of Visalia through the St. Johns River. A levee on the south side of that channel protects the city, but has failed in a number of floods. See the section of this document that describes the St. Johns Levee — Condition in Recent Years, for a summary of the challenges that the city and county face in maintaining that levee and keeping it from failing.

## *Success Dam*

This 142-foot-high[250] earthen dam is on the Tule River. The watershed drainage area of the Tule as measured from above the dam is 393 square miles.[251] The dam was completed and storage began in November 1961. It is located off Highway 190 between Porterville and Springville.

Success Dam was built with a gross-pool capacity (aka usable storage capacity) of 85,400 acre-feet.[252, 253] USACE has since revised the area-capacity curve for the reservoir. The reservoir is now rated as having a gross-pool capacity of 82,291 acre-feet (generally rounded to 82,300).[254] That amount of gross storage can hold 45% of the 118-year average runoff (1894–2011) for the Tule River. That gross storage percentage is important primarily from the standpoint of irrigation, not flood control.

Success Reservoir has a gross-pool capacity of 82,300 acre-feet, of which 75,760 acre-feet is reserved for a flood-control pool.[255] That leaves 6,540 acre-feet available for a conservation pool in the winter.

The dam is operated to reduce floodflows in order to achieve a maximum objective flow immediately downstream of the dam to 3,200 cfs. Dam operators also try to minimize floodflows into the Tulare Lakebed.[256]

Success Dam is capable of providing a downstream level of protection of 100-200 years. That is, it can catch that level of rain flood while keeping flows downstream of the dam within stated channel capacity / maximum objective flow.

The above level of protection estimate has not been reviewed and approved by appropriate USACE personnel and represents a draft value only. It is based on the existing dam design. The estimate reflects only hydrology, not seismic or geotechnical risk. It assumes that the full flood-control pool is available at the beginning of the flood. There are no recent data available on sedimentation rate or on the amount of sedimentation that has occurred in the reservoir.

At full pool, Lake Success covers 2,477 acres and extends 35 miles back from the dam. Its gross pool elevation is 652.5 feet.[257]

In the science fiction novel "Lucifer's Hammer," fragments of a comet hit the lake and destroy the dam. Although less romantic, the earthen dam has been found to have seepage problems and to be at risk of failure in the event of an earthquake. After studying alternative solutions, the USACE chose a preferred solution: constructing a 350-foot extension downstream and the replacement of the dam's core. However, the USACE did further study and announced in April 2012 that the risk of catastrophic failure was not as great as formerly thought.

The full pool elevation of Lake Success is 652.5 feet, equivalent to 82,291 acre-feet. In 2004, the USACE set restrictions on the maximum amount of water that the lake could hold, lowering the maximum water level to 620 feet elevation (equivalent to 29,183 acre feet). This pool restriction was felt necessary because at the time:

1. The USACE thought that there was a large risk due to earthquakes.
2. They thought that there was a bigger seepage risk with water running through the earthen dam.

In April, 2012, the USACE came to the conclusion that there was not as much seismic risk as they had thought. When the dam was evaluated a few years earlier, concerns were raised that sandy soil beneath the dam might be so prevalent that if a sizable earthquake occurred, the soil would settle, causing the dam above it to settle lower than its current height. That could cause large amounts of water to spill over the dam. Depending on the volume and speed at which the water spilled out, it could further eat away at the dam and make for a worse flood.

As a result of the reevaluation, the official pool restriction was increased to 640 feet elevation (equivalent to 56,084 acre-feet), but encroachment is allowed to 645 feet elevation (65,473 acre-feet). Whether the higher water level remains in effect or if the restrictions are completely removed isn't clear yet. Sensors are monitoring seepage rates in and under the dam as the amount of water in the lake increases, and so far no problems have been found. The USACE still has to look at the risks. The previously identified risks have not been eliminated; they have just come down.[258] The USACE is reevaluating the water level (i.e., the pool restriction and allowable encroachment level) while they evaluate the preferred structural solution for the dam.

This pool restriction largely impacts recreation and irrigation users. It does not affect the dam's ability to control floods. The flood-control pool remains unrestricted and the dam's ability to provide a downstream level of protection remains unaffected.

## *Isabella Dam*

This 185-foot-high[259] earthen dam is on the Upper Kern River, just below where the North Fork and South Fork of the Kern merge. That is about 34 miles northeast of the city of Bakersfield. The watershed drainage area of the Kern as measured from above the dam is 2,074 square miles.[260] The watershed drainage area as measured from above the First Point of Measurement gage (located just upstream of the city limits of Bakersfield) is 2,407 square miles.[261] The entire watershed drainage area of the Kern as measured from above the Buena Vista Lakebed is 3,612 square miles.[262] The dam began operation on April 15, 1954. Storage in the reservoir prior to that date was negligible.

Isabella Dam was built with a gross-pool capacity of 568,100 acre-feet.[263] That amount of gross storage can hold 49% of the 118-year average runoff (1894–2011) for the Kern River. That gross storage percentage is important primarily from the standpoint of irrigation, not flood control.

Isabella Reservoir has a gross-pool capacity of 568,100 acre-feet, of which 169,760 acre-feet is reserved for a flood-control pool.[264] That leaves 398,340 acre-feet available for a conservation pool in the winter.

The dam is operated to reduce floodflows in order to achieve a maximum objective flow at the First Point of Measurement gage (located just upstream of the city limits of Bakersfield) to 4,600 cfs. Dam operators also try to minimize floodflows into the Tulare Lakebed.[265]

Isabella Dam is capable of providing a downstream level of protection of 50-100 years. That is, it can catch that level of rain flood while keeping flows downstream of the dam within stated channel capacity / maximum objective flow.

The above level of protection estimate has not been reviewed and approved by appropriate USACE personnel and represents a draft value only. It is based on the existing dam design. The estimate reflects only hydrology, not seismic or geotechnical risk. It assumes that the full flood-control pool is available at the beginning of the flood. There are no recent data available on sedimentation rate or on the amount of sedimentation that has occurred in the reservoir. There are several uncertainties involved in modeling flow downstream of the dam.

At full pool, Lake Isabella Reservoir covers about 11,500 acres,[266] making it one of the larger reservoirs (by surface area) in California. It is nearly twice as large as the reservoir formed by Pine Flat Dam.

At full pool, Lake Isabella's gross pool elevation is 2,605.5 feet.[267]

The USACE is conducting a wide-ranging study of the main and auxiliary dams at Lake Isabella. The study was undertaken as a result of seismic concerns as well as water seepage detected in 2006. Scientists looked at core samples of the rock below the dams and dug several deep trenches to look for movement along the Kern Canyon Fault, which runs directly under the western edge of the auxiliary dam. Once thought to be inactive, the fault is now believed to be capable of causing a 7.5 magnitude earthquake, large enough to rupture the auxiliary dam.

The studies may lead to the need to repair the dams. It could take 10–15 years before the anticipated repairs can be completed. In the meantime, the USACE has reduced the amount of water that can be safely stored in the reservoir. As originally designed and constructed, the lake's capacity was 568,000 acre-feet. But until the dam is once again certified safe to hold that volume of water, the fill-limit has been set at 360,000 acre-feet, about 63% of the total capacity of the reservoir. That amount of gross storage can hold 49% of the 118-year average runoff (1894–2011) for the Kern River.

This 360,000 acre-feet pool restriction largely impacts recreation and irrigation users. It does not affect the dam's ability to control floods. The flood-control pool remains unrestricted and the dam's ability to provide a downstream level of protection remains unaffected.

Historically, the Kern River divided in its lower reaches. The distributary point was located just west of Bakersfield, a few hundred yards east of the present-day Stockdale Bridge, at a place where the historic wooden Bellevue Weir was built across the river. That weir is now gone, but there is a rock spillway across the river at the same location.

At that point, the Kern split into two parallel channels, both of which flowed eventually north toward Tulare Lake. The main channel flowed along the west side of the San Joaquin Valley, connecting with Buena Vista Lake and Buttonwillow Swamp. Between Buena Vista Lake and Tulare Lake, this channel was known as Buena Vista Slough. The other channel, known as Goose Lake Slough, flowed along the east side of the valley. Interstate 5 occupies the high ground between those two channels.

Goose Lake Slough typically only carried water during flood periods. Bull Slough was the northern extension of Goose Lake Slough and was located north of Goose Lake. Goose Lake Slough and Bull Slough have not carried Kern River floodwaters since 1983. For a more complete description of these channels and their associated wetlands, see the section of this document: General Notes on Kern, Buena Vista, and Goose Lakes.

With the completion of Isabella Dam in 1954, the risk of Kern River water reaching the Tulare Lakebed was greatly reduced. During floods, all four federal reservoirs coordinate their operations to minimize inflows to the lakebed. Still, that has proved insufficient in a major flood such as 1969.

Now it is possible to route a large portion of Kern River floodwaters into the California Aqueduct rather than into Buena Vista and/or Tulare Lakes. This is done using the Kern River Intertie and Cross Valley Canal. Once the water enters the California Aqueduct, it is pumped over the Tehachapi Mountains and sent to the Los Angeles area. The Kern River Intertie is located just north of the Taft Highway (Highway 199). It was completed in 1977 and has a capacity of 3,500 cfs.

## *Storage Capacity in the Tulare Lake Basin*

As shown in Table 7, the total storage capacity of the four southern federal reservoirs (Pine Flat, the expanded Kaweah, the reduced Success, and the reduced Isabella) is 1,608,000 acre-feet. If and when Success and Isabella are restored to full capacity, the total storage capacity of these four reservoirs will be about 1,835,900 acre-feet.

Table 7. Reservoir storage capacity in the Tulare Lake Basin.

| Reservoir | Original Capacity (acre-feet) | Current Capacity (acre-feet) | Planned Capacity (acre-feet) |
|---|---|---|---|
| Pine Flat | 1,000,000 | 1,000,000 | 1,000,000 |
| Kaweah | 150,000 | 185,600 | 185,600 |
| Success | 82,300 | 65,473 | 82,300 |
| Isabella | 568,300 | 360,000 | 568,000 |
| Total | 1,800,300 | 1,608,073 | 1,835,900 |

The 1,608,073 acre-feet in current capacity is based on pool restrictions at Lake Success and Lake Isabella that largely impact recreation and irrigation users. These restrictions do not affect the ability of those dams to control floods. The full 1,835,900 acre-feet is available for flood control purposes without restriction, and the dam's ability to provide a downstream level of protection remains unaffected.

People who live below the reservoirs tend to think that they're safe, that the reservoirs are so big that they can catch and hold the floodwaters of the biggest events. Those reservoirs have been very effective at protecting downstream communities since their construction, but they do have their limits. For comparison:

- The 1,608,073 acre-feet in combined current capacity can hold 54% of the 118-year average runoff (1894–2011) of the four rivers.
- The combined runoff of the four major rivers in the Tulare Lake Basin in 1983 was 8,746,222 acre-feet (see Table 66 and Figure 13). For comparison, that is 5.4 times the combined current capacity of the federal reservoirs on those four rivers.

## Variation in Runoff over Past 118 Years: 1894–2011

Figure 13 illustrates how full natural flow runoff has varied for each of the four major river drainages in the Tulare Lake Basin over the last 118 years (1894–2011). The average runoff of the four rivers during this period was 2,986,578 acre-feet.

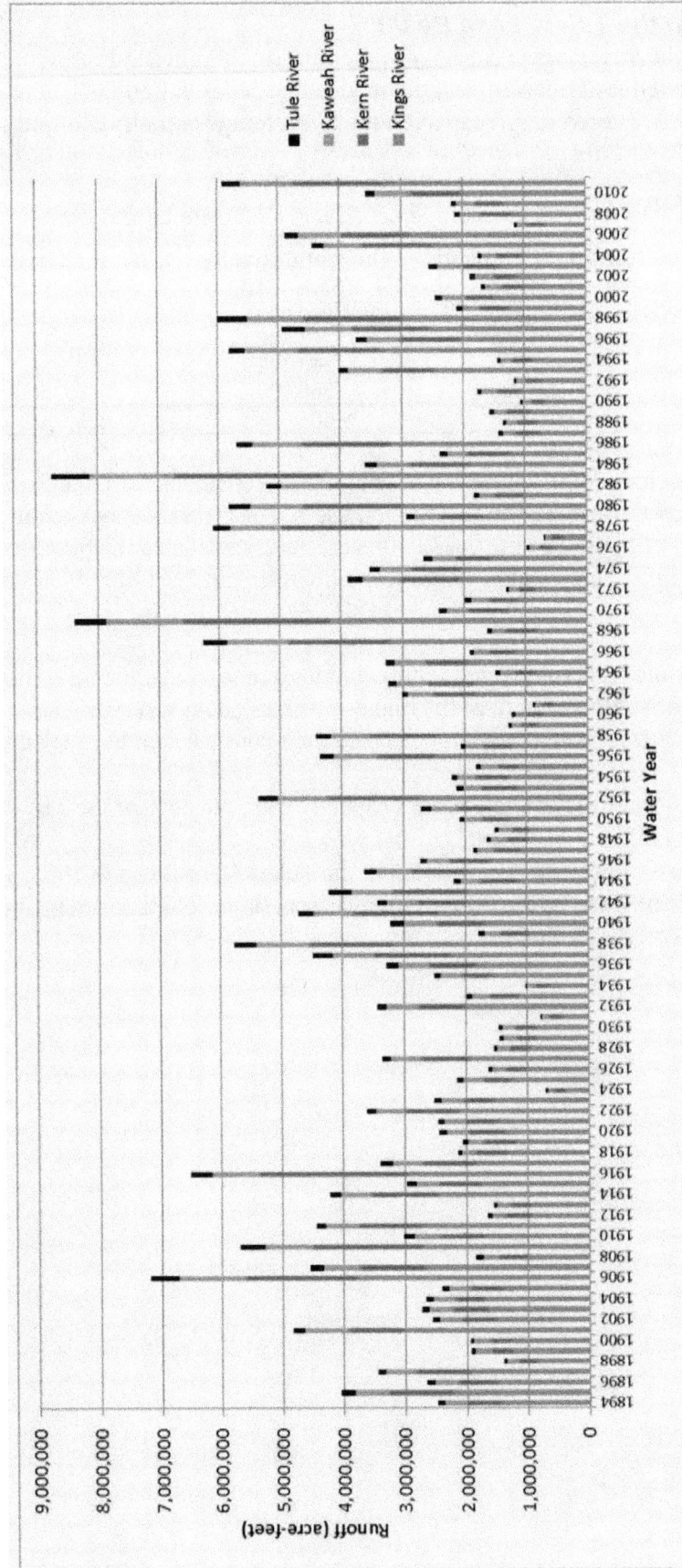

Figure 13. Variation in runoff over past 118 years: 1894–2011.
Source: Data compiled from DWR California Data Exchange Center[268] and USBR which obtained it from USACE[269]

## General Drought and Temperature Notes

### *What Constitutes a Drought*

The term "drought" is commonly used in two different ways in the Tulare Lake Basin. Traditionally, the term "drought" has been used to reflect precipitation that is significantly less than average. That is generally how the term drought is used in this document — a period of significantly less than average precipitation. Typically this condition has to last for at least two years before it is recognized as a drought. This type of drought ends when precipitation returns to approximately average conditions. The 1976–77 drought is an example of this type of drought.

The other use of the term "drought" refers to conditions when, for various reasons, we don't have all the water that we feel we are entitled to. In this kind of drought, 1) available water supply fails to meet demand, and 2) we perceive that water that is rightfully ours is being used somewhere else. If "they" would just let us have our water, we would be better able to meet our needs. This type of drought has two components: hydrologic and socio-political. The Tulare Lake Basin has had a lot of this sort of drought in recent years. It is more challenging to mark the end of this type of drought. Precipitation can return to average or even above-average conditions, but there still isn't enough water to meet demand. The latter part of the 2007–09 drought is an example of this type of drought.

In the 2007–09 drought, water years 2007 and 2008 really were drought years by the traditional definition. The runoff was so low in those years that the state's water year index rated those years as critically dry. The 2007–09 drought was California's first drought for which a statewide proclamation of drought emergency was issued.

That turned out to be critical. When precipitation returned to near-average or above-average, it was hard politically for the governor to declare an end to the drought. There clearly wasn't enough water to go around. It wasn't until March 30, 2011, after an incredibly wet winter, that the state of emergency was finally rescinded. That was long after the end of the hydrologic drought. The 2007–09 drought had morphed from the traditional type of drought (below-average precipitation) into the socio-political type (we're entitled to more water than we're getting).

Unlike floods, droughts are not clearly defined. Identifying periods of drought is a matter of subjective interpretation, even in retrospect. The droughts noted in this document are generally those that meet both of the following criteria:
- recognized at a statewide level or identified in a peer-reviewed scientific publication
- appear to have impacted the Tulare Lake Basin in some way, such as being reflected in the annual flows of the rivers within that basin

Most water users in California are cushioned to some degree from drought. We have developed a number of conveyance and storage sources and have invested in redundant systems in many areas. But it is generally not cost-effective to provide this level of secondary storage for rural areas. As a result, rural areas have to put a greater reliance on rain for their principal water supply. That leaves them very vulnerable to drought.

Impacts of drought are typically felt first by those most reliant on annual rainfall — ranchers engaged in dryland grazing, rural residents relying on wells in low-yield rock formations, or small water systems lacking a reliable water source. (Criteria used to identify regional and state drought conditions generally do not address such localized impacts.)

It comes down to the cost of developing secondary sources for such rural areas and the memory of those that are now alive. John Steinbeck grew up in the Salinas Valley, just west of the Tulare Lake Basin. He captured this view well in his greatest novel, *East of Eden*:

> *I have spoken of the rich years when the rainfall was plentiful. But there were dry years too, and they put a terror on the valley. The water came in a thirty-year cycle. There would be five or six wet and wonderful years when there might be nineteen to twenty-five inches of rain, and the land would shout with grass. Then would come six or seven pretty good years of twelve to sixteen inches of rain. And then the dry years would come, and sometimes there would be only seven or eight inches of rain. The land dried up and the grass headed out miserably a few inches high and great bare scabby places appeared in the valley. The live oaks got a crusty look and the sagebrush was gray. The land cracked and the springs dried up and the cattle listlessly nibbled dry twigs. Then the farmers and the ranchers would be filled with disgust for the Salinas Valley. The cows would grow thin and sometimes starve to death. People would have to haul water in barrels to their farms just for drinking. Some families would sell out for nearly nothing and move away. And it never failed that during the dry years the people forgot about the rich years, and during the wet years they lost all memory of the dry years. It was always that way.*

*East of Eden* was published in 1952, but Steinbeck's message is still resonant today. Human memories are short; we forget our rainfall and drought history all too fast.

To help with this problem, the California Department of Water Resources maintains two indicators of annual surface water supplies for the Central Valley:[270]
- the Sacramento Valley Water Year Index
- the San Joaquin Valley Water Year Index

Those indices measure unimpaired natural runoff from 1906 to the present for the Sacramento River Basin and from 1901 to the present for the San Joaquin River Basin. As part of those indices, DWR categorizes each year compared to average runoff.

The San Joaquin Valley Water Year Index is based on the combined flows of four rivers: the Stanislaus, Tuolumne, Merced, and San Joaquin Rivers. For the combined flow of those four rivers, the hydrologic index classification (water year type) is calculated as:

- wet (equal to or greater than 3.8 million acre-feet)
- above normal (between 3.1 and 3.8 million acre-feet)
- below normal (between 2.5 and 3.1 million acre-feet)
- dry (between 2.1 and 2.5 million acre-feet)
- critical (less than 2.1 million acre-feet)

The results are illustrated in Figure 14.

Table 8 summarizes the number of years in each of the hydrologic index classifications illustrated in Figure 14. For example, the San Joaquin River Basin has experienced 19 critically dry years during the past 111 water years (1901–2011).

Table 8. Frequency of different water year types in the San Joaquin River Basin.

| Water Year Classification | Number of Years | Percent |
|---|---|---|
| Wet | 37 | 33% |
| Above normal | 22 | 20% |
| Below normal | 18 | 16% |
| Dry | 15 | 14% |
| Critical | 19 | 17% |
| Total | 111 | 100% |

Figure 14 illustrates the San Joaquin Valley Water Year Index during the past 111 water years: 1901–2011. That index is a measure of unimpaired runoff in the San Joaquin River Basin.

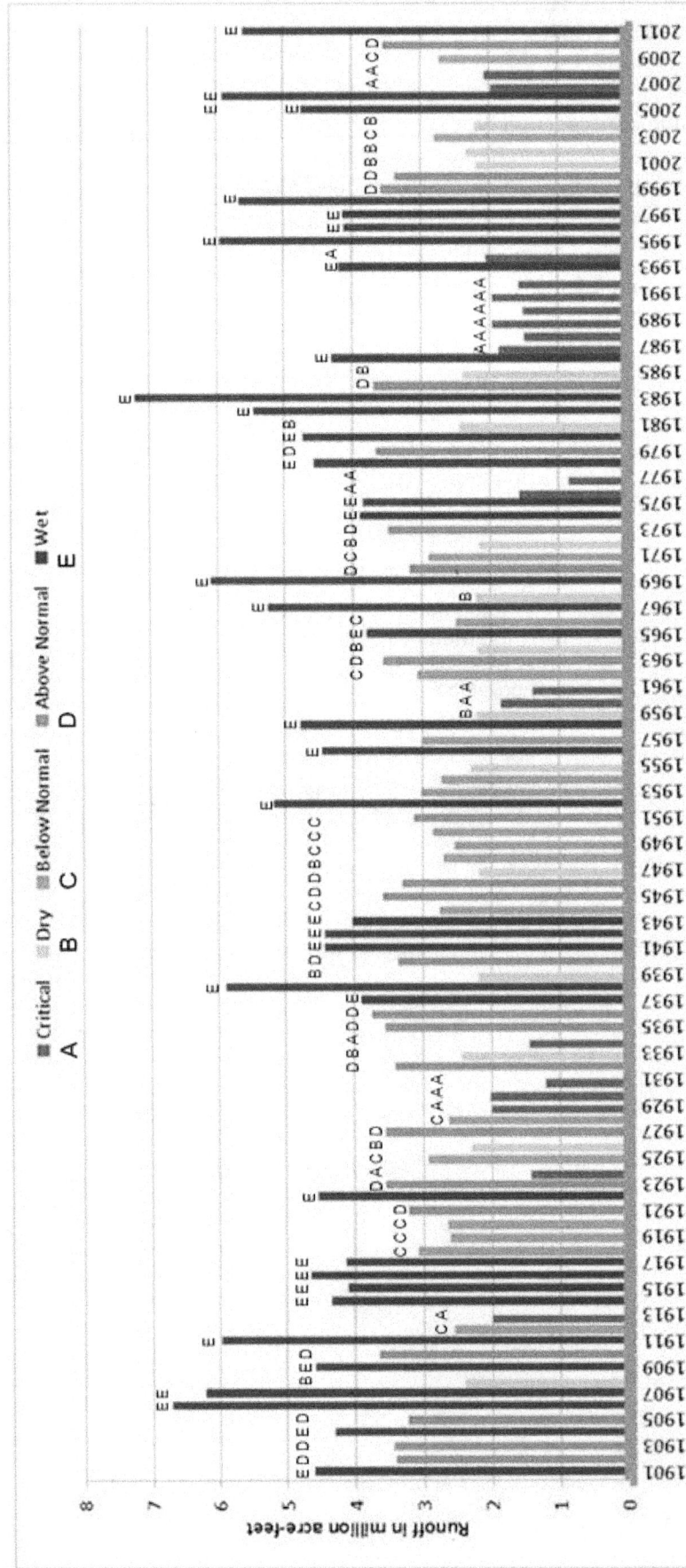

Figure 14. San Joaquin Valley Water Year Index for past 111 years: 1901–2011.
Source: California Department of Water Resources

The Sacramento Valley Water Year Index is also based on the combined flows of four rivers: the upper Sacramento, Feather, Yuba, and American Rivers. Although directly applicable to the Sacramento River Basin, the Sacramento Valley Water Year Index is widely used throughout the state as a measure of drought severity.

One dry year does not constitute a drought in California. Droughts occur slowly, over a multi-year period. There is no universal definition of when a drought begins or ends. As a general rule of thumb, if DWR's San Joaquin Valley Water Year Index is dry or critical for two years in a row, then that constitutes a drought.

## *Notes on Runoff Reconstructions*

Christopher Earle and Harold Fritts used tree-ring data to reconstruct annual runoff for the Sacramento River Basin for 1560–1980 from the analysis of tree-rings.[271] Dave Meko and others used tree-ring data to push that reconstruction back to A.D. 869.[272]

Lisa Graumlich used tree-ring data from subalpine conifers in the Southern Sierra to reconstruct temperature and precipitation back to A.D. 800.[273] That study found that summer temperatures were warmer than late twentieth-century values from about 1100–1375, corresponding to the Medieval Warm Period. It also found a period of cold temperatures from about 1450–1850, corresponding to the Little Ice Age. Precipitation during the 1,000± year record varied, but generally averaged less than twentieth-century levels regardless of the variation in temperature.

Edward Cook and others used tree-rings to reconstruct the Palmer Drought Severity Index for North America.[274, 275] Their studies found that the Sierra was in a period of exceptional and extended drought from the late 800s to about 1300. The driest two periods in western North America were centered on the mid–1100s and the mid–1200s; those are both reflected in the Sierra.

A 3,000-year record from giant sequoia trees in Sequoia National Park was reconstructed by Tom Swetnam, Tony Caprio, and others.[276] Their study found that the western Sierra was droughty and often fiery during the Medieval Warm Period (i.e., from 800–1300). This time period had the most frequent fires in the 3,000 years studied. During that period, extensive fires burned through parts of Giant Forest at intervals of about 3–10 years. Any individual tree was probably in a fire about every 10–15 years.

Tree-rings can only be used to reconstruct the precipitation record for a few thousand years at most. Scott Mensing and others analyzed a set of sediment cores extracted from Pyramid Lake that had been deposited over the past 7,630 years.[277] They used the ratio of moisture-loving *Asteraceae* to drought- and salt-tolerant *Chenopodiaceae* in that lakebed as a proxy for drought. This allowed the authors to reconstruct a drought record for the western Great Basin. Since Pyramid Lake gets most of its water from the Sierra, this drought record can — with caveats — be applied to the Sierra as well. The authors documented multiple droughts in the region that each lasted 150–200 years. Their work suggested that variable solar activity may well be the major factor in determining the hydrological state of the region.

There have been two important studies of sediments in the Tulare Lakebed.
- In 1999, Owen Kent Davis at the University of Arizona provided a record of late Quaternary climate for the Tulare Lake region based on the palynology (pollen study) of a depocenter core. A depocenter is that part of a basin where the greatest subsidence occurred. Davis's study found that the vegetation of the southern San Joaquin Valley used to resemble that of the contemporary Great Basin, including abundant greasewood. He also found that giant sequoia was widespread along the Sierra Nevada streams draining into Tulare Lake prior to 9,000 year B.P. (Before Present). The end of Great Basin plant assemblages 7,000 B.P. coincided with increased charcoal (i.e., fire frequency in the woodland and grasslands). Davis's study also included conclusions regarding relative lake levels throughout the Holocene.[278]
- In 2006, Rob Negrini and his associates at CSU Bakersfield built on Davis's results with improved constraints on absolute elevations and ages of past lake levels from two trench sites at higher elevations in the Tulare Lakebed.[279]

## *Notes on Temperature Changes*

Popular perception is that North American temperatures have been relatively stable for the last thousand years or so and have recently begun an unprecedented climb.

However, Lisa Graumlich, Tom Swetnam, and other researchers have found evidence that temperature anomalies corresponding to Europe's Medieval Warm Period (approximately 800–1300) and Little Ice Age (approximately 1450–1850) occurred in the Tulare Lake Basin.

These two different views of long-term temperature changes have been presented by others as the Battle of the Graphs:

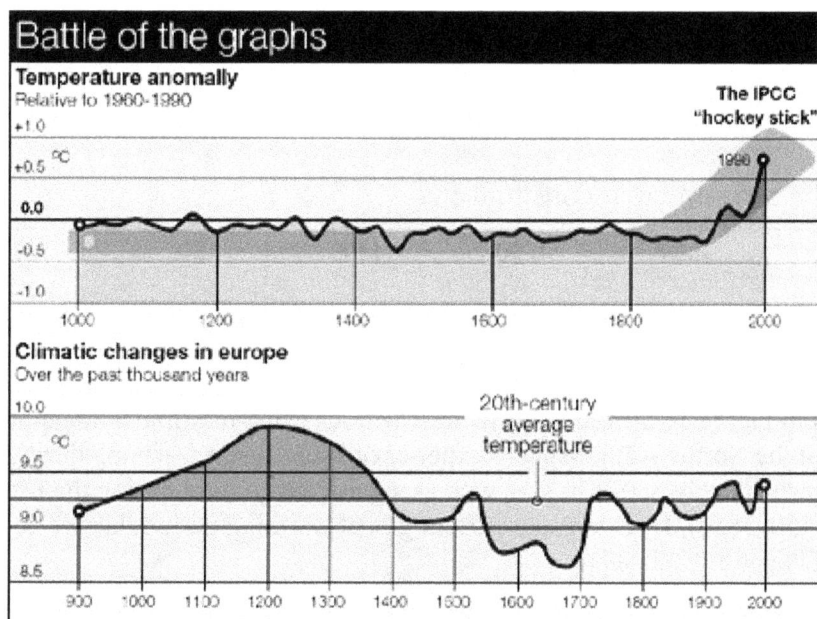

Figure 15. Comparison of two temperature reconstruction graphs.

Although the concept of the lower graph in Figure 15 may be correct in general concept, it exaggerates and greatly oversimplifies the situation. For one thing, it is based just on European conditions and probably overstates the magnitude of the temperature anomalies even there, especially during the Medieval Warm Period. Figure 16 is a more reliable and peer-reviewed presentation of the available data.

116

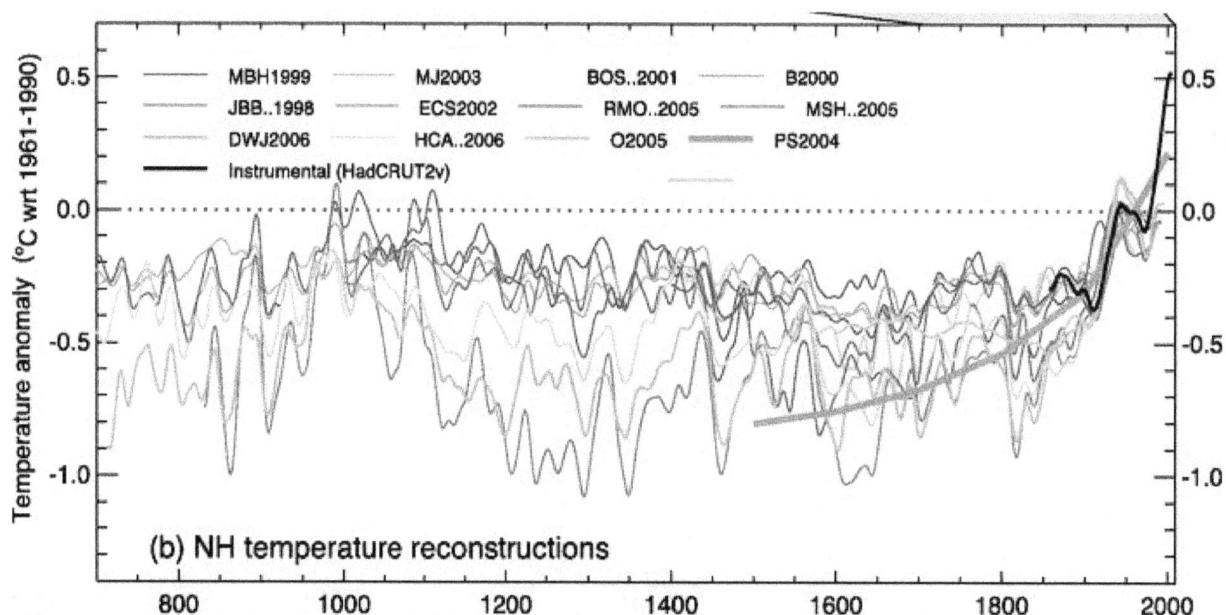

Figure 16. Northern Hemisphere temperature reconstructions for the last 1,300 years.
Source: Intergovernmental Panel on Climate Change 2007[280]

The temperatures in Figure 16 are based on reconstructions using multiple climate proxy records. They all show that the Northern Hemisphere experienced significant warming during the Medieval Warm Period (approximately 800–1300) as well as significant cooling during the Little Ice Age (approximately 1450–1850). However, the current global warming period that we're experiencing is unprecedented in the past 1,300 years.

The year 2012 continued this trend in rising temperatures. For example, there were 3,215 daily high-temperature records set or tied during June. Among those records, 1,748 of them were for temperatures of 100 degrees or higher.[281]

July 2012 was the hottest month on record for the contiguous U.S., breaking the record set in July 1936. The warm July temperatures contributed to the warmest 12-month period that the nation has experienced since record-keeping began in 1895.[282]

October 2012 marked the 332th consecutive month (27.7 years) with a global temperature above the 20th-century average.[283]

In 2012, Adrian Das and Nate Stephenson at the USGS field station at Sequoia at Kings Canyon National Parks performed an analysis of long-term historical temperature records in the vicinity of the national parks. They analyzed data from the Lemon Cove, Ash Mountain, Independence, Bishop Airport, Grant Grove, and Lodgepole weather stations.

Their analysis showed that the long-term temperature records for the above six stations, taken as a whole, have generally followed patterns seen statewide: there has been an increase over the periods of record.

117

## *California Snow Conditions during the Little Ice Age*

In 1542, Juan Rodríguez Cabrillo led the first European expedition to explore what is now the West Coast of the United States. The Cabrillo expedition sailed out of the port of Navidad, Guatemala (near modern day Manzanillo) on June 24, 1542. He sailed as far north as the Monterey Peninsula. Returning south on November 18 of that year, he noted the snow-capped Santa Lucia Mountains in southern Monterey County. That is not a place where we expect to see such conditions today.

Sebastian Vizcaino led an expedition that departed Acapulco on May 5, 1602, and arrived in the bay that he named Monterey on December 14 of that year. That expedition occurred during the height of the Little Ice Age (1450–1850).

Vizcaino spent a very cold Christmas in the area. He recorded that on Christmas Day, 1602, the mountains near the port were covered with snow and that on New Year's morning the water holes were frozen to the depth of a palm. The expedition encountered no Native Americans, but did find a deserted village. Vizcaino speculated that the inhabitants had taken refuge in the interior to escape the biting cold. (Today, the average low temperature in Monterey in December and January is a much more pleasant 43°.)

Vizcaino wanted to attract colonists to his new bay. So although he was numb with cold, he wrote a glowing report in which he said the area's climate was like that of Seville's.[284] So untrue was the picture that he painted, that when Captain Gaspar de Portolá and Father Juan Crespi arrived with their colonizing party 167 years later, they failed to recognize the fabled port. Food was so scarce that they were reduced to eating seagulls and pelicans. After snow began to cover the hills on November 30, 1769, the survivors decided to return to San Diego.[285]

Zenas Leonard was a fur trapper, a mountain man. He left St. Louis on April 24, 1831, with a fur trapping company. He returned to civilization some 4½ years later on August 29, 1835. He left us a highly readable account of his adventures: the Narrative of the Adventures of Zenas Leonard.

In July of 1833, Leonard joined with the Joseph Walker party in exploring the unknown country from the Great Salt Lake to the Pacific Ocean. With great difficulty, they traveled down the Humboldt River, crossed the Sierra in October 1833, followed the San Joaquin River downstream, and arrived in the vicinity of Half Moon Bay on November 20, 1833.

Their route down the Humboldt River took them to Carson Lake and the vicinity of present-day Carson City, Nevada. How they actually crossed the Sierra and reached the San Joaquin River is less clear. They are traditionally thought to have crossed north of the Merced River, becoming the first whites to discover Yosemite Valley. However, that route would have been well south of the Carson City area, and that would seem inconsistent with Leonard's account.

Furthermore, Leonard described the section of the Sierra that the Walker Party passed through. Based on his research, George Durkee is convinced that the area Leonard described wasn't Yosemite. Scott Stine's research indicates that the Walker Party's probable route was well north of Yosemite: over Ebbetts Pass, and generally along the route of present-day Highway 4.

In any case, Leonard described encountering a lot of old and consolidated snow as they crossed the Sierra in October 1833:

*In some of these ravines where the snow is drifted from the peaks, it never entirely melts, and may be found at this season of the year, from ten to one hundred feet deep. From appearance it never melts on the top, but in warm weather the heap sinks by that part melting which lays next the ground. This day's travel was very severe on our horses, as they had not a particle to eat...but the most of the distance we this day traveled, we had to encounter hills, rocks and deep snows. The snow in most of the hollows we this day passed through, looks as if it had remained here all summer, as eight or ten inches from the top it was packed close and firm — the top being loose and light, having fell only a day or two previous.*

The Walker Party encountered snow that was persisting from year to year. They were crossing the Sierra near the end of the Little Ice Age. There is no longer persistent snow in that area; conditions have changed dramatically in the last 180 years.

## Groundwater Overdraft

The groundwater aquifer that underlies the San Joaquin Valley stretches from the Sierra Nevada to the Coast Ranges and from the Tehachapi Mountains to the western edge of the Sacramento–San Joaquin River Delta. The story of this aquifer and the history of its use have been told in several USGS publications.[286, 287, 288, 289]

A groundwater aquifer can be thought of as an underground lake. The "underground lake" under the San Joaquin Valley is a large porous body of loosely packed alluvium (clay, silt, sand, and gravel) that is saturated with water.

It averages about 2,400 feet in thickness, but increases from north to south to a maximum thickness of more than 9,000 feet near Bakersfield. The aquifer has two zones: a mostly unconfined upper zone, separated from a confined lower zone by the Corcoran Clay. The water below the Corcoran Clay generally has distinctly different water chemistry from the water above the clay.[290] The main source of groundwater in the San Joaquin Valley is the upper 1,000 feet of basin-fill deposits.[291]

The water that fills this alluvium comes largely from mountain rivers and streams on the east side of the San Joaquin Valley. These are the streams we are all familiar with that flow down out of the Sierra.

However, some of the ephemeral streams on the west side of the valley can also have large flow events in certain years. These are the streams that flow down out of the Coast Ranges. Some of these events in the ancient past resulted in huge volumes of water. Flows from these west-side streams have contributed to groundwater recharge on the upper end of the alluvial fans and have created groundwater bodies of distinct chemical types related to the soils and geology of the upper watershed. An example is Arroyo Pasajero (aka Los Gatos Creek) in western Fresno County. Both Coalinga in the Pleasant Valley and the lands east around Interstate 5 have used groundwater from that system. The quality was not great, but usable for agriculture and industry until a better supply came along: the California Aqueduct.

Some of the water in the alluvium flowed down from the mountains quite sometime ago. In that sense, some of the water being pumped up now is being mined from the distant past.

Very little of the runoff soaks into the ground as the rivers flow through the mountains and foothills; those areas tend to be underlain with granitic, metamorphic, and other relatively impermeable rocks. However, once the rivers emerged out onto the valley floor, they historically flowed over the highly permeable sand and gravel of the alluvial fans that soaked up the water. An alluvial fan is a distributary system made of multiple channels that allow for large areas of shallow inundation. This water then became part of the groundwater aquifer and continued moving between the sand grains.

This situation changed dramatically with the coming of irrigated agriculture to the San Joaquin Valley, beginning in about the 1870s. Settlers dug canals to tap the streams and rivers, spreading that water onto fields and orchards. This spreading of water over the surface of the ground permitted a steady agricultural growth that required more and more water. As more water was spread, less reached the valley's wetlands and less was returned to the groundwater aquifer. Then the four federal reservoirs began operation during the 1954–61 period, allowing large quantities of water to be stored for use later in the year.

Today, rivers are largely cut off from the natural process of the routine spring flooding of the alluvial fans with the most permeable materials. Groundwater recharge has become largely dependent on irrigation events during the summer irrigation months. The amount that can be percolated is substantially less than the natural process because the crop uses some of the water, and the most permeable areas don't receive as much total water for the length of time that the wet year inundation (overland flow) would cause.

There are many areas of the San Joaquin Valley that cannot contribute significantly to the groundwater aquifer. The heavy Corcoran Clay layer is distributed throughout the central and western San Joaquin Valley. It varies in thickness up to 160 feet under the Tulare Lakebed.[292] Ken Schmidt has found places in the valley where clay is vertically continuous for 1,000 feet; no decent water-bearing zones to be found. Similarly, older hardpan soils in many areas of the valley have little contribution to the groundwater.

There are several projects currently underway to attempt to reconnect the more permeable alluvial soils with more water for groundwater recharge. A recent state law (AB 359) promotes the management and protection of the state's groundwater supplies by requiring local water agencies to map groundwater recharge areas and to submit those maps to local planning agencies. Investments in utilizing such lands for recharge are underway in some areas. Some of the more promising projects seek to work with farmers to have them use their land in wet years with the promise to avoid or mitigate any damages to crops.

Many water districts have constructed shallow ponds (aka recharge basins) to put excess water into the groundwater aquifer. These recharge basins somewhat mimic the intermittent wetland habitat that once existed on the alluvial fans historically. As an example, the Kaweah Delta Water Conservation District has 40 recharge basins covering approximately 5,000 acres.[293]

Recharge basins such as these are typically small and somewhat widely distributed. They are generally located over the alluvial fans because that is where the rivers are. The water that has been stored in the groundwater aquifer is retrieved during dry years by individual wells operated by agriculture and cities.

Water banking takes this concept one step further. It combines wells with the recharge basins; all of which are concentrated in one place and operated by the water district. The idea is that the water placed in the groundwater aquifer won't move significantly; it will be there to be retrieved when desired. It uses the aquifer as the equivalent of an underground reservoir. Kern County is the lead in water banking. Virtually all of the water districts in the San Joaquin Valley portion of Kern County are involved in water banking at some level.

The biggest water bank in the San Joaquin Valley is the Kern Water Bank. It covers about 20,000 acres of the Kern River's sandy alluvial fan southwest of Bakersfield and has about 7,000 acres of recharge basins. Up to 72,000 acre-feet per month can be recharged at the beginning of a recharge program, a rate that gradually declines as the program progresses.[294, 295]

When Secretary of the Interior Bruce Babbitt visited the Kern Water Bank in January 2000, he hailed it as "the most effective groundwater storage program in the United States, probably the whole world." Water recharged by the Kern Water Bank comes primarily from the State Water Project via the California Aqueduct, but some flood flows from the Kern River and the Central Valley Project have also been recharged.

In the days prior to Euro-American settlement, the rivers and the groundwater aquifer had a sort of working agreement. In dry years when the rivers were low, the groundwater aquifer still contained so much water that it spilled out into numerous springs.

When irrigated agriculture started in the San Joaquin Valley, this relationship began to change. The early wells in the southern San Joaquin Valley were hand-dug pits. As well-drilling technology improved, wells could go deeper. This was one of the keys to tapping the Artesian Belt (see Artesian Wells — Discovery of the Artesian Belt). The valley's artesian wells continued to flow naturally until they all stopped for uncertain causes during the first decade of the 20th century. The probable cause seems to be because the water table dropped as the wells went deeper.

The deep well turbine was introduced by Bryon Jackson in 1901.[296] It got to the Westside Subbasin of the San Joaquin Valley fairly soon after that. The town of Tranquillity (30 miles west of Fresno) had such wells for drinking water in 1920.

Around 1930, the development of an improved deep-well turbine pump and rural electrification enabled additional groundwater development for irrigation.[297] This was right when the valley was suffering through some of the worst of the 1922–34 drought. In the 1930s, the hand-dug pits in the Poplar area (northwest of Porterville) began to run dry as the groundwater table dropped.[298] The post-WWII period brought a tremendous increase in the amount of pumping for irrigation, resulting in an ever-increasing decline in the level of the groundwater aquifer.[299]

California's groundwater has been described as one of the least-regulated, least-monitored aquifers in the U.S. While possibly an exaggeration, that gives a sense of the challenges involved in trying to get a handle on the situation with the Tulare Lake Basin's aquifer. Water managers typically don't have current status and trend data on that aquifer; they have had to make decisions based on assumptions that sometimes turn out much later to be incorrect.

In 2009, the USGS released a report on a five-year study of groundwater availability of the Central Valley aquifer.[300] After 1900, when large-scale farming began in the Central Valley, water tables dropped significantly as wells were drilled to feed crops. Groundwater aquifers in parts of the western Tulare Lake Basin eventually dropped by more than 400 feet compared with pre-1900 levels.[301] Water-head levels in the area of greatest overdraft had been above ground level prior to 1900. But by 1961, they were well below sea level. Until 1968, irrigation water in those areas was supplied almost entirely by groundwater. As of 1960, water levels in the deep aquifer system were declining at a rate of about 10 feet per year.

The severity of the groundwater overdraft in the San Joaquin Valley was part of the impetus for building the state and federal canal systems in the 1960s that divert water from the water-rich northern half of the state to the arid southern half. Huge pumping stations located in the Sacramento–San Joaquin River Delta are used to export a portion of the northern waters to the south via these state and federal systems.

The California State Water Project, commonly known as the State Water Project (SWP), is the world's largest publicly built and operated water and power development and conveyance system. The SWP is operated by the California Department of Water Resources. The major feature of the SWP is the 702-mile-long California Aqueduct.[302]

The original purpose of the SWP was to provide water for arid Southern California, whose local water resources were insufficient to sustain that region's growth. Construction began in the late 1950s. The SWP provides drinking water for more than 23 million people. However, roughly 80% of the water carried by the project is used for agriculture, primarily in the San Joaquin Valley.[303]

The Central Valley Project (CVP) is a federal water project operated by the Bureau of Reclamation. It was devised in 1933 in order to provide irrigation and municipal water to much of the Central Valley — by regulating and storing water in reservoirs in the northern half of the state, and transporting it to the San Joaquin Valley and its surroundings by means of a series of canals, aqueducts, and pump plants, some shared with the SWP.[304] Two of the major features of the Central Valley Project are the Friant–Kern Canal and the Delta–Mendota Canal.

Beginning in 1950, water was diverted through the Friant–Kern Canal from below Millerton Reservoir to the east side of the San Joaquin Valley. In 1951, surface water deliveries along the northwest side of the San Joaquin Valley began through the Delta–Mendota Canal. In 1967, surface water deliveries to farms along the west side and near the southern end of the San Joaquin Valley began through the California Aqueduct. The availability of imported surface water following the construction of these canals resulted in a decrease in groundwater pumpage.[305]

The State Water Project reported record high-water exports from the Delta; 4.90 billion cubic meters of water, in water year 2011, the highest export rate recorded since 1981. The federal Central Valley Project exported 3.13 billion cubic meters of water in 2011, an increase from exports in 2008–2011, but comparable to exports from 2002–2007. Translated into acre feet, the total exports via the state and federal Delta pumps was 6,520,000 acre-feet in 2011 — 217,000 acre-feet more than the previous record of 6,303,000 acre-feet set in 2005.[306]

For perspective, that 6.5 million acre-feet is nearly twice as great (1.9 times to be precise) as the combined annual flow of the two largest rivers in the southern San Joaquin Valley. The San Joaquin River produces a long-term average annual flow (measured at Millerton Dam) of 1.8 million acre-feet per year. The Kings River has an average annual flow (measured at Pine Flat Dam) of 1.7 million acre-feet. Together, these two rivers produce an average of 3.5 million acre-feet of water.

Switching farms to this new surface water supply allowed groundwater aquifers to recover somewhat. For a while, it had been assumed that water tables had stabilized after about 1970 because the groundwater overdraft was thought to have largely stopped. Unfortunately, we now know that this was not a safe assumption; the groundwater table did not stabilize at this time.

Between 1961 and 2003, the period covered by the 2009 USGS study, groundwater storage in the Tulare Lake Basin showed a steep and fairly steady overall rate of decrease. The report found that the San Joaquin Valley experienced a net loss of 59.7 million acre-feet of groundwater storage during the 41-year period (1961–2003) covered by the study. For comparison, that is over 37 times the combined current capacity of the four federal reservoirs in the Tulare Lake Basin.

While the northern and western parts of the San Joaquin Valley saw water level recovery during this period, the report found that "overall, the Tulare Basin part of the San Joaquin Valley still is showing dramatic declines in groundwater levels and accompanying increased depletion of groundwater storage."

The groundwater aquifer levels recovered somewhat after 2003. However, during the three-year 2007–09 drought, farmers relied so heavily on groundwater that it brought the groundwater aquifer down to near the historic low.[307]

Some recovery occurs after each wet year or sequence of wet years, but overall the trend has been down. For example, the California Water Institute at CSU Fresno recently produced a map of the San Joaquin Valley showing the change in groundwater depth from 1983–2009. The only areas of first water that did not decline during this period were generally in the Westside Subbasin of the San Joaquin Valley where salty perched water resides on the first clay lens below the surface. ("First water" is the depth to the first fully sustained saturated zone in the subsurface, regardless of quality.)

Groundwater is like a bank account. Over the long run, we can't afford to withdraw any more water from the underground aquifer than flows back into the ground. In that regard, it's helpful to have a general accounting of some of the ways in which surface water leaves the Tulare Lake Basin. Surface water that leaves the basin has no opportunity to soak back into the aquifer.

Another way of looking at this issue is that despite the Tulare Lake Basin's huge import of water from up north, we have had a major net loss of water. Our basin has sprung a huge leak. So it's worth getting a handle on the accounting; where is the leak?

The Tulare Lake Basin is an essentially closed basin; it functions as an inland sink. Because of the Tulare Lake sill, water from the various rivers in the basin generally remains within the basin. That water would normally only make it to the Sacramento–San Joaquin River Delta if Tulare Lake were to overflow the sill; that hasn't happened since 1878. Although about 90,000 acre-feet of lake water was pumped over the sill in 1983 in an attempt to drain the lakebed.

There are only two significant ways for water to leave a closed basin like ours:
1. Exports of water outside the basin (aka inter-basin transfers).
2. Evapotranspiration. This is the sum of evaporation and plant transpiration from the earth's land surface to the atmosphere. Evaporation accounts for the movement of water to the air from sources such as the soil, canopy interception, and water bodies. Transpiration accounts for the movement of water within a plant, and the subsequent loss of water as vapor through stomata in its leaves. Evapotranspiration in the Tulare Lake Basin is estimated to have doubled when we began irrigating crops on the valley floor during the summer.[308] Most of this water vapor condenses and falls back within the basin as rain or snow. But it is possible (perhaps probable) that a good bit of this moisture has been leaving the basin unseen. However improbable, that seems to be the most plausible route for how water is leaving our basin; that is how we have sprung our huge leak. We haven't found anybody who has studied this issue.

We have a relatively good handle on exports because they are visible and can be gaged:
- Diverting Kings River floodwaters to the San Joaquin River in order to minimize flooding in the Tulare Lakebed. This is done using the North Fork / Fresno Slough / James Bypass channel. The James Bypass began operation in 1872, and the capacity of the associated system has since been increased several times. Prior to about 1872, all of the Kings River water flowed into Tulare Lake. See the section of this document on Pine Flat Dam for a more detailed description of the James Bypass. Water that is sent through this system winds up in San Francisco Bay; it is essentially a loss from the point of view of Tulare Lake Basin water users. With the construction of Pine Flat Dam in 1954, the need to divert water through this system was greatly reduced. Even

so, diversions through this system have occurred in 38% of the years since the dam was completed.[309]

- Diverting Kern River floodwaters into the California Aqueduct rather than into Buena Vista and/or Tulare Lakes. This is done using the Kern River Intertie and Cross Valley Canal. Once the water enters the California Aqueduct, it is pumped over the Tehachapi Mountains and sent to the Los Angeles area. (Los Angeles is always happy to receive high-quality water from the Tulare Lake Basin, especially when they only have to pay shipping costs.) The Kern River Intertie was completed in 1977. Prior to that, a big flood on the Kern would first fill Buena Vista and/or Goose Lakes, and then spill into Tulare Lake. The Kern Intertie was used to make large diversions to the Los Angeles area during the 1983 and 1998 floods. Lesser diversions may have been made in other floods. The shipping costs to get water to Southern California are not insignificant. We all see the big pumping plant and pipes west of Interstate 5 when we drive the Grapevine. The State Water Project (SWP) is the largest single consumer of energy in California with a net usage of 5.1 million mWh.[310] The SWP consumes so much energy because of where it sends its water. To convey water to Southern California from the Sacramento–San Joaquin River Delta, the SWP must pump it 1,926 feet over the Tehachapi Mountains, the highest lift of any water system in the world. Pumping one acre-foot to the region requires approximately 3,000 kWh. Southern California's other major source of imported water is also energy intensive: pumping one acre-foot of Colorado River Aqueduct water to Southern California requires about 2,000 kWh.[311] It requires an average of more than 9,000 kWh to move a million gallons of water to Southern California.[312] The Metropolitan Water District of Southern California estimates that the amount of electricity used to deliver water to residential customers in Southern California is equal to one-third of the total average household electric use in Southern California.[313]

- Transferring water from the Kings, St Johns, and Tule Rivers into the Friant-Kern Canal in order to minimize flooding in the Tulare Lakebed. This is done by using pumps at the point where each of those rivers crosses the canal. Once the river water enters the canal, it flows by gravity to the canal's terminus near Bakersfield. There it is emptied into the Kern River. The water is then routed to the Los Angeles area using the Kern River Intertie and Cross Valley Canal as described above. A combined total of over 472,000 acre-feet of floodwaters was pumped into the canal during the years 1978, 1980, 1982, 1983, 1986, 1995, 1997, 1998 and 2006. (The total amount may have been a good bit more than this; records are incomplete.) Transfers may have been made in later years as well. Including all sources (four rivers), exports to the Los Angeles area occurred in 30% of the years since the Kern River Intertie began operation in 1977.[314]

- Selling water for use elsewhere, such as in the Los Angeles area. That is outside the scope of this document.

We don't have anything like a full grasp of just how large the total exports were for most years. However, Table 9 gives examples of what the numbers add up to for three of the larger years for water exports from the Tulare Lake Basin.[315]

Table 9. Examples of large water exports out of the Tulare Lake Basin: 1969–1998.

| Source: | Kings River | Kings/Kaweah/Tule | Kern River | Total Exports |
|---|---|---|---|---|
| Route: | James Bypass | Friant-Kern Canal | Kern Intertie | |
| To: | SF Bay | LA | LA | |
| Year | (million acre-feet) | (million acre-feet) | (million acre-feet) | (million acre-feet) |
| 1969 | 1.6 | not an option | not an option | 1.6 |
| 1983 | 2.3 | | .8 | 3.1 |
| 1998 | 1.0 | .2 | .1 | 1.3 |

Total exports of at least 700,000 acre-feet also occurred in 1978, 1980, 1986, and 2006. There are many years for which no data are available.

That isn't to say that exports (or inter-basin transfers) are necessarily a bad thing; just that we need to recognize that such diversions have consequences to the groundwater aquifer. Every time that water is transferred out of the Tulare Lake Basin, it's that much less water available for use or for groundwater recharge. The pressure to make certain types of diversions may increase in the future.

In December 2009, a NASA/UC Irvine study concluded that for the period from October 2003 through March 2009, the groundwater supplies of the San Joaquin Valley were depleted by an average of over 2.8 million acre-feet per year. The data covered a period consisting of one very dry year, two moderately dry years, and two wet years, an average mix in our region. The lead author of the 2009 study was Jay Famiglietti, the director of Hydrologic Modeling at the University of California, Irvine.[316]

For perspective, that 2.8 million acre-feet overdraft is nearly as great (80%) as the combined annual flow of the two largest rivers in the southern San Joaquin Valley. The San Joaquin River produces a long-term average annual flow (measured at Millerton Dam) of 1.8 million acre-feet per year. The Kings River has an average annual flow (measured at Pine Flat Dam) of 1.7 million acre-feet. Together, these two rivers produce an average of 3.5 million acre-feet of water.

Bill Tweed described the situation in a column that he wrote for the *Visalia Times-Delta*.[317] In the San Joaquin Valley, water for agriculture, cities, rivers, wildlife refuges, etc. comes from three sources:
1. The most sustainable and local of these three is the Sierra, mainly in the form of water from the Tuolumne, Merced, San Joaquin, Kings, Kaweah, and Kern Rivers.
2. The second source of our water is exports from the Sacramento–San Joaquin River Delta, which is the immediate source of water that we import from Northern California via the Delta-Mendota Canal and California Aqueduct. That water, moved south at considerable expense, is increasingly fought over and hard to get.
3. The third source is what we pump from the groundwater aquifer, much of which is never replaced. A century ago, much of the valley had groundwater almost to the surface; artesian wells were common. Now, many areas have been mined for water to a depth of several hundred feet. The groundwater situation in Tulare County is particularly well studied and understood.

When surface supplies are inadequate to meet the demand, we pump out of the groundwater aquifer. When surface supplies allow, water districts work to recharge the groundwater aquifer. Such efforts form an important part of water management in the San Joaquin Valley.

In recent years, it has become increasingly apparent that we are withdrawing more than we are returning, we have a groundwater overdraft. Our demand for water is dramatically higher than the surface supply, so we are mining the underground supply. By that definition, we are effectively in drought conditions most of the time, even when precipitation is above average.

Thanks to the NASA/UC Irvine study, we now know how badly we are failing to replace the water that we pump for agricultural and urban use. We would have to divert most of the water from our two biggest rivers in order to cover the shortfall.

The implications of this are hard to escape. Even if our snowpack were to remain stable, the groundwater table in the San Joaquin Valley will continue to drop due to the overdraft. One of the consequences of this is that the wells will continue to go ever deeper and the cost of pumping will continue to rise. Eventually the pumping will be limited by supply and demand. Agriculture (the valley's single biggest water user) will be forced to reduce its reliance on groundwater.

In 2012, Bridget Scanlon and her colleagues at the University of Texas produced the highest-resolution picture yet of how groundwater depletion varies across space and time in California's Central Valley and the High Plains of the central U.S.[318] The authors of that report used water level records from thousands of wells, data from NASA's GRACE satellites, and computer models to study groundwater depletion in the two regions.

The 2012 study built on the 2009 NASA/UC Irvine study. Both studies identified the southern areas of the Central Valley — particularly the Tulare Lake Basin — as facing the most dire groundwater issues.

Scanlon's study paints a stark picture of how much water has been removed from the area's groundwater aquifer. In the mostly drought-defined years of 2006–09, water users in the Tulare Lake Basin used enough groundwater to fill Lake Mead, the nation's largest man-made reservoir.[319]

The consequences of overdrawing the groundwater aquifer aren't just financial. Subsidence in the San Joaquin Valley is one of the great changes that human activity has imposed on the environment. The San Joaquin Valley has the largest vertical subsidence (a maximum of 29.7 feet, or more than 28 feet, depending on the source) and the largest areal extent (5,400 square miles, or 5,200 square miles, depending on the source) of subsidence in the world because of groundwater withdrawal.[320, 321]

The 5,400 square miles of the San Joaquin Valley that has experienced subsidence represents about half of the total valley floor (10,000 square miles). The most severe subsidence (29.7 feet) is near Mendota. The maximum rate of subsidence occurred before the California Aqueduct and Delta-Mendota Canal were built, in part to reduce the amount of groundwater overdraft in that region.

Those measurements were as of 1970 when the last comprehensive surveys of land subsidence were made. For a while, it had been assumed that subsidence had largely stopped after the big canals were completed in about 1970. That assumption turns out to have been based on wishful thinking and a lack of monitoring. It is now recognized that further significant subsidence has occurred along the western side of the Tulare Lake Basin since 1970, particularly in Kern, Kings, and Fresno Counties. It has been speculated that some areas in those counties have now subsided upwards of 50 or 60 feet.[322] USGS is currently conducting a three-year study (2009–12) to measure the actual amount.

Recent subsidence has been detected in the San Joaquin River Basin in western Madera County. In the last few years, the land there has subsided almost three feet (photograph on file in the national parks).

Subsidence began in the Tulare Lake Basin in 1935 or a few years earlier.[323, 324] One of the best places to see an example of that early subsidence is in the community of Three Rocks, along Highway 33 on the west side of Fresno County.[325] Subsidence has continued off and on in that community since then. Three Rocks is located in the Westside Subbasin (aka Westlands Water District) which extends roughly from Firebaugh on the north to Kettleman City on the south.

The San Joaquin Valley groundwater system is the largest storage reservoir south of the Sacramento–San Joaquin River Delta and therefore has a large potential and interest from a statewide perspective. The future strategy of Delta water exports will be one of opportunistic flows, rather than dependable steady-state flows. Making the best use of this resource will require unprecedented cooperation and coordination of surface water and groundwater users south of the Delta. That will be challenging. But given the right mix of infrastructure and institutional arrangements, the San Joaquin Valley groundwater system conditions could improve significantly compared to present conditions.

The groundwater overdraft in the Tulare Lake Basin is arguably due to a combination of pumping too much, not recharging enough, and not conserving enough. Many groups are working on different aspects of these various problems.

It isn't clear how to summarize the overall groundwater overdraft situation. Perhaps the most optimistic viewpoint is that the situation is grave, and no practical basin-wide alternative exists except for failure and reduction of groundwater use by agriculture and some cities. However, the effects will be locational even in the Tulare Lake Basin; some areas will remain sustainable.

The more pessimistic view is that this is not just a localized groundwater issue. The entire Tulare Lake Basin is losing water at a dramatic rate; the system has essentially sprung a massive leak. Therefore, the entire basin will be forced to significantly reduce its reliance on groundwater use.

.

# Summary of Droughts

## Summary of Past Mega-Droughts

California was subject to prolonged, severe droughts from about A.D. 900–1400. The first epic drought lasted more than two centuries before the year 1112; the second drought lasted more than 140 years before 1350.

A reconstruction of the Sacramento River streamflow developed from tree-rings found that the most intense 20-year drought of about the last 1,000 years was between 1140–1160.

## Summary of Droughts in the Central Valley: 1579–1900

The droughts in Table 10 were compiled from a variety of sources, including tree-ring reconstructions, Tulare Lake elevations, and settler accounts. It probably captures most of the larger droughts, but may miss some of the lesser ones.

The drought length given is the number of years that the drought was active somewhere in the state. It may not have been active each of those years in the Central Valley. The first drought (1566–1602), which is known from the Southern Sierra, overlaps with the next two droughts which are known from the Sacramento River Basin.

Table 10. Droughts in the Central Valley: 1566–1900.

| Drought | Length (years) |
|---------|----------------|
| 1566–1602 | 36 |
| 1579–1582 | 4 |
| 1593–1595 | 3 |
| 1618–1620 | 3 |
| 1651–1655 | 5 |
| 1719–1724 | 6 |
| 1735–1737 | 3 |
| 1755–1761 | 6 |
| 1776–1778 | 3 |
| 1793–1795 | 3 |
| 1827–1829 | 3 |
| 1839–1841 | 3 |
| 1843–1846 | 4 |
| 1855–1857 | 3 |
| 1860–1861 | 2 |
| 1862–1864 | 3 |
| 1870 | 1 |
| 1873–1882 | 10 |
| 1887–1888 | 3 |
| 1897–1900 | 4 |

## Summary of Droughts since 1901

Table 11. Summary of droughts in the Tulare Lake Basin since 1901.

| Drought | Water Year | San Joaquin Basin Water Year Classification |
|---------|-----------|---------------------------------------------|
| 1912–13 | 1912 | Below normal |
|         | 1913 | Critically dry |
|         |      |                |
| 1917–21 | 1917 | Wet |
|         | 1918 | Below normal |
|         | 1919 | Below normal |
|         | 1920 | Below normal |
|         | 1921 | Above normal |
|         |      |                |
| 1922–27 | 1922 | Wet |
|         | 1923 | Above normal |
|         | 1924 | Critically dry |
|         | 1925 | Below normal |
|         | 1926 | Dry |
|         | 1927 | Above normal |
|         |      |                |
| 1928–34 | 1928 | Below normal |
|         | 1929 | Critically dry |
|         | 1930 | Critically dry |
|         | 1931 | Critically dry |
|         | 1932 | Above normal |
|         | 1933 | Dry |
|         | 1934 | Critically dry |
|         |      |                |
| 1943–51 | 1943 | Wet |
|         | 1944 | Below normal |
|         | 1945 | Above normal |
|         | 1946 | Above normal |
|         | 1947 | Dry |
|         | 1948 | Below normal |
|         | 1949 | Below normal |
|         | 1950 | Below normal |
|         | 1951 | Above normal |
|         |      |                |
| 1959–62 | 1959 | Dry |
|         | 1960 | Critically dry |
|         | 1961 | Critically dry |
|         | 1962 | Below normal |
|         |      |                |
| 1976–77 | 1976 | Critically dry |
|         | 1977 | Critically dry |
|         |      |                |
| 1987–92 | 1987 | Critically dry |
|         | 1988 | Critically dry |
|         | 1989 | Critically dry |
|         | 1990 | Critically dry |
|         | 1991 | Critically dry |
|         | 1992 | Critically dry |
|         |      |                |
| 2007–09 | 2007 | Critically dry |
|         | 2008 | Critically dry |
|         | 2009 | Below normal |

# Summary of Floods

## Summary of Past Mega-Floods

Botanic and geomorphic evidence indicates that a very large flood occurred in Northern California in about 1600. As detailed in Table 12, geomorphic evidence indicates that mega-floods occur in Southern California on approximately a 200-year cycle. This bicentennial flooding was skipped only three times since 212 and never twice in a row.

The last skip was in the early 1800s. Even that skip may have only been a skip from the perspective of Southern California. It's possible that the huge flood that the Central Valley experienced in 1805 belongs in this series.

Table 12. Mega-floods in Southern California.

| Approximate Date of Flood |
|---|
| 212 |
| 440 |
| 603 |
| 1029 |
| 1418 |
| 1605 |

It is uncertain whether the Northern California mega-flood of about 1600 was one and the same as the Southern California mega-flood of about 1600. In any case, it's tempting to think that there would also have been floods in the Tulare Lake Basin during the 1600–1610 time period since the regions to our north, east, and south were experiencing immense precipitation events at this time. The Tulare Lake Basin would have been under the same general storm tracks.

## Summary of 19th Century Flood History

Table 13 presents what we know about those floods that occurred between 1800–1849. Information on floods from this period is very incomplete. The information about these floods typically came from Native Americans and from the earliest settlers such as John Sutter.

Table 13. Partial list of major floods in the Central Valley: 1800–1849.

| Date of Flood | Type of Flood |
|---|---|
| 1805 | Rain |
| 1826 | Rain |
| 1847 | Rain |

High-water marks observed in the San Joaquin Valley and attributed to the 1805 flood were some six feet higher than the huge 1861–62 flood reached.

Extensive settlement in California began around 1850 following the discovery of gold. Settlement of the Tulare Lake Basin began about that time as well. Between 1850 and 1900, a number of great floods occurred in the Central Valley. For this time period, we have much better data for the San Joaquin River Basin than for the Tulare Lake Basin.

Table 14 lists 12 of the major floods in the San Joaquin River Basin during the period 1850–1900. It excludes floods that clearly don't appear to have been major floods in the Tulare Lake Basin. We have fairly decent to very good descriptions of five of those 12 floods: 1850, 1852–53, 1861–62, 1867–68, and 1890. The others were included in this table because they were identified by the USGS and/or USACE as having been a major flood in the San Joaquin River Basin.[326, 327] The effect of these floods on the elevation of Tulare Lake is reflected in Figure 10.

Table 14. Selected major floods in the San Joaquin River Basin: 1850–1900.

| Date of Flood | Type of Flood |
|---|---|
| May–June 1850 | Snowmelt |
| December 1852–February 1853 | Rain |
| 1861 | Unknown |
| January–February 1862 | Rain |
| December 1867 – January 68 | Rain |
| 1869 | Unknown |
| 1872 | Unknown |
| 1878 | Unknown |
| 1884 | Unknown |
| 1886 | Unknown |
| 1889–90 | Rain |
| 1893 | Unknown |

In the northern part of the Central Valley, the 1861–62 flood is the flood-of-record on most rivers. The 1861–62 flood affected a very wide area stretching from the Columbia River to the Mexican border. However, the 1867–68 flood was especially severe on Sierra Nevada streams tributary to the southern part of the Central Valley. During recorded history, the 1867–68 flood was one of the greatest in the Tulare Lake Basin. Peak stages in that region during December 24–25 were the highest of record.

The 1867–68 flood also resulted in the deepest flood depths ever on the streets of Visalia. The 1861–62 flood put a maximum of 24 inches on Main Street while the 1867–68 flood put 5–6 feet.

## Selected Floods in the Tulare Lake Basin Since 1905

Table 15 lists floods that generally had very high flood flows on at least one of the major rivers in the Tulare Lake Basin since 1905. Selecting floods to include in this list was somewhat arbitrary, just as our concept of what constitutes a flood is arbitrary.

Table 15. Peak flood flows for selected floods: 1905–2011.

| Date of Flood | Type of Flood | Peak Flood Flows (Average Daily Full Natural Flow) | | | |
| | | Kings River at Pine Flat (cfs) | Kaweah River at Terminus (cfs) | Tule River at Success (cfs) | Kern River at Bakersfield (cfs) |
|---|---|---|---|---|---|
| March 1906 | Rain | | 8,861 | | |
| May–June 1906 | Snowmelt | 24,900 | 7,260 | | 9,500 |
| January 1909 | Rain | | 9,578 | | |
| December 1909 | Rain | | 8,226 | | |
| January 1914 | Rain | | 10,275 | | |
| January 1916 | Rain | | 10,540 | | |
| February 1937 | Rain | | 13,520 | | |
| February 1938 | Rain | | 11,232 | | |
| March 1943 | Rain | | 9,714 | | |
| February 1945 | Rain | | 9,890 | | |
| November 1950 | Rain | | 16,640 | | |
| January, 1952 | Rain | | 5,918 | | |
| May–June 1952 | Snowmelt | 15,500 | 5,170 | 860 | 8,360 |
| December 1955 | Rain | 72,589 | 44,512 | | 12,787 |
| February 1963 | Rain | 34,612 | 18,405 | 6,100 | 15,612 |
| December 1966 | Rain | 64,564 | 53,280 | 40,085 | 72,787 |
| January 1969 | Rain | 40,513 | 22,437 | 12,822 | 22,359 |
| September 1978 | Rain | 19,205 | 3,890 | 337 | 3,868 |
| January 1980 | Rain | 33,283 | 16,933 | 8,676 | 13,036 |
| April 1982 | Rain | 48,909 | 18,514 | 6,690 | 8,638 |
| September 1982 | Rain | 30,415 | 6,308 | 586 | 6,673 |
| December 1982 | Rain | 24,682 | 8,325 | 3,680 | 4,236 |
| May 1983 | Snowmelt | 24,218 | 6,671 | 2,036 | 13,812 |
| February 1986 | Rain | 25,060 | 9,428 | 5,650 | 7,528 |
| March 1995 | Rain | 26,970 | 8,369 | 2,413 | 7,347 |
| January 1997 | Rain | 50,217 | 17,948 | 9,676 | 18,780 |
| November 2002 | Rain | 10,969 | 9,436 | 4,906 | 10,306 |

The flow data shown above are the peak daily flows (the average hourly flow for the peak day of the flood). Flows are expressed in cubic feet per second (cfs). Where necessary, these flows have been adjusted to remove the effects of dams upstream of the gage. Source: USACE, Sacramento District, and USGS.

Table 16. Flood exceedence frequencies for selected floods: 1905–2011.

| | Type of Flood | Flood Exceedence Frequencies | | | |
| --- | --- | --- | --- | --- | --- |
| | | Kings River at Pine Flat | Kaweah River at Terminus | Tule River at Success | Kern River near Bakersfield |
| March 1906 | Rain | | 16% | | |
| May–June 1906 | Snowmelt | | | | |
| January 1909 | Rain | | 14% | | |
| December 1909 | Rain | | 18% | | |
| January 1914 | Rain | | 13% | | |
| January 1916 | Rain | | 13% | | |
| February 1937 | Rain | | 10% | | |
| February 1938 | Rain | | 12% | | |
| March 1943 | Rain | | 14% | | |
| February 1945 | Rain | | 13% | | |
| November 1950 | Rain | | 8% | | |
| January, 1952 | Rain | | 25% | | |
| May–June 1952 | Snowmelt | | | | |
| December 1955 | Rain | 1% | 1.2% | | 8% |
| February 1963 | Rain | 5.3% | 6.4% | 20% | 5% |
| December 1966 | Rain | 1.4% | 0.6% | 0.5% | 0.3% |
| January 1969 | Rain | 4% | 4% | 1.5% | 1.8% |
| September 1978 | Rain | 12% | 29% | 93% | 30% |
| January 1980 | Rain | 5.5% | 8% | 9% | 8% |
| April 1982 | Rain | 2.9% | 6% | 13% | 13% |
| September 1982 | Rain | 6.5% | 25% | 84% | 16% |
| December 1982 | Rain | 10% | 18% | 30% | 25% |
| May 1983 | Snowmelt | | | | |
| February 1986 | Rain | 9.5% | 15% | 17% | 15% |
| March 1995 | Rain | 9% | 18% | 40% | 15% |
| January 1997 | Rain | 2.5% | 7% | 8% | 4% |
| November 2002 | Rain | 30% | 15% | 20% | 10% |

The flood exceedence values shown in the above table reflect the chance of a flood of a certain size or larger occurring in any given year. The exceedence frequencies were calculated by combining the peak daily flow rates from Table 15 with the appropriate flood frequency curve.

The rain flood frequency curves were calculated by the USACE based on observed rain floods for the following periods of record:

- Kings River below Pine Flat Dam: water years 1955–1978
- Kaweah River below Terminus Dam: water years 1905–2004
- Tule River below Success Dam: water years 1953–1988
- Kern River below Isabella Dam: water years 1953–2008

Table 17 presents the flood recurrence intervals for selected floods. Although that's the term used in this document, this concept will be found referred to elsewhere under a wide variety of terms, including:

- exceedence interval
- occurrence rate
- 50-year return period, etc.
- 50-year flood event, 50-year-flood, etc.

Table 17. Flood recurrence intervals for selected floods: 1905–2011.

| | Type of Flood | Flood Recurrence Intervals | | | |
| --- | --- | --- | --- | --- | --- |
| | | Kings River at Pine Flat (years) | Kaweah River at Terminus (years) | Tule River at Success (years) | Kern River near Bakersfield (years) |
| March 1906 | Rain | | 6 years | | |
| May–June 1906 | Snowmelt | | | | |
| January 1909 | Rain | | 7 years | | |
| December 1909 | Rain | | 5 years | | |
| January 1914 | Rain | | 8 years | | |
| January 1916 | Rain | | 8 years | | |
| February 1937 | Rain | | 10 years | | |
| February 1938 | Rain | | 8 years | | |
| March 1943 | Rain | | 7 years | | |
| February 1945 | Rain | | 8 years | | |
| November 1950 | Rain | | 13 years | | |
| January, 1952 | Rain | | 4 years | | |
| May–June 1952 | Snowmelt | | | | |
| December 1955 | Rain | 100 years | 85 years | | 13 years |
| February 1963 | Rain | 19 years | 16 years | 5 years | 20 years |
| December 1966 | Rain | 70 years | 170 years | 200 years | 333 years |
| January 1969 | Rain | 25 years | 25 years | 67 years | 56 years |
| September 1978 | Rain | 8 years | 3 years | 1 year | 3 years |
| January 1980 | Rain | 18 years | 13 years | 11 years | 13 years |
| April 1982 | Rain | 35 years | 17 years | 8 years | 8 years |
| September 1982 | Rain | 15 years | 4 years | 1 year | 6 years |
| December 1982 | Rain | 10 years | 6 years | 3 years | 4 years |
| May 1983 | Snowmelt | | | | |
| February 1986 | Rain | 11 years | 7 years | 6 years | 7 years |
| March 1995 | Rain | 11 years | 6 years | 3 years | 7 years |
| January 1997 | Rain | 40 years | 14 years | 13 years | 25 years |
| November 2002 | Rain | 3 years | 7 years | 5 years | 10 years |

The recurrence intervals in the above table were calculated by combining the peak daily flow rates from Table 15 with the appropriate flood frequency curve. An alternative — and equally valid — approach would have been to combine the peak *hourly* flow rates with the corresponding flood frequency curve. This can result in noticeably different results. (See the January 1997 flood on the Kaweah for a comparatively extreme example. Using the daily flow rate, that flood had a recurrence interval of 14 years; but using the peak hourly flow rate, it had a recurrence interval of 25 years.)

One of the big lessons learned from preparing this document is that our rivers have been relatively quiet of late. Table 17 shows that the Tulare Lake Basin hasn't really experienced any big floods in over 40 years. There haven't been any 50-year floods or 100-year floods. The Kaweah and Tule Rivers haven't even seen any 20-year floods during this time period.

That probably doesn't mean anything; it's just a statistical coincidence. The problem is more psychological. When we don't experience a big flood for a while, we tend to forget just how big our floods can be. We have come to think of the federal reservoirs and our levees as protecting us from the effects of big floods, and that isn't necessarily realistic when we consider our flood history.

Those of us who live and work above the federal reservoirs have a greater responsibility to become involved in planning for floods than those who live below the dams. The last really big flood in the Tulare Lake Basin was the December 1966 flood. It's sobering to reflect back on the experience of that flood. Fifteen-foot waves were reported to have been common on the mainstem of the Kaweah in Three Rivers. Today we think of Dry Creek below Terminus Dam as not much more than a quiet foothills stream. However, in the 1966 flood, Dry Creek carried 44% more water than the Merced River in Yosemite Valley during the much more famous January 1997 flood.

The take-away message is that it would be prudent to prepare for big floods, much bigger than we've been experiencing during the last 40 years. This is particularly important for those of us who live and work in areas that aren't protected by a federal reservoir.

# Summary of Floods and Droughts since 1849

This document describes what we know about approximately 183 floods and 31 multi-year droughts that have occurred during the last 2,000 years. The floods of 1849–50 are the first for which there are fairly accurate historic descriptions. Figure 17 illustrates the 173 floods and 16 multi-year droughts that have occurred since 1849.

This document sometimes groups floods from adjacent years together. This can occur when floods wrap over the winter months as in 1955–56. This can also occur when two years of floods run together as in 1982–83. In Figure 17, the floods are shown by the first year of the two-year flood grouping.

Droughts are generally illustrated for the number of years that the drought was active somewhere in the state. The drought may not have been active each of those years in the Tulare Lake Basin. See the section of this document that describes each drought to learn what we know about local conditions during those years.

This graph illustrates how often floods occur during multi-year droughts. That was one of the lessons learned from preparing this document. Floods occur at all manner of times. When they occur varies so widely because there are such a variety of causes for floods.

Figure 17 illustrates far more individual flood events in recent years than in earlier years. Certainly 1982–83 was a period of exceptional flooding events. But in general, this document reflects more flood events from recent years because of data availability. In the last few decades, there has been an explosion in information that society records and puts on the Internet. In many ways, this graph of flood frequency reflects that explosion in information availability.

This document describes 10 floods that occurred before 1849; the earliest of those was dated about A.D. 212. See the section of this document that describes the California mega-floods for more about that event. Half of the droughts described in this document, 15, occurred before 1849; the earliest began in about A.D. 900. See the section of this document that describes California's mega-droughts for more about that drought. In general, the farther back that you look, the bigger the flood or drought has to be in order to be detectable.

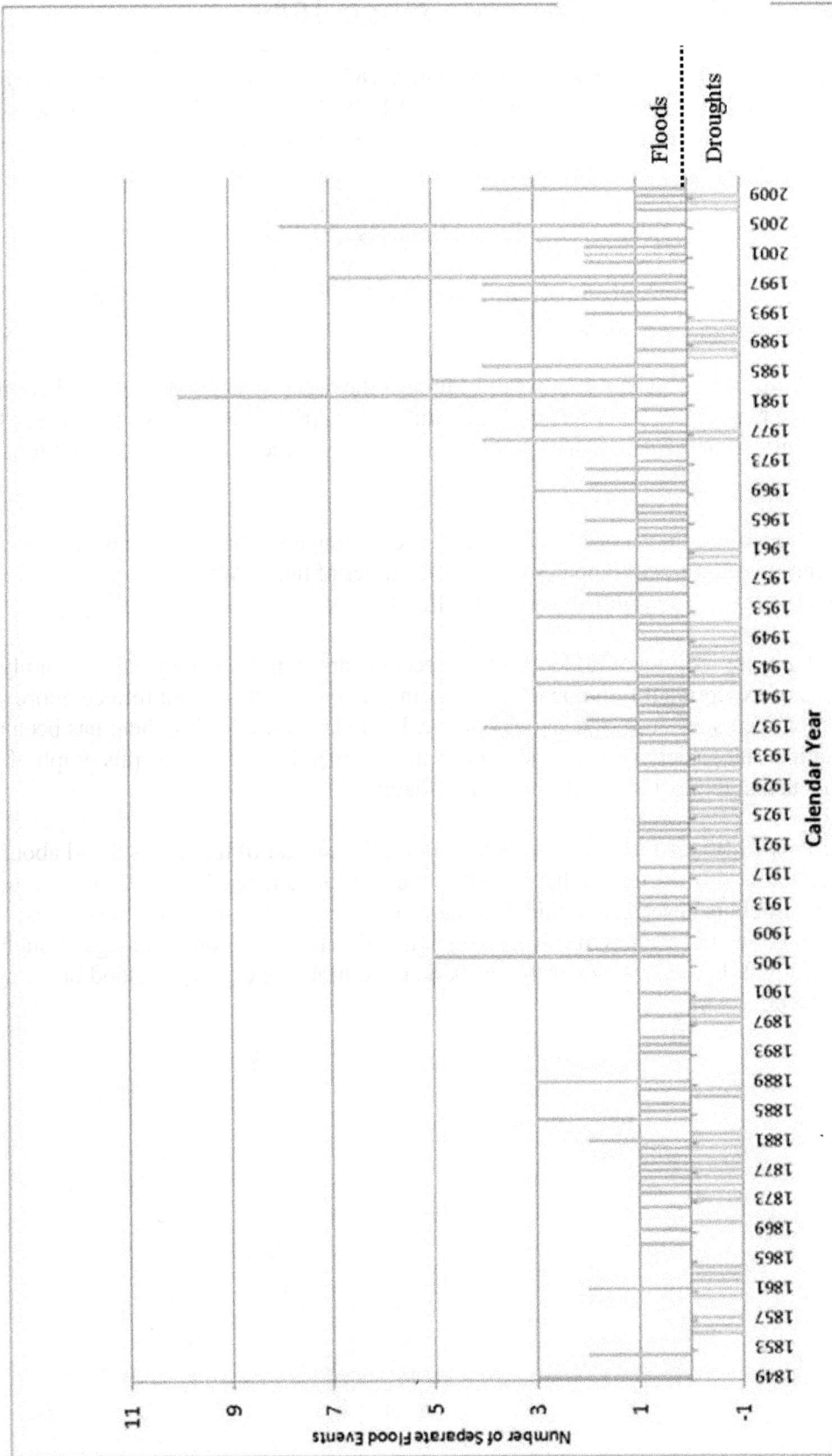

Figure 1. Known floods and multi-year droughts for past 163 years: 1849–2011

# Specific Floods and Droughts

## California's Two Mega-droughts (A.D. 900–1400)

California, including the Sierra, was subject to prolonged, severe droughts from about A.D. 900–1400.[328]

The driest two periods in western North America were centered on the mid–1100s and the mid–1200s; these are reflected in the Sierra.[329, 330]

A 1994 study of relict tree stumps rooted in present-day lakes, rivers, and marshes showed that California sustained two epic drought periods, extending over more than three centuries.[331, 332] The first epic drought lasted more than two centuries before the year 1112; the second drought lasted more than 140 years before 1350. In that 1994 study, Scott Stine used drowned tree stumps rooted in Mono Lake, Tenaya Lake, West Walker River, and Osgood Swamp in the Central Sierra. Scott concluded that runoff from the Sierras during those periods was "significantly lower than during any of the persistent droughts that have occurred in the region over the past 140 years."

At Fallen Leaf Lake (located near the southern end of Lake Tahoe), trees up to 82 feet in height have been found rooted upright 118 feet below the current lake surface. Those trees became established and grew during this mega-drought.

By the early 1980s, the Los Angeles Department of Water and Power's diversions had caused Mono Lake to drop to elevation 6,372 feet, roughly 50 feet below its natural elevation. This drawdown exposed thousands of acres of lake bottom along with many cottonwood and Jeffrey pine stumps. Scott Stine sampled those stumps and had them radiocarbon dated. They fell into two groups: one that had died in about 1100 and the other in about 1350. Thus there had been two mega-droughts in the area that brought Mono Lake down far enough to allow trees to become established and grow to considerable size.

Scott found a similar situation at Tenaya Lake in Yosemite National Park, where a dozen trees protrude from this glacial lake. Some individual trees are standing in nearly 70-foot-deep water. When Scott had those trees radiocarbon dated, he found that they fell into similar groups as at Mono Lake. One group had died in about 1100 and the other in about 1350. A count of the annual rings showed that the first drought (the one ending in about 1100) had lasted at least 140 years, and the second drought (the one ending in about 1350) had lasted at least a century. Phil Catarino and Eric Henningsen, the divers who worked at Fallen Leaf Lake, have confirmed that the bases of the Tenaya trees are rooted in growth position.

Scott found relic stumps of Jeffrey pines in West Walker River on Highway 108, the Sonora Pass road.[333] Since the root systems of Jeffrey pines can tolerate complete inundation for only a couple weeks at most, he reasoned that there must have been an extended drought in the area while these trees were growing. Sure enough, they dated to the same periods as the two mega-droughts at Mono and Tenaya Lakes.

During much of the 20th century, Owens Lake has been reduced to a playa due to water diversions by the Los Angeles Department of Water and Power. In the 1990s, Scott went out into the Owens Playa as part of an archeological assessment. The search turned up not only archeological materials

140

(some of them at very low elevations on the playa), but a rooted, tufa-encrusted shrub stump as well. Radiometric analysis of the shrub stump indicated that it comported in age with the circa 1100 mega-drought sites from farther north.

Susan Lindstrom, while diving in Independence Lake north of Tahoe, discovered stumps that may have been comparable to the mega-drought that ended around 1350.[334]

From reviewing studies such as Scott Stine's, DWR concluded that California is subject to droughts far more severe and prolonged than anything witnessed in the historical record.[335]

Steve Wathen is an ecologist/forester from UC Davis. He said that a mega-drought occurred from 1150–1225 in the Northern Sierra.

In a 1996 study, Lynn Ingram and others concluded from their analyses of variations in the isotopic composition of fossil shells in the San Francisco Bay, that "alternate wet and dry (drought) intervals typically have lasted 40 to 160 years" in the San Francisco Bay drainage area (Sacramento River and San Joaquin River Basins) over the past 750 years (since about 1250).[336]

In addition to these very long duration mega-droughts, there were other severe droughts of shorter duration occurring during the 12th and 13th centuries.

A reconstruction of the Sacramento River streamflow developed from tree-rings found that the most intense 20-year drought of about the last 1,000 years was between 1140–1160.[337]

George Durkee (national park wilderness ranger) collected a core from a stump above the Crabtree Ranger Station and sent this to Scott Stein who dated it. Radiocarbon dating showed that the tree died in the 1100s, so it may have been contemporaneous with the 1140–1160 drought. That is, that stump apparently became rooted during that drought.

Tony Caprio and Linda Mutch studied how the Mountain Home Grove of giant sequoias responded to a fire that occurred in 1297.[338] Their research suggested that this fire event was of unusually high severity, not equaled over the last 2,000 years. It appeared to result in the mortality of most non-sequoias and a considerable number of giant sequoias.

Time series of precipitation, reconstructed from tree-rings, showed that a long drought occurred between 1292–96 throughout the Southern Sierra. Tony and Linda speculated that this drought was a contributing factor in creating the conditions for the Mountain Home fire that occurred in 1297.

The last severe mega-drought that impacted the Great Basin and adjacent parts of California ended about 1350. Just as it was ending, another drought was getting started in the Central Valley and the Sierra. A reconstruction of the Sacramento River streamflow developed from tree-rings found that the driest 50-year period in the last 1,100 years was between about 1350–1400.[339] In addition to the Central Valley, this drought affected most or all of the Sierra.

Steve Wathen said that this drought lasted from 1350–1500 in the Northern Sierra. It may have also affected the Rockies.

## 1566–1602 Drought

This 36-year drought affected at least the Southern Sierra and is known from tree-ring reconstruction.[340] It ended when the climate of North America changed dramatically during the 1600–1610 time period. See the section of this document that describes the California mega-floods for more about that event. This drought in the Southern Sierra overlaps the following two droughts that are known from tree-ring reconstruction in the Sacramento River Basin.

## 1579–82 Drought

This four-year drought affected at least the Sacramento River Basin and is known from tree-ring reconstruction.[341]

By some measures, 1580 is the drought year of record in the Southern Sierra. The tree-ring for that year is barely detectable, demonstrating the severity of the drought; it is even more severe than the 1976–77 drought.

However, from other perspectives, the pair of six-year droughts that lasted from 1929–34 and 1987–92 are considered to be longer, more severe droughts than any other drought since at least 1560. See the sections of this document that describes those droughts for the difference between them.

## 1593–95 Drought

This three-year drought affected at least the Sacramento River Basin and is known from tree-ring reconstruction.[342]

This drought's claim to fame is that it has the lowest average runoff (9.3 million acre-feet) on the Sacramento River of any drought since 1560. See Table 70 in the section of this document that describes the 1987–92 drought for a comparison of selected multi-year droughts.

## California's Six Mega-floods (A.D. 212–1605)

Arndt Schimmelmann did research based on sediment cores taken in the Santa Barbara Basin and published his results in two scientific papers.[343, 344] Schimmelmann found evidence for a huge flood striking Southern California approximately every 200 years, centered on the years shown in Table 18. This cyclic mega-flood is generally described as a Southern California event because that is where the research was done. However, the flood presumably strikes Northern and Central California as well.

Table 18. Mega-floods in Southern California.

| Approximate Date of Flood |
|---|
| 212 |
| 440 |
| 603 |
| 1029 |
| 1418 |
| 1605 |

The bicentennial flooding in Southern California was skipped only three times since 212 and never twice in a row. The quasi-periodicity of approximately 200 years for Southern California floods recorded in the Santa Barbara Basin matches the approximate 200 year periodicities found in a variety of high-resolution paleoclimatic archives. More importantly, it matches the roughly 208-year cycle of solar activity (the Suess Cycle) and inferred associated changes in atmospheric circulation. The last skip was in the early 1800s, leading the authors of one of the Schimmelmann papers to conclude that "we foresee the possibility for a historically unprecedented flooding in Southern California during the first half of the 21st century."

The skip in the early 1800s may have only been a skip from the local perspective of the Santa Barbara Basin. The Central Valley experienced a huge flood in 1805, one that was even bigger than the huge 1861–62 flood. Perhaps the storm track that year just didn't extend far enough south to be recorded in the Santa Barbara Basin.

The 1605 event was measured with a precision of ±5 years. The dating of that flood event is consistent with tree-ring evidence for a wet and cold paleoclimate elsewhere in the region. The depth of the silt layer deposited in the Santa Barbara Basin in the 1605 event implied an intensity of precipitation and flooding of the Ventura and Santa Clara Rivers unmatched in the last 1,000 years.

We tend to think of floods as relatively short-term events, lasting for no more than a few months. However, judging from what we know of the 1605 flood, mega-floods appear to be a much longer-term type of event. Once such an event begins, it can last for up to 10 years. During that period, multiple episodes of flooding and extreme runoff can occur as well as other unusual climatic events. The 1605 flood was associated with a large-scale change in climate that affected the Northern Hemisphere from roughly 1600–1610.

The decade 1600–1609 stands out as the coldest in a 570-year (A.D. 1400–1970) comprehensive record of summer temperatures across the Northern Hemisphere, based on tree-ring and ice-core data.[345] The years 1601 and 1605 produced unusually narrow tree rings in the Sierra, suggesting very cold growing seasons.[346]

The 16-year period from 1597–1613 in the Sacramento River Basin had the maximum reconstructed riverflow for the 420-year (1560–1980) time period.[347] There was a major flood on several Northern California rivers, including the Salmon and Klamath, in about 1600. Those rivers wouldn't see another flood that big until the December 1964 flood.

The year 1602 was the end of a 36-year drought (1566–1602) drought in the Sierra.[348] Mono Lake filled to record levels from approximately 1600–1650. Those were the highest levels of the past millennium.[349] The Mojave River terminates at the Silver Lake playa in the Mojave Desert. (That playa is located along Interstate 15, just north of the town of Baker.) At very infrequent times, the Mojave River delivers so much water to the playa that it forms a perennial lake. The last time that this happened was likely during approximately the 1600–1610 time period.

Along the coast of California in the Santa Barbara area, 1604 was the fourth wettest year, and 1601–1611 was the third-wettest 11-year period in a 620-year (A.D. 1366–1985) reconstruction of precipitation.[350] There was a major flood on the Santa Ana River in Orange County in about 1600. That river wouldn't see another flood that big until the January 22, 1862 flood. Severe flooding occurred around Mexico City in 1604 and 1607.

It's tempting to think that there would also have been floods in the Tulare Lake Basin during the 1600–1610 time period since the regions to our north, east, and south were all experiencing immense precipitation events at this time. The Tulare Lake Basin would have been under the same general storm tracks. However, we haven't found evidence of such floods. If that evidence exists, it is presumably available from sediment samples taken in the Tulare Lakebed. (See the section of this document on Runoff Reconstructions).

Regional and global climate evidence indicates that much of the world experienced rapid, intense cooling around 1605. That cooling might have been brought on by a cluster of volcanic events, but that wouldn't explain the apparent 200-year cyclical nature of the mega-floods in Southern California.

An alternative explanation is based on the observation that there was a sharp minimum in the Carbon-14 record about 1605. (Carbon-14 is a naturally occurring radioactive isotope of carbon.) This suggests that low solar activity might have been responsible for both the minimum in the Carbon-14 record and for the global climate change.

The authors of one of the Schimmelmann papers speculated that the 1605 flood might have represented a double whammy. The low in the 208-year cycle of solar activity created a period of preexisting cooling conditions across the Northern Hemisphere. By coincidence, a cluster of volcanic events may have then combined with the preexisting cooling to intensify the flooding conditions.

In any case, it is presumed that the cooling in the climate was probably accompanied by a shift of prevailing wind patterns and associated storm tracks toward the Equator during the time period 1600–1610. That appears to be the unifying link in the paleoclimatic evidence listed above and from other sources.

The authors of the two Schimmelmann papers suggested that after 1600, frequent winter/spring outbreaks of cold polar air, called ''nortes,'' originating from a much colder North American continent, picked up moisture on their southbound path, and brought intense precipitation as far south as central Mexico.

In addition to the above research, USGS has apparently done sediment research in San Francisco Bay that also showed evidence of past mega-floods. This may be a reference to the work done by Lynn Ingram at UC Berkeley. The details of that research are unknown.

## 1618–20 Drought

This three-year drought affected at least the Sacramento River Basin and is known from tree-ring reconstruction.[351]

## 1651–55 Drought

This five-year drought affected at least the Sacramento River Basin and is known from tree-ring reconstruction.[352]

## 1719–24 Drought

This six-year drought affected at least the Sacramento River Basin and is known from tree-ring reconstruction.[353]

## 1735–37 Drought

This three-year drought affected at least the Sacramento River Basin and is known from tree-ring reconstruction.[354]

## 1755–61 Drought

This six-year drought affected at least the Sacramento River Basin and is known from tree-ring reconstruction.[355]

## 1776–78 Drought

This three-year drought affected at least the Sacramento River Basin and is known from tree-ring reconstruction.[356]

## 1793–95 Drought

This three-year drought affected at least the Sacramento River Basin and is known from tree-ring reconstruction.[357]

## 1805 Flood

Based on research done in the Santa Barbara Basin, California experiences a mega-flood approximately every 200 years. (See the section of this document that describes the California mega-floods.) The last such flood to occur in Southern California was in about 1600. The 1805 flood (a Central Valley flood) may have been the next one in that series.

In any case, 1804–05 was an unusually wet winter throughout the Central Valley. There was heavy runoff the following year.

Histories of early settlements state that California Indians spoke of a great flood, which was supposed to have occurred about the beginning of the 19th century and to have drowned thousands of them. This reference may have been to the flood of 1805, which is said to have covered the entire Sacramento River Valley except the Sutter Buttes.[358, 359, 360] The Sutter Buttes is a volcanic plug that rises about 2,000 feet above the valley floor near the center of the Sacramento Valley.

High-water marks observed in the San Joaquin Valley and attributed to the 1805 flood were some six feet higher than the huge 1861–62 flood reached.[361]

## 1826 Flood

1825–26 was reported to have been an unusually wet winter throughout the Central Valley. (One source said that the wet winter was 1824–25, but that was almost certainly an error.) There was heavy runoff and flooding the following year.[362]

According to the Yuba County history, the Native Americans said that the Sacramento Valley had a large flood in the winter of 1825–26. One trapping party was compelled to camp in the Marysville Buttes because of high water. Those hills were full of grizzlies, elk, antelope, and smaller game that had taken refuge there. Native Americans recalled the flood of 1826 as a devastating one.[363, 364]

We have no records of what was occurring in the Tulare Lake Basin at the time of this flood. The flood may have extended this far south, but we really don't know that.

## 1827–29 Drought

This three-year drought affected at least part of the state.

Annie Mitchell wrote that Native Americans said that Tulare Lake went dry about 1825.[365] Possibly that was associated with the 1827–29 drought

## 1839–40 Flood

County histories and journals of pioneers mention floods in the lower Sacramento River Basin during the 1839–40 season.[366]

## 1839–41 Drought

The extent of this three-year drought is not well understood because the state was so lightly populated at the time. What is known comes from scattered accounts:
- 1839–41 was a three-year drought in the Sacramento River Basin based on tree-ring reconstruction.[367]
- John Bidwell came to California via the California Trail in 1841. He recalled that 1841 was an extremely dry year in the Sacramento area.

## 1843–46 Drought

The extent of this four-year drought is not well understood because the state was so lightly populated at the time. What is known comes from scattered accounts:
- 1843–46 was a four-year drought in the Sacramento River Basin based on tree-ring reconstruction.[368]
- Drought ruined the crops at Sutter's Fort in 1844.
- John Bidwell recalled that 1843 and 1844 were extremely dry years in the Sacramento area.

## 1847 Flood

1846–47 was reported to have been an unusually wet winter. (One source said that the wet winter was 1845–46, but that was almost certainly an error.) There was heavy runoff the following year.[369]

The winter of 1846–47 was the one that trapped the Donner Party in the Sierra. A severe blizzard during the last week of October 1846 buried the upper elevations of the Sierra and blocked the trail into Northern California. During the winter of 1846–47, the snowline east of Sutter's Fort (located in present-day Sacramento) was typically about 3,000 feet, indicating a severe winter.

According to the Yuba County history, the early settlers spoke of floods in the winter of 1846–47, which did but little damage, simply because there was not much to be injured. John Sutter described the area near present-day Sacramento as a vast expanse of water.[370]

One account reported that the Stanislaus River, at a point about 1½ miles upstream from its mouth, overflowed the country for miles beyond its channel, and that the San Joaquin River was about three miles wide at the crest of the flood. This is the earliest flood mentioned in historical accounts of the settlement of the San Joaquin River Basin.[371] As in the Sacramento River Basin, the extent of the overflow in the lower reaches of these rivers in the early days provides little indication of the discharge, as the minor floods would have spread beyond the normal channels in many places almost as far as the major floods.

This flood is known from its effect on rivers farther north in the Central Valley. Rivers in the Tulare Lake Basin may have flooded as well, but no settlers were living here to record the event.

## 1849–50 Floods (3)

There were three floods in 1849–50:
1. December 1849 – January 1850
2. May–June, 1850
3. December 1850

The floods of 1849–50 are the first for which there are fairly accurate historic descriptions, and the one of January 1850 undoubtedly was of major proportions. However, it would be exceeded at Sacramento in 1852.[372]

1849–50 was reported to have been an unusually wet winter. In December 1849, John Benjamin Hockett and a party of immigrants from Arkansas camped on the Tule River when the whole valley appeared a vast sheet of water.[373]

According to the Yuba County history, the winter of 1849–50 was a wet one, and the streets of Marysville were for a time so muddy that they were almost impassable. The miners along the river were compelled to work in the creeks and ravines in the hills until the water subsided.[374] There was heavy runoff throughout the Sierra in 1850, including in the Tulare Lake Basin.

Sacramento was hit by a major flood on January 7–8, 1850. This was the first major flood to inundate that city's waterfront since Euro-American settlement. Fundraising began for building levees on the Sacramento and American Rivers. Some very low-standard levees would be built in March 1850.

It seems likely that there was also flooding on the rivers in the Tulare Lake Basin, but we have no records of that. There were no missions in the San Joaquin Valley, nor were there any American settlements of note south of Sacramento. Fort Miller wouldn't be established on the San Joaquin River until May 1851. Fort Visalia wouldn't be established on the Kaweah Delta until November 1852.

In May 1850, Lieutenant George H. Derby of the U.S. Army's Topographical Engineers explored the Tulare Lake Basin. His assignment was to find a practical location for a wagon road to the Kings River and to find a location for a military post to control the Native Americans.[375]

On May 7, Derby observed that the Tule River was 100 yards wide, 12–20 feet deep, and very rapid. Two days later, he came to the Kern River, which he described as very broad and deep, and with a 6 mph current. It was running so full that it couldn't be crossed by his mules. It was discharging into Buena Vista Lake which was 10 miles long and 4–6 miles wide.

Returning north, Derby's party reached the Kaweah Delta on May 14. Including the main river, there were five distinct channels. Derby described four of those channels as being much wider than the Tule River. All five appeared to be at their height, and all were deep and rapid. Derby would later conclude that this was still several weeks before the peak of the runoff.

(When Lt. R.S. Williamson of the U.S. Army's Topographical Engineers mapped the Tulare Lake area three years later in 1853, the Kaweah Delta was already known as the Four Creeks Country. That name stayed in popular currency for several decades, although there was never general agreement as to which of the various creeks in the area were *the* four creeks.)[376]

Derby crossed the Kings River by boat on May 18. It was about 300 yards wide, rapid, and as cold as ice. While exploring farther downstream, the Native Americans told him that the Kings was higher than they had ever seen it. Derby then turned west, cutting across the swampy portion of the San Joaquin Valley. He discovered that all of the water of the Kings was flowing toward Tulare Lake. In addition, a large amount of overflow from the flooding San Joaquin was also flowing toward Tulare Lake with a strong current.

When Derby finally reached the outlet for Tulare Lake (what we now call the Fresno Slough), he discovered that it had only an extremely slow current flowing toward the San Joaquin River. Derby's party became entrapped in the Fresno Slough area by the rising waters and barely escaped with their lives.

Clearly Derby found extremely high water when he encountered the Kern, Tule, Kaweah, and Kings Rivers during the runoff of May–June, 1850. The Native Americans said that the Kings was higher than they had ever seen it. We have encountered no reports of subsequent spring snowmelt floods that resulted in such high water. Spring snowmelt floods take place at the onset of hot weather after a wet winter has built up a larger-than-average snowpack in the mountains. The May–June 1850 flood may have been one of the largest such floods to occur in historic times.

S.T. Harding researched the total runoff for water year 1850.[377] He estimated the total runoff for that year to be 1,420,000 acre-feet. A glance at Table 5 will show that since Derby's visit, there have been 19 years with more than three times that much total runoff. But that was still a huge snowpack, and could have generated a very large flood given the right temperatures.

Since the late 1800s, spring runoff waters have largely been diverted onto irrigated lands. For an explanation of how this came to be, see the section of this document: Why is there no lake in the Tulare Lakebed today? The May–June 1850 flood was one of the last great snowmelt floods before these irrigation diversions largely captured the runoff.

Harding found that Tulare Lake's lowest level in water year 1850 was about elevation 208.0. The runoff that year was sufficient to raise the lake to a maximum elevation of 211.5 feet. Derby's experience suggests that Tulare Lake was at a relatively low level prior to the flood and was now being refilled when he was crossing the distributaries of the Kings and San Joaquin Rivers. Although the lake was one foot above the delta sill at the beginning of the water year (elevation 208 – 207 feet), C.H. Lee found that significant outflow didn't really start until the lake reached an elevation of 210 feet. That was because the delta sill was so heavily vegetated with tules. That could help to explain why Derby first observed rivers flowing toward Tulare Lake, then flowing back toward the Sacramento–San Joaquin River Delta.

148

Panoche/Silver Creek west of Mendota flooded at sometime in 1850.[378]

In December 1850, a major flood struck Sacramento; it was bigger than the flood which had struck that city in January. The levees that had been built in March of that year failed. The flood destroyed most of Sacramento. It seems likely that there was also flooding on the rivers in the Tulare Lake Basin, but we have no records of that.

## 1852–53 Floods (2)

There were at least two floods in the Central Valley in 1852–53:
1. December 1852
2. March–April 1853

Flooding was widespread in the Sacramento and San Joaquin Valleys in the winter of 1852–53. Based on fragmentary accounts, it is possible that the December to April period might best be viewed as a more or less continuous series of flood events rather than as individual floods. The December and March–April floods cited above were just the best documented events.

The December 1852 flood in Sacramento was bigger than the January 1850 flood had been. An article in a Red Bluff newspaper in 1861 described the 1852 floods as being the highest known to the oldest residents prior to December 1861.[379]

That may have been true on the Upper Sacramento. However, on the lower river at Sacramento, the March–April flood 1853 flood was larger than the December 1852 flood.[380] On March 29, Sacramento residents watched the river rise 12 feet in 24 hours. This time the city stayed submerged for six weeks.[381]

The floods were closely followed on the East Coast. Newspapers in Oregon, San Francisco, Sacramento, and Nevada would write up the latest news. Those newspapers would be carried by steamer to the Pacific side of the Isthmus of Panama. From there, they would go overland to the Caribbean side, then by steamer to New York. The articles would then be reprinted in the *New York Times*, about two weeks after the original event had occurred.[382, 383] News from the Sandwich Islands (Hawaii) and Australia was relayed in a similar manner.

The USACE identified the 1852–53 flood as a major flood in both the Sacramento and the San Joaquin River Basins.[384] Heavy snows, flooding, property damage, and hardship were also reported from Oregon and Nevada. The Willamette River had a major flood in January 1853.

1852–53 was an unusually wet winter, the third such winter in eight years (1846–47, 1849–50, and 1852–53). The San Francisco *Alta California* newspaper said that this was the most severe winter since California had been inhabited by the Americans.

At Gilroy's Ranch (site of the present-day town of Gilroy), the waters were higher than they had been known for 25 years. (Trivia: John Gilroy, a Scotsman, was the first non-Spanish settler in Alta California to be legally recognized by the Spanish crown. He arrived in Monterey in 1814, 25 years before John Sutter established his fort in what would become Sacramento.)

There were heavy snows in the higher portions of the mining districts. The *New York Times* reported that snows were over 10 feet deep by the end of December, and numerous roofs collapsed. Snow was within six miles of Fort Miller (now covered by Millerton Lake).

The mud, floods, and heavy snows made travel very difficult; many roads were virtually impassable even within Sacramento. Provisions were scarce to nonexistent in the mining communities, and prices quickly soared to exorbitant levels. Some miners retreated from the worst-hit areas so that the remaining provisions could be shared by those who couldn't get out.[385]

Immense rains combined with the melting snows to cause widespread flooding, loss of life, and property damage in the mining districts. Many bridges were washed out and ferries swept away.

At Marysville (north of Sacramento at the confluence of the Yuba and Feather Rivers), the waters were reported to be two feet higher in February 1853 than they had been in March 1852. The Feather River was carrying such an immense quantity of logs and driftwood that it looked like it was full of porpoises.

Sacramento experienced severe flooding in both the December and March–April floods. This prompted the city to raise the main levee.[386] One source said that some of the streets in the business district (now Old Town Sacramento) were raised by up to five feet at this time. They would be raised again by a much larger amount between 1862–1869.

Four floods hit Marysville during the winter of 1852–53, and the surrounding country was more or less under water the whole season. The rains commenced early in November 1852, and toward the latter part of the month the water was as high as it had reached the season before. Again, a week or two later, the water rose 6½ inches higher than at first. The waters then subsided, but the last week in December was one of continued rain, and on December 31 the water from the Feather and Yuba Rivers began to come into the city. The next day the water was 20½ inches higher than at the last flood, and was from 6–10 inches deep on the floors of the buildings about the plaza. There had been a grand ball planned at the Merchants' Hotel on New Year's Eve, 1853; but when the hour arrived, the hotel was surrounded by water. Several young gentlemen, loath to lose the anticipated pleasure, proceeded to the hotel in boats, and with a number of ladies residing there, danced merrily until morning.[387]

All the low and bottom lands were completely submerged by this flood. As it was the first experience that the new ranchers had of this kind, they lost very heavily in stock, crops, etc. Communication from the city with the outside world, and among the farmers, had to be maintained by boats. People were compelled to come to the city in boats in order to obtain supplies, and trading with the mines was effectively blockaded for some time. The continuous rains and almost impassible muddy roads were such an impediment to freighting that a great shortage of supplies was caused in the mines.

At the earliest possible moment, a number of energetic and enterprising men started out with trains of supplies, hoping to reach the destitute regions before the markets were supplied, and thus reap a bountiful harvest of gold to reward them for their labor. Those who reached the mines first were amply rewarded for their exertions, and were able to secure any price that their conscience would permit them to ask, such as one dollar per pound for flour, and twenty cents per pound for hay.

The fourth and last flood of the season in Marysville commenced on Saturday, March 25, 1853; and on Tuesday the water reached a point eight inches higher than in January. The country on all sides of

Marysville and Yuba City was under water. By Saturday, the waters had subsided sufficiently to permit the pack trains to leave the city.

One source said that Sacramento was virtually wiped out by the 1852–53 flood, just two years after the devastating flood of December 1850. In any case, it was heavily damaged. The warm rain that struck the city on December 29–30 was described as being of unprecedented severity. Not even the new brick houses provided a place of refuge. Most of the foundations settled due to the saturated ground. Those buildings were roofed with tin, and the storm rolled the tin up like parchment. In many houses, the occupants were obliged to go out into the streets to seek shelter from the rain. There were only two brick buildings in the entire town that didn't leak during the storm.[388, 389]

(Sacramento would rebuild after the flood, and in 1854 it would become the fourth and final capital of the state. In 1852–53, the twin towns of Vallejo and Benicia were serving as the capital of the newly formed state. The initial capital had been San Jose.)

There was extensive flooding and property damage in the region of Colusa in the lower Sacramento Valley.

The *Alta California* newspaper said that floods were widespread in both the northern and southern mining districts. Ferries were destroyed on a number of rivers. Bridges were washed away on the Stanislaus, Calaveras, and other rivers. Flood levels were higher than in the memorable winter of 1849–50.[390]

The Sacramento and San Joaquin Valleys formed a "world of water." The bottoms on the San Joaquin River were under one vast sheet of water, estimated to be some 20 miles wide. At the mouth of the Merced, the owners of the ferry took up residence on their boats. Food supplies were low for the 200 settlers on the San Joaquin.

A Belgian gold miner, Jean-Nicolaus Perlot kept a diary of events in the Mariposa area:[391]

> *Never in my life have I seen it rain more heavily or for a longer time. From the sixth of December (1852) to the first of March (1853), the rain didn't stop for as much as three hours, unless it was during my sleep, which is hardly probable; how many times, during those three mortal months, how many times I awakened at some hour of the night! And always I heard the monotonous sound of the rain falling on the roof of our house.*

The flood of 1852–53 raised Tulare Lake by 11.5 feet. At that point, the lake had an elevation of almost 216 feet and a depth of about 37 feet at its deepest point (216–179 feet). There was 9 feet of water in the outlet channel flowing over the delta sill (216–207 feet). From there, the water connected through the Fresno Slough to the San Joaquin River and flowed on to San Francisco Bay.

Panoche/Silver Creek west of Mendota flooded sometime in 1852, probably in December.[392]

Tulare Lake would reach that size only twice more: in 1862 and 1867. Over the next eight years (1853–61), the lake dropped 16 feet. This became the pattern over subsequent decades. Floods would abruptly raise the level of the lake, after which it would gradually shrink as shown in Figure 10.

In 1852, Nathaniel Vise and others settled in the Four Creeks Country (the area that we now call the Kaweah Delta). Their timing was less than ideal; they were caught in the high water of the 1852–53 flood.[393] The *New York Times* reported that the news from Four Creeks was dreadful. The 500 settlers there were living on beans.

The White River is the next river south of the Tule. 1852 was also a major flood in the mining district on the White.[394] (The village that supported the mines is located 10 miles east of Delano. It was then known as Tailholt but was later renamed White River.)

Gordon's Ferry was established on the Kern River just north of present-day Bakersfield College in the spring of 1852. Eight months later, rain fell for three weeks across California. An observer wrote: "The rivers have been swelled to such an extent as to inundate all the low lands, causing immense damage, destroying stock and agricultural products." According to José Jesús López, early pioneers said that the Kern River swept Gordon's "perfectly bare of all signs of improvements."[395]

Below Gordon's Ferry, the Kern River flowed through Kern and Buena Vista Lakes on its way to Tulare Lake. Tejón Creek flowed into the southeast end of Kern Lake in a channel two feet deep and ten feet wide. In the 1852–53 flood, Tejón Creek overflowed its channel for more than two months.

During the height of the 1852–53 flood, some sailors jumped ship in San Francisco. They stole a whaleboat, hoisted the sail, and headed inland. Taking advantage of the prevailing winds, they sailed south up the San Joaquin River, through the Fresno Slough, and entered Tulare Lake. This is the first of five documented trips between that lake and San Francisco Bay to occur in historic times. (The other four were in 1868, 1938, 1969, and 1982–83.)

But the sailors didn't stop in Tulare Lake: they continued south up the Kern River to Buena Vista and Kern Lakes. And since Tejón Creek was in flood, they kept going up that creek (east) another 15 miles or so until they were about two miles north of a Native American village. That village was located where the old Tejon Ranch headquarters would later be built (at the end of present-day Sebastian Road, due east of where Interstate 5 and Highway 99 diverge). There the sailors beached their boat and walked over to the village.

After staying with the villagers for about two weeks, the sailors returned to their boat to go back downstream. However, by then, Tejón Creek had gone down so much that the water was back in the channel, and they almost didn't make it. About 10 Tejón Indians helped them. Santiago Montez was one of them, and he later recalled the event:[396]

> *Them sailor pretty smart. When water not deep, they put boat across creek and sit on boat. That make dam. That back water up. They all jump out of boat, grab boat by sides and run ahead of water fast they can. Then boat go maybe hundred yards and stick again. They do that lots time before they get back in Kern Lake.*

The sailors stayed on in Kern and Buena Vista Lakes where they trapped beavers and otters. At least some of them married Native Americans and raised their children in the local schools. The whaleboat remained in use until it was apparently swept away in the 1867–68 flood. Perhaps it washed down to Tulare Lake or even back to San Francisco Bay. If that boat could talk, what a tale it could tell.

## 1855–57 Drought

This three-year drought affected at least part of the state. Tulare Lake continued its steady decline during this period. Floyd Otter said that Robert Glass Cleland documented the effects of this drought.[397] During the drought, cattle grazed everything that they could reach and then died by the tens of thousands.

## 1860–61 Drought

This two-year drought does not seem to be recognized at the state level. Tulare Lake continued its steady decline during this period. Floyd Otter said that Robert Glass Cleland documented the effects of the drought of 1860–61.[398] This was to be a virtual replay of the 1855–57 drought which had resulted in the deaths of so many cattle. During the 1860–61 drought, cattle once again grazed everything that they could reach and then died by the tens of thousands.

Large numbers of cattle starved to death during the droughts of 1855–57, 1860–61, and 1862–64. A Mussel Slough woman vividly recalled life on the shore of Tulare Lake during this period:[399]

> *The country was nothing but a dry, barren desert with bands of wild roving cattle that would come out of the timber along the river in the morning and go out to the lake to feed. Where the water of the lake had receded a little grass would spring up and they would get a little feed. The poor things were almost starved…so we could not blame them for eating the hay we had stacked for our horses. The settlers all dug ditches for fences to keep them out but without much effect.*

It's possible that this drought affected much of the state, but we just haven't found records to that effect. In any case, an article in the *Red Bluff Independent* reported that summer of 1861 was the hottest, driest season since California became a state in 1850. The fall rains were late in coming, and cattle starved to death in large numbers in early November.[400, 401]

The drought would end decisively when a series of epic storms moved into the state at the beginning of December, unleashing the great 1861–62 flood.

## 1861 Flood

This flood occurred during the 1860–61 drought.

Relatively little is known about this flood. It is different from the much more famous 1861–62 flood. The USACE identified it as a major flood in both the Sacramento River and the San Joaquin River Basins.[402] How it affected the rivers within the Tulare Lake Basin is unclear. Runoff during water year 1861 caused Tulare Lake to rise 2.3 feet in elevation.

## 1861–62 Flood

Flooding occurred from December 1861 through January 1862.

For reference, 1861 was the first year of the Civil War. General Beauregard fired on Fort Sumter on April 12, 1861, igniting that war. California managed to participate in the Civil War in various ways despite conflicted loyalties and the effects of the disastrous flood of 1861–62.

1861–62 was an incredibly wet winter. Extremely wet winters in California are often associated with an El Niño weather pattern. However, research done at Oregon State University indicates that was not the case with the 1861–62 storms.[403] The polar jet stream apparently swept up and down the West

Coast during that winter, causing the temperatures to vary wildly, from very warm to very cold. The warm storms brought the typical rain-on-snow events. However, the cold storms brought snow down nearly to sea level in the Sacramento Valley. San Francisco recorded nine days with temperatures below freezing in January alone. On January 28, San Francisco registered 22 degrees, a full 5 degrees colder than any temperature ever measured in the modern era in that city.

The atmospheric mechanisms behind the storms of 1861–62 are unknown; however, the storms were likely the result of an intense atmospheric river, or a series of atmospheric rivers.[404] Immense quantities of water were delivered during this storm event.

The mining community of Sonora received 102 inches of rain (8½ feet) in a 74–day period (November 10, 1861 – January 23, 1862).[405]

San Francisco recorded 28.25 inches in 30 days. This was 6.48 standard deviations above the average rainfall for 30 consecutive days. The associated recurrence interval is 37,000 years.[406]

Most of the states of Oregon and California were affected by the flooding. Record stages resulted on the major rivers throughout those two states. It almost certainly had a recurrence interval greater than 100 years.

The 1861–62 flood was a huge event. One source said that it stretched from Canada to Mexico, but that overstates the case. The storm tracks responsible for the flood were generally aimed at the southern part of Oregon. As a result, the northern part of Washington received less than average precipitation during the period of this flood. Overall, the 1861–62 flood had minimal impact on Washington.

The story was very different in Oregon. November brought one storm after the other, resulting in a marked excess in precipitation over most of the state. In the Cascades, temperatures were cold enough to result in well above-average accumulations of snow. December turned warm, and the rains melted much of that snow. There were a series of major floods throughout the state. The meteorological conditions of the Pacific Northwest during the winter of 1861–62 were summarized by Edward Lansing Wells, the chief meteorologist for many years at the Portland office of the U.S. Weather Bureau.[407]

The storms that struck Oregon in November moved in rather far south. However, the one that hit at about the beginning of December seems to have passed just far enough north to produce strong, warm southerly winds with extremely heavy rains that reached into eastern Oregon.

Of all the floods to hit Oregon that winter, the most impressive occurred on the Willamette River. (Based on fragmentary evidence, that was also the one that was most closely followed in the new town of Visalia, far to the south.) A Belgian gold miner, Jean-Nicolaus Perlot, left the California gold fields to settle in Portland, Oregon in time to witness the flood on the Willamette:[408]

> *The peaceful Willamette became, by the fifth of December, an impetuous torrent; leaving its bed, it upset and carried away the establishments which bordered its banks. It was, for two days, a curious and heart-rending spectacle: the river was covered with strays of all kinds, trees, animals, fences, provisions, houses, sawmills, flour mills all that was floating pellmell, and passed before Portland with a speed of three leagues an hour.*

Some 353,000 acres were inundated; "the whole Willamette valley was a sheet of water." It was the largest flood on that river in recorded history. Many towns were damaged or destroyed.

The Willamette peaked at Oregon City on December 4, 1861 at 635,000 cfs, 35% greater than the average flow of the Mississippi River. Oregon City sits at the base of Willamette Falls, the largest waterfall in the Pacific Northwest (based on volume). One night during the flood, the residents of that town watched a number of houses come over the falls, with lights still burning inside. Then, on December 5, they watched as the side-wheel steamer St. Clair ran the 40-foot-high falls "with great rapidity."[409]

December 1861 to January 1862 constitutes one of the greatest flood periods in the history of California. The 1861–62 flood period was remarkable for the exceptionally high stages reached on nearly every stream, for repeated large floods, and for the prolonged and widespread inundation in the Sacramento River and San Joaquin River Basins. Rainstorms were heavy in the lower elevations and snowfall continuous in the upper elevations throughout the two basins.[410]

The summer of 1861 was the hottest, driest season that the northern Central Valley had experienced in over a decade. The fall rains were late in coming and cattle deaths were high in November. However, by December 10, the drought was over in the Red Bluff area and flood damage from the Sacramento River was extensive.[411, 412] The settlers wanted an end to the drought, and they got it.

Northern California experienced record-setting precipitation and flooding. The initial flood began late in November 1861 when storms brought rains to the lowlands and covered the mountains with up to 20 feet of snow. This was followed by warm rains in the mountains which melted the accumulated snow. This pattern was repeated several times through the winter. The storms kept coming right through January.

Flooding was severe in the North Coast of California. Flooding on the Klamath River was particularly impressive. A 500-foot-long wire suspension bridge spanned that river in a canyon below Weitchpec (east of Trinidad). That bridge was 99 feet above low water and thought to be safe from any possible flood. But the Klamath overtopped that bridge and swept it away. Most of the Native American ranches on the Klamath Indian Reservation were located along the Klamath River. The Klamath destroyed all of those ranches that were within 25 miles of the river mouth.

Sacramento was built where the American and Sacramento Rivers meet. It had experienced severe flooding in the 1852–53 flood, causing the city to raise its streets and strengthen the main levee. Despite these precautions, Sacramento was one of the hardest hit cities in the 1861–62 flood. It was flooded about five times during that winter. The first inundation occurred at 6 a.m. on the morning of December 9, 1861, when the American River breached the east levee. By 10 a.m., many houses were floating or overturned. To drain the city, engineers directed a prison chain gang to cut through the R Street levee between Fifth and Sixth, and the water rushed through the opening to the Sacramento River, sucking about 25 floating houses through the gap.

Leland Stanford was elected governor in November 1861 and was just taking office when the flood hit. The Sacramento River was flooding so badly that Stanford had to crawl out the window of his home and row himself to his inauguration.

But this was just the beginning of the flooding. The Sacramento area got another 25 inches of rain during the following two months, almost four times the average rainfall. The Sacramento River surged at three times its average seasonal flow of 285,000 cfs, inundating the Sacramento–San Joaquin River Delta region. Sacramento was under water for three months. After much debate about appearances and propriety, the State Legislature voted on January 23 to abandon the state capital and move to San Francisco. The California Supreme Court also moved its operations to San Francisco, but it never moved back.[413]

The flooding prompted Sacramento to raise the streets of its business district (now Old Town Sacramento) by up to 15 feet between 1862–1869. The tunnels under present-day Old Town Sacramento are reminders of the original downtown buildings and streets.

In the Sacramento River Basin there was a succession of floods starting on December 8, 1861 and continuing into March 1862. Many reports published during the period described the lower Sacramento and San Joaquin valleys as one vast sea of water. Probably as much as 5,000 to 6,000 square miles of the valley floor were submerged.[414]

In 1860, the State of California had hired Josiah Whitney and William H. Brewer, both Yale graduates, to conduct a long-term, in-depth investigation of the state's resources. They were just two years into their studies when the great flood of 1861–62 bankrupted the state and soon thereafter terminated their project. Brewer was a botanist and an agriculturist. He was also a compulsive diarist — keeping detailed notes of his experiences from 1860–64. They were mostly letters to his brother, which were assembled into a book: *Up and Down California*.[415] This book was printed by the Yale University Press in 1930. It provides detailed accounts of both the 1861–62 flood and the ensuing drought of 1862–64.

A defining feature of the flood in the Central Valley was that the rains came down far faster than the Sacramento and San Joaquin Rivers could drain the floodwaters to San Francisco Bay. Brewer was there to describe the resulting enormous lake that swelled up in the Central Valley. Nothing remotely like it has ever been seen since. The prolonged period of flooding in the lower Sacramento Valley lasted from December 13, 1861 to about February 1, 1862.[416]

Brewer wrote from San Francisco on January 19, 1862:[417]

> *The amount of rain that has fallen is unprecedented in the history of the state....The great central valley of the state is under water — the Sacramento and San Joaquin valleys — a region 250 to 300 miles long and an average of at least twenty miles wide, a district of five thousand or six thousand square miles, or probably an area of three to three and a half millions of acres!*

Brewer wrote of the Central Valley on February 9, 1862:[418]

> *Nearly every house and farm over this immense region is gone. There was such a body of water — 250 to 300 miles long and 20 to 60 miles wide, the water ice cold and muddy — that the winds made high waves which beat the farm homes in pieces. America has never before seen such desolation by flood as this has been, and seldom has the Old World seen the like.*

That lake that formed in the Central Valley in 1862 was roughly three times larger than the Great Salt Lake (5,500 square miles versus 1,700). The low-elevation lakes and wetlands in the Tulare Lake Basin were part of that big lake.

156

The Sacramento Valley was so inundated that steamers ran back over the ranches fourteen miles from the river, carrying stock, etc., to the hills. Approximately 100,000 cattle died in the valley.

Brewer reported that "It is supposed that over one-fourth of all taxable property of the state has been destroyed." Brewer kept in touch with the State Treasurer and news of the dwindling state government income because he was having long delays in being paid for his work.[419]

The magnitude of flooding in Sacramento was due in part to the sediment coming from the hydraulic mining. Brewer reported that the river was choked with sediments before the flood and that the riverbed was raised by at least six feet during the flood. Sacramento responded by once again strengthening the levees and raising the elevation of the streets by as much as 10 feet.

According to the Yuba County history, long and incessant rains ushered in the rainy season, and the Feather River started to rise rapidly on Saturday, December 7, 1861. All day Sunday the rain poured down, and that night the city was nearly under water. Early Monday morning several buildings, undermined by the water, fell crumbling to the ground, creating great consternation. The floors of the Merchants' Hotel fell through to the basement, carrying with them the sleeping occupants, several of whom were severely injured by the fall, although no one was killed. A great many frame houses were floated from their positions, and some of them were carried down the stream. The steamer Defiance, playing tunes on her calliope, made its way through the streets giving assistance to those who were rescuing the unfortunate.[420]

The condition of the country was described in the Marysville *Appeal*:[421]

> *Westward one vast water level stretched to Yuba City, where a kindred inundation was raging; the entire town site being under water. Beyond this to the foothills of the Coast Range there appeared to be no dry land. Northward the plains were cut up into broad streams of running water, which were swiftly coursing toward the great sheet of water stretching between the Yuba and Feather rivers, up as far as the residence of Judge Bliss, unbroken except by the upper stories of houses, trees and floating debris. Southward the whole plain toward Eliza was one sheet of water, dotted with trees, roofs of houses, floating animals and wrecks of property of every description. Where Feather River sweeps past Eliza, stock of every kind could be seen constantly passing downstream, some alive and struggling and bellowing or squealing for life. Hare and rabbits were destroyed by thousands.*

The people in the country had to leave everything and flee to high ground for safety; many who were too late for this, climbed trees and remained perched among their branches until rescued by friends. Nearly all the bridges on the Yuba and Bear Rivers were carried away, and drift timber and saw-logs came down the streams in great quantities, some of which were left in gorges thirty feet high when the water fell. A deposit of sand up to six feet thick was left on the bottomlands when the waters retreated.

On January 11, 1862, the water raised six inches higher than before, but the warning of the previous flood had caused the merchants and farmers to move everything perishable beyond the reach of danger. The loss of stock that winter and the next summer was very great, and in Sutter County was estimated to be three-fourths the entire number. Few animals escaped except those able to reach the Marysville Buttes, and the cold weather nipped the grass, causing large numbers of the cattle to die from starvation.

For a week, the tides at the Golden Gate did not flood; rather, there was continuous and forceful ebb of brown, fresh water 18–20 feet deep pouring out above the salt water. A sea captain reported that his heavily laden ship foundered in the Gulf of the Farallones off of San Francisco due to the layer of fresh water. Tule islands floating across the bay and out to sea were crawling with rattlesnakes. Some of these islands came to rest under the San Francisco wharves where the snakes were a menace for months. Freshwater fish were caught in San Francisco Bay for several months after the peak of the flood. Such events have not happened since.

The 1861–62 flood is known as a flood of Northern California because that is where the population of the state largely lived at the time. However, the flooding from that event was also quite severe in Central and Southern California.

In the San Joaquin River Basin, there were extreme, successive floods during December 1861 and January 1862. By early January, snow had accumulated to unusual depths in the Sierra. Much of this snow deposit was melted by warm rains and helped to swell the flood volume.

Out of 100,000 head of cattle in San Joaquin County, only 10,000 survived.[422]

William Knight, a fur-trader, came to California in 1841. Caught up in the gold fever of 1849, he was heading south when he was stopped by the wide Stanislaus River. Seeing a business opportunity, he began a ferry operation using an old whaling vessel. That ferry was later replaced with an open-truss bridge. Legend has it that the plans for that bridge were drawn up by Ulysses S. Grant. In any case, the 1861–62 flood washed away that first bridge and all but one of the other bridges on the Stanislaus. (A new covered bridge was constructed at the same site in 1863. That bridge has withstood many subsequent floods and is in remarkably good condition. However, it was closed to vehicular traffic in June 1981.)

At daylight on Friday, January 10, the crest of the Stanislaus hit the town of Knights Ferry like an avalanche, rising rapidly until it covered the business section, which was built on a flat above the river. At dusk, the river fell about four feet, and the residents thought that the worst was over. Then at 2 a.m. Saturday, the Stanislaus rose again and carried off all the remaining buildings on the flat. Only the buildings on the hill remained.

Every bridge on the Tuolumne River was washed away except the one at Steven's Bar.

The Mariposa area was hit by a heavy storm in late December 1861, resulting in dramatic flooding on the Merced River and at the mining community of Coulterville. The tale was told in the January 7, 1862 issue of the *Mariposa Gazette* as reprinted in the February 6, 1862 issue of the *Visalia Delta*:

> *It was the hardest storm, particularly that part of it occurring Thursday night, Dec. 26th, that has ever swept these mountains within the recollection of that very respectable individual "the oldest inhabitant." The Merced River rose fearfully high, sweeping off every bridge upon it, tearing out dams, etc. In Coulterville, the gale of Thursday night was terrible, accompanied by a heavy rain. That night was the most hideous we have ever known. It was worse there than further south. The town, it might be said, was afloat. The rear portion of all the establishments bordering on (Maxwell) Creek, went along with its turbulent waters. Bell's Saloon was flooded, and Cashman & Co.'s barn, a large building, raised anchor and took a notion to sail. R. McKee, Esq., however, with characteristic intrepidity, hitched it to the liberty pole.*

158

The Merced River, downstream from the mouth of its canyon, flooded the town of Snelling. The flood widened and changed the course of the Merced River channel. Reports stated that the whole country surrounding lower Mariposa Creek and the Fresno and Chowchilla rivers, as seen from the foothills, was one vast sheet of water.[423]

During December 1861 and January 1862, the San Joaquin River rose from 24,000 cfs to approximately 133,000 cfs, a fivefold increase. The city of Stockton and the surrounding country were inundated for many miles. Floating farmhouses broke the telegraph wires on the outskirts of Stockton. A steamboat ran through the back wall of the Russ Hotel in the town of Hill's Ferry (northwest of Los Banos). The flood destroyed nearly all the bridges, mills, and other structures along the channels of the San Joaquin River and its major tributaries.[424]

Panoche/Silver Creek west of Mendota flooded in December 1861 and January 1862.[425]

In the Tulare Lake Basin, there was an exceptionally great flood on January 11, 1862. The Kings, Kaweah, and Tule Rivers brought down tremendous quantities of timber from the Sierra and deposited them on the plains.[426]

The 1861–62 flood on the Kings River began the formation of Cole Slough, cutting the head of that slough. (See the section of this document on Pine Flat Dam for a more detailed description of the formation of Cole Slough and associated waterworks.) The entire town of Scottsburg was washed away by the Kings during this flood and was subsequently rebuilt at a safer location.

In the winter of 1861–62, Joseph Hardin Thomas had just completed construction of a sawmill on Mill Flat Creek, downstream from present-day Sequoia Lake. It was a double-circular sawmill with a 40 horsepower steam engine, one of the two primary sources of lumber for Visalia. Thomas's mill was destroyed by 30-foot-deep floodwaters resulting from a debris slide in January 1862. The flood also destroyed Feggan's Mill, an older sawmill located six miles farther downstream nearer the Kings River.[427]

From the number of large trees washed down from the mountains by the floods on the Kings, Kaweah, Tule, and White Rivers, the settlers inferred that this was the greatest flood for many years.[428]

The 1861–62 flood brought plenty of destruction, but it also brought opportunity. Thomas Flaxman, the owner of the 80-foot-long sternwheeler Alta, decided to use the flood to bring the Alta to Tulare Lake. The crew was composed of men familiar with freighting from Firebaugh to Stockton on the San Joaquin River. The route that Captain Giddings chose was through the San Jose Slough to Summit Lake and from there to Tulare Lake. Upon entering the San Jose Slough, Giddings took aboard several vaqueros to act as pilots. The first day went fine, but that evening the Alta got into the wrong channel.

The captain resorted to kedging: carrying the anchor ahead, dropping it, and then pulling the larger vessel ahead by reeling in the anchor cable. It was arduous work. At midnight, the weary crew lay down to sleep until morning, when they were to resume their task. However, when they woke, the floodwaters had gone down, leaving the vessel high and dry. The boat was stranded about four miles southeast of present-day Burrel. The following years were ones of drought. There the Alta sat, just two miles from Elkhorn Station. A strange sight indeed to stage passengers who passed on the dusty road nearby. The Alta was gradually picked apart for lumber and firewood. In 1875, her engine, boiler, and pilot house were salvaged for a side-wheel steamboat — the *Mose Andross* — that A.J. Atwell was having built at Buzzards Roost on the northeastern shore of Tulare Lake.[429, 430, 431]

In the Visalia area, rain started early in October 1861. By the end of that month, the ground was wet down to a depth of eight inches. By the end of November, sufficient rain had fallen to wet the ground down to a depth of 2½ feet. The rain started again in mid-December and continued to fall for several weeks. January brought a week of warm gentle rain which filled the creeks to the banks. Heavy rains continued until March 1862.[432] The wind remained southerly from mid-December throughout most of January.[433]

Prior to the 1861–62 flood, the Kaweah River waters spread at high stages over what was known as the Kaweah River Swamp (aka Visalia Swamp). The swamp commenced at present-day Terminus Dam and extended southwesterly about 9 miles with a width of 1–3 miles. The spreading waters were reunited in various channels in and below the swamp. Channel capacity was inadequate to pass floodwaters, not only within the swamp, but also below it.

The Shipp Cut had been made in 1854; it was a small drain ditch from the Kaweah River Swamp near Rocky Ford (north of present-day Kaweah Oaks Preserve) west to Canoe Creek. The 1861–62 flood cut a new channel along the northern border of the swamp. Shipp Cut and a section of Canoe Creek were enlarged by the floodwaters and became a part of this new channel, and finally a connection was established with the Cross Creek channel downstream of Visalia, creating what is now known as the St. Johns River.[434]

This rerouting of the floodwaters to the north of the Kaweah Delta may have reduced the flooding in Visalia. In any case, surprisingly little water came down Mill Creek.

Mill Creek flooded downtown Visalia three times during the 1861–62 flood:
- evening of January 11–12 (22 inches deep on Main Street)
- January 17–18 (20 inches deep on Main Street)
- night of January 20 – January 23+ (24 inches deep on Main Street)

Most of the wells in Visalia were contaminated by the floodwaters, so drinking water was hard to obtain.[435] Since rain continued to pour down during the flood, some people caught rainwater for their drinking water. There was one pump in town (at the corner of Encina Avenue and Oak Street), and it was the duty of the young boys in many families to fetch the unpolluted water from that pump, carrying their loads in boats.

The floodwaters caused significant property damage in the Visalia area as well. The flood destroyed many irrigation ditches, a lot of fencing, and four bridges in and around town.[436] The flood in the Visalia area was described in some detail in the January 23, 1862 issue of the *Visalia Delta*.[437]

Some 42–46 homes as well as some businesses were destroyed in Visalia during the flood.[438, 439] Many homes of this period were made of adobe. As the water came up about these, they began melting and sinking, necessitating immediate departure of their inhabitants. At the time of the flood, there were only a few brick structures in Visalia; most mercantile buildings were made of wood or adobe. The floodwaters melted away the foundations of the adobe buildings and toppled them over, so to speak, on their heads.[440] However, not a single wood or brick building came down in Visalia

160

during the flood.[441] It was a hard-learned lesson for the town, rather like the moral of the *Three Little Pigs and the Big Bad Wolf*. After the 1861–62 flood, most of the homes and businesses in Visalia were constructed of either brick or wood. Adobe was a building material reserved for high ground.

This was the first major flood to come into Visalia since settlement was begun, and there was significant property damage. However, the floodwaters were shallow and the flood damage was trivial compared to what was happening farther north and south. The residents of the town realized their good fortune. The *Visalia Delta* reflected on this in late January 1862:[442]

> *The more we learn of the late terrific storm, both at home and abroad, the more do we find matter for congratulation that Tulare (County) has escaped so cheaply. The loss to the mass of citizens is absolutely nothing, as compared with less fortunate localities.*

People who lived around Tulare Lake were also keenly aware of the benefits that the floodwaters brought. The *Visalia Delta* highlighted one example:[443]

> *The recent high water has had the good effect of stocking all the creeks and small streams with an abundance of lake trout. In such vast numbers did they ascend that we are informed that in Antelope Valley (vicinity of present-day Elderwood) and at the Cottonwoods, the receding flood has left wagon loads of them standing high and dry. It has been so long since the creeks have been full enough to allow them to ascend, that they had nearly disappeared from the smaller streams in this vicinity.*

By February 1862, the flood was over. In an effort to allay the fears of new immigrants to Visalia, the *Visalia Delta* wrote:[444]

> *Many boats are still observable lying high and dry at the gates and steps in front of the residences of some of our citizens. Put them away, gentlemen, there is no chance of their being wanted very soon again and their presence might give strangers the impression that floods were a regular institution in Tulare (County).*

Visalia was built on the alluvial fan of the Kaweah River: the Kaweah Delta. An alluvial fan is a distributary system made of multiple channels that allow for large areas of shallow inundation. Because the flooding in the 1861–62 flood was shallow, no horse, cow, or even full grown hog was known to have drowned in the country around Visalia. In fact, the Tulare area had thousands of fat hogs ready to be driven north to where prices for all manner of foodstuffs had just skyrocketed. The cry went out: *Drive up the hogs*.[445]

The 1861–62 flood deposited considerable quantities of drift, silt, and sand on the Kaweah Delta. A portion of the channel of the Lower Kaweah River was obstructed by these deposits, significantly reducing its ability to carry flows. During low-flow periods, that section of the Kaweah would now be dry; it would only serve to carry water during flood periods. The St. Johns would become the principal channel of the Kaweah River across the delta to Tulare Lake for the next 15 years.

Prior to the 1861–62 flood, the Kaweah had divided into its distributaries (the Four Creeks, if you will) at a point more or less in the middle of the delta. However, after the flood, the primary distributary point moved up near the top of the delta, well up into the swamp, about a mile below present-day McKay's Point. The four channels continued to come back together to reform into a single river channel on the lower side of the delta.

There was a fifth channel that came directly out of the foothills south of the Kaweah (presumably that followed the course of present-day Yokohl Creek), flowed along the south side of the Kaweah Delta (apparently following a course similar to present-day Outside Creek) and then joined the Kaweah in the marshy ground near its junction with Tulare Lake.

W.B. Cartmill recalled what the 1861–62 flood was like. The Cartmill Ranch was located north of Tulare on present-day Cartmill Road (Ave 248), about four miles west of Highway 99. It was located on Packwood Creek, well back from Tulare Lake. The Cartmill family arrived at the ranch on October 26, 1861. It started raining shortly after their arrival and continued raining almost every day until about Christmas. The flood came on Christmas Eve. The Cartmill parents struggled all that night to keep the floodwater out of their cabin, but to no avail. When W.B. and his sister woke on Christmas morning, they were surprised to find that the water had entered their cabin and was nearly up to the bed. For a 4½ year old boy, this was an excuse for fun, jumping from the bed into the water. However, the family was compelled to abandon their cabin that morning and move in with a neighbor. When they left the cabin, they had to wade several hundred yards to reach dry land. W.B. recalled that the water was up to his armpits and running so strong that he had to hold onto his mother's dress. It was two weeks before they could return to their cabin.[446]

The flood caused the Tule River to change its course in January 1862. Prior to the flood, the Tule ran northward between what are now Second and Third Streets in Porterville, turned west past the foot of Scenic Hill and thence in a northwesterly direction across the plain.[447] After the flood, the river continued directly west as it left the foothills in a new channel some distance south of the settlement. The settlement that came to be called Porterville was relocated as a result of the flood.[448]

The Butterfield Overland Mail route crossed the Tule River at the foot of Scenic Hill (at the junction of present-day Main St. and Henderson Ave). The Tule River stage stop was located at that point, and was operated by Porter Putnam for a time. This portion of the stage route was discontinued in March 1861 due to Civil War fighting in the South. However, Porter stayed to found the town named for him.[449]

In the Tulare Lake Basin, reports stated that a damaging flood on the Tule River overflowed farms to a depth of several feet. The lowlands along the tributaries of Tulare Lake were probably flooded continuously from the middle to the end of January 1862.[450]

After the Tule changed course, the area between the old and new channels was declared swampland under terms of the Swamp Land Act of 1850, which provided for the reclamation of such land. To satisfy the terms of the law, applicants for such land were said to have loaded a rowboat into a wagon and, sitting in the boat, driven over the land which they were interested in claiming. They could then swear to an affidavit that they had gone over the land in a boat and that it now had been reclaimed. It is likely that the federal authorities in Visalia knew of the little joke, but settlers were wanted, and the land was valued at only $1.25/acre anyway.[451]

The White River had a major flood in the gold mining district. The river rose 5 feet higher than it had in the 1852 flood. There was great property damage.[452]

Poso Creek flooded on January 18 with a rush of logs and water 60 feet high.[453]

The 1861–62 flood was a major flood on the Kern, causing huge property damage in the mining country. Virtually all the bridges, dams, and mills were destroyed. The river was 45 feet higher than ever previously recorded.[454]

The rain and flooding on the Kern lasted for two months. There were only a few settlers living in the lower portions of Kern River Island (present-day city of Bakersfield), and for the few years that they had lived there, the rising winter runoff had spared their tule and adobe homes.

Christmas Day, 1861, though, was not like the past light floods that had occurred regularly. The floodwaters rose during the night. Within a few hours, every home in the low-lying areas was washed away. The plight of the settlers was described in the *Visalia Delta*:[455]

> *The settlement known as Alkali City or Kern River Island is also ruined. They have lost all — stock, grain, and everything else — scarcely escaping with their lives. Several of the inhabitants were forced to remain on a very small island ten or twelve days, with nothing to eat except half rations of roasted corn.*

The flood of December 25 changed the river channel at the site of the present-day city of Bakersfield and inundated all except the higher knolls in the vicinity. It seems certain however, that the flood of 1861–62 flood was not as great as the 1867–68 flood would be.[456]

Gordon's Ferry (aka Gale's Ferry) was located just north of present-day Bakersfield College. The Sinks of the Tejón was the first Butterfield Overland Mail stop north of Fort Tejon. It was located at the intersection of present-day David and Wheeler Ridge Roads, roughly 10 miles northeast of where Interstate 5 and Highway 99 diverge. When the Kern River came out of the canyon, it created one vast sea of water from Gordon's Ferry to the Sinks of the Tejón. Somewhere within that huge sheet of water was Kern Lake. Buena Vista Lake backed up to within 12 miles of Fort Tejon.[457]

Before the 1861–62 flood, the Kern River channel ran where the Kern Island Canal now runs in Bakersfield: by the Beale Library (between Chester and Union Ave) on its way to Kern Lake. The flood shifted the river to the west. The new channel began at Gordon's Ferry and passed through what is now Old River and into the Las Palomas slough system between Kern Lake and Buena Vista Lake on its way to Tulare Lake (see Figure 9). Not only did that new channel bypass Kern Lake, but one source said that it also bypassed Buena Vista Lake, meaning that those lakes would only get water during years with very high runoff. In any case, the river would shift even farther northwest in the 1867–68 flood.[458, 459]

A major debris slide formed on the South Fork of the Kern in January 1862. This was described in the *Visalia Delta*:[460]

> *The cause of this disaster is owing to a slide from the mountain, filling up the bed of the stream, the water forming in the immense reservoir above, and after forcing its way through the obstruction, forbid all opposition…The crumbling of the mountain is described by those who saw it as a grand and terrific sight. Huge masses of rock were hurled from their base, trees uprooted, were sent whirling through the air, and this mass of matter gathering force, as it went, came down the steep mountain declivity, with wild and terrific confusion, indescribable. Mr. Jacob Macomb and family, residing on the South Fork of Kern, were awaked at the midnight hour, by the water and had barely time to leave their house before it fell.*

Reed Tollefson thinks that the above landslide may have occurred fairly far upstream on the South Fork Kern, perhaps on land that is now within Sequoia National Forest. The South Fork Valley has very flat topography. However, just above these private lands on the national forest, the river enters a deep gorge for many miles. Presumably that's where the slide occurred.

Tulare Lake had been at a very high stage after the 1852–53 flood, the second highest ever recorded. After 1853, there was a gradual shrinkage of the lake until the fall of 1861. Over those eight years, the lake dropped about 13 feet in elevation.

There were multiple causes of this:
- The maximum elevation in 1853 (215.5 feet) was higher than the elevation of the delta sill (207 feet). The water above this elevation simply flowed out of the lake and connected through the Fresno Slough to the San Joaquin River, and from there it flowed on to San Francisco Bay.
- Normal evaporation in our hot valley summers.
- Eight years with only low or average runoff and no floods.
- Two years of drought (1856 and 1857).
- Diversion for irrigation was just getting underway.

The 1861–62 flood raised the lake by 15.7 feet to elevation 216, the highest that the lake has been during historic times. At elevations above 207 feet, the lake over-topped the lowest point on the delta sill. At the lake's highest stage, about 9 feet of water was flowing in a broad expanse northerly over this ridge (216–207 feet). From there, the water flowed into the Fresno Slough and the San Joaquin River. At the height of this flood, the lake was about 37 feet deep at the deepest point (216–179 feet). The surface area increased from about 350 square miles in 1861 to about 790 square miles in July 1862.

C.E. Grunsky estimated that more than 5,000,000 acre-feet of water flowed into Tulare Lake in the single season 1861–62. For comparison, that is 3.1 times greater than the combined current capacity of all four of the federal reservoirs in the Tulare Lake Basin. Using more precise data, S.T. Harding later recalculated the total inflow as being 6,290,000 acre-feet of water, the equivalent of 3.9 times the combined current capacity of our present-day reservoirs.[461]

The 1861–62 flood was a record flood in the Tulare Lake Basin not just because of its volume. The force of the flood was such that all four of the major rivers (Kings, Kaweah, Tule, and Kern) cut new channels. Given that, one can only wonder that Visalia received only the most minor of flooding, and that essentially no livestock drowned in the vicinity of the Kaweah Delta.

One source said that thousands of cattle were drowned in the Tulare Lake Basin during the 1861–62 flood. No details were provided to support that claim. Livestock deaths on the Kaweah, Tule, and White Rivers appear to have been minimal. The Kern did experience serious flooding, so perhaps some cattle drowned in that area.

The winter storms engulfed all of Southern California. Beginning on Christmas Day, 1861, the Los Angeles area had 15 days of gentle rain followed by 28 continuous days of heavy rain. During the course of the 1861–62 season, Los Angeles received over 66 inches (5½ feet) of rain.

The Mojave River rose 20 feet above average in present-day Oro Grande. Lakes formed in the Mojave Desert.

Planes were cut by gulches and arroyos from Ventura to San Luis Rey. (Mission San Luis Rey de Francia is located just south of Marine Corps Base Camp Pendleton in San Diego County.)

The Santa Ana River flooded catastrophically on the night of January 22, 1862, sweeping away the village of Agua Mansa (literally "Gentle Water"), located on the Santa Fe Trail just south of present-day Colton.

Hearing the roar of the river that night, Father Borgotta frantically rang the church bell, sounding the alarm. The inhabitants of the village ran or swam to high ground. "The gentle Santa Ana River became a raging torrent which washing, swirling and seething, swept everything from its path." One writer reported that there were "billows fifty feet high." Peter C. Peters of Colton recalled that "when morning came — (there was) a scene of desolation." Only the church and a house near it survived.

USGS reconstructed the cross-section of the flood and determined that the normally placid Santa Ana had been flowing at approximately 320,000 cfs that night. That was three times greater than any subsequent flow in the area, even the 1938 flood.[462]

The Santa Ana flood formed two large lakes south of that river — one in the Inland Empire and another in the floodplain of Orange County. The lake in Orange County lasted about three weeks with water standing four feet deep up to four miles from the river. Sediment cores suggest that the last time the area saw a flood that big was in about 1600.

In San Diego, over seven inches of rain fell in January alone. All of Mission Valley was under water, and Old Town was evacuated.[463]

The 1861–62 flood is the flood-of-record for much of Southern California.[464]

## 1862–64 Drought

This three-year drought followed on the heels of the huge 1861–62 flood. It affected at least the Central Valley and possibly a larger area.

During this drought, Sacramento received less than half its average rainfall and the San Joaquin Valley was even drier. In the first year of this severe drought, William Brewer described the San Joaquin Valley as "a plain of absolute desolation." On February 27, 1864, the *Stockton Independent* reported that there had been no rain of consequence for 11 months. By the end of March, the wheat and barley fields around Stockton were dried out completely.

Cattle died in large numbers during 1864, bringing their numbers down dramatically. Sheep came through the drought in better shape, their numbers increasing. The double whammy of the 1861–62 flood followed by the 1862–64 drought is one of the main reasons that the Sierra grazing business changed from primarily cattle to primarily sheep.

Table 19 uses the livestock censuses to illustrate the dramatic change that occurred between 1860 and 1870.

Table 19. Livestock censuses of the San Joaquin Valley.

| Year | Cattle | Sheep |
|------|---------|---------|
| 1860 | 226,248 | 78,568 |
| 1870 | 288,483 | 901,892 |

William Brewer wrote about the drought and the effect of sheep grazing in 1864. Brewer's party arrived in Visalia on June 6 of that year. A few days later, they undertook what has become one of the great steps in American mountaineering, entering the Sierra at Big Meadow. They finally emerged at Mariposa at the end of the summer. When Brewer eventually got around to counting up his miles of travel, he found that he had ranged over 15,000 miles within the state — but that summer of 1864 spent exploring the crest of the Sierra had to be among the best.[465]

In the Tulare Lake Basin, 1864 is remembered as the most severe year of the 1862–64 drought by far. Most of the streams had dried up, the feed either did not mature or withered, and there was not even sufficient water for drinking.

W.G. Cartmill recalled how severe the drought was on the lower part of the Kaweah Delta. Tulare Lake served as a gigantic watering hole. Thanks to the huge 1861–62 flood, the lake was brimful when the drought set in. Vast herds of cattle would spread over the country for miles, traveling as far back from the lake as they could go without water in search of the scant grasses. Then they would rush back to the shore each day to quench their thirst. By 1864, grass had disappeared completely on the plains, with oak leaves, acorns, and salt grass serving as fodder of last resort. Cattle died in large numbers, their stench filling the air.[466] Their hides were taken, but the meat was left to rot. It's easy to visualize the scene with huge flocks of turkey vultures and California condors, reminiscent of the Serengeti or the Pleistocene.

David Campbell, an early Tulare County pioneer, recalled that 1864 was so dry that the grass did not even get started that year.[467]

After the Kaweah River reaches the area of the present-day Terminus Dam, it flows onto its delta and divides into distributaries. The 1861–62 flood deposited considerable quantities of drift, silt, and sand on the Kaweah Delta. A portion of the channel of the Lower Kaweah River was obstructed by these deposits, significantly reducing its ability to carry flows. During low-flow periods, this section of the Kaweah was now dry; it only served to carry water during flood periods.

In addition, the 1861–62 flood had created a new distributary, which came to be called the St. Johns River, along the north side of the delta. The St. Johns would remain the principal channel of the Kaweah until 1877 when the Fowler Cut would reopen the old Kaweah River channel. This worked out great for north-side farmers, but not well at all for farmers on the south side of the delta.

As the low-water period of 1862 approached, irrigation water was scarce for those who depended upon Deep Creek, Packwood Creek, and Visalia Creek for their supply, and many projects were proposed for relief. (Visalia Creek was originally known as Tiber Creek. We now know it as Mill Creek.) During 1862 and the years immediately following, a number of ditches were constructed from the St. Johns River near Rocky Ford, southwesterly to the original main channel of the Lower

Kaweah River. All of those ditches were intended to increase the flow in the delta streams in the Visalia area and on the south side of the delta.

After much contention between settlers on these several streams as to the apportionment of their scant supply of water, they finally reached agreement. In 1867, a gate was constructed in the head of Visalia Creek (presumably where the Lower Kaweah River ends, just north of the Ivanhoe turnoff (Road 156/158) on Highway 198). The timing was not good; that gate would last less than a year before being swept away by the 1867–68 flood.

The drought, especially with its near total failure of winter pasture grasses, forced ranchers to look elsewhere for previously unused rangelands. What resulted was the first major utilization of the Sierra for large-scale livestock feeding. During the drought years, hungry cattle from the lowlands swarmed over the Sierra foothills and forests while the high country suddenly found itself assaulted by huge herds of domestic sheep. Within a few years, much of the herbaceous vegetation of the Sierra had either been destroyed or replaced. In the foothill grasslands, annual Eurasian grasses replaced the grazing-sensitive native perennial species. In the high country, entire basins were so thoroughly denuded that parties traveling on horseback lamented the almost total lack of feed for their animals.[468]

In 1859, Paschal Bequette, Sr. brought his family to Visalia and became a cattle and horse breeder. He recalled that they saved their horses during the 1862–64 drought by taking them up the South Fork trail to Hockett Meadows where there was good feed and water.[469]

Floyd Otter said that valley ranchers also drove their hogs into the high country during drought years.[470] In July 1864, at the height of the worst drought year, Clarence King took a trip from Visalia to the Mt. Whitney country. In a letter to Josiah Whitney (Chief of the California Geological Survey), he wrote:

> I rode until nine in the evening, when we came to the "Hog Ranch," two acres of tranquil pork, near a meadow in the most magnificent forest in the Sierras.

That "ranch" was probably a temporary drover's camp on the South Fork of the Kaweah. Floyd Otter thought that it might have been in or near Hockett Meadows. King later described this pig-herd in the words of its owners as "The pootiest hogs in Tulare County — nigh three thousand." One can only imagine what 3,000 "half-wild boars, sows, and pigs" could do in a summer on the Hockett Plateau.

The route that Paschal Bequette, Clarence King, and the hog drovers took was presumably along the newly constructed Hockett Trail. 1864 was the first year that this trail from the Kaweah Delta area to the Cerro Gordo silver mines was open for use. The trail was built by John Benjamin Hockett and his partners under a grant by the California Legislature, and remained a private toll trail until the state regained ownership of the trail by purchase.

According to Samuel Thomas Porter's history of the Mineral King mining rush, Pleasant Work (what a nice name) was also known to run hogs in Hockett Meadows before 1867.[471]

The 1862–64 drought was viewed as something of a blessing by the folks trying to drain the swampland around Bakersfield.

The drought finally ended in November 1864, and three years of average precipitation followed.

## 1867–68 Flood

Flooding in 1867–68 occurred primarily in December 1867. On the lower parts of the Sacramento — San Joaquin River systems, the floods carried over into January 1868.[472]

The winter of 1866–67 had been rough on the Chinese laborers constructing the Transcontinental Railroad over Donner Summit. Avalanches had wiped out two of their work camps. The winter of 1867 proved equally challenging. Sub-tropical storms deluged the region with more than 40 inches of rain in December 1867, causing extensive flood damage.

In the northern part of the Central Valley, the 1861–62 flood is generally the flood-of-record. In the Sacramento River Basin, the main river and its lower tributaries were at extreme flood stages between December 22, 1867 and January 2, 1868. However, the floods of 1867–68 are believed to have been generally lower in discharge and volume than those of 1861–62. At some points the American and lower Sacramento Rivers were reported at higher stages in 1867–68, but it is probable that these high stages were caused by aggradation of stream beds or by channel contraction due to levee building. In the Sacramento River Basin upstream from the Feather River, the floods of December 1867 were definitely secondary to those of 1861–62.[473]

The 1861–62 flood affected a very wide area stretching from the Columbia River to the Mexican border. However, the 1867–68 flood was especially severe on Sierra Nevada streams tributary to the southern part of the Central Valley. In the foothills, the flood on the San Joaquin River exceeded considerably any other known flood and was probably higher than any known flood at all points upstream from the mouth of the Merced River. However, the San Joaquin River stages downstream from the mouth of the Stanislaus River were not as high in 1867 as in 1862.

During recorded history, the 1867–68 flood was one of the greatest in the Tulare Lake Basin. Peak stages in that region during December 24–25 were the highest of record. Major floods occurred on all the main tributaries in the Tulare Lake Basin.[474] The Kings. Kaweah, Tule, and Kern Rivers carried flood flows in 1867–68 that are believed to be the greatest known, exceeding those of the 1861–62 flood.[475]

Flooding occurred throughout the San Joaquin Valley in December 1867, extending barely into January 1868. The San Joaquin Valley was described as looking like an ocean. Unlike the 1861–62 flood, this flood lasted only weeks rather than months.

The preceding multi-year drought had ended in November 1864. One account said that the high country then experienced two consecutive years of heavy snows with virtually no summer between. This supposedly resulted in a huge accumulation of snow in the Sierra.

Whether or not that was true, there are multiple accounts that rain and snow began in mid-November 1867 in the Kaweah River Basin and came down almost continuously through December. One account said that the snowline was at about 5,000 feet until December 20, at which point the weather turned warmer. Presumably similar weather conditions were happening throughout the Central and Southern Sierra.

As on the other rivers in the Tulare Lake Basin, flooding in 1867 on the Kings was greater than the 1861–62 flood.[476] It is considered to be the greatest flood on the Kings since at least the flood of 1805.[477] The 1867–68 flood completed the formation of Cole Slough. (See the section of this document on Pine Flat Dam for a more detailed description of the formation of Cole Slough and associated waterworks.) Tremendous quantities of timber were brought down from the Sierra and deposited on the plains.

The Kings River engulfed the newly rebuilt town of Scottsburg. (Even though that townsite had been selected because it was thought to be safe from flooding.) That community was then rebuilt at an even safer location and renamed Centerville.

As on the other rivers in the Tulare Lake Basin, flooding in 1867 on the Kaweah was greater than the 1861–62 flood. It is considered to be the greatest flood on the Kaweah since at least the flood of 1805.[478, 479]

Again, as with the Kings, the Kaweah brought tremendous quantities of timber down from the Sierra and deposited them on the plains. Smith Mountain is about a mile east of Dinuba. By some accounts, flooding was so extensive in the 1867–68 flood that one could have ridden in a boat from Smith Mountain to the Tule River.[480]

According to the book *Floods of the Kaweah*, the Kaweah rose 17.5 feet above the average low-watermark in the foothills.[481] Despite considerable searching, Valerie McKay was unable to find the source of that figure or where the watermark was measured. The earliest gage on the Kaweah was a USGS staff gage (USGS gage #11-2105) that was located near Three Rivers just above the junction with Horse Creek; that gage was established in October 1903 and was read twice daily. Harry H. Holley established the measuring station at McKay's Point in October 1916 for the River Association. Perhaps the watermark reported above was measured in a foothill canyon such as the present-day Terminus Dam area.

The flood deposited considerable quantities of drift, silt, and sand on the Kaweah Delta, raising its elevation. The head of the Lower Kaweah River channel had been partially closed by the 1861–62 flood. The 1867–68 flood further obstructed that channel by depositing still more drift and silt in it.

The flood refilled and otherwise destroyed some of the ditches that had been constructed to bring water from the St. Johns River back into the old channel of the Lower Kaweah River. It washed out the new gate in the head of Visalia Creek (what we now know as Mill Creek). It partially closed the head of Packwood Creek (just north of the Ivanhoe turnoff (Road 156/158) on Highway 198). It also partially closed the head of Deep Creek (northeast of the present-day Kaweah Oaks Preserve).

The 1867–68 flood further enlarged the St. Johns River that had been created in the 1861–62 flood. It eroded a new head of the St. Johns River about a mile farther upstream, farther into the swamp. This new point of separation of the two channels became known as McKay's Point. That was the original spelling, but it is now sometimes written as McKays Point (as on the Woodlake USGS quad) and McKay Point (the preferred spelling used by the Kaweah Delta Water Conservation District and at least sometimes by the U.S. Army Corps of Engineers). This document uses the original spelling: McKay's Point.

McKay's Point is located about a mile northwest of present-day Lemon Cove and three miles below Terminus Dam. A variety of structures would later be built at this location in an attempt to control the Kaweah and split the flow between the Lower Kaweah River and the St. Johns River. The river's natural tendency to relocate this point has been actively resisted, similar to the much bigger struggle to keep the Mississippi from following its preferred course down the Atchafalaya.

In 1870, a brush and rock diversion weir was built at McKay's Point. Presumably this weir served to allocate the water between the St. Johns and the various minor distributaries (aka creeks and ditches) that still connected at that point. It would remain that way until 1877 when the Fowler Cut (built by Samuel Fowler under contract to the Kaweah Canal and Irrigation Company) would reopen the Lower Kaweah River channel. The brush and rock diversion and weir at McKay's Point was also rebuilt in 1877. Once the Lower Kaweah River channel was rewatered, then the irrigation ditches attached to that channel got reliable water for the first time in 15 years.

No doubt the brush and rock diversion weir at McKay's Point had to be frequently repaired and it apparently had to be completely reconstructed in 1884 and 1897. The first concrete weir was built at McKay's Point in about 1909. It had to be replaced after the 1937 flood. That diversion weir was destroyed or at least bypassed in the 1955 flood. The current concrete weir diversion at McKay's Point is maintained and operated by the Kaweah Delta Water Conservation District.

Stringtown was a settlement of five families living in a line south of present-day Woodlake along the Kaweah River, east of Bravo Lake. (This reference to Stringtown's location is somewhat unclear. Bravo Lake was actually located adjacent to the St. Johns River.) The 1867–68 flood came into several of their houses; only one of which was on sufficiently high ground to survive completely intact. After the flood, the other four families decided to move. This gives the 1867–68 flood the often cited reputation of having destroyed Stringtown.[482, 483]

In 1864, the newly created People's Ditch Company had built 12 miles of ditch to what is now Farmersville. That system failed during the 1867–68 flood. Logs from the Sierra reached the Farmersville area. Some of those logs were apparently giant sequoias from the South Fork of the Kaweah. In the years after the flood, ranchers would use this wood to build fences.

During the 1867–68 flood, all the streams in Tulare County were reported to be on a rampage with great loss of property. The Kaweah and St. Johns Rivers made a vast expanse of waters. The Visalia area was awash with water; boats were widely used for transportation, and there was significant loss of property.

There are multiple accounts of the flooding which clearly relate to the general flooding that was occurring throughout the area. David Campbell, an early Tulare County pioneer, recalled that the floodwaters formed almost a solid sheet of water from Porterville to Visalia.[484]

Another account was about three Visalia families who lived near where Packwood Creek crosses the present-day Mineral King Road (about two miles east of Lovers Lane on Highway 198). All of those families gathered at A.H. Broder's place during the day because his was situated on higher ground. They then built a three-foot-high embankment, enclosing about half an acre of ground. The siding from the barn was removed, and a raft built. If the river continued to rise, they planned to move to a still higher sand knoll which lay to the southwest. By 9:00 the following morning, Broder, who had been keeping tabs on the water level by means of sticks, reported that it had receded half an inch and that it would not be necessary to move.

About 200 Native Americans took refuge on the same high mound as Broder and his neighbors. They made a gala festival of the predicament. Squirrels and rabbits in great numbers were caught and hung on lines to dry; the flood affording both amusement and provender.

Another account that has survived was about the residence of the Evans family, which was located on high ground near present-day Tulare Avenue in the general vicinity of Ben Maddox or Santa Fe.[485] (Later this location was known as the Evansdale Orchard.) The water had risen previously, and then it rose again suddenly during the night. It surrounded their home and almost engulfed some of their neighbors' homes. The Prothero family lived on the Bentley place and there the water ran through the windows. They moved to the Evans home for shelter.

Then came a call for help from the home of Mrs. Williams, who lived adjoining. This was about 1:00 in the morning, pitch dark and the swirling waters icy cold. Mrs. Williams had a baby but four or five days old and was unable to walk. Samuel and James Evans waded over, and placing her in a rocking chair, carried her to safety. Tom Robinson, with his wife and family, also took refuge with the Evanses, making a total of 25 gathered there. The barn, several hundred yards away, half full of hay, provided the only place for sleeping quarters for so many people.

Between it and the house, the water ran two or three feet deep. Luckily, a boat had previously been constructed in which to go to Visalia, and so the half-dried refugees cuddled around the stove in the Evans's kitchen were enabled to get to bed without again getting wet. Jim Evans, acting as gondolier, conducted his guests to their hay mow lodgings. This nighttime flooding event may have occurred on the night of December 23–24.

As on the other rivers in the Tulare Lake Basin, flooding in 1867 on the Kern was greater than the 1861–62 flood. It is considered to be the greatest flood on the Kern since at least the flood of 1805. In the Kern River Basin, the flood was at high stage from December 25, 1867, to January 1, 1868. A remarkable feature of the flood was the large quantities of logs from the Sierra, including cedar and giant sequoia, that were deposited on the overflowed lands of Kernville and Bakersfield.[486]

All of the streams in the southern part of the Central Valley reached peak stages during December 24–25.[487] Three witnesses told Walter Fry that the flood on the Kaweah arrived in Visalia late on the evening of December 23.

The 1867–68 flood resulted in the deepest flood depths ever on the streets of Visalia. The 1861–62 flood put a maximum of 24 inches on Main Street. When the peak of the 1867–68 flood arrived in Visalia, it flooded the development along Mill Creek 5–6 feet deep as measured at the grist mill. That mill was on Mill Creek, on the southeast corner of Main and Santa Fe Streets.

Walter Fry recorded the experiences of several families who lived in the countryside around Visalia. One of those, Betty Townsend, lived near Cutler Park on the St. Johns. On the evening of December 23, 10 people sought refuge in her house and stayed there for a week. Her account included:

> *Our Christmas dinner, in part, consisted of a turkey feast. The turkey was captured by one of the party from a bale of hay which was being swept down the torrent. A pig was similarly rescued and consumed.*

The Tule River spread all over the Poplar and Woodville sections.

Deer Creek and the White River merged their waters in their lower courses.

The Kern River had rerouted to the west in the 1861–62 flood. The 1867–68 flood moved the river channel even farther north to its present location, ending in the Buena Vista Slough a few miles north of Buena Vista Lake and entered that lake from the northwest as shown in Figure 9.[488]

Tulare Lake gradually declined in elevation after the 1861–62 flood. In the summer of 1867, the lake level was 200.7 feet. However, the 1867–68 flood raised it by 14.7 feet, bringing it back to a maximum elevation of 215.4 feet. At the height of this flood, Tulare Lake was almost 37 feet deep at the deepest point. The lake has not been that deep since.

The onset of the flooding in the lakebed swept in at night, catching people by surprise. Jack Phillips (son-in-law of early pioneer Dan Rhoades) and four other guys had their hogs together and were camped near each other on the lakeshore below present-day Stratford. When they went to sleep at night, there was no water in their camp. By morning, the hog camp was going under water and a hurried retreat was made toward high ground. The water was at their heels all the way to the Dan Rhoades adobe in spite of the best time they could make. The same morning, Doc Vaness came out of the lakebed with his family just in time to avoid drowning.[489]

Based on limited information, the flood described above most likely occurred on Christmas Eve, 1867. Peak stages on all the main tributaries in the Tulare Lake Basin occurred during December 24–25.[490] The flood on the Kaweah arrived in Visalia on the evening of December 23. Those floodwaters would probably have reached the Tulare Lakebed on the morning of December 24 (having traveled roughly 30 miles at about 2.5 m.p.h. = 12 hours). The much larger volume carried by the Kings River would have had a longer route to travel, so those floodwaters presumably took somewhat longer to reach the lakebed. That would fit with a possible arrival time of the evening of December 24.

In any case, flooding was eventually so extensive that boats bearing supplies were reported to have passed freely from Visalia to places in Kings and Fresno counties. One account said that the flooding made the San Joaquin Valley a continuous lake of water from Buena Vista Lake to the San Joaquin River; that seems quite plausible given the height of Tulare Lake.

In 1868, Richard Smith loaded a 16-foot scow with a one-ton cargo of honey and made the 170-mile journey from Tulare Lake to San Francisco Bay.[491] That remains the only recorded commercial trip ever made from the lake to the bay. (There were four *non-commercial* trips: 1852, 1938, 1969, and 1982–83.) Apparently Smith was able to make the return trip back through the tules to Tulare Lake.

How different people viewed the 1867–68 flood was a matter of perspective. And their perspective was determined in part by the elevation where they were living when they experienced the flood.
- Joseph Palmer lived at an elevation of perhaps 1,000 feet. So he experienced the flood from a canyon perspective; a brief event that passed by causing him no harm (see description below).
- Visalia was about 331 feet in elevation. The new settlers there experienced the flood as the Kaweah spread across its delta, causing destruction.
- Tulare Lake was about 215 feet in elevation at its peak. For people who lived in the vicinity, it was lake flooding. The lake was a major resource that was sustained by such periodic flooding.

In addition to the widespread flooding described above, a very rare landscape-wide event occurred. After some six weeks of steady rain, the ground had become saturated to a considerable depth. Slopes across the Southern Sierra became unstable, resulting in an apparently large number of landslides during a nine-day period in December 1867. A small portion of those slides dammed flooding rivers, and then failed when those dams were overtopped.

The Southern Sierra was very sparsely populated in 1867, so the chance of any one of those landslide dam failures being noted and recorded, and that record surviving into the 21st century is slim. Through good fortune, we have been able to reconstruct four of those landslide dam failures to varying degrees.

Such events don't add any water to the flood that is already occurring down below. But they withhold the water for a while, and then release it all at once, resulting in a wall of water, a pulse.

### Landslide Dam Failure #1: South Fork of the Kaweah

One of those landslide dam events occurred on the north side of Dennison Ridge Peak on the night of December 20, 1867.

This is the largest landslide to have occurred in the national parks in historic time. This event is included in a USGS report of documented historical landslide dams from around the world.[492] The event was analyzed by Walter Fry.[493]

It had been raining in the Three Rivers area almost steadily for some 41 days, with heavy snows above the 5,000 foot elevation level. All the rivers were very high. The weather turned warm on December 21, and a hard rain fell all day, even at high elevation.

The soil involved in this landslide was described as a sandy loam. If so, it would have had lots of voids that could hold water. After 41 days of steady rain, it would have been saturated. The storm on December 20 would then have been the triggering mechanism. That was the way that the somewhat similar Mill Creek Landslide began on the South Fork of the American River on January 24, 1997.

When the slope became unstable, a mass of dirt and vegetation broke loose from near the crest of Dennison Ridge. The head of that landslide began on a 45 degree slope. It swept 2½ miles down into the canyon of the South Fork of the Kaweah. The landslide had a total estimated mass of about 580,000 cubic yards (445,000 cubic meters).[494]

The landslide stripped the steep hillside of a thick forest of giant sequoia, pine, and fir in a path of devastation that ranged from 1,500–4,000 feet in width. This included the westernmost 300–400 acres of Garfield Grove. Walter Fry calculated that 350 million board feet of timber came down in that slide.

Most of the big landslides in the Southern Sierra have contained a large component of rock, including huge boulders. However, this landslide was markedly different. It consisted largely of trees, a thick layer of sandy loam, and relatively small rocks. Many of the pines, firs, and sequoias were quite large, including sequoias up to 30 feet in diameter. There were relatively few large rocks in this debris to provide structural stability. Therefore, this landslide dam failed much more catastrophically than other large landslide dams, and it left relatively little evidence of its presence immediately adjacent to the river. A large chunk of the dam washed out with the first failure, but it apparently took several days to completely wash out the dam.

Walter Fry analyzed the site in 1931, 64 years after the event occurred. Presumably there were still large logs and other debris present to allow him to determine that the top of the dam had been over 400 feet high at its highest point. Some of that material may still be there today.

Fry described the top of the dam and its volume, but he did not describe its shape. We can conjecture that it had a relatively steep slope and was much higher on the side adjacent to Garfield Grove. It may have flattened out near the bottom of the slide, typical of other landslides. If that were the case, then the amount of water impounded might have been only on the order of 200 feet or so; we have no way of knowing the low point on the dam. However, there is no apparent evidence of the dam on the opposite side of the river (the Ladybug Trail side); presumably all of those logs were removed by the flood.

Although new growth has disguised most signs of the landslide, its effects are still dramatically apparent in the vicinity of Snowslide Canyon, where dense sequoia forest ends abruptly at a landslide boulder field. The absence of large, old-growth trees is apparent throughout the disturbed area. There appears to be an even-age stand of giant sequoias covering some of the lower part of the landslide's path, presumably having germinated or taken root in the freshly disturbed mineral soil.

The landslide would have exposed a lot of mineral soil in and immediately adjacent to a giant sequoia grove. Significant sequoia germination would have been expected following this event. Walter Fry conducted a detailed survey of the area some 60 years later, but did not mention seeing sapling sequoias. Nevertheless, those even-age stands of relatively young sequoias are there today. Dating these trees could quantify how the grove responded to this large-scale disturbance.

The landslide occurred just before midnight on December 20. When it came to rest, it formed a landslide dam that was ½ mile wide and over 400 feet high at its highest point. The South Fork Kaweah was presumably running at flood or near-flood stage because of all the previous rain and snow. It didn't take the river long to fill the temporary reservoir. The dam failed about 1:00 a.m. on December 22, just 25 hours after the slide occurred.

The collapse of the landslide dam produced a flood surge about 40 feet deep that rushed down the South Fork Canyon.

Joseph Palmer was a homesteader who lived in that canyon, several miles below the slide. Just before midnight on December 20, he

> *was aroused by a heavy rumbling sound such as I had never heard before, and which lasted for an hour or more. Then a great calm set in, and even the roaring of the river ceased. On leaving my cabin in the morning, I found that despite the heavy rain the river was low. From this I knew that a great slide had blocked the canyon above and that later the dam would give way and cause a flood...About 1:30 a.m. I was aroused by a tremendous thundering and rumbling sound...I jumped out of bed, grabbed my clothing, and ran for safety up the mountain side some 200 yards from the river. In a few minutes the flood came along with a crest of water some 40 feet in depth that extended across the canon, carrying with it broken-up trees which were crashing end over end in every direction with terrific force and sound. The river remained high for several days, and all the while timber was going down and being swept clear out to the valley.*

The bursting of the landslide dam at 1:30 a.m. on the morning of December 22 let loose a great flood, and the impounded water spilled and smashed its way down the South Fork Canyon, carrying everything before it, including giant sequoia logs. Some of those sequoia logs can still be found along the South Fork. A particularly good place to see them is the peninsula formed by the junction of Grouse Creek and the South Fork (the old Hat Maxon Ranch; Hat was Kirk Stiltz' great uncle). That peninsula was right in the path of the flood and must have gotten hammered. The land has since revegetated to some degree, but sequoia logs lie scattered about, some of them upwards of 40 feet above the level of the riverbed.

The flood swept past what is now Three Rivers, 15 miles below the landslide. Possibly this is the flood that deposited the big rocks along Cherokee Oaks Drive. The flood moved rapidly through the steep canyons, but slowed dramatically when it emerged from the canyon and spread out onto the gentle slopes of the Kaweah Delta.

Harry H. Holley, the Kaweah Watermaster for many years, said that water takes about 6 hours to travel the 15 or so miles from McKay's Point to Visalia.[495] Presumably the flood was traveling at roughly the same speed, about 2.5 mph, as it moved across the delta.

Traveling at that speed, the flood would have arrived in the small town of Visalia (42 or so miles downstream from the landslide) toward the end of the day on December 22, roughly 12–18 hours after the landslide dam had failed.

Many sources attribute the widespread flooding in the Tulare Lake Basin to the failure of the South Fork of the Kaweah landslide dam. That represents a misunderstanding of this event. The landslide dam failure, while dramatic, did not cause the flooding. All that it did was to impound the flow on one tributary for 25 hours and then release that water in a surge. It didn't contribute any additional water to the flooding that was already occurring throughout the basin.

There is no credible record that anybody in Visalia even took note of the increased volume of water; it probably wasn't all that much. (It would have been spread out many miles wide by then, so its height would have been greatly reduced.) But they definitely noticed the huge increase that occurred when the peak of the flood occurred on the following evening, December 23.

No doubt the mainstem of the Kaweah at peak flood (the biggest flood in that river's recorded history) had the power to pick up and carry many sequoia logs out onto the delta, far more than had been moved on December 22. It's easy to see how it became folklore that the flood of December 23–25 was caused by the spectacular landslide dam failure on the South Fork of the Kaweah. However, that flood was really caused by the same events that caused the flooding occurring on the other rivers in the Tulare Lake Basin. It was just another of our rain-on-snow events.

When the floodwaters subsided, a huge number of logs were left scattered widely about the Kaweah Delta. One big sequoia log came to rest right beside the grist mill at Main and Santa Fe. The trees lasted for years and appeared in numerous pioneer tales. Some accounts indicate that some of the trees were sawed for lumber. One portable sawmill was ordered specifically to see if it were feasible to mill the logs that had been left in the upper part of the Packwood Creek swamp.[496] However, others recalled that so much sand and rock was imbedded into the trunks that the trees could not be sawed for lumber. When the 1874 No Fence Law made stockmen liable for the damage caused by their trespass cattle, many of the sequoia logs were split and used to fence the open range.[497]

### *Landslide Dam Failure #2: San Joaquin River*
The story of this dramatic landslide dam failure is known from two sources:
- Lilbourne Alsip Winchell's 1920 *History of Fresno County*.[498] This document has proved quite a challenge to track down. Fortunately, its account of the event was reprinted in Gene Rose's *San Joaquin: A River Betrayed*.[499]
- Floyd Otter's *The Men of Mammoth Forest*.[500]

The town of Millerton was created near Fort Miller, on the banks of the San Joaquin River. It was the original county seat of Fresno County, formed in 1856.

The fall rain came early and became more frequent by November, when the rains turned to snow in the Sierra. By December, record amounts of snow had been observed in the mountains. Then the temperature moderated and warm rains began sending the San Joaquin surging. In breathless prose, Winchell described the flood from the viewpoint of the residents of Millerton.

> *The San Joaquin at Millerton steadily grew in volume and height. Day after day the rains came. Anxiously the people awaited abatement of the storms. Each hour the angry stream reached higher and higher. The occupants of the buildings along the lower street began moving their most valued possessions, yet hoping for relief from the merciless encroachment.*
>
> *Nightfall came — black under the overcast skies. It was Christmas Eve; but there were no devotional offerings. The harassed people were beyond joyous expression; though, from the women, there may have been silent prayers for mercy. There was universal vigilance and excited effort, and concern for community safety. Despairingly, as the black night measured the hours, they watched the unceasing advance of the surging torrent. Lanterns gleamed through the street; lights shone in all the upper houses; and the rain fell, and splashed in sheets in the frowning earth!*
>
> *At eleven o'clock that night — Christmas Eve — the river was higher than the white men had ever seen it. Suddenly, crashing, roaring, sounds came to the ears of the wakeful villagers. Rushing with appalling speed and force a high wall of water, bearing on its surface an overwhelming tangle of broken and twisted trees from the forests of the high mountains. The whole blossom of this avalanche flood was thickly covered with the smashing, grinding, tearing logs — trunks, tops, roots were whirled along with the destroying speed of a tornado. Greater than the combined blows of all the batter rams and catapults of old, the massed projectile struck the town. Nothing in its tracks resisted it. In a few moments the awful work was done. Millerton was wrecked.*

Floyd Miller's account said that there were multiple slides on December 24, temporarily damming the San Joaquin. It was the failure of that landslide dam or dams that made the event so catastrophic, resulting in the destruction of Millerton.

The San Joaquin River passes through a narrow granite gorge above Millerton. The San Joaquin is a huge river, and it was at flood stage. However, the large quantity of debris that slid off the mountains combined with the narrowness of the gorges allowed the landslide dam(s) to block the river. Eventually the river overtopped and breached the dam, sending a tidal wave of water and debris down the canyon.

At Millerton, Jones's trans-river ferry was swept all the way to Sycamore Point. The same thing was repeated downstream at Hill's Ferry, where the ferry had been destroyed during the 1861–62 flood. Debris from the 1867 siege damaged paddlewheel steamers plying the river. The steamers that were not damaged chugged around the inland sea plucking those residents lucky enough to have a second story home to which they could escape. Much of the port city of Stockton was inundated; floods had long been part of the Stockton scene. Boats were torn from their moorings and left as derelicts below. Stumps and debris from the Christmas catastrophe were seen as far away as Suisun Bay.[501]

Some Millerton residents rebuilt, some moved. However, as a result of the flood, the county seat was soon relocated to Fresno. The townsite of Millerton was inundated after Friant Dam was completed in 1942, forming Millerton Lake.

### Landslide Dam Failure #3: Mill Flat Creek

There are several "Redwood Mountains." The particular Redwood Mountain referred to in this story appears to have been the mountain that the Big Stump Grove is located on within present-day Giant Sequoia National Monument.

Forest Mill was a sawmill constructed on Mill Flat Creek, perhaps 5–10 miles downstream from present-day Sequoia Lake. This is the same location where Feggan's Mill had been destroyed after a flood resulting from a January 1862 landslide dam failure. (It's easy to confuse Mill Flat Creek with the similar-sounding Mill Creek. Mill Creek intersects the Kings River about two miles below present-day Pine Flat Dam. Mill Flat Creek intersects the Kings about two miles above the reservoir.)

In 1867, Forest Mill was owned by Jasper (Barley) Harrell and D.V. Robinson. It was powered by a 26-foot overshot water wheel. The following story of the debris slide and resulting flood was told in an 1881 issue of the *Visalia Weekly Delta*.[502]

> *'Twas on Christmas night 1867. It had rained all day, as it had done for about a week. The clouds were low; the day was dreary and lonesome; the night was one of those intensely dark, stormy nights that occasionally come in the pine forest, that one has to see in order to realize. Sometime in the fore part of the night, quite a tract of land with heavy timber, on the side of Redwood Mountain, slid into the creek, forming a dam which collected a large head of water, then giving way started down the creek crashing the timber before it. The first habitation it came to was an old log house inhabited by S.B. Corderoy. He heard it coming and caught his clothes and ran for life. The water just caught him as he reached high ground. The next place reached was (Michael) Hart's. He heard the noise, thought it was a tornado, tore up the floor, and put his family under it and took his gun and stood in the door to await his doom, but the water did not reach his house. Next it came to the house of a man by the name of Root. He heard a great noise as of many waters, and jumped out of bed into it knee deep, where he stood fishing after his clothes and trying to convince his wife that it was better to lie a bed than to get out. Next it struck the Forest Mill, leaving it a complete wreck, and rose to the doorstep of the house where D.V. Robinson dwelt with his family, then passed on (to the Kings River) doing no more damage.*

Floyd Otter said that this flood destroyed another sawmill in addition to the Forest Mill.[503] This flood is apparently documented in detail in a 1906 *Lumbering in Tulare County* report prepared by H. Barton for the Fresno office of California's Department of Forestry. However, CalFire has been unable to locate that file.

## *Landslide Dam Failure #4: North Fork of the Kern*

While the South Fork of the Kaweah landslide dam flood was happening in Visalia, a similar drama was about to unfold on the North Fork of the Kern River. The location was about three miles below the present-day Kern Ranger Station, just upstream from Little Kern Lake. At the time, Kern Lake did not exist. The event occurred within what is now Sequoia National Forest.

The Kern Canyon at that point is about 3,000 feet high. The canyon walls were already saturated from the rain and the weight of the snowpack, which further burdened the slopes. The winter of 1867 had exceptionally high precipitation that helped to mobilize the slide material. When the slope on the west wall (one source incorrectly said that it was the east wall) became unstable, a mass of dirt, rock, trees, and snow broke loose and swept down into the river canyon. The slide occurred on or about December 28, 1867. There were no witnesses to the event. The nearest settlers were some 30 miles down-canyon in the mining community that we now call Old Kernville.

The mountainside that slid into the upper Kern River was supposedly about a square mile in size, according to a pack train that passed the location a month later while in route to Death Valley. One source said that the pack train was using the Jordan Trail, but the Hockett Trail would seem more likely.

This landslide occurred about ½ mile upstream from where a massive slide off the east wall had created Little Kern Lake at some earlier time.

The temporary lake that formed behind this huge landslide was estimated by one source to have had a depth of about 1,000 feet and extended 12–16 miles upstream. The reliability of that estimate is highly suspect; it seems improbably large. If it were really this big, the lake would have backed up to the vicinity of Kern Hot Spring. Failure of such a huge landslide dam would have presumably done even more damage to Kernville and Bakersfield than was experienced. It seems more likely that the dam was on the order of several hundred feet high rather than 1,000 feet. A 400-foot dam would have backed the lake up 7–8 miles to about Rattlesnake Creek (elevation 6,600 feet). A 200-foot dam would have reached to about Lower Funston Meadow (elevation 6,469 feet).

The Kern River would have been running at about flood stage after the previous six weeks of rain and snow. One source suggested that it took the river two days to fill the temporary reservoir. Another source said that the riverbed downstream in Kernville was only dry for about 24 hours. In any case, the Kern River appears to have filled its new reservoir in less than two days and then overtopped the landslide dam. Because it was winter, there were no witnesses to this event. But evidently the dam failed catastrophically. Judging from when the flood reached Bakersfield, this failure occurred on or about December 30.

One source said that the failure of the Kern Lake dam caused the Little Kern Lake dam to fail as well. The reliability of that claim is unknown. We haven't found any reliable record of how intact the Little Kern Lake dam was prior to the 1867–68 flood.

In any case, the floodwaters poured into Little Kern Lake and over its dam. The flood continued down-canyon about 30 miles to Kernville, where it destroyed many homes. According to Harvey Malone (whose father-in-law's family was living in the Kernville area at the time of the flood), the floodwaters backed up when they reached the head of the narrow section of the canyon (where Isabella Dam would later be built). Water was about 50 feet deep in present-day Weldon near the Kern River Preserve. None of the accounts that were passed down speak of any fatalities in

178

Kernville. Presumably that was because the residents realized that when the river went dry, a flood would soon be following.

From Kernville, the floodwaters turned west and poured into Kern Canyon, down toward the village that would become Bakersfield. The massive wall of water, ice, uprooted trees, soil and rocks scoured the canyon walls. By the time it reached the mouth of the canyon, the floodwaters were 200 feet high.

It had rained continuously in Bakersfield for many weeks, having stopped only a few days earlier. One source said that some of the residents were living in thatched tule houses. The better homes like that of Colonel Thomas Baker were constructed from adobe, but with roofs made from thatched tules. The tule thatch was coated on the outside with mud. All the houses had dirt floors. Six weeks or so of rain trickling through the dirt on the roofs had created a miserable, muddy mess.

By Christmas Day, 1867, the Kern River had risen near the top of the levees, but no flooding had occurred. Volunteers constantly patrolled the river banks protecting against any signs of a breech in the levee system. The river then gradually receded until near the year's end, when the river flow supposedly stopped totally for two days. Apparently this was on or about December 30.

The river presumably didn't stop entirely, just decreased greatly in volume. Even with the landslide dam on the North Fork of the Kern, there would still have been significant water coming down from the smaller South Fork of the Kern. Presumably some of the residents took alarm over this unexpected reduction in the river's volume and realized its significance, just as they had in Kernville.

On New Year's Day, 1868, residents were awakened to loud roaring sounds that were accompanied by the earth trembling. When the 200-foot-high flood came out of the canyon, it spread out and dropped in height; but it was still quite impressive. The trees, boulders, ice and brush in the forefront of the flood created a 50-foot-high logjam near where the Chester Avenue Bridge now exists. This towering logjam dammed the channel and diverted the river north around the village and onward to Buena Vista Lake. Although the village was flooded by a foot of water on what is present-day Chester Avenue, the village escaped the majority of the flooding. By some measures, this was the most spectacular disaster to occur in Bakersfield prior to the 1952 earthquake.

Immediately east of Bakersfield, the flood left an uprooted cedar wedged into the boulders high upon the north wall at the canyon's mouth, about 200 feet above the riverbed (photograph on file in the national parks). That tree was about 40 feet long and four feet in diameter and was badly scoured and scraped from its long trip. It was deposited about 20 feet above the end of the powerhouse tunnel that would be built 30 years later. Local fishermen and hikers climbed the cliff to inspect that cedar log for years until the powerhouse's wooden flume was constructed in 1894. At that time, the cedar was cut up for use in that waterway.

After the flood, hundreds of thousands of logs were scattered over the south end of Kern County as far as Buena Vista and Tulare Lakes. The mesa where the Kern River Golf Course is now situated was covered with hundreds of large uprooted trees. Small logs covered an area half a mile square on Tom Barnes's ranch (later known as the Canfield Ranch) east of Elk Hills. Barnes had been born in North Carolina in 1827 and moved to what is now Kern County in 1859.

Two sources said that Colonel Baker built a sawmill to mill the large number of trees that had been washed down by the flood. The availability of lumber apparently greatly improved the quality of housing in the area. Another source questioned that because the sand in those logs would have caused problems with the saw blades.

In any case, Tom Barnes is known to have built a log cabin on his ranch in 1868. The walls were made from logs left by the flood. The lumber for the roof was milled wood from the mountains south of the San Emidio Ranch (located at the base of the mountains south of Bakersfield). The milled wood was hauled through Fort Tejon to his ranch by ox team. Apparently Bakersfield did not yet have a sawmill. Barnes's house is now an attraction at Kern County Museum's Pioneer Village.

Thanks to the logjam formed when the flood came out of the canyon, the river cut a new channel north of town. As a result, the town that would become Bakersfield then had to be relocated to be near that new channel. The present Kern River channel is the one that was cut during the 1867–68 flood.

Andrew C. Lawson visited the site of the two slides (Little Kern and Kern Lakes) in 1903 as part of a geological study of the Upper Kern.[504] Lawson talked to W.T. Grant who had visited the area in 1867 before the Kern Lake slide and then returned in 1868 after that slide had occurred.

According to the account in James Moore's book on the geology of the national parks, Lawson observed that the trees growing on the Little Kern Lake dam looked like they were less than 100 years old.[505] That seems reasonable. If the oldest trees had germinated in 1868 after the flood, they would have been 35 years old when Lawson visited the site in 1903.

Dating of those trees and the ones farther down the canyon might provide some interesting information. All the trees in areas that were exposed to the full force of the flood (typically the valley floor and the lower canyon walls) should date from 1868 or later. However, there should be a point on the side of the canyon wall — corresponding to the high-water mark of the flood — where those young trees would abruptly meet the older trees that survived the flood.

Steve Moffit, Sequoia National Park's former trails supervisor, traveled the section of the Kern Canyon below Kern Lake for over 20 years. He recalled that the evidence of the flood still looked fresh and near biblical in size. One of the most impressive sites is an area just north of the Grasshopper Flat campsite in Sequoia National Forest, a couple of miles below Little Kern Lake.

Some guidebooks refer to this area as being a floodplain. It is a stretch of relatively treeless land with long windrows of large, very uniform-size river boulders. In some places, these windrows are 12 or more feet tall and a hundred or more yards long. They are mostly parallel to the canyon, and therefore to the trail. As you ride through this maze of river rock, it is very impressive as being big, very, very different, and covering a large area.

When Steve first encountered these strange windrows, he thought that perhaps they were some type of odd parallel moraine. But after he learned of the Kern Lake flood story, it all fell together for him. If you stick a garden hose in a pile of dirt and gravel, it will tail out in windrows of uniform-size gravel according to the rate and volume of water flowing at that point and time. That appears to be the Grasshopper Flat floodplain on an enormous scale. Perhaps geologists have developed a different explanation, but that was Steve's deduction.

Presumably Andrew C. Lawson covered this flood and the Grasshopper Flat windrows in one of the two geologic studies he did on the Kern.[506, 507] Grasshopper Flat is a remote area of Sequoia National

Forest and has apparently not been visited in recent years by agency staff. It may be that no hydrologist has visited this site since Lawson was there over a century ago.

## 1869 Flood

Relatively little is known about this flood. It was not a major flood in the Sacramento River Basin. The USACE identified it as a major flood in the San Joaquin River Basin.[508] We have not found any record of how the flood affected the rivers within the Tulare Lake Basin in detail. Runoff during water year 1869 caused Tulare Lake to rise a very impressive 9.7 feet in elevation.

## 1870 Drought

The 1870 drought is not recognized at the state level. It was a severe drought in the Tulare Lake Basin. It is unknown when this drought began; it may have been a multi-year drought. Many cattle died. The losses to local cattlemen were so great that Tulare County supervisors asked the state board of equalization to make general reductions in the taxes of Tulare County. In August, the county supervisors made dramatic reductions in the assessed valuations of many farmers as a result of the drought.

## 1872 Flood

Relatively little is known about this flood. The USACE identified it as a major flood in the San Joaquin River Basin.[509] Moderate flooding was reported near Sacramento in 1871–72, but it was not considered a major flood in the Sacramento River Basin.

Water year 1872 was a very large runoff year, delivering an estimated 2.6 million acre-feet of water to Tulare Lake. Tulare Lake was at essentially full pool (elevation 207 feet) when the flood started. The runoff raised the lake 5.3 feet to elevation 212.3 feet.

Despite the huge volume, we have found no record of flooding along any of the rivers in the Tulare Lake Basin. Perhaps the record of this old flood has just been lost. Or perhaps it has to do with how the flood is viewed. Maybe the high runoff did little damage to Visalia and the other settlements on the delta areas. But down in the lakebed, floods brought the volume of water necessary to sustain the lake through periods of drought. From that perspective, the flood wasn't damaging at all, it was a boon.

## 1873–82 Drought

This drought was active somewhere in the state from 1873–82. In the Tulare Lake Basin, it is often referred to as beginning in 1877. There is general consensus that conditions didn't really start to improve until 1883.[510] Based on fragmentary data, the drought seems to have waxed and waned over the 1873–82 time period in the Tulare Lake Basin:
- A severe drought was reported for the Visalia area in 1873–75.
- A severe drought was reported for the Kern County area in 1876–77.
- David Campbell, an early Tulare County pioneer, recalled that 1877 was so dry that the grass did not even get started that year.[511]
- Kenzie Whitten "Blackhorse" Jones lived on the west side of Fresno County.[512] His valley farm failed during the terrible drought of 1877.[513]

- Morgan Blasingame of Millerton recalled the conditions on the San Joaquin: *The drought of 1877 was the reason the cattlemen started going to the mountains. A few sheep and cattlemen had been going earlier, but in 1877 conditions were so desperate that you either moved your cattle to the mountains — or they died.*[514]
- *Vanishing Landscapes* refers to the protracted drought of 1877–79.[515]
- Total precipitation in Visalia for water year 1878 was 104% of that city's long-term average (1878–1972).
- David Campbell recalled that 1879 was so dry that the grass did not even get started that year.[516]
- Total precipitation in Visalia for water year 1879 was only 39% of that city's long-term average.
- Total precipitation in Visalia for water year 1880 was 127% of that city's long-term average.
- Total precipitation in Visalia for water year 1881 was 116% of that city's long-term average.
- Total precipitation in Visalia for water year 1882 was only 67% of that city's long-term average.
- Total precipitation in Visalia for water year 1883 was 83% of that city's long-term average.
- On the legal side, lawsuits asserting riparian rights were fought out in Fresno courts in 1883 because of a dispute over Kings River irrigation water.
- On the not-so-legal side, a canal brush dam was blown up in 1883 because of a dispute over Kings River irrigation water.[517]

One source said that there were 18 diversions from the Kaweah River in 1875.

Referring to the Tule River sequoia forests after his hike through that area in 1875, John Muir later wrote:[518]

*All the basin was swept by swarms of hoofed locusts, the southern part over and over again, until not a leaf within reach was left on the wettest bogs, the outer edges of the thorniest chaparral beds, or even on the young conifers, which unless under the stress of dire famine, sheep never touch.*

A Sierra Club Bulletin from this era recommended carrying lemons on wilderness trips to hide the taste of sheep piss in mountain streams and lakes, specifically in the East Lake/Reflection area of what is now Kings Canyon National Park.

Forest fires were widespread in 1875. In the summer of 1875, the *Visalia Weekly Delta* carried the news that "Heavy fires are raging in the mountains east of here, (giving the appearance at night) of immense lanterns suspended from the heavens."[519] Tony Caprio, the national parks' fire ecologist, commonly finds burn scars for 1875. The fires extended at least as far south as the Green Horns, presenting what was described as "a grand sight at night."[520] Floyd Otter attributed the majority of those fires to sheepherders.[521]

During the period of this extended drought, there were three major floods (1874, 1876, and 1878) and up to five small to moderate size floods (1875, 1877, 1879, 1880, and 1881).

Table 20 presents inflow to the Tulare Lakebed for each year of the 1873–1882 drought.

Table 20. Inflow to the Tulare Lakebed during the 1873–1882 drought.

| Water Year | Inflow to Tulare Lake (thousand acre-feet) |
|---|---|
| 1873 | 400 |
| 1874 | 3,130 |
| 1875 | 115 |
| 1876 | 2,595 |
| 1877 | 315 |
| 1878 | 1,032 |
| 1879 | 58 |
| 1880 | 410 |
| 1881 | 120 |
| 1882 | 80 |

## Artesian Wells — Early Attempts

The first intentional attempt to dig an artesian well took place in the center of the intersection of Main and Court Streets in Visalia in 1859. That effort didn't hit artesian water, but it produced very good water and became the town well. Since the water table in much of the Tulare Lake Basin was high and ordinary wells were inexpensive, no further attempt was made to find artesian water for many years.

In 1877 the Southern Pacific Railroad brought in a successful artesian well at a depth of 310 feet, two miles south of Tipton on the arid west side of Tulare County. The railroad created a 40-acre oasis that came to be known as the Tree Ranch. An artificial lake was created, stocked with carp, and a boat was provided for the enjoyment of travelers along the dusty west side. Thousands of trees were grown there as was nursery stock that was then planted elsewhere along the railroad right-of-way. It was quite a novelty.

## Artesian Wells — Discovery of the Artesian Belt

Finally, in 1881, the intentional search for wells resumed. That attempt took place on the Paige & Morton Ranch, three miles west of Tulare. The Water Professor, A.P. Cromley, used a water witch to locate the well. Water "in grand abundance" was struck at 330 feet, just as the professor had prophesied. It produced a flow of about 800,000 gallons per day.

> *The Board of Supervisors visited the sparkling and wonderful fountain, and many people from the country round assembled, toasts were drunk and speeches made in its honor.*

The success of that well generated much excitement, inducing many others to search for artesian water. Some were successful, some were not. By trial and error, an area that came to be called the Artesian Belt was identified within a few years. It covered over one million acres.

The water from those wells was quite warm, warm enough to ward off frost. Mrs. Henry McGee, a pioneer of Buzzards Roost (on the northeast shore of Tulare Lake), recalled that the warm water made the top of the pipe a favorite play area for children, especially for little girls. And it served as a grand steam laundry.

The wells continued to flow until they all stopped for uncertain causes during the first decade of the 20th century. The reasons proposed for that stoppage included:

- Some said that the water table suddenly dropped. This does seem to be the probable cause. See the section of this document that describes Groundwater Overdraft for a discussion of why the water table was dropping in this area at this time.
- Draining of Tulare Lake.
- Underground rivers that suddenly ceased to be, or other less fanciful subterranean hydrologic changes. The 1906 San Francisco Earthquake was proposed as a possible causative factor of these unseen changes.

The 1892 Thompson Historic Atlas Map and a *Los Tulares* bulletin provide a much more detailed account of the fascinating history of artesian wells in the Tulare Lake Basin.[522, 523]

## 1874 Flood

This flood occurred during the early stages of the 1873–82 drought. It is relatively common in the Tulare Lake Basin for floods to occur during multi-year droughts.

Water year 1874 was a very large runoff year, delivering an estimated 3.1 million acre-feet of water to Tulare Lake. The lake was just below full pool (elevation 206.5 feet) when the flood started. The runoff raised the lake 6.0 feet to elevation 212.5 feet. Tulare Lake has not been this high since.

As with the 1872 flood, we have found no record of flooding along any of the rivers in the Tulare Lake Basin. It could be that little damage was done to Visalia and the other settlements on the delta areas, or it could be that we just haven't found the record of this old flood.

A map drawn in 1874 shows Tulare Lake as being nearly 700 square miles in size. This apparently reflected the condition when the lake was roughly 212 feet in elevation.

The 1874 flood in the Tulare Lake Basin is different from the huge storm that struck the San Francisco area on November 22–23, 1874. That storm dropped over 18 inches of rain on Fort Ross in 24 hours, with a recurrence interval of 32,000 years.[524]

## 1875 Flood

This flood occurred during the 1873–82 drought.

Hydraulic mining contributed to the flooding and shoaling of rivers in the Sacramento Valley. Marysville, located at the confluence of the Yuba and Feather Rivers, experienced a number of devastating floods from 1852 through 1875. Debris from the Malakoff mine choked the Yuba until the river bottom was higher than the town.

According to the Yuba County history, the 1875 flood was the greatest and most destructive flood to hit Marysville, even bigger than the 1861–62 flood. As a result of the previous floods, the city had spent an immense sum of money to surround itself with a seven-mile-long ring levee. Trusting to the levee, the residents did not take the precautions that they had in the previous floods. This proved their undoing. When the flood came, it swept everything before it. Even goods that were placed upon platforms above the traditional high-water mark were lost as the river continued to rise.[525]

For a week, heavy and incessant rain and snow storms prevailed, accompanied in some instances by thunder and lightning, an unusual phenomenon in the valley. Warm rain on a heavy snowpack is the typical recipe for valley floods. On the morning of January 19, the flood was threatening the city's levee. Despite frantic efforts to raise the low spots, the levee was overtopped before the end of the day. Many houses were abandoned that evening as the residents sought safety in large houses, churches, the courthouse, etc.

By noon on January 20, the water was three to ten feet deep in the streets. A strong current ran down the F Street slough to the Yuba River. The whole valley, including the city, was one vast sheet of water on a level with the rivers.

Enormous damage was done to the residences and to even the most substantial buildings of the city. The railroads were badly damaged, and in the country there was a great deal of destruction of stock and farm property. John Muir provided an eloquent description of the storm in *Flood-Storm in the Sierra* in the June 1875 issue of the *Overland Monthly*.[526]

The levee was later strengthened and Marysville has only been flooded three times since 1875. Another major rebuilding of the levees in the area was conducted from 2004–2008.

In the Tulare Lake Basin, rain began on January 15. By January 20, Visalia had received three inches of rain and the foothills east of town had received about 10 inches. The streams in the area were running higher than they had at any time since the 1867–68 flood, but no real damage had occurred as of that date.[527]

During the spring runoff of 1875, there was high water and minor flooding on the Kaweah.[528] Visalia incurred only minor to moderate flood damage from this flood. Presumably there was flooding on other rivers within the Tulare Lake Basin.

On the other hand, inflow to Tulare Lake in water year 1875 was relatively low, only 115 thousand acre-feet. That suggests that the spring runoff didn't last very long.

## 1876 Flood

This flood occurred during the 1873–82 drought.

The winter of 1876–77 was a strong El Niño event. That was presumably the cause of the 1876 flood since strong El Niño conditions typically result in an abundance of moisture in the Tulare Lake Basin. However, the 1876–77 El Niño affected the climate far beyond our area: it was a global event of major significance.

The extreme weather produced by this El Niño gave rise to the most deadly famines of the 19th century. Australia, New Zealand, and the South Pacific were in drought conditions during 1876–77. The drought was so severe in the Northern Hemisphere that up to 10 million died in India and 20 million died in China from malnutrition and drought-related diseases.[529]

Water year 1876 was a very large runoff year in the Tulare Lake Basin, delivering an estimated 2.6 million acre-feet of water to Tulare Lake. The lake was just below full pool (elevation 206.3 feet) when the flood started. The runoff raised the lake 5.4 feet to elevation 211.7 feet.

As with the 1872 and 1874 floods, we have found no record of flooding along any of the rivers in the Tulare Lake Basin. It could be that little damage was done, or it could be that we just haven't found the record of this old flood.

## 1877 Flood

This flood occurred during the 1873–82 drought.

During the spring runoff, there was high water on the Kaweah and probably on other rivers within the Tulare Lake Basin. The levee built to protect Visalia gave way, causing flooding.

That levee was presumably on the south bank of the St. Johns River. This is the earliest record we have of that levee failing. It would not be the last. See the section of this document that describes the St. Johns Levee — Condition in Recent Years, for an account of some of its continuing problems.

## 1878 Flood

This flood occurred during the 1873–82 drought.

The USACE identified it as a major flood in both the Sacramento River and the San Joaquin River Basins.[530]

The Zalda Canal had been constructed on the north side of the Kings River Delta in 1872. The 1878 flood enlarged that canal to form what is now known as the North Fork of the Kings River.

The 1878 flood filled Tulare Lake, bringing it to elevation 207.5 feet, causing it to spill through Summit Lake and into Fresno Slough and the San Joaquin River. "Eating" Smith chose that opportunity to bring the 32-foot-long schooner *Water Witch* (formerly the *Alcatraz*) from San Francisco to Fresno Slough. From there he had it loaded onto wagon beds and hauled overland to Tulare Lake.[531, 532]

Tulare Lake has never filled again since the 1878 flood. The *Water Witch* appears to have been the last boat of any significant size to have made it to the lake. Since 1878, the Tulare Lake Basin has functioned essentially as a closed basin, an inland sink without a regular outlet to the ocean.

## 1879 Flood

This flood occurred during the 1873–82 drought.

According to one source, there was minor flooding on the Kaweah River in 1879.[533] However, inflow to Tulare Lake in water year 1879 was very low, only 58 thousand acre-feet. If flooding actually occurred, it must have been localized and of a relatively minor nature.

## 1880 Flood

Flooding in 1880 occurred in April. This flood occurred during the 1873–82 drought.

The storm or storms and resulting floods appear to have affected much of the Central Valley.

Donner Summit set the U.S. record for the snowiest April ever with 298 inches (24.8 feet) of snow falling in one month.

A low-pressure area came ashore west of Red Bluff. The heaviest rainfall was located in a west-to-east band extending from Mt. St. Helena (near Napa) to Nevada City. Sacramento received 8.37 inches of rain during the two-day storm event of April 20–21. That was 5.79 standard deviations above the average with a recurrence interval of about 3,500 years.[534] Flooding was reported near Sacramento and likely occurred elsewhere in the Central Valley.

W.G. Pennebaker recalled that the Kaweah peaked on April 12, flooding Visalia and part of the surrounding countryside. It seems probable that this flood occurred at the same time as the big flood in Sacramento. So perhaps there was an editing typo at sometime, and the flood really peaked on April 21.

## 1881 Floods (2)

These floods occurred during the 1873–82 drought.

In the Sacramento River Basin, there were two floods:
1. one in mid-January
2. one that lasted from late January through early February

Outstanding flood peaks occurred in the upper part of the Sacramento River valley between January 14 and February 4. In the San Joaquin River and Tulare Lake Basins, the floods of 1881 were not of major proportions, although flood conditions were reported at some points in the San Joaquin River basin. There is evidence that the Sacramento River at Red Bluff and the Pit River reached peak discharges greater than in the 1861–62 flood. It is believed that the record-high stages on the lower Feather, Yuba, American, and Sacramento River reflected changed channel conditions, and that these stages were not the highest discharges of record.[535]

Based in part on research by the USGS, the USACE identified the 1881 flood as being a major flood in both the Sacramento River and the San Joaquin River Basins.[536] However, it was not a major flood in either the upper part of the San Joaquin River Basin or in the Tulare Lake Basin.

We have found no historical accounts of flooding on the rivers within the Tulare Lake Basin during 1881.

Inflow into Tulare Lake in water year 1881 was 120 thousand acre-feet, similar to water year 1875. That suggests that the flooding in 1881 might have been on the order of the relatively minor flooding that occurred in 1875.

## 1884 Floods (3)

There were at least three floods in 1884:
1. February
2. October
3. December

1883–84 was reported to have been an unusually wet winter. There was heavy runoff the following year throughout the San Joaquin River Basin including the Tulare Lake Basin. The USACE identified it as a major flood in the San Joaquin River Basin.[537] It was not a major flood in the Sacramento River Basin.

Floods threatened Fresno every winter because it was located in the sink of four creeks:
- Dry Creek ran just to the north of town.
- Red Banks and Dog Creek merged in the flat lands to the east.
- Fancher Creek ran nearby.

The center of Fresno was the confluent point for these four creeks.

On February 16, floodwaters covered every street in Fresno. All basements and ground floors were flooded. The only means of transportation within the city was by boat. The national parks files contain three photographs of this flood in Fresno:
1. This photograph depicts the situation on J street looking northwest from Fresno Street.
2. This photograph depicts the situation on M Street, south of Merced Street. The Fresno County Courthouse is seen in the background to the left. The church seen to the right is the Southern Methodist Church at the intersection of L and Fresno Streets.
3. This photograph depicts the situation on K Street looking north from Fresno Street. The Mill Ditch is visible in the foreground on Fresno Street. The steeple of the Methodist Church can be seen in the distance.

In the spring, the Kaweah flooded lowland farms on the delta. A brush and rock diversion weir had been constructed at McKay's Point in 1877. The flood damaged that structure so extensively that it had to be reconstructed in 1884.

A cloudburst occurred in Yokohl Valley at sometime in 1884. The valley was relatively well populated at the time. The Peter Stewart family lived about 10 miles up the Yokohl Valley from Merriman Station. In a few minutes, the storm turned the usually placid Yokohl Creek into the proverbial raging torrent. The flood swept away their small house, drowning Peter and his wife, his mother, their two children, and seriously injuring "Rat" Weisner. Some of the bodies were found as far away as Merriman.[538]

Parts of the Central Valley experienced a major storm in December 1884. The highest rainfall measurements of this storm were for Bowman Dam in the Tahoe National Forest where 33.8 inches was reported in six consecutive days. This is 4.94 standard deviations above the average with a recurrence interval of 1,900 years. This storm delivered over half of the average annual rainfall at that station during those six days.[539] This was a big and powerful storm, so it might have extended as far south as the Tulare Lake Basin. Panoche/Silver Creek west of Mendota flooded in December 1884.[540] Other than this record, we haven't found any account of flooding on area rivers.

Inflow to Tulare Lake in water year 1884 totaled 1.5 million acre-feet. This raised the level of the lake by 7.6 feet (from elevation 188.0 to 195.6).

## 1885 Flood

There were high floodwaters during the winter of 1885 along Cottonwood Creek on the route between Visalia and Badger. This was described by a Mrs. Lizzie in a talk that she gave at a picnic at Big Stump on August 27, 1950.[541] Presumably there were high floodwaters in other streams and rivers in the Tulare Lake Basin that winter as well.

Inflow to Tulare Lake in water year 1885 totaled 483 thousand acre-feet. This raised the level of the lake by 5.6 feet (from elevation 188.0 to 193.6).

## 1886 Flood

Relatively little is known about this flood. The USACE identified it as a major flood in the San Joaquin River Basin.[542] It was not a major flood in the Sacramento River Basin. How it affected the rivers within the Tulare Lake Basin is not known in detail. Runoff during water year 1886 caused Tulare Lake to rise 4.5 feet in elevation.

## 1887–88 Drought

This two-year drought is recognized at the state level. It definitely affected the Tulare Lake Basin.

In 1878, Tulare Lake had been at least 30 feet deep. The *New York Times* reported in its August 11, 1889, issue that the lake was less than three feet deep at the deepest part. (S.T. Harding would later estimate the minimum depth in 1889 to be 4.5 feet (183.5–179 feet).[543]

The Kings River dried up in 1889 below the mouth of the Fresno irrigation system. Those two events were presumably due in part to the 1887–88 drought.

## 1888 Flood

This flood occurred either during or right at the end of the 1887–88 drought.

We know nothing about this flood except for two photographs from Fresno (on file in the national parks):
1. The flooded Union Ice Company warehouse located just south of the Southern Pacific Railroad tracks. The photographer was on G Street looking northeast toward Kern. The sign in the background says "Planing Mill," probably Madary's Planing Mill located on the corner of H Street and Kern.
2. Shows the damage along Inyo Street. The Arlington Hotel is seen on the right at the intersection of Inyo and J streets. In the distance is probably a freight warehouse of the Southern Pacific Railroad.

## 1889 Flood

Relatively little is known about this flood. Based on research by the USGS, the USACE identified it as a major flood in both the Sacramento River and the San Joaquin River Basins.[544] While this may have affected the San Joaquin River Basin, it does not appear to have had much of a presence in the Tulare Lake Basin. It had no net effect on the elevation of Tulare Lake.

## 1889–90 Floods (2)

There were two flood events in 1890:
1. a big rain flood from December 1889 through February 1890
2. a moderate snowmelt flood during May and June

The USACE identified the January event as a major flood in both the Sacramento River and the San Joaquin River Basins.[545] Accounts of the May/June flooding episode are somewhat vague.

Statewide, the two wettest water years during historic times were 1890 and 1983. The heavy rainfalls of water year 1890 (the winter season of 1889–90) were confined to the northern half of the state.[546] It was reported to have been an unusually wet winter throughout the Sierra.

The winter season of 1889–90 was remarkable for exceptionally heavy precipitation in the Central Valley Basin which produced floods of considerable magnitude from December 1889 – February 1890. Heavy snowfall in the Sierra resulted in unusually high runoff from melting snow during May and June, 1890. The floods were relatively greater in the San Joaquin River Basin, and they are considered to rank as the largest in that area for the period between the floods of 1867–68 and those of 1907.[547]

The Transcontinental Railroad had been completed in 1869. In January 1890, a relentless barrage of blizzards and a derailed train shut down Donner Pass for 15 days. In addition to an armada of snowplows and railroad crews, nearly 5,000 snow shovelers were hired to help clear the tracks, but the 66 feet of snow that fell on the pass that winter overwhelmed their efforts. This nearly stymied the attempt by journalist Nellie Bly to circumnavigate the globe in less time than novelist Jules Verne's fictional voyage *Around the World in 80 Days*. But Nellie was a very determined woman. With the help of the Central Pacific, she made it back to New York City on January 25, 1890, having traveled 72 days, 6 hours, and 11 minutes in her epic, planet-circling journey.

The winter of the 1890 water year was remarkable for the exceptionally heavy and widespread precipitation that produced floods of considerable magnitude throughout Northern California in January and February 1890 and moderate floods at other times from December 1889 through May 1890. The winter season of 1889 to 1890 featured an exceptionally heavy snowfall in the mountains, and the snow runoff period was one of the heaviest and longest of record. Lowlands in the lower Sacramento River Basin were flooded for many weeks. In December 1889, the Sacramento River reached flood stages from Tehama to Sacramento. The peak stages on the river at Colusa and Sacramento were the highest yet observed. However, these high stages were primarily due to reclamation work along the river.

There were many breaks in the levees from Colusa downstream, and considerable damage was done to grain lands. A large break on the right bank levee of the Sacramento River below Sacramento helped to reduce subsequent flood stages. In January 1890, the tributaries of the Sacramento River were again at high stages. Stony and Putah creeks were reported to have been at the highest stages known to local residents. Considerable overflow from Cache Creek near Yolo flooded farms and caused washouts along the railroad. In February 1890, a flood occurred on the upper Sacramento River. The Sacramento River at Redding washed out part of a bridge.[548]

The January flood resulted from a warm rain falling on this heavy snowpack. The San Joaquin River flooded, enveloping Stockton and other valley communities.[549] Large floods occurred throughout the San Joaquin River Basin during the latter part of January. The upper San Joaquin River possibly reached an extremely high stage. The Merced, Stanislaus, Tuolumne, and Mokelumne rivers were at dangerously high stages, and some of the foothill tributaries of these rivers were reported to have been at the highest known stages to date. Several towns were flooded and railroad and highway structures washed out. The maximum stage of the season, however, was reached, at least on the lower San Joaquin River, during the snow runoff period in May 1890.[550]

The levee protecting Visalia had not received regular maintenance in a number of years. (This was presumably the south levee on the St. Johns River.) As a result, that levee failed near the end of January, resulting in widespread flooding in Visalia that lasted for part of one day.[551] The flood wiped out railroad tracks near the town. Photographs of Main and Court Streets indicate that floodwaters were running about a foot deep through town. Travel on Main Street was said to have been by boat.

It seems probable that the Kings and Kern Rivers were flooding at the same time, but no accounts of that have yet been found.

There was heavy runoff in 1890 on all the rivers within the Tulare Lake Basin. Presumably this happened during the April-July snowmelt period, but we haven't seen any documentation as to the actual timing. Inflow to Tulare Lake in water year 1890 totaled 2.0 million acre-feet. This raised the level of the lake by 12.0 feet (from elevation 183.5 to 195.5).

## 1893 Flood

Flooding in 1893 occurred in February.

Relatively little is known about this flood. The USACE identified it as a major flood in the San Joaquin River Basin.[552] It was not a major flood in the Sacramento River Basin. How it affected the rivers within the Tulare Lake Basin is unclear.

Apparently flooding occurred throughout the Tulare Lake Basin, although the reports are fragmentary.

One source said that major flooding occurred along the Kaweah in February. Visalia was certainly flooded. Although the 1893 flood may have been perceived as a major flood in Visalia and elsewhere on the Kaweah Delta, it wasn't a heavy runoff year. The level of Tulare Lake increased only 2.9 feet that year. So by that measure, it was not a major flood. It appears that the total runoff for the year was only average, but there was a rain event in February that caused flooding on the delta.

Presumably there was flooding on the Kings in February, but we haven't discovered any reports to that effect.

The Kern River overran its banks on February 10 due to melting snow and heavy rains. It didn't quite get into downtown Bakersfield, but it came a block from 19th and Chester Ave. The Kern Valley Nursery was one of the areas damaged.

Troop B, Fourth Cavalry arrived in Three Rivers on June 20. The snow had been so heavy that the Mineral King Road was still blocked with snow. Many of the streams were running so high as to be impassable.

## 1894 Flood

Flooding in 1894 occurred in February. It was a major flood in some portions of California. It may have affected the Tulare Lake Basin, but we haven't seen any reference to it.

The cavalry reported that the winter season ended exceptionally early. As a result, sheep had entered General Grant National Park and the northern part of Sequoia National Park in considerable numbers. At least 300 of those sheep perished in a late snowstorm.

## 1895 Flood

Flooding in 1895 occurred in January. It was a major flood in some portions of California. It may have affected the Tulare Lake Basin, but we haven't seen any reference to it.

Troop I, Fourth Cavalry arrived in Three Rivers on May 23. Many of the streams were running so high as to be impassable. On June 4, the runoff was still so high as to prevent the cavalry from going up the South Fork of the Kaweah to Hockett Meadows. It was July 10 before the Mineral King Road became passable.

A brush and rock diversion weir had been reconstructed at McKay's Point on the Kaweah River after the 1884 flood. It had to be reconstructed again in 1897. Some flood, perhaps the 1895 flood, had evidently damaged it so extensively that repairs alone were insufficient.

Tulare Lake rose four feet in 1895, indicating that there was higher than average runoff.

## 1897–1900 Drought

This four-year drought is recognized as affecting at least part of the state, and it certainly affected the Tulare Lake Basin.

Fresno received 0.30 inch of rain on April 1, 1897, but then the rains stopped. It was a cruel April Fools' joke. That would set a record as the earliest occurrence of the last measurable rain for the water year.[553] In 1924, the *Visalia Morning Delta* referred to 1897–1898 as being a dry year in Tulare County.[554] Floyd Otter said that valley ranchers drove their hogs to the high mountains in the drought years of 1897–99.[555]

In 1878, Tulare Lake had filled (elevation 207.5 feet) and spilled through Summit Lake and into the Fresno Slough and the San Joaquin River. But in 1883, just five years later, the lake had reached its lowest level in memory. With many minor fluctuations (and one big flood, the flood of 1890), the lake gradually dwindled in size over the next two decades.

The *New York Times* reported in August 1898 when Tulare Lake dried up completely.[556] This was the first time that had happened in historic times. It had gone from full-pool to bone-dry in just 20 years (1878–98). See the section of this document on the Chronology of Tulare Lake for the various causes that contributed to this drying.

The lakebed would remain dry through 1900.

## 1898 Flood

Flooding in 1898 occurred in September. This flood occurred during the 1897–1900 drought.

The storm occurred on September 26 and was centered on the town of Tulare. It was apparently an isolated low-elevation event. The atmospheric mechanism behind the storm is unknown. It could have been a thunderstorm embedded in a tropical storm remnant.

Tulare received 3.89 inches of rain during the storm event. That was 5.66 standard deviations above the average with a recurrence interval of 2,700 years. Several other stations in the vicinity had over 3 inches of rain on September 26. It was the wettest day ever at Dinuba.[557]

This storm almost surely resulted in localized flooding, but we haven't seen any accounts to that effect.

## 1901 Flood

We haven't discovered any account of flooding this year. However, flow for the water year was 181% of the 1894–2011 average for the Kings, 170% for the Kaweah, 125% for the Tule, and 119% for the Kern. This suggests that some flooding in the delta areas seems possible, especially in the northern portion of the Tulare Lake Basin.

The Tulare Lakebed had dried completely for the first time in August 1898 and remained dry through 1900. The 1901 flood brought the lake back to life, if only modestly. Inflow to Tulare Lake in water year 1901 totaled 408 thousand acre-feet. That left the lake six feet deep (elevation 185.5 feet) at its deepest point.

## 1906 Floods (5)

Depending on how you count them, there were at least five floods in 1906:
1. January
2. March (2)
3. May
4. June

The January to March period might best be viewed as a more or less continuous series of flood events rather than as individual floods.

Likewise, the May to June period might also best be viewed as an extended snowmelt runoff flood with multiple peaks.

The winter of 1905–06 was a strong El Niño event. As shown in Table 21, precipitation during water year 1906 was heavier than average throughout the valley.

Table 21. Precipitation during water year 1906.

| Reporting Station | Total Precipitation (inches of rain) | Percent of Mean |
|---|---|---|
| Fresno | 13.54 | 140% |
| Visalia | 13.85 | 143% |
| Hanford | 11.65 | 142% |
| Tulare | 14.78 | 178% |
| Porterville | 17.82 | 170% |
| Bakersfield | 8.72 | 148% |
| Kernville | 16.26 | 158% |

A massive snowpack accumulated in both the San Joaquin River and Tulare Lake Basins during the winter of 1905–06. There were only a limited number of weather stations in the snow country at that time to measure the snow directly. The California Cooperative Snow Surveys wouldn't begin until 1930.

The winter of 1905–06 appears to be the snowpack of record in the Southern Sierra. After the heavy winter of 1951–52, the USGS reexamined available records of snowfall at stations in the San Joaquin River and Tulare Lake Basins from the winter of 1905–06.[558] See the section of this document that describes the 1952 flood for more detail about the results of that study.

Based on the limited data available, the USGS study concluded that the 1952 snowpack appeared to equal or exceed the snowpack that caused the great snowmelt floods of 1906. However, the only way to be certain was to wait until the snowpack melted and ran off. The results turned out to be quite clear, and rather surprising. In no case where the period of record included the year 1906, did the 1952 snowmelt runoff exceed that of 1906 on any river. That confirmed that all those watersheds had a bigger snowpack in 1906 than in 1952.

The difference was particularly remarkable in the Kaweah River Basin. The winter of 1951–52 set a modern-day snowfall record at Lodgepole, one that would last until the winter of 2010–11. However, as shown in Table 38 on page 257, the snowmelt runoff in the Kaweah River Basin in 1906 was 38% greater than in 1952.

Table 22 gives the total runoff in 1906 for each of the major rivers in the Tulare Lake Basin.

Table 22. Snowmelt runoff in 1906.

| River | Total Runoff April 1 – July 31 (thousand acre-feet) | Maximum daily flow Date | (cfs) |
|---|---|---|---|
| Kings River | 2,980 | June 20, 1906 | 24,900 |
| Kaweah River | 814 | May 28, 1906 | 7,260 |
| Tule River | 192 | | |
| Kern River | 1,390 | June 21, 1906 | 9,500 |

The first two big winter storms of 1906 that we know of occurred on January 12–13 and February 9. Heavy rain fell on Friday and Saturday, January 12 and 13. The rain was reported to be exceedingly heavy in the foothills and mountains east of Visalia.[559] Table 23 provides the total precipitation for the January 12–13 storm event for selected reporting stations.

Table 23. Precipitation during the January 12–13, 1906 storm event.

| Reporting Station | Storm Total (inches of rain) |
|---|---|
| Visalia | 2 |
| Eshom Valley | 7 |
| Kaweah #1 Powerhouse in Three Rivers | 8 |

According to official NWS records, it initially looked like the February 9 storm was a storm of near-epic proportions in Bakersfield, dropping a total of 4.1 inches on the city that day. However, the National Climatic Data Center (NCDC) has since questioned the 1906 records for that city. Several transcription errors have been noted, including the use of non-standard annotations. For example, the entry for February 9 was written as 4-10 instead of 4/10. So it appears that a monster storm didn't occur in Bakersfield on February 9. That explains why there wasn't a major flood reported in that month.

Bakersfield still experienced a storm on February 9, and it was the biggest storm of the month. However, total precipitation in the city that day was 0.4 inches. It's reasonable to believe that the storm dropped a lot of snow up in the conifer zone, but we have little data for making an estimate of the amount.

Likewise, it originally looked like Bakersfield set a record with an outstanding total of 8.70 inches of precipitation during the month of February. However, that record turned out to be due to the non-standard annotations. The corrected total was a much more modest 0.70 inch, 57% of which had fallen in the big February 9 storm.

There was no weather station in the national parks in 1906. The Fourth Cavalry detachment wouldn't arrive in the park until June. However, there were four civilian rangers employed year-round. Walter Fry was the chief of those rangers. One of the other rangers, Charlie Blossom, kept a dairy, and his diary sheds some light on weather conditions in the parks during 1906.

Charlie's patrol district included the South Fork of the Kaweah. He noted many days of heavy rain from January into May. The El Niño conditions resulted in an unprecedented snowpack in the Southern Sierra. On June 26, Charlie traveled from Hockett Meadows to where the South Fork Campground is today. That was a 24-mile day, 6 miles of which were over packed snow still 6 to 8 feet deep.

Those were far from normal trail conditions for the end of June. Something truly aberrant had happened. It was almost as if the Southern Sierra had been thrown back into the Little Ice Age.

That was just a hint of the phenomenal snowpack that had accumulated in the Southern Sierra that winter. There was no formal weather gaging station to make direct measurement of the national parks' snowfall, but the parks do have measurements of the compacted snowpack.

Walter Fry reported that the winter of 1905–06 brought the heaviest snowfall in the sequoia groves of both Sequoia and General Grant National Parks ever recorded to that date. The snow was 29 feet (equivalent to roughly 348 inches) on the level in Giant Forest on February 25, 1906. Even by June 25, the snowpack in Giant Forest had only melted down to about 12 feet on the level.[560]

Conditions like that have not been seen since. Based on Fry's report, the winter of 1905–06 far exceeded the record-setting winter of 2010–2011 or the winters of 1951–52, 1968–69 and 1982–83.

There is some precedent for this much snow falling in the Sierra. Donner Summit received 298 inches (24.8 feet) in April 1880.

Walter Fry is generally regarded as a very reliable source, but his report of 29 feet of snow in February is astounding. (Not to mention the 12 feet remaining on June 25.) It does seem generally consistent with the USGS finding that the Kaweah River Basin's snowpack in 1905–06 was 38% bigger than the huge snowpack of 1951–52. And it fits with the big snowmelt floods that occurred down in the valley in 1906.

But it is still difficult for us to comprehend the sheer magnitude of the snowpack that accumulated in the winter of 1905–06. The depth of snow in the national parks was double any that has occurred since. The deepest the snow has ever been measured at Giant Forest (including snow surveys) was 151 inches on March 16, 1952.

From Fry's report, we know that there was roughly 348 inches of snow in Giant Forest in the Southern Sierra. Usually there would be a greater — or at least a comparable — depth of snow in the Northern Sierra. As a comparison, the snow survey site at Lower Lassen Peak (which is easily, year after year, the deepest snow in California) has only exceeded 300 inches once (in March 1983). But that was not the case in 1906; there was a complete disconnect between the Southern Sierra and the Northern Sierra that winter.

Thanks to the USGS study, we know that the record-setting snowpack that formed in the winter of 1905–06 affected most or all of the watersheds of the San Joaquin River and Tulare Lake Basins. However, it apparently did not affect the Northern Sierra, and it did not affect the area immediately west of Tahoe. Bob Meadows researched the available data at some of the other sites in the Sierra in 1906, but none of them showed highly elevated snow depths. At the two sites west of Tahoe (Blue Canyon and Emigrant Gap), the snowpack was 30 inches or less at the end of February 1906.

A phenomenal amount of snow had been delivered to the Southern Sierra by the end of February 1906. How to account for the 29 feet of snow in Giant Forest? Based on what we know, it appears that a substantial portion of it might be accounted for by just one huge storm. The January 12–13 storm brought "exceedingly heavy rain" to the foothills east of Visalia, including 7 inches of rain at Eshom Valley and nearly 8 inches in Three Rivers (see Table 23).

That storm would have delivered significantly greater levels of moisture to the conifer belt of General Grant and Sequoia National Parks. Based on a study conducted by USGS, Giant Forest averages about 2.4 times as much moisture from winter storms as Three Rivers, at least in November (see Table 34 on page 245).

If that ratio held true for this storm, then the 8 inches of rain that Three Rivers received might have resulted in 19 inches or so of moisture at Giant Forest, which would have been upwards of 19 feet of snowfall. (Snowfall to rain ratios vary from about 6:1 to 15:1 in the Southern Sierra.) Gary Sanger said that if the core of the storm moved over Giant Forest, snow production could have been greater than this. Otherwise, it would likely have been less, especially in a convective storm.

The snowpack would have peaked in April. Based on available information, the Sierra has never seen a greater snowpack in historical times. An interesting related detail is that the San Francisco Earthquake occurred on April 18. At what was very close to the moment of peak snowpack during this very exceptional year, the Sierra was shaken by a very powerful earthquake (magnitude 7.8). That event was centered north of San Francisco, but was strongly felt in the Tulare Lake Basin. The earthquake apparently triggered a very strong cycle of avalanches in our local mountains. Historical records document that many of the buildings remaining in the Mineral King Valley from the 1870s silver rush, including the Smith Hotel, were destroyed by those avalanches.

As incredible as the winter of 1905–06 was in the national parks, an even bigger surprise would be coming the following winter, some 200 miles to the north. The year after the national parks got its phenomenal snow dump, the winter of 1906–07 would bring 884 inches (74 feet) of snowfall to Tamarack, California (that's snowfall, not snowpack). That station is located at 8,000 feet elevation, about 20 miles northeast of Calaveras Big Trees State Park. The Tamarack station was only established at the beginning of the 1906–07 season. Think what it might have recorded if it had been established a year earlier.

The January 1906 flood was mild enough, just a nice way to break the drought. It resulted from the very heavy rain that fell in the foothills on January 12–13. The resulting floodwaters arrived suddenly, but didn't last very long. (The short duration of the flood suggests that precipitation at the mid- to upper elevations probably came as snow and didn't contribute to the flood.) Virtually all the streams and rivers in the Kaweah and Tule River Basins flooded. Hundreds of acres were inundated. Adjacent watersheds presumably flooded as well.

The heaviest damage occurred on the St. Johns River north of Visalia. At 8:00 a.m. on Sunday, following the two days of rain, the St. Johns was still a dry sandy streak. But one hour later, the river was running bank-full. The rapidly growing flood soon proved too big to pass through the culvert under the Santa Fe track. The bed of the track gave way, creating a break of thirty feet. (Trains had to be detoured via Hanford until repairs could be made.) All of the residences east of the Santa Fe were surrounded by a small sea of water.[561]

Visalia's China Town was also flooded. China Town generally encompassed an area bounded by East Main on the south and Oak Street on the north, Santa Fe on the east and Bridge Street on the west. Newspaper accounts of the day said that the Celestials moved their possessions from their basements up above flood level.

("Celestial" was a term used to describe Chinese emigrants to the United States, Canada and Australia during the 19th century. The term was widely used in the popular mass media of the day. It was not a disparaging term. It was an attempt to translate a classical term in Chinese by which the emigrants referred to themselves in dealing with the non-Chinese. China was often called the "Celestial Kingdom.")

The Tulare Irrigation District's large flume across the St. Johns River near Venice Hill was swept away.

The Santa Fe track two miles south of Visalia was under water for a considerable length of time. The land near the power company's substation east of Visalia was under water. Considerable damage was done to the Southern Pacific's bridge across the Tule River south of Porterville. In general, this heavy rain was considered a great blessing for farmers and stockmen; it had been dry for way too long. But if they had been praying for an end to the dry season, their wish was about to be granted.[562]

Charles Tollerton, a Dinuba pioneer, recalled that it rained every day in February. (Anecdotal accounts such as this should generally be taken with a grain of salt. Bob Meadows checked the weather logs for the nearby town of Reedley and discovered that it actually only rained on eight days in February for a total of 2.28 inches.)

Sand Creek was a mile wide in the Orosi district for weeks. People in the northern section of the country couldn't get to Visalia. A railroad engine bogged down in Monson (southwest of Orosi) for two or three days, and everything along Sand Creek and the Cottonwood Creek channel was a lake.

Flooding occurred multiple times from March through June. It was a major flood on the Kings, Kaweah, and Kern Rivers. In the Tulare Lake Basin, the flood of 1906 would be recalled as the big one for the next three decades (until the December 1937 flood).

Troop B, Fourth Cavalry, arrived in the village of Kaweah on June 3. The preceding winter had been exceptional in the amount of snow and rainfall, resulting in considerable damage to roads and trails. The Colony Mill Road was impassable due to landslides. Most of the national parks' trails were impassable due to snow and fallen trees.

The channel of Mill Creek through Visalia had been deepened, straightened, and covered with planks in 1891. That had been more or less adequate to carry runoff until the 1906 flood. But the Mill Creek channel overflowed four times that year:
1. March 16–20. Floodwaters reached a maximum depth of four feet in parts of the city.
2. March 25–28. Less severe than the previous flood.
3. May 29. Caused by snowmelt. Lasted one day and covered only a small portion of the city.
4. June 13. Resulted from an unusually hot spell. Pronounced cooling saved the city from this being an even greater flood.

The Kaweah's peak natural flow occurred at McKay's Point on March 15: 12,749 cfs. (That was the peak hourly flow; the peak average daily flow, as reflected in Table 15, was 8,861 cfs.) Based on the flood exceedence rates in Table 16, this had a recurrence interval of 6 years for the Kaweah.

The June flood was apparently the most impressive of the four. Water flowed down both Center and Oak Streets in what the *Visalia Times-Delta* described as a river running past the Tulare County Courthouse (located on Court Street between Oak and Center).

Eastman had just popularized the camera, so this was the best documented flood yet in Visalia (multiple photographs on file in the national parks). A steam engine of the Southern Pacific Railroad was photographed sitting in about a foot of water on Oak Street near Church Street. Boats became a commonplace means of transportation in the downtown area. One account of this flooding episode read:

> *In 1906, Visalia had a bad flood. My father and mother lived out at Venice Cove here*
> *and he and his brother built a boat. They went to put in the river and then right where the*

*break was, up about where Ben Maddox is, they got out and went down Main and Court
Street. And the only place they got stuck was about a block from Center Street. Somebody
threw a horse line to them and pulled them over this low spot and they went on down
Main Street in a boat.*

As a result of the 1906 flooding, Visalia explored two options for diverting a portion of the
floodwaters from Mill Creek into Packwood Creek. It isn't apparent that either of those options was
pursued. The main response to the flood was the construction of a new and bigger, half-mile-long,
concrete aqueduct/conduit in Visalia in 1910.

South of Mill Creek are the drainages of Packwood and Cameron Creeks. Waukena was a small
farming community on Cameron Creek (a distributary of Deep Creek) along the northeast shore of
Tulare Lake. This area was inundated in the 1906 flood, killing some of the orchards.[563]

The North Fork Bridge (aka the Three Rivers Bridge) was constructed across the mainstem of the
Kaweah in Three Rivers after the flood. Presumably one of the 1906 floods (probably the June flood)
destroyed whatever bridge had previously crossed that river. The new bridge was a post and timber
bridge anchored by four steel cylinders filled with concrete. It remained in use until washed away in
the December 1937 flood.[564]

The Tule River is the next drainage south of Cameron Creek. The Tule caused serious flooding in
Porterville during the 1906 flood. One woman recalled going to the Opera House to fetch the tables
and chairs, only to find them floating near the ceiling.[565]

Orlando Barton was the superintendent of the Devil's Den Oil Company and traveled by bicycle
between Devil's Den and Visalia. He recorded a vivid and detailed account of the flooding on the
lower section of the Kern River. The floodwaters reached Sand Ridge on June 7, 1906. But instead of
passing through the natural gap in the ridge as expected, the floodwaters began pooling into Ton
Taché, the southern part of Tulare Lake. Three weeks later, that lake was 7 miles wide and 15 miles
long. Finally, on June 20, the floodwaters began to break through the gap and flow on toward what
we now think of as Tulare Lake proper.[566]

As Tulare Lake filled up, Orlando continued his observations like a field naturalist:[567]

*Hay will float on water. Two stacks arrived at George Scherin's ranch on the south shore of
the lake last week. One of them has about 20 tons of barley hay in it. Neither of the stacks are
much out of shape after their cruise from the north shore.*

Total flow for water year 1906 was 225% of the 1894–2011 average for the Kings, 256% for the
Kaweah, 336% for the Tule, and 252% for the Kern. It was the third wettest year ever recorded for
the Kings, Kaweah, and Tule, and the fourth ever for the Kern.

Tulare Lake was virtually dry when Hobart Whitley visited it in 1905. However, the high runoff of
the 1906 flood brought the lake back. The flood left the lake about 12 feet deep at the deepest point
(elevation 191 - 179 feet), submerging 300 square miles. (This compares to 790 square miles at its
maximum in 1862 and 1868.) As a relative measure of the volume of the runoff, that was the biggest
increase in the lake's depth since the 1890 flood.

One authoritative source gives the combined runoff of the four rivers in the Tulare Lake Basin during water year 1906 as 7,360,000 acre-feet, the second largest runoff of record. However, that is based on outdated information. The total runoff for the four rivers in water year 1906 was 7,195,240 acre-feet. (The runoff in water years 1969 and 1983 would be substantially larger: 8,379,585 and 8,746,222 acre-feet, respectively.) The total floodwater entering the Tulare Lakebed in water year 1906 was about 1,530,000 acre-feet. As detailed in Table 5, this inflow exceeded that of any year since that time.[568]

Many levees were constructed in the Tulare Lakebed between 1903 and 1905 when lake levels were low. Unfortunately, those levees were light and poorly constructed. As a result, they failed when the high flows of 1906 entered the lake. The failure of those levees resulted in large financial losses, as almost 175,000 acres of wheat and barley had been planted that year. Most of that land was flooded before the crops could be harvested.

Although the 1906 flood was a disaster for lakebed farmers, others saw a silver lining. A number of recreation pleasure craft boats were quickly built or purchased for use on Tulare Lake that summer. The boats had gasoline engines with one owner bragging that his "lightweight" engine would weigh only 110 pounds.[569]

The series of flooding events that occurred in the Tulare Lake Basin in 1906 was unrelated to the intense storm that occurred just to our north on December 11, 1906. That storm covered a narrow band extending in a northeasterly direction from Monterey to Ione in the Sierra foothills north of Sonora. On December 11, Forest Lake on the 17-Mile Drive in Pacific Grove recorded 6.07 inches during that storm event. This was 6.38 standard deviations above the average with a recurrence interval of 9,000 years.[570]

## 1907 Floods (2)

There were two floods in 1907:
1. March
2. July

This was the second wet winter in a row. On January 17, 1907, Yosemite Valley reported a 60-inch snowpack, the greatest ever measured on the ground.[571]

Tamarack, California (about 20 miles northeast of Calaveras Big Trees State Park), set a seasonal snowfall record for the Sierra, recording 884 inches (74 feet) of snow falling during the 1906–07 season.

The March 1907 event was a very destructive flood in the Sacramento River Basin. It was caused by a severe rain from March 16–20 followed by a period of comparatively high runoff. Stages were exceptionally high throughout that basin. On the Feather River at Oroville, the flood height was the greatest ever observed, although it was believed that the riverbed at that location had been raised since 1862 by deposition of mining debris. Flooding on the Sacramento River system was so extensive that this event has been cited as the inspiration for construction of a 1,100-mile system of levees and dams for flood control.

This flood was also significant in the San Joaquin River Basin. Only a moderate rise on the upper San Joaquin River was observed during this flood, but there were exceptionally high stages on the large tributaries in the lower part of the basin. From the Merced River to the Mokelumne River, stages peaked on March 19, and were followed by high stages for several days. The San Joaquin River downstream from Mendota was at or above flood stage from the middle to the end of March.

The California Water Plan calls out the 1907 flood as being one of the major floods in the San Joaquin River Basin. Presumably that was in reference to the March flood.

It is not clear to what extent the 1907 floods affected the Tulare Lake Basin. Most of what we know about the flood impacts comes from the valley lakebeds. Based on Tulare Lake elevation data, the March flood appears to have delivered far more water to the lake than the July flood did.

Total flow for water year 1907 was 159% of the 1894–2011 average for the Kings, 139% for the Kaweah, 149% for the Tule, and 144% for the Kern.

On July 3, the levee that constrained Buena Vista Lake failed. This event was written up in the *New York Times*. The resulting flood inundated 25,000–30,000 acres south and west of Bakersfield including the old bed of Kern Lake. It damaged twelve miles of the Sunset Railroad. The levee that failed was supposedly built in 1866–67. If so, it would presumably have been built by Colonel Baker.

Tulare Lake had come back to life in 1906 and it expanded even more in the 1907 floods. Total inflow to the lake in water year 1907 was 977,000 acre-feet, raising the elevation 6.7 feet.

In 1907, a massive levee was built around the four sides of Tulare Lake, containing it to a fraction of its full natural size. Ripley's *Believe It or Not* featured the "Square Lake" in its syndicated cartoon. The lake was now harnessed, and the former lakebed was declared safe for growing orchards.

## 1909 Flood

There were at least two floods in 1909:
1. January–February
2. December

The heavy rainfalls associated with the storm sequence of January 1–20, 1909 extended in a southwest to northeast direction from Fort Ross near San Francisco to Greenville in the Feather River Basin. Nine stations reported their highest-ever rainfall totals for 20 consecutive days. La Porte in the Feather River Basin had 57.41 inches during the 20-day storm event which was 5.38 standard deviations above the average. The associated recurrence interval is 12,000 years. The Sacramento River at Red Bluff responded to the heavy rainfall with a flood crest of 30.5 feet on February 3, 1909, one foot higher than the previous record of 29.5 feet on February 4, 1881.[572]

The January flood was written up in the *New York Times*. It was a major flood in both the Sacramento River and San Joaquin River Basins. The flood of 1909 is believed to have been as great as that of March 1907, at least in the Sacramento River Basin.[573]

During January 1909, flooding occurred at several places in the Sacramento River Valley from Red Bluff to the mouth of the Sacramento River. The Sacramento River reached high stages at Red Bluff in January and continued to rise into the beginning of February. The Sacramento River at Red Bluff reached a peak stage that was the highest yet observed. The lower river at Sacramento reached the maximum stage of record in the middle of January, and exceptionally high stages were recorded on nearly all the main tributaries to the river. Flood conditions prevailed in the lower basin through the end of the month. However, damaging floods occurred again in the beginning of February. The floods of 1909 were the most disastrous of any for which there is an authentic account, although it is believed that the flood discharge from the Sacramento River Basin in 1862 was probably far greater than the discharge from the floods of 1907 or 1909.[574]

It was a major flood along the Kaweah and the Tule Rivers and presumably on the Kings and Kern as well. Levees failed at both Visalia and Porterville.

The Kaweah's peak natural flow occurred at McKay's Point on January 21: 12,227 cfs. (That was the peak hourly flow; the peak average daily flow, as reflected in Table 15, was 9,578 cfs.) Based on the flood exceedence rates in Table 16, this had a recurrence interval of 7 years for the Kaweah.

The brush and rock diversion weir at McKay's Point on the Kaweah River had been reconstructed in 1897. It was apparently severely damaged in the 1909 flood and required reconstruction yet again. This would be the fourth time in less than forty years that the structure had to be replaced. This time it was rebuilt using concrete instead of brush and rock. The replacement structure would last until the 1937 flood.

The levee protecting Visalia (presumably the one on the south bank of the St. Johns) broke on the afternoon of January 14, and the floodwaters swept into town. The northwestern part of the town was quickly submerged.

Porterville also flooded on January 14. Twenty-five families living in the lower part of town were rescued by citizens with rafts.

Ernest Clayton Northrop recalled the extensive flooding that occurred during the winter of 1908–09. At the time, he was living on Bear Creek, a tributary of the North Fork of the Tule River, downstream from present-day Mountain Home State Forest in the general vicinity of the SCICON school. He said that it rained for many days and nights, followed by extensive flooding. The flood washed down sequoia logs which his family later made into fenceposts. Looking out from a point near their farm, they saw Tulare Lake spreading over most of the valley; there was water as far as they could see.[575]

There was a major flood on Garza Creek on the west side of Fresno County in 1909. This flood may have happened in January, or it may have happened on October 9. In any case, this was apparently the flood that killed Kenzie Whitten "Blackhorse" Jones. He died while leading a horse across that creek.[576]

Troop G, Fourteenth Cavalry, arrived in Three Rivers on May 7. They reported that there had been a great fall of snow during the preceding winter. This prevented them from reaching their outpost camps until June 15. Buck Canyon had snow for so much of the summer that they were never able to establish their usual outpost camp there. Because of the heavy snow, tourists didn't begin arriving until about July 1.

We only know about the December flood from the gaging stations. We haven't found any anecdotal reports of it.

The Kaweah's peak natural flow occurred at McKay's Point on December 9: 14,108 cfs. (That was the peak hourly flow; the peak average daily flow, as reflected in Table 15, was 8,226 cfs.) Based on the flood exceedence rates in Table 16, this had a recurrence interval of 5 years for the Kaweah.

Total flow for water year 1909 was 163% of the 1894–2011 average for the Kings, 186% for the Kaweah, 278% for the Tule, and 241% for the Kern.

Tulare Lake had high water, bringing it even higher than after the 1906 and 1907 floods. Total inflows to the lake in water year 1909 were 1,175,000 acre-feet, raising the elevation 4.7 feet.

## 1911 Flood

Flooding in 1911 occurred in January.

The winter of 1910–11 was a moderate to strong La Niña event.

The extreme rainfalls of the storm of January 9–11, 1911 extended in a southwest to northeast line between Los Gatos and Galt. Los Gatos recorded 17.34 inches, resulting in a recurrence interval of 800 years. Thirteen stations reported their highest-ever six-day rainfalls during this storm.[577]

This was one of the greatest floods of the 20th century in the lower San Joaquin Valley. During this flood, the upper San Joaquin River near Friant reached high stages at the end of January. The flood was higher downstream near Newman at the mouth of the Merced River. The peak stage of 1911 set a record. High stages were also reached on the Calaveras, Mokelumne, Stanislaus, Tuolumne, and Merced rivers. The floods on these tributaries combined to raise the San Joaquin River to a record-breaking stage. Reports estimated that 75,000 acres of land were flooded from the overflow of the San Joaquin, Mokelumne, and Calaveras rivers.[578]

The Kaweah also flooded.[579] The other rivers within the Tulare Lake Basin may have flooded as well.

At the least, it was a year with high runoff. Total inflows to Tulare Lake in water year 1911 were 724,000 acre-feet.

## 1912–13 Drought

Table 24 shows how the San Joaquin Valley Water Year Index categorized the drought years in that basin.

Table 24. Rating of drought severity during the 1912–13 drought.

| Water Year | San Joaquin Basin Water Year Classification |
|---|---|
| 1912 | Below normal |
| 1913 | Critically dry |

1901–11 had generally seen average to well above-average water flows for the rivers within the Tulare Lake Basin. In contrast, 1912–13 saw those rivers deliver only about 50% of average flows.

While this may not have been have a severe drought, it represented a significant change from previous years.

## 1913 Flood

The "Black Flood" happened in Coalinga on July 22–23 (multiple photographs on file in the national parks). Presumably this was caused by an intense rainstorm. We haven't found an explanation for why it was called the Black Flood. Perhaps something swept up in the path of the storm turned the floodwaters black.

This storm apparently covered most or all of the Tulare Lake Basin. Panoche/Silver Creek west of Mendota flooded sometime in 1913, probably in July.[580]

Fresno and Bakersfield each received 0.33 inches of rain on July 22, making that the wettest July day on record for both of those cities.

## 1914 Flood

Flooding in 1914 occurred in January.

An intense rainstorm struck Fresno and Coalinga on January 25. It apparently covered much or most of the Tulare Lake Basin. Portions of Coalinga were flooded (photograph on file in the national parks).

Panoche/Silver Creek west of Mendota flooded sometime in 1914, probably in January.[581]

The Kings River peaked near Piedra on January 25 at 59,700 cfs.[582] By then, the bridge over the Kings near Reedley was awash (photograph on file in the national parks), and the Kings was reported to be at its highest point since the 1867–68 flood. Major damage was done to the county bridges. The flood did much damage to buildings and killed many animals. One source said that the Kings wouldn't see another flood of this magnitude until 1950, but the 1937 flood was in the same category.

The national park's superintendent reported that the rainfall and snowfall of the preceding winter was greater than usual, resulting in heavy damage to roads and trails. The January 25 storm was presumably a major contributor to that damage.

The North Fork Kaweah peaked on January 25: 7,400 cfs. This was the largest flow on that river since record-keeping began in 1910. This would remain the flow of record until the 1937 flood.

The mainstem Kaweah's peak natural flow occurred on January 25: 13,899 cfs. (That was the peak hourly flow; the peak average daily flow, as reflected in Table 15, was 10,275 cfs.) Based on the flood exceedence rates in Table 16, this had a recurrence interval of 8 years for the Kaweah.

Flow on the Kaweah for January was almost 10 times greater than for the previous month. The town of Lemon Cove suffered major damage, washing away a small resort and hotel. Thousands of acres of valley farms were flooded. There was widespread flooding between Visalia and Exeter, halting highway travel. (The first Model T began production in 1908. Some of this highway traffic presumably would still have been non-motorized in 1914.) Water was neck-deep in some parts of Exeter.

Visalia had constructed a new, half-mile-long concrete aqueduct/conduit in 1910 to carry Mill Creek underground through town. The 1914 flood provided the first major test of that structure. It worked as designed; Visalia reported no significant damage.

The Kern River peaked near Bakersfield on January 26 at 18,300 cfs.[583] It was the biggest flood that city had seen since record-keeping began in 1893. It would remain the flood-of-record until the 1937 flood.

Corcoran was incorporated in 1914, the year that the 1912–13 drought ended. It seemed to end rather dramatically; floodwaters from the Kings, Kaweah (via Cross Creek), and Tule all reached the lakebed in January. An article in the *Hanford Journal* said that farmers in the lakebed were "doomed." However, an article in the *Corcoran Journal* said that assessment was premature; it was too soon to say how great the damage would be.[584] As it turned out, the flood was short-lived. A record barley and wheat harvest was brought in that summer, apparently having suffered relatively little from the flood.[585]

## 1916 Flood

Flooding in 1916 occurred in January.

The winter of 1915–16 was a La Niña event.

1916 was a year of vigorous weather systems throughout California. There was a very rare snowstorm on the west side of the southern San Joaquin Valley at the end of December 1915. A man-made forest of 2,300 oil derricks occupied the west side of the San Joaquin Valley in 1916. Half of those derricks, which ranged in height from 70–130 feet, were destroyed in two big windstorms that occurred on January 17 and 27, 1916.[586]

River flooding occurred in January and continued in the Tulare Lakebed for about four months thereafter. That flooding resulted from two Pacific storms. The first storm lasted from January 14–20 and covered an area that extended at least from San Diego north to the Kern River Basin. The next storm struck on January 24. Although the second storm extended to the Canadian border, it may not have produced as much precipitation in the Tulare Lake Basin as the earlier storm did.

The first storm was unusual in covering such a large area and extending so far north. The entire water year was an anomaly in that the Kern River Basin received twice as much precipitation that year as did any of the other watersheds in the Tulare Lake Basin. The Kern River Basin presumably received the brunt of this mega-storm because it was south-facing and was the southern-most watershed in the Tulare Lake Basin.

The Kings River peaked at Piedra on January 17 at 45,400 cfs.[587]

The Kaweah's peak natural flow occurred at McKay's Point on January 17: 15,362 cfs. (That was the peak hourly flow; the peak average daily flow, as reflected in Table 15, was 10,540 cfs.) Based on the flood exceedence rates in Table 16, this had a recurrence interval of 8 years for the Kaweah.

The Tule River must have also flooded, since it sent significant floodwaters into Tulare Lake.

The Kern River peaked near Bakersfield on January 19 at 18,000 cfs.[588]

Total flow for water year 1916 was 174% of the 1894–2011 average for the Kings, 177% for the Kaweah, 249% for the Tule, and 337% for the Kern. This was the highest runoff ever recorded for the Kern.

In 1907, a massive levee had been built around four sides of Tulare Lake to constrain its growth in times of flood. The lake had been harnessed, and the lakebed declared safe for growing orchards. However, the 1916 flood would bring the lake back to life and put an end to those hopes, at least temporarily.

The Tulare Lakebed was dry on January 1, 1916. Inflows from the Kings, Kaweah, and Tule Rivers began in January. Inflow from the Kern River began in March. The Kern River Basin received a huge amount of precipitation during January, some of it falling as rain and some as snow. Apparently the initial runoff from those storms went into filling Buena Vista Lake. Subsequent melting of the snowpack during the March–June period was presumably responsible for the Kern's contribution to Tulare Lake's inflows in 1916.

The 1916 flood is worthy of note in that it was a very large flood with especially heavy inflows from the Kern River. As illustrated in Table 25, it was also the first flood in which measurements of inflow were sufficient to determine the contribution from each river basin.[589]

Table 25. Inflow to the Tulare Lakebed during water year 1916.

| Stream | Total Lakebed Inflow (acre-feet) | Percent Contribution |
|---|---|---|
| Kings River | 186,100 | 18% |
| Kaweah River | 212,500 | 20% |
| Tule River | 96,000 | 9% |
| Kern River | 547,100 | 53% |
| Total | 1,041,700 | |

This inflow left Tulare Lake about 11 feet deep at its deepest point (elevation 190 - 179 feet). It would be 19 years before the Kern River or Tulare Lake would experience another flood this big.

Paso Robles received 15 inches of rain during the January storm, as much rain as that area receives in an average year. That remains the wettest January ever in Paso Robles.

The 1916 flood was a major event throughout Southern California. We have to look there to understand what happened in the southern part of the Tulare Lake Basin. This storm story begins in the San Diego area in the summer of 1915.[590, 591]

For four years, San Diego had been experiencing below-normal flows in the local rivers. It was feared that the area was entering another prolonged drought. Memories of past droughts were still all too fresh; there seldom seemed to be enough water. To make matters worse, the fall rains didn't start on schedule. A dry November sharply reduced the supply in the city's reservoirs.

Charles Mallory Hatfield was a rainmaker — an expert in pluviculture. He was originally from Fort Scott, Kansas, but was by then living near Los Angeles. His supporters believed that Hatfield was the real McCoy.

Under pressure from the San Diego Wide Awake Improvement Club and others, the San Diego Common Council (the equivalent of today's city council) approached Hatfield and voted 4–1 to pay him to fill the Morena Reservoir for a $10,000 fee, payable by the inch. They figured that they had nothing to lose; it was a fee-for-service agreement. The council instructed the city attorney to draw up a contract. By New Year's Day, 1916, Hatfield had set up his moisture accelerator near the Morena Dam and was releasing strange and secret vapors into the atmosphere.

The first of two Pacific storms struck January 14–20. Cuyamaca (in eastern San Diego County, source of the headwaters of the San Diego River) received 18 inches in that six-day storm, an amount equal to nearly half of its average annual rainfall. Major flooding resulted downstream in the San Diego area. After the first storm, Morena Reservoir still wasn't full, so Hatfield continued producing his vapors; he intended to earn his fee. One man caught up in the initial flooding in Mission Valley recommended: "Let's pay Hatfield a $100,000 fee to quit."

After just a four-day break, the second storm arrived on January 24. It added more than 14 inches of rainfall at Cuyamaca. Nearby Descanso received a total of 27.79 inches of rain during January 14–28. This represents a recurrence interval of about 6,400 years. A total of 25 stations recorded their highest-ever rainfalls during the two storms. There were 8 stations that received 10 inches of rain or more on January 17.[592] The total for the month at Dorman's Ranch in the San Bernardino Mountains was 57.91 inches (4.8 feet). The ground was already saturated from the first storm and could hold no more. When the second storm hit on January 24, raging torrents of water raced down the rivers and creeks.

The spillway of the Sweetwater Dam was designed with a capacity of 5,500 cfs. At the peak of the first flood on January 17–18, it had a flow of 45,500 cfs.[593] The dam was overtopped by 3.7 feet of water, causing a large section of the south abutment dike to fail catastrophically.

On January 26, the city dynamited the dam in Switzer Canyon in the south portion of Balboa Park. It had been cracked and weakened over the years; blowing the dam was apparently considered preferable to letting it fail during the height of the flood. Two houses on 16th Street were overturned as the water rushed down to San Diego Bay.

On the evening of January 27, Lower Otay Dam was also overtopped. Water filled the observation shafts on the downstream side of that dam's steel core, and the pressure blew out the rock that provided the dam's structural stability. The steel core then swung out like a gate, releasing the full depth of water, which created a flood wave in the canyon of gigantic proportions. The dam was about 130 feet high, and the depth of the wave in the canyon a short distance below the dam site was about 100 feet high. As the lower canyon widened, the wave height decreased to approximately 20 feet, which was still devastating to the people living in the valley below. It required only 2½ hours for 13 billion gallons to empty out of that reservoir.

The San Diego River reached a crest six feet higher than in any previous flood. The city's concrete bridge at Old Town was the first to go. Next was the Santa Fe Railway bridge, even though it was weighed down with loaded freight cars. In Mission Valley, the San Diego River peaked at an estimated 75,000 cfs. (For comparison, the more famous 1980 flood peaked in Mission Valley at just

25,000 cfs.) All but one of the large bridges in San Diego County was destroyed. The Southern Pacific lines were severely damaged, and train service to Southern California was discontinued. The only way to travel from San Diego to Los Angeles was by boat.

Hatfield talked to the press on February 4, saying that the damage was not his fault and that the city fathers should have taken adequate precautions. He had fulfilled the requirements of his agreement with the city: the Morena Reservoir was full.

The city attorney pointed out to Hatfield that although a contract had been drawn up, it had not been signed. Hatfield had been working without a written contract; he had effectively entered into a gentleman's agreement with the Common Council. Hatfield countered that he was then the owner of the water that he had added to the Morena Reservoir, which was valued at $400,000. It was a nice try, but the city still refused to pay.

Hatfield filed a lawsuit on December 2 to force the city to pay their bill. The city later offered to negotiate; they'd pay him his $10,000 fee if Hatfield would accept liability for the $3.5 million in claims that had been filed against the city as a result of the flood. Hatfield declined the offer. His lawsuit never came to trial and was eventually dropped. The court did rule in related suits that the flood was an act of God, and therefore the city was not liable for the damages.

The San Diego Common Council never did pay Hatfield the fee that they had agreed to. But the flood had been good for Hatfield. His fame continued to grow, and he received more contracts for rainmaking.

Total damage in San Diego County was nearly $8 million and about 15 people died.

Orange County received 11.5 inches of rain during the period January 17–28. The Santa Ana River overflowed, sending a wall of mud through farmland and streets. Almost all the bridges in the county were destroyed. The state highway and virtually all the roads in the foothill and mountain areas were washed out. Four people drowned in the county, two in a cottage floating down the Santa Ana River.[594]

The Santa Ana River also flooded upstream in Riverside County as did the San Jacinto River. The cities of Indio, Coachella, and Mecca were completely inundated; 9 inches of rain fell in the Coachella Valley. Lake Elsinore rose very quickly. All rail traffic was halted in the county due to landsides or tracks washing out.[595]

Los Angeles experienced flooding January 14–19 and 25–30. The Los Angeles River ran 3 miles wide.[596]

San Bernardino County experienced flooding from January 17–28, 40 bridges were destroyed. All roads in Cajon Pass were washed out. The Santa Ana River ran 2 miles wide. It was one of the largest floods ever on the Mojave River. Two drowned in the county.[597]

## 1917–21 Drought

This drought affected the entire state except the Central Sierra and North Coast and has a possible recurrence interval of 10 to 40 years. It was most extreme in the north.

Table 26 shows how the San Joaquin Valley Water Year Index categorized the drought years in that basin.

Table 26. Rating of drought severity during the 1917–21 drought.

| Water Year | San Joaquin Basin Water Year Classification |
|---|---|
| 1917 | Wet |
| 1918 | Below normal |
| 1919 | Below normal |
| 1920 | Below normal |
| 1921 | Above normal |

The drought was comparatively mild in the Southern Sierra. Water year 1917 was exceptionally dry up to the middle of February 1918; less than an inch of precipitation had fallen in Fresno. But then the situation changed dramatically, with more than eight inches of rain falling between mid-February and the end of March. As a result, the rivers within the Tulare Lake Basin had slightly above-average flows in water year 1917. From 1918–21, those rivers experienced generally 50%–80% of average flows.

Tulare Lake became dry on April 30, 1919 and would remain dry for the next three years (1919–21). Repairs to the lake levees, which had been breached in the 1916 flood, could finally be made.

## 1918 Flood

Flooding in 1918 occurred in September. This flood occurred during the 1917–21 drought.

The 1918 flood occurred during the El Niño of the winter of 1918–19. That El Niño was the subject of a recent study.[598] It was one of the strongest El Niño events of the twentieth century, comparable in intensity to the prominent El Niño events of 1982–83 and 1997–98.

Although a strong El Niño event typically causes flooding in the Tulare Lake Basin, it can have very different effects in other parts of the world. The 1918–1919 El Niño was likely responsible for the severe drought that took place in India in 1918. That was one of the worst droughts that country experienced in the 20th century. There was famine and a lack of potable water, resulting in a compromised population. The drought coincided with an influenza pandemic that was sweeping the globe at that time. The 1918 influenza pandemic killed about 18 million people in India and between 50 to 100 million globally. The authors of the 1918–1919 El Niño study speculated that it might have been linked to the influenza pandemic, especially in India.

In addition to this strong El Niño, 1918 also produced an unusual hurricane. When a Pacific hurricane degrades, it usually makes landfall in Southern California or in Mexico. The year 1918 was the only instance in historic times in which the remnants of a hurricane are known to have come inland as far north as Central California. It isn't clear what effect this storm had in the Tulare Lake Basin. The Coast Ranges would have gotten heavy rains and flooding. Storms in the Coast Ranges typically spill over into the drainages of western Fresno County and Kings County. We haven't found any records to indicate whether that happened in this storm or not. Possibly this storm and flood was just outside the Tulare Lake Basin.

But even if it were, the story of the 1918 hurricane merits inclusion in this document because it serves as a model for two unusual storms that occurred in the Tulare Lake Basin. Both of those were robust cyclic storms which vigorously entered rain shadow areas to the northeast, resulting in a deluge in normally dry areas:[599]

- The February 1978 storm. That storm produced large rainfalls on the windward slopes of Ventura County and then continued over into the rain shadow area in the Buena Vista Lake region.
- The March 1995 storm. That storm produced devastating rainfalls on the windward slopes of the Coast Ranges. It was still quite energetic as it moved into the rain shadow area to create further devastating floods. That was the storm that washed out the Interstate 5 bridges near Coalinga.

On September 11–12, 1918 the remnants of a hurricane tracked to the north-northwest off the coast of Baja California and Southern California, generating only light amounts of rain in the coastal mountains of Southern California.

The storm system apparently moved onshore near Monterey Bay. The town of Antioch, east of Stockton, is on the lee side of the Coast Ranges. Antioch received 6.59 inches during September 12–14 with a recurrence interval of 2,200 years. A total of 12 stations reported rains with return periods in excess of 100 years.[600]

The storm moved north to the Red Bluff area before dissipating. Typically such storms bring a surge of moist warm tropical air that triggers thunderstorms. Red Bluff had 1.19 inches of rain in 30 minutes, 3.72 inches in 2 hours, and 6.12 inches in 24 hours on September 13. Red Bluff received a total of 7.12 inches during the storm event, more than any other station.[601]

## 1922 Flood

Flooding in 1922 occurred in May.

This flood is known only from its effect on Tulare Lake. The lake had dried up on April 30, 1919. One source said that the 1922 flood on the Kaweah was sufficient to leave Tulare Lake eight feet deep at its deepest point. The reliability of that measurement is unknown; it seems improbably deep.

S.T. Harding was aware that the Tulare Lakebed received some inflows from both the Kings and the Kaweah Rivers in 1922. However, he was under the impression that the quantity was relatively small and that the water was quickly absorbed by the soil or used directly for irrigation of crops growing in the lakebed.[602]

Mae Weis's account of Tulare Lake history from that era said that the floodwaters arrived in May 1922, and that by June a total of 23,680 acres of lakebed cropland was flooded. A little more water was added from heavy rains during the winter of 1923–24, but the lakebed was completely dry again early in 1924.[603]

Total flow for water year 1922 was 129% of the 1894–2011 average for the Kings, 107% for the Kaweah, 102% for the Tule, and 115% for the Kern.

## 1922–34 Droughts (2)

This drought had two components:

- 1922–27
- 1928–34 (recognized as 1929–34 at the state level)

This drought was statewide, although not all areas were affected equally. The first part of the drought (1922–27) affected much of the entire state except the Central Sierra and has a possible recurrence interval of 20–40 years. The first two years of the drought (1922–23) had no effect whatsoever in the Tulare Lake Basin.

The drought was simultaneously in effect for the entire state in 1924, and it was particularly severe that year.

The first part of the drought ended in 1927, and the last part isn't generally considered to have begun until 1929. The year 1928 is not considered a drought year at the state level. But in the Tulare Lake Basin, there are indications that the second part of the drought may have begun in 1928. In particular:

- On April 3, 1928, Fresno began a 214-day stretch without measurable rain, the longest such streak on record for that city.[604]

The second part of the drought (1929–34) is considered the longest, most severe drought in the state's history. Parts of the state, especially the northern quarter, were in drought from 1928–37. This drought is sometimes referred to as the drought of 1928–37. Because of the extended duration, the second part of the drought accumulated the largest deficiency in runoff of any drought in the state's history.

The six-year drought of 1929–34 is unequaled in the historical record of the Sacramento Valley Water Year Index dating back to 1872; this indicates that the drought had a recurrence interval of more than 100 years.

A 420-year reconstruction of Sacramento River runoff from tree ring data was made for the California Department of Water Resources (DWR) in 1986 by the Laboratory for Tree-Ring Research at the University of Arizona.[605] The tree ring data suggested that the 1929–34 drought was the most severe in the 420-year reconstructed record from 1560–1980. This indicates that the 1929–34 drought has a possible recurrence interval of more than 400 years.

The 420-year reconstruction also suggested that few droughts prior to 1900 exceeded three years, and none lasted over six years, except for one period of less than average runoff from 1839–46. See the section of this document that describes the 1987–92 drought for a comparison of the multi-year droughts in the Sacramento River Basin.

Table 27 shows how the San Joaquin Valley Water Year Index categorized the drought years in that basin.

Table 27. Rating of drought severity during the 1922–27 drought.

| Water Year | San Joaquin Basin Water Year Classification |
|---|---|
| 1922 | Wet |
| 1923 | Above normal |
| 1924 | Critically dry |
| 1925 | Below normal |
| 1926 | Dry |
| 1927 | Above normal |

Total flow for water year 1924 was 23% of the 1894–2011 average for the Kings, 24% for the Kaweah, 18% for the Tule and 26% for the Kern. Flow for 1931 would be nearly as low. The Tulare Lake Basin wouldn't see flows this low again until 1977 (see Figure 13 on page 109).

By some measures, the two-year period 1923–24 was the worst Kings River drought on record.[606]

Water year 1931 was the lowest runoff (184,130 acre-feet) experienced on the Kern River since record-keeping began in 1894.

The drought was active somewhere in the state from 1922–34. But within the Tulare Lake Basin, it was primarily active from 1924–34. Based on the San Joaquin Valley Water Year Index, only two years during that period (water years 1927 and 1932) received average or better flows. The Kern didn't have average flows even in those years. When Walter Fry wrote Bulletin #8 in November 1931, his editor noted the despondency among valley residents that had been caused by the recent cycle of dry years.

W.E. Bonnett, meteorologist with the U.S. Weather Bureau in Fresno, said that only 0.06 inch of rain fell in March 1923; it was the driest March in 42 years. However, April 1923 was the wettest April in 42 years.[607]

The drought brought grazing on the ranges almost to a standstill; ranchers were reduced to feeding hay. It was standard practice for cattle from Arizona and Texas to be shipped into Tulare County around the first of the year; 15,000 had already arrived at the beginning of 1924. The foothills had a little grazing left at that time due to recent rain, but the plains were already bare.[608]

1924 was a very small tree-ring year in the Sierra.

The August 1924 national park monthly report said that the drought continued unrelieved. There were fires in the national forests and erroneous reports that travel to the national parks was prohibited or unsafe. Many park streams and springs dried which had never failed in the memory of the oldest inhabitants. Fish were dying in streams. There was ample water at Giant Forest only because of a new water system that had been installed in 1923.

The September 1924 national park monthly report said that virtually no rain had fallen in the park since March, only the lightest showers in April and May. Winter horse feed at lower levels didn't start until the fall rains, so the park thought that it might have to make emergency purchase of fodder. The September report concluded:

> *But as the hart panteth for the water brook so do all residents of this part of California await the breaking of the Great Drought.*

Table 28 illustrates the runoff in the Tulare Lake Basin during water year 1924.

Table 28. Runoff in the Tulare Lake Basin during water year 1924.

| Watershed | Total Runoff (acre-feet) | % of average (1894–2011) |
|---|---|---|
| Kings | 391,920 | 23% |
| Kaweah | 101,650 | 24% |
| Tule | 24,700 | 18% |
| Kern | 190,810 | 26% |
| Total | 709,080 | 24% |

The combined runoff of the four rivers in the Tulare Lake Basin during 1924 was only 709,080 acre-feet, the second lowest since record-keeping began in 1894, only 1977 would be lower (see Table 58 and Figure 13).

Bill Tweed recalled being told by old-timers in the 1970s that the mainstem of the Kaweah stopped running in 1924; it was only a series of disconnected pools. This may well be true, or it may be a slight exaggeration, colored by the passage of time. There's no way to know for sure. Anecdotal accounts such as this should be taken with a grain of salt. For comparison, total runoff in the Kaweah River Basin in water year 1977 was 8% less than in 1924. However, the Kaweah did not quite stop flowing in 1977.

The drought initially appeared to end in October 1924. Sequoia National Park's monthly report for October began with the statement:

> *The principal event of the month was the end of the Great Drought. Heavy rains fell on October 6th, the first for over six months. Rains and snows continued during the month so that precipitations at Giant Forest for the month was 6.30 inches as compared with only 1.07 inches last year. This is an encouraging start for the heavy winter which California so badly needs.*

On November 9–10, 1924, the national parks were deluged with an exceptionally heavy rainstorm; rain came down more heavily than had ever been recorded. December would bring abundant moisture to all of California. For a more complete write-up of that winter, see the section of this document that describes the 1924 flood.

Despite this relief during the winter of 1924–25, the drought would continue for another nine years (1926–34).

213

Table 29 shows how the San Joaquin Valley Water Year Index categorized the drought years in that basin.

Table 29. Rating of drought severity during the 1928–34 drought.

| Water Year | San Joaquin Basin Water Year Classification |
|---|---|
| 1928 | Below normal |
| 1929 | Critically dry |
| 1930 | Critically dry |
| 1931 | Critically dry |
| 1932 | Above normal |
| 1933 | Dry |
| 1934 | Critically dry |

Tulare Lake dried up in 1924. The lake did get some inflows from the Kings and the Kaweah Rivers during several of the drought years. However, most of the quantities were small and S.T. Harding believed that much of the water was quickly absorbed by the soil or used directly for irrigation of crops growing in the lakebed. Tulare Lake would not reappear as a large lake until 1937.

Limited floods were welcomed by the grain growers in the Tulare Lakebed since a certain amount of irrigation water was needed, especially during the drought years. On December 6, 1927, practically all of the lakebed reclamation districts joined in forming the Tulare Lake Basin Water Storage District, setting aside 18 sections in the lowest portion of the lakebed for a reservoir. (Yes, you read that right. The drought was so severe that they were creating a reservoir in the lakebed.) Since it was being freely predicted that Tulare Lake was a thing of the past due to the increasing use of the water from the tributary streams for irrigation purposes, the idea behind this scheme was to conserve as much as possible of the Kings River runoff for irrigation purposes. The reservoir would afford a certain amount of flood protection; but in 1927, amid the scramble for irrigation water, the flood control angle received scant consideration.[609]

In August 1926, the Kaweah Fire burned 86,000 acres in the drainage of the North Fork of the Kaweah. In August 1928, the South Fork Fire burned 22,000 acres in the drainage of the South Fork of the Kaweah. The national parks haven't seen fires of this magnitude since. It may not have been a coincidence that these huge fires occurred during one of the more intense and extended droughts that the Tulare Lake Basin has experienced in historic times.

In 1927, California was in the sixth year of one of the longest droughts since Euro-American settlement. However, the state's papers would surely have been covering the great flood that was besieging the lower Mississippi River Valley that year. The Mississippi River broke out of its levee system in 145 places, inundating 16 million acres and destroying 130,000 homes. In places, the river swelled to 80 miles wide.[610, 611] Not until May 2011 would a flood of comparable magnitude come down the Mississippi. That must have seemed like a different world, when viewed from the perspective of a state so mired in drought.

Around 1930, the development of an improved deep-well turbine pump and rural electrification enabled additional groundwater development for irrigation.[612] The early wells in the southern San Joaquin Valley were hand-dug pits. In the 1930s, the pits in the Poplar area (northwest of Porterville) began to run dry as the water table dropped.[613] See the section of this document that describes

Groundwater Overdraft for a discussion of why the water table was dropping at this time. This marked the transition to the current practice of irrigation using deep-well pumps.

The winter of 1932–33 was an El Niño event. It was remarkable in the national parks for heavy low-elevation snows. Sequoia National Park's monthly report for January 1933 began with the note:

> *The only matter of special interest is the weather, which for the last two weeks of January resembled that of Alaska rather than California. All records for snowfall were broken, and old-timers with over fifty years' experience can recall no such precipitation. The damage to live oak and other trees between the 1,500 and 4,000 foot levels is tremendous; and the damage to park telephone lines, roads, campgrounds, etc. cannot yet be estimated. We shall be hard put to keep the park open and repair damage before summer. But a splendid water supply is piling up in the mountains, and our slight inconveniences are as nothing in the general scheme.*

Following the longest dry winter in the history of the region, a series of storms began on January 15, 1933, that continued with only a two-day break through the end of the month. Precipitation at the national parks' reporting stations jumped from the least on record to nearly average. Giant Forest received 60 inches (5 feet) of snow in 24 hours on January 19, setting a record for California that would stand until 1982. Giant Forest received 181.5 inches of snow during the month (including 15 feet in 15 days). This huge snowfall was the more remarkable because it occurred during such a severe drought. For the Giant Forest station, both total snowfall and the amount of snow on the ground rose from below average to the greatest ever recorded at the end of January.

The national parks had taken delivery of a new Snow King rotary snowplow just before the storm, and it worked wonderfully. Even so, snow fell faster than equipment could remove it. The Generals Highway closed on the night of January 19 when a slide buried a plow and a pickup. High winds drove the snow into drifts up to 30 feet deep on the steeper sections above Deer Ridge. Equipment continued to be operated 24 hours a day, and additional equipment was rented. Even so, the snow was 6½ feet deep on the roadbed at Granite Springs at the end of the month. The section of road from Deer Ridge to Slide Spring averaged 10 feet deep with many slides 12–15 feet deep. And it was still snowing. It would be February before the storm broke and the Generals Highway could be reopened.

Although less dramatic, the lower elevations of the national parks also got hit hard in the January storm. Ash Mountain received a total of 23 inches of snow, and snow lay on the ground from January 16–30, a record for that elevation. An emergency purchase of 20 tons of hay was required to feed the parks' pack and saddle stock for the last half of the month. The barn at Clough Cave Station collapsed under the weight of the snow. The water intake at Hospital Rock was demolished by snow slides. Sleet storms wrecked all of the lower telephone lines. Heavy rains at the lower elevations caused considerable damage and slides.

Despite this relief during the winter of 1932–33, the drought returned and continued through the fall of 1934.

Water year 1934 was the driest ever in both Fresno and Bakersfield. Fresno received a total of 4.4 inches of rain that year while Bakersfield got only 2.2 inches. That water year had the next-to-lowest runoff experienced on the Tule River since record-keeping began in 1894, only 1977 would be lower.

As illustrated in Table 29, water year 1934 was categorized as a critically dry year in the San Joaquin River Basin.

The national parks' Ash Mountain headquarters development (originally known as Alder Creek) began operation in 1921. Alder Creek was the sole source of water for that development until auxiliary water sources were developed starting in 1950. With the loss of the river pump in 1997, Alder Creek has once again become the headquarters' primary water supply, but this was particularly true in the early years. Because the flow of that creek is relatively low, there has long been concern about how to get through drought years. In 1934, the park constructed the 75,000-gallon Alder Creek Reservoir; the dam for that reservoir is still largely intact.

Sequoia National Park's monthly report for October 1934 began with the following note:

> *A hopeful feeling pervades the park and adjacent Valley because of better than usual early fall rains. The fire hazard has ended and young grass is coming up in the foothills. But the horse ranges have been badly damaged by drought and overuse, and we may have to kill some stock as we cannot get authority to purchase hay and grain.*

The year 1934 deserves special mention. It was the end of the longest, most severe drought in California's history (1929–34). But 1934 was also right at the beginning of the Dustbowl drought in the Great Plains (1933–40). Based on the Palmer Drought Index, 80% of the contiguous 48 states was experiencing at least moderate drought by the end of June 1934; 63% was in severe to extreme drought. By either metric, 1934 remains the greatest drought year in our country since record-keeping began in 1895.[614]

The period 1927–34 was a serious drought for the Great Basin as well. In December 1934, Lake Tahoe reached its lowest elevation since record-keeping began. A group of stumps was exposed by the receding waters along the south shore of the lake. Since then, these and other submerged stumps in the area have been intensively studied by Susan Lindstrom and others. The stumps measure as tall as 10 feet and up to 3½ feet across. Based on radiocarbon dating of these stumps, the lake reached a low stage on one or more occasions between about 2250–3560 BC. During that period, trees became established and grew for some 100–350 years before the lake raised enough to drown them. The takeaway message is that what we think of as a long-term drought today is mild compared to earlier mega-droughts.

## 1923 Flood

Flooding in 1923 occurred in April. This flood occurred during the 1922–34 drought, although the drought really hadn't taken effect in the Tulare Lake Basin yet.

This was either a rain or a rain-on-snow event. April 1923 was the wettest April in 42 years.[615]

The Kaweah's peak natural flow occurred at McKay's Point on April 6: 6,333 cfs. (That was the peak flow; the peak average daily flow was 4,410 cfs.)

The Tule River had a maximum daily discharge near Porterville on April 6 of 3,820 cfs. (The term "maximum daily discharge" is presumably the same as "peak average daily flow"). That was the highest flow since record-keeping began in 1901. It would remain the flow of record for over three decades. Not even the flood of 1950 would exceed this record.[616]

### 1924 Flood

Flooding in 1924 occurred in November. This flood occurred during the 1922–34 drought.

The winter of 1924–25 was a La Niña event.

The November 1924 national park monthly report said that an exceptionally heavy rain occurred over almost the entire park on November 9–10. The rain came down more heavily than had ever before been recorded. Giant Forest received 6½ inches of rain in 24 hours and other points received 2–3 inches in less than an hour. There was severe erosion of park roads, many culverts were washed out, and two small bridges were destroyed. There was a big slide on the Giant Forest Road above Cedar Creek. (The headwaters of Cedar Creek were near the Colony Mill Ranger Station, but possibly this was referring to one of the tributary branches of that creek.)

The December 1924 national park monthly report said that:

> *Overshadowing everything else in the park as in California is the abundant rain and snowfall of this winter. There were ten rainy days during December while the month was colder and more gloomy than usual. Six inches of snow were on the ground at Alder Creek park headquarters one morning and it remained for several days.* (Alder Creek was the original name for Sequoia National Park's headquarters development, now known as Ash Mountain.)

By December 31, 1924, Giant Forest had received 108 inches of snowfall, compared with just 31 inches by the same date in 1923. The increase in precipitation was less marked in the valley.

Despite this relief during the winter of 1924–25, the drought would continue for another eight years.

### 1931 Flood

This flood occurred during the 1922–34 drought. Flooding occurred on the Kaweah River at sometime in 1931.[617] Possibly other rivers within the Tulare Lake Basin also flooded.

Little is known about this flood. Two possible clues:
- 1.02 inches of rain fell on Bakersfield on December 8, 1931, setting the record for the wettest December day ever in that city.[618]
- A total of 54 inches (4½ feet) of snow fell in Yosemite Valley in December 1931, setting the record for the snowiest December ever on record there.[619]

### 1932 Flood

Flooding in 1932 occurred in September. This flood occurred during the 1922–34 drought.

This flood was selected by the National Weather Service forecast office in Hanford as one of the top Central Valley weather stories of the 20th century, more noteworthy even than the December 1955 flood.

The subject matter expert on this flood is believed to be Jon Hammond, the editor of the *Tehachapi News*. Jon has written up this flood in his newspaper. Unfortunately, it has proved impossible to obtain a copy of that document. The flood is also supposed to be described in detail in the book

*Three Barrels of Steam* by James E. Boynton. The following write-up is based on newspaper accounts from the time of the flood and on other sources.[620, 621, 622, 623]

The remnants of an unnamed Pacific hurricane moved up into the Gulf of California and came ashore near Mexicali on September 29. (Pacific hurricanes didn't start getting names until 1960.)

The storm traveled into the lower desert without much resistance only to break up in the Tehachapi Mountains. Tehachapi received 7.11 inches of rain from September 28 – October 1. The recurrence interval for that event at Tehachapi was 200 to 500 years. Apparently the rains greatly exceeded that in the surrounding mountains.

Reports of this storm are long on details of the effects, but short on quantitative details of the rain amounts. The USACE analyzed the streamflow associated with this storm. They found that the peak runoff rate was 3,815 cfs/sq mi on the 3.5 square miles of the Cameron Creek watershed. This runoff rate (3,815 cfs/sq mi) means that the rain was coming down at an average rate of 6 inches an hour over the entire watershed. That is a rather stunning rate. (The obscure conversion rate used was 1,000 cfs/sq mi = 1.55 inches/hr.)

This is another rare occurrence of extremely large rainfalls on the lee side of an orographic barrier (i.e., the rain shadow of a mountain). For three other examples, see the section of this document that describes the 1918 hurricane.

This is the first time that we know of in historic times in which a Pacific hurricane caused flooding in the Tulare Lake Basin. The 1918 hurricane may have caused some flooding along the northwest side of the basin, but we haven't found records to document that.

Tehachapi received 4.38 inches of rain in seven hours on September 30, the most extreme rainfall ever recorded in that city. For a time that day, the town of Tehachapi was under three feet of water, with a torrent tearing through the streets and sweeping furniture out of houses. The nearby community of Monolith was also flooded.

Some of the floodwaters flowed north into the Mojave Desert, forming a large lake. The town of Mojave was under two feet of water. However, most of the water poured south down Tehachapi Creek which is the southern fork of Caliente Creek. Caliente Creek drains into the San Joaquin Valley near Arvin, southeast of Bakersfield.

The rain was so intense that it brought Santa Fe Engine No. 3834 to a stop. That train was waiting out the storm atop a new concrete trestle over Tehachapi Creek, ½ mile east of Woodward Station (about a mile upstream from the village of Keene). This was during the Great Depression. In addition to the crew, the Southern Pacific estimated that there would typically have been up to 50 hobos on a freight train such as this. In the middle of that train was a helper locomotive assisting it up the grade to Tehachapi Pass. The Santa Fe train was sitting on a siding. Sitting next to it on the mainline was Sunset freight train No. 829 of the Southern Pacific.

As the floodwaters poured down Tehachapi Creek, they encountered six railroad bridges. At each bridge, debris snagged and created unstable debris dams which held back floodwaters long enough to create temporary reservoirs of runoff. These dams broke apart as water built behind them, creating surges of floodwaters that exasperated the flooding problems. Walls of floodwater, some 40 feet high, raced down Tehachapi Creek as each bridge gave way.

The floodwaters first hit a KAAD service station at Woodford where 15–19 men had taken refuge from the storm. (One source said that this station was at Keene.) Those inside the building were caught up in the floodwaters and some may have drowned. In a separate incident, a family of four in Woodford was drowned when the flood swept away their creek-side house.

Some 20 road camp workers were camped between Keene and Tehachapi. The initial emergency report (before the phone went dead) was that the flood swept down on their camp. Apparently most or all of those workers escaped, but that would not be known until county rescuers could reach the scene a day or two later.

The Kern County Tubercular Sanatorium at Keene was right next to Tehachapi Creek. The flood swept away the pump house, just a few yards from the main building. Three patients were drowned, but the rest survived.

The raging floodwaters piled up 50 feet deep against the trestle that the Santa Fe train was sitting on, undermining it. The trestle gave way directly in the center, collapsing with a roar that could be heard above the deafening noise of the storm. The helper locomotive in the center of the train plunged into the torrent, pulling seven freight cars with it. The Santa Fe locomotive also plunged in, but the Southern Pacific train remained on the mainline track, witnessing the horrifying event. (Southern Pacific passenger train No. 52 had passed only three minutes before the torrent hit the trestle.)

By the time the floodwaters reached Caliente, Tehachapi Creek was flowing at 37,000 cfs. All railroad crossings and 31 miles of track had been undermined, and 600 feet of track were washed out. The cost to the railroad for track repair was $600,000. Huge sections of the state highway through Caliente Canyon (the route that we now know as Highway 58) had been washed out, and at least nine highway bridges were destroyed. In addition to Monolith and Tehachapi, four communities in Tulare Lake Basin were flooded: Woodford, Keene, Caliente, and Arvin. Flooding in Caliente resulted in the death of a telegraph operator and her two-year-old niece.

High winds accompanying the storm tore down telegraph and telephone lines, isolating the mountain communities. The stricken area was largely cut off from contact with the San Joaquin Valley.

First responders immediately set out from Bakersfield with ambulances to check on the road camp workers and the other areas in the path of the flood. But they were stopped by the washed out bridges, 20 miles short of the road camp. They had to hike over two mountain ranges to get to the scene of the devastation.

One of the first reports came from Harry W. McGee, a United Air Lines pilot. When he arrived at United Airport in Burbank on October 1, he reported that Tehachapi seemed to have been inundated. He flew over that village in route from San Francisco with 10 passengers, flying out of his way to avoid the worst of the recurrent storms. He reported that mud and debris were visible in the Tehachapi streets.

Total property damage was about $1 million and resulted in 15–26 deaths. Among the dead were the engineer and brakeman of the wrecked train. Two unidentified bodies were assumed to be hobos from the train; there was no way to know how many more remained buried under the mud.

The floodwaters rolled the Santa Fe engine far downstream and buried it under 10 feet of silt and rocks. It took five days to even find the severely damaged engine and a month to free it. All rail traffic over Tehachapi Pass, the inland route between the San Joaquin Valley and the Los Angeles area, was halted by the destruction of the Southern Pacific Railroad track. For the next 14 days, all rail traffic had to be rerouted along the coast.

Following the cloudburst at Tehachapi, another downpour fell near Lebec, near the summit of the Ridge Route (the precursor of the Grapevine or Interstate 5 route over the Tehachapis). Great stretches of that state highway were damaged by the flood and traffic was halted. Most serious was a large rock and mudslide that occurred between Oak Glen and Camp Tejon.

The highway was opened to light traffic the following day, October 1, but guards of highway patrolmen warned motorists that the road was barely passable and prohibited trucking and heavy traffic entirely. County tractors assisted the cars over the stretch of highway between Oak Glen and Camp Tejon.

## 1935 Flood

Flooding in 1935 occurred in April.

The 1922–34 drought had finally ended. The first half of April was stormy. Daily temperatures at both Ash Mountain and Giant Forest were considerably lower than in 1934. Snow fell below Ash Mountain on one occasion. The snowpack in the national park at the end of April was heavier than it had been for many years. But April was remarkable over and above that. Precipitation in the park for April was much above average. In valley towns, all-time records of rainfall were exceeded in April, Fresno receiving over 16 inches.

The Kaweah River flooded and other rivers within the Tulare Lake Basin may also have flooded. The flood is known from two photographs (on file in the national parks):
- one of Elk Creek flowing across the Generals Highway
- one of the Kaweah in flood adjacent to the Kaweah Hatchery

The culvert at Elk Creek overflowed, washing out 260 yards of the Generals Highway. The Kaweah experienced above-average flows for April, although not nearly as high as April flows would be in the years 1936–38.

The state had constructed a fish hatchery on the Kaweah River in 1919. It was located directly across from the turnoff for the Mineral King Road. The purpose of the hatchery was to stock the streams of Fresno and Tulare Counties and a portion of Kern County. The hatchery survived a number of floods over the next three decades, some of which caused severe damage. The hatchery would eventually be removed after sustaining major damage in the 1950 flood.

## 1936 Flood

Flooding in 1936 occurred in February.

This was apparently a rain-on-snow event.

The Kaweah's peak natural flow occurred at McKay's Point on February 13: 8,360 cfs. (That was the peak flow; the peak average daily flow was 5,366 cfs.)

The Tule River peaked near Porterville on February 13 at 12,000 cfs.[624]

Tulare Lake had been dry since 1924. No flood flows reached the lakebed from any river during 1936.

Total flow for water year 1936 was 109% of the 1894–2011 average for the Kings, 113% for the Kaweah, 122% for the Tule, and 107% for the Kern.

## 1937 Floods (4)

There were four floods in 1937:
1. February (this was the biggest of the four floods on the Tule and Kern Rivers)
2. June
3. July
4. December (this was the biggest of the four floods on the Kings and Kaweah Rivers)

Photographs taken during the 1937 and 1938 floods show widespread flooding in the Kaweah Delta. The areas north and east of Visalia looked much as they would in the 1945 flood. Visalia was flooded in one of the 1937 floods, probably the December flood.

The winter of 1936–37 had the heaviest precipitation recorded in the national parks until then. (Record-keeping of weather data began in the parks in 1920. The winters of 1861–62 and 1905–06 were probably heavier, but there was no system for recording precipitation in those years.)

During the last week of December 1936, 75 inches (6¼ feet) of snow fell at Giant Forest, one of the heaviest snowfalls on record up until that time. This was the winter when an avalanche swept away the 125-foot-long Hamilton Gorge Suspension Bridge.

In Giant Forest, 11.96 inches of warm rain fell on six feet of snow between February 5–7, resulting in flood conditions unknown since 1916. There was considerable damage to the Generals Highway and to the Colony Mill Road. The Kaweah River rose 11 feet in 13 hours. Another 7½ inches of rain fell the following weekend, bringing the river to within one foot of its previous high mark. Flood conditions were widespread throughout Central and Southern California.

The Generals Highway was closed for nearly a month (until February 27) by storm damage at a score or more places including major damage at "Deer Creek" (possibly that was a typo and was supposed to say Deer Ridge) in February. Apparently that was referring just to the section of the Generals Highway between Grant Grove and Giant Forest. The section of road below Giant Forest did not reopen until many months later.

Bill Tweed recalled that there was a huge road failure just above Deer Ridge. That slide took out more than just the road; the mountainside virtually disintegrated. The Colony Mill Road had been closed for at least a few years prior to 1937. However, the landslide of February 1937 was so bad that the Colony Mill Road was reopened and oiled so that it could it be used as a detour for much of the following summer. This was the last time that the Colony Mill Road saw significant public use. This may have been one of the last times that the Generals Highway has been closed for more than a couple of weeks. (The highway would be closed for most or all of 1956 and 1967 for replacement of the Marble Fork Bridge near Potwisha.)

Ward Eldredge found a 1937 park monthly report that included the bid for a Deer Ridge bin wall complete with a photograph identifying the location. Manuel Andrade said that this would almost surely have been a wood bin wall, not a metal one. Bill Tweed recalled that a wood bin wall above Deer Ridge was replaced with a metal bin wall about 1980. Perhaps that was the same bin wall that was installed after the 1937 road failure.

The heavy culverts and trash cans that now exist adjacent to the Marble Falls Trail resulted from a failure along the Generals Highway in the vicinity of Deer Ridge and Eleven Range Overlooks, 1,200 feet above. Whether that was from the failure that occurred in 1937, 1952, or 1966 is unknown.

The Kaweah's peak natural flow occurred at McKay's Point on February 6: 19,751 cfs. (That was the peak hourly flow; the peak average daily flow, as reflected in Table 15, was 13,520 cfs.) Based on the flood exceedence rates in Table 16, this had a recurrence interval of 10 years for the Kaweah.

Specific damage in the national parks (including CCC work areas) from the February flood included:
- Salt Creek Truck Trail (outside the national parks but managed by the parks). One section washed out with gullying 2½–3 feet deep (photograph on file in the national parks).
- Washout on the Colony Mill Road. Water flowed across oiled surface to shoulder, causing undermining and loss of over ½ mile of road (photograph on file in the national parks).
- Progress Gulch on the Ash Mountain–Advance Truck Trail (what we now call the Shepherd's Saddle Road). Over 1,000 feet of un-oiled road washed out at one location. Maybe this was where the water tank is near Rattlesnake (photograph on file in the national parks).
- Slide and washout on the Generals Highway at Station 162+79 (photograph on file in the national parks).
- A section of the Generals Highway washed out, tearing out a 36-inch culvert. This happened on the 10 mph curve just below One Shot Rock, about two miles above the 3,000-foot elevation sign (photograph on file in the national parks).

National park CCC crews were employed in the emergency to save life and property in the Three Rivers area.

The river gage at McKay's Point (three miles below Terminus Beach) had been installed in October 1916. The Kaweah peaked there at 16,000 cfs on February 14, 1937. Two days earlier, the flood had set a record for the highest average daily flow since record-keeping began there in 1916.[625]

The Tule River peaked near Porterville on February 6, 1937 at 12,000 cfs.[626] By coincidence, the Tule had peaked the previous year on February 13, 1936, at 12,000 cfs.

The Kern peaked near Bakersfield on February 7, 1937, at 20,000 cfs.[627] Bakersfield narrowly escaped inundation in this event. The levee along the south bank of the Kern came within one foot of being overtopped. An emergency flood-fight helped to protect the levee from overflow. (A similar emergency stand would be required on this levee in the 1950 flood.) The Fruitvale and Fairhaven areas near Meadows Field were flooded, and 16 people had to be rescued by boat in those areas. Over 50 people were evacuated, and all of their homes were destroyed or badly damaged.[628]

The 1937 flood was an impressive event; it was the outstanding early flood in the Kern River Basin. However, when the settlers looked around, they saw high-water marks at much higher elevations. About two miles below the confluence of the North Fork and the South Fork of the Kern, the older marks were 40 feet higher than those of the 1937 flood. Those marks had been left by the 1867–68 flood.[629] That had to be a sobering thought. The reason that the 1867–68 flood had been so high in that area was because of the massive landslide dam failure that occurred on the North Fork in December 1867. See the section of this document that describes the Landslide Dam Failure #4: North Fork of the Kern.

The second flood of 1937 happened on June 4–7. It did serious damage down in the valley, but there was no damage of note in the national parks.

July brought numerous thunderstorms to higher elevations of the national parks, a fairly typical situation. However, a cloudburst on the evening of July 24 did considerable damage in the Mineral King area. The road was washed out in one spot and fabric automobile tops were riddled by hailstones. The East Fork Kaweah rose two feet in 20 minutes.

And then came the fourth flood of the year. During December 9–12, a single, intense storm moved rapidly from the North Pacific across California. In most places, this storm was recorded as a two-day event: December 10–11. The storm was warm, and precipitation in the Sierra fell primarily as rain instead of snow. There was little snow on the ground at the start of the storm, so snowmelt was not a major factor.

This storm was a high-elevation event centered in the northeast corner of the state. A total of 21 stations reported their highest-ever two-day rainfall during that storm. The highest intensity part of this storm was in a zone between Inskip Inn (northeast of Chico) to Alturas (northeast of Redding). Alturas had 5.08 inches of rain with a recurrence interval of 22,000 years.

This storm resulted in the highest-ever rainfalls at 80 river gaging stations from the Trinity River in the north to the Kaweah River in the south. Five stations reported over 10 inches of rain on December 11. Hobergs (south of Clear Lake) received a total of 20.50 inches during the two-day storm event. Felton (north of Santa Cruz) and Los Gatos Summit both reported their highest-ever two-day rainfall, with over 14 inches during the storm event.

Table 30 shows the elevation and precipitation for the December 9–11 storm event. The greatest precipitation was received on December 10.[630]

Table 30. Precipitation during the December 9–11, 1937 storm event.

| Reporting Station | Elevation (approximate) | Storm Total (inches of rain) | Drainage Basin |
|---|---|---|---|
| Crane Valley | 3,500 | 15 | San Joaquin |
| Auberry | 2,050 | 9 | San Joaquin |
| Huntington Lake | 7,000 | 10 | San Joaquin |
| Big Creek research facility | 1,950 | 12 | Kings |
| Balch Camp | 1,300 | 10 | Kings |
| General Grant | 6,600 | 15 | Kings |
| Giant Forest | 6,358 | 16 | Kaweah |
| Ash Mountain | 1,708 | 11 | Kaweah |
| Springville | 1,050 | 14.5 | Kaweah |

The most extreme flood-peak discharges were in parts of the Northern and Central Sierra. The December 1937 flood was widespread over the northern two-thirds of the state. It had a recurrence interval that was greater than 100 years on some rivers. There were several peaks of record in the Northern and Central Sierra. Damage was $15 million.

In coastal streams there was extensive flooding from the Russian River south to the Santa Clara Valley. There was extensive flood damage in the Feather River Basin. Main Street in Chester (east of Red Bluff) was washed away. A new record-high river stage of 31.95 feet occurred on the Sacramento River at Red Bluff on December 11. That was 1.5 feet higher than the previous record high of 30.5 feet set in 1909. The Yosemite Valley Highway was flooded by the Merced River. Extensive flooding occurred in the Tulare Lake Basin.[631]

The Kings River peaked at Piedra on December 11 at 80,000 cfs. That was the biggest flood since record-keeping began in 1895. That would remain the flow of record on the Kings until the 1950 flood. However, just as on the Kern, there were reminders of an earlier and bigger flood. At Pine Flat, the high-water marks of an early flood (believed to be the 1867–68 flood) were seven feet higher than the December 1937 flood. The 1867–68 flood remains the greatest flood on the Kings since at least the flood of 1805.[632]

The 1937 flood caused severe damage to the state's Kings River Hatchery. That hatchery was located on the South Fork of the Kings River, upstream from the junction with the North Fork.

The U.S. Forest Service mapped the high-water line of the 1937 flood in the vicinity of where the Cedar Grove Bridge would later be built. That line coincides reasonably well with the modeled 50-year flood event. That puts it in a category with other Cedar Grove floods of the past 70 years (1937, 1950, 1955, 1966, 1969, 1978, 1982, 1983, and 1997) that rise to the level of the modeled 50-year flood: that is, a flood event that occurs about every eight years on average. See the section of this document that describes Cedar Grove Flooding.

Giant Forest received 16.28 inches during the December 9–12 storm event, 14 inches of which fell during December 10–11. Ash Mountain received 9 inches during the December 10–11 period.

The Kaweah peaked near Three Rivers (USGS gage #11-2105) on December 11 at 33,300 cfs.[633] This was the highest flow since that gage was installed in 1903. This was believed to be the highest level since 1867. As on the Kings River, this would remain the flow of record until the flood of 1950.

The national parks' records said that the 1937 flood took out or damaged 13 of the 14 bridges that spanned the Kaweah's various branches; six of those bridges were in the parks. The Marble Fork Bridge near Potwisha was severely damaged (photograph on file in the national parks).

The records don't specify which bridge survived the flood unscathed, but it's tempting to think that it was the Oak Grove Bridge on the East Fork Kaweah. That bridge appears to be relatively flood-proof and we have found no record of it being damaged in any flood.

A concrete arch culvert under the Generals Highway at Dorst Creek was so badly damaged that it had to be replaced. The flood also washed out three trail bridges in the parks.

The national parks were closed for two days until the worst of the road damage was repaired. It took 10 days to rebuild the Marble Fork Bridge so that it was once again passable to traffic.

The approaches to the Pumpkin Hollow Bridge (Bridge #46-29) were destroyed, but the bridge survived. The parks' approach would be washed out again in the 1955 and 1966 floods, but the original bridge is still in use.

The Kaweah Hatchery near Hammond was severely damaged during the 1937 flood. The hatchery was repaired after that flood and operations continued.

The Dinely Bridge washed away.

Jim Barton recalled that warm rain came down for three days straight. It rained even at Lodgepole, so there was no ice to skate on at the skating rink. The mainstem of the Kaweah crossed the North Fork Drive above the Barton ranch and flooded their pasture. It flowed down North Fork Drive until it turned back toward its original channel across from present-day Flora Bella Farm.[634]

The North Fork Kaweah peaked on December 11 at 8,290 cfs. This was the largest flow on that river since record-keeping began in 1910. This would remain the flow of record until the 1950 flood.[635]

The Upper North Fork Bridge survived.

Jim Barton and his family gathered at the North Fork Bridge (aka the Three Rivers Bridge) to see how it would withstand the flood. That bridge was a wooden truss structure built in 1906. The post-and-timber bridge was anchored by four steel cylinders filled with concrete. It was the only public bridge across the mainstem of the Kaweah River. As Jim and his family watched, two young men — Orlen "Baldy" Loverin and Fred Gimm — drove onto the threatened structure from the North Fork side in Loverin's 1934 Chevrolet coupe. Upon reaching the Highway 198 side, they realized that the approach was gone, so they backed the car back toward the North Fork side. But now the water was too deep at that approach, and they were stranded. The men were able to get off the bridge by hanging onto a barbed wire fence while fording the raging water, but the car had to stay. The river rose several more feet, completely inundating the car. Eventually the bridge and car washed away. The bridge and car were found the next morning on the Thorn Ranch below where the present-day

225

North Fork Bridge stands. The car was pretty banged up, but had landed upright back on the bridge, its tires still on the runner planks.[636]

The two cylinders on the highway side of the bridge washed completely away and came to rest, along with some other pieces of the bridge, in a swimming hole (across from what would later be Pat O'Connell Towing) but was then known as the "Old Twenty." That, Jim said, was the end of that swimming hole.[637]

The Airport Bridge survived. At the time, there was no Kaweah River Drive. The road ended at the Taylor Ranch, just beyond the Three Rivers Airport, which had opened just two years before. At the Taylor Ranch, there was a private bridge that connected that area with Highway 198. It was built in two sections, connecting at an island in the middle of the river. That bridge also washed out, leaving North Fork residents stranded. Jack Hill, a county road foreman who lived on what is today the Anjelica Huston ranch, took his bulldozer to the Taylor Ranch and scratched out a road up to Dinely Drive. That route would later become Kaweah River Drive.[638]

It took a while to replace the North Fork Bridge so that cars could get across the mainstem of the Kaweah. Initially, people were accommodated by a cable and trolley system. A person would sit on a swing board, place their possessions in an orange box dangling from it, and pull themselves across and over the river. Jim Barton recalled that this system was in place in time for Three Rivers School's Christmas program as he remembers being on it in the dark with his entire family.[639] A cable trolley would again be used at this location after the North Fork Bridge was destroyed during the December 1955 flood.

A temporary bridge was put in place for 1½ years until a permanent North Fork Bridge could be completed. That new permanent bridge would be located next to the present-day Three Rivers Market and remained in use until it washed away in the 1955–56 flood.[640]

Highway 198 (later known as Old Three Rivers Road) crossed the South Fork Kaweah via a concrete bridge. The South Fork undermined that bridge, causing it to collapse in the center during the flood. A temporary plank bridge was installed until a new bridge could be built in the same location.[641]

Sometime after 1938, Highway 198 was realigned and a new bridge across the South Fork was constructed where Kaweah Park Resort is today.

The December flood severely damaged Terminus Beach. The Kaweah peaked at 35,000 cfs at McKay's Point on December 11. That was the highest flow since the gage was installed in 1916.[642] The flood apparently destroyed or otherwise overwhelmed the concrete weir at McKay's Point.

The Tule River peaked near Porterville on December 11 at 11,300 cfs.[643]

The Kern River peaked on December 11, 1937. It was a significantly smaller flood than the February 6, 1937, flood had been.[644]

Total flow for water year 1937 was 135% of the 1894–2011 average for the Kings, 157% for the Kaweah, 219% for the Tule, and 167% for the Kern.

The Kern River sent floodwaters into Tulare Lake for the first time since 1916. Tulare Lake reappeared on February 7, 1937, for the first time since 1924. American white pelicans, waterfowl, and shorebirds reappeared almost instantly and in incredible numbers. (See the section of this document that describes the Chronology of Tulare Lake for a description of this remarkable biological event.) By the end of water year 1937, Tulare Lake was about 13 feet deep at its deepest point (elevation 191.9 - 179 feet).

### *Big Creek Debris Flow*

In addition to flooding, the December storm caused a major debris flow in the lower Kings River Basin. Many debris flows in the Sierra are never recorded. We know about this debris flow in large part because it occurred on a USFS research station: the Pacific Southwest Research Station. That facility was known at the time as the California Forest and Range Experiment Station. The watershed is located immediately north of Pine Flat Reservoir. The event was analyzed and summarized by Jerry DeGraff, a geologist for the USFS.[645]

The storm began at 5:00 p.m. on December 9. Precipitation fell mainly as rain and ended at 7:00 p.m. on December 11. The rainfall included two high intensity periods of 2 inches for one half-hour and 1.5 inches for one hour in the Kings River Basin. In the Big Creek Basin, a tributary to the Kings River, the experiment station maintained weir and gage instrumentation on eight small subwatersheds ranging in size between 4 and 15 hectares. Total rainfall in the Big Creek Basin from the storm was 12.3 inches. The rainfall occurred when only eight inches of snow was present on the summits. Structures housing streamflow instrumentation near the mouths of the subwatersheds were destroyed or severely damaged by flooding which carried considerable debris.

While the damage to the dams and instrumentation on the subwatersheds in Big Creek was attributed to flooding, Jerry concluded that it was actually the result of a debris flow. Land slumps occurred in the upper parts of the subwatersheds. Photographs document shallow slope movements which lead into the channels. In the channel above one gaging station, the passage of the debris down the channel appears to have followed its own course rather than remaining strictly confined to the channel banks. Other photographs show the channels scoured to bedrock. The bulk of the debris was described as having originated in the channel bottoms and sides. The small dams were pounded terrifically by large boulders, some weighing as much as 5,000 pounds, which literally battered out the centers of three of the dams. At one of the subwatersheds, the passage of the debris left a four-foot-high debris levee which redirected flow and reduced damage to the dam. The presence of debris levees, the large size of transported boulders, the channel scouring, the variance of the path or track of the flow relative to the channel banks, and the shallow landslide sources in the upper watershed are all evidence of a debris flow rather than floodwaters.

### 1938 Floods (2)

There were two floods in 1938:
1. February–March
2. December

The winter of 1937–38 was a heavy snow year, the second such year in a row. Yosemite recorded 61.09 inches of precipitation in 1938, setting a record that would last until 1983.[646] The winter of 1938–39 was a moderate to strong La Niña event.

Following on the heels of the four 1937 storms, another heavy storm and flood event hit Central and Southern California just three months later on March 2–3, 1938. It was a 50–80 year flood event. The storm of March 2, 1938 produced some of the largest stream flows ever recorded in much of

Southern California. (The 1861–62 flood was much worse, but very few Anglos were living in Southern California at the time.) Bakersfield set a 24-hour precipitation record for the month on March 3. Sixteen stations (most of which were in Los Angeles and San Bernardino Counties) reported 10 or more inches of rain on March 2. It resulted in ⅓ to ½ of the average annual rainfall at those stations in that one day. Records were set by the resulting flood that wouldn't be broken until the 1969 flood. The flood totaled $79 million in damages and resulted in 87 deaths.[647]

In February and March, 1938, heavy storms flooded the San Joaquin Valley.

Panoche/Silver Creek west of Mendota flooded sometime in 1938, probably in February or March.[648]

The Kaweah River crested in Visalia on the night of February 26.

The late February part of the 1938 flood was centered in the Tulare Lake Basin. The early March part of the flood was a major event and affected all of Southern California.[649]

Migrant laborers suffered the most from the flooding. John Steinbeck came to Visalia in February 1938 to help relieve their suffering. This experience was apparently the motivating factor in his writing *The Grapes of Wrath*. Horace Bristol photographed the migrant encampment in Visalia and elsewhere. Steinbeck based the central characters in his masterpiece on the farm workers that he and Bristol encountered in Visalia that winter. The book's climatic flood was based on what he witnessed in Visalia. Bristol's photographs were used to cast the movie and were later published in *Life* magazine.

The Kaweah's peak natural flow occurred at McKay's Point on December 11: 34,799 cfs. (That was the peak hourly flow; the peak average daily flow, as reflected in Table 15, was 11,232 cfs.) Based on the flood exceedence rates in Table 16, this had a recurrence interval of 8 years for the Kaweah.

A great flood struck Los Angeles County on March 3, resulting in $45 million dollars in damages and 113 deaths. A total of 5,601 homes were destroyed and another 1,500 were severely damaged. Thousands of people had to be evacuated and thousands more were left homeless. Two CCC camps were destroyed and over 300 relief workers had to be rescued. A total of 91 highway and railroad bridges were destroyed or badly damaged. With all the rail lines out of service, there was no mail delivery, so mail was taken by the U.S. Coast Guard between L.A. and San Diego. The peak flow of the Los Angeles River at Long Beach exceeded the average flow of the Mississippi River at St Louis.

Orange County also experienced a great flood in early March. It was the most destructive in the county's history, resulting in 19 deaths and 2,000 homeless. Eyewitness accounts say that an 8-foot-high wall of water swept out of the Santa Ana Canyon. At the peak of the flood on March 3, the Santa Ana was flowing at an estimated 100,000 cfs. Near the mouth of the river, the Santa Ana overflowed its banks and covered an area 15 miles long and 5 miles wide.

San Bernardino County experienced major flooding from March 1–5 due to a series of storms which resulted in very heavy rainfall. This flooding event seems to have been centered in the upper Santa Ana River Basin. Some areas in that watershed received over 30 inches of rain during the event. Over 100 bridges were destroyed and 800 miles of roads were lost. Over 150 homes were destroyed and many more flooded, leaving over 1,000 homeless. Most USGS gaging stations were destroyed. Cajon Pass was closed to traffic due to miles of road destruction, bridges washed out, rail lines destroyed, and dozens of landslides. All communications were cut off; the only routes left open were by foot or air. The Mojave River experienced a major flood; 22 homes were swept away in Victorville. The peak discharge from the 1938 flood exceeded any flood since the 1861–62 flood which is considered the flood-of-record for this area. Damage in the county exceeded $11 million dollars, and 22 people died.

Riverside County was extensively damaged by the flood of March 1–3. The northern section of Riverside was inundated and many people were forced from their homes. Men, women, and children had to be rescued from trees as they were unable to reach higher ground when their homes became imperiled. Fairmont Park saw great destruction when the dam at Lake Evans was ripped out by floodwaters. Livestock of all sorts was lost to flooding in the Santa Ana River. Damage to roads, bridges, and rail lines in the county was extensive.

December precipitation at both Giant Forest and Ash Mountain was the greatest since record-keeping began in 1920. Between December 9–12, 16 inches of rain fell in Giant Forest, bringing the Kaweah to flood condition and causing unprecedented damage to roads and trails. Six bridges were destroyed. In addition, the Marble Fork Bridge near Potwisha was badly damaged, and the arched culvert over Dorst Creek was so badly damaged that it had to be replaced. Damage and bridge destruction immediately outside the national parks was even greater than damage within the park.

Total Giant Forest precipitation during the calendar year was 66 inches (5½ feet), the greatest since record-keeping began in 1920.

The Kern River peaked near Bakersfield on March 3 at 14,600 cfs.[650]

Total flow for water year 1938 was 189% of the 1894–2011 average for the Kings, 202% for the Kaweah, 254% for the Tule, and 187% for the Kern. This was the first time since record-keeping began in 1894 that the Kaweah River had two back-to-back years with flows that were over 150% of average.

The 1938 flood caused major flooding in the Tulare Lakebed. When the elevation of Tulare Lake reached 192 feet, one of the main levees in the lakebed broke and the lake spilled over 49 square miles of land. The lake continued rising, eventually cresting at 195 feet. By June, 135,600 acres of the lakebed were underwater. That was the maximum acreage flooded since the 1906 flood. Tulare Lake has not been this big since.[651] By the end of water year 1938, Tulare Lake was about 16 feet deep at its deepest point (elevation 195.0 - 179 feet).

In the lakebed, the barley harvest would normally have started in mid-May, but had been delayed by the after-effects of the December 1937 flooding followed by the heavy spring rains. Because of those delays, the harvest got underway just as the rivers went on a rampage, tearing into the levee systems. The farmers found themselves in a race to bring in the harvest before the various lakebed levees failed. As harvesters worked around the clock, earthmovers and an army of shovel-wielding men fought on the levee banks. If the harvesters won and got the grain pulled over the levee in tractors and eased down into the next block, then the levee would be blown up. But if the levee broke before the harvesters were finished, then the harvesters stayed in the field, working just ahead of the slowly spreading water and, at the last minute, were jerked out of that field and into the next.[652] The USBR estimated that 126,000 acre-feet of water came into the Tulare Lakebed during water year 1938.[653]

While the high lake levels of 1938 were a disaster for the lakebed farmers, others saw their opportunity. Near the height of the flood, Frank Latta and three boys took a 15-foot homemade motor boat from Bakersfield to San Francisco. They left Pioneer Weir on the Kern River on June 18, 1938. The trip ended 14 days later at the wharf on the south end of Treasure Island.[654, 655]

Treasure Island had been built specifically for the Golden Gate International Exposition (aka World's Fair). The exposition would open to the public on February 18, 1939. Building of the exhibits was well underway when Latta and the boys were there, so they billed their trip as a visit to see the exposition. This was the third documented trip between Tulare Lake and San Francisco Bay. (The other trips were in 1852, 1868, 1969, and 1982–83.)

## 1939 Flood

Flooding in 1939 occurred in June.

A very intense summer storm struck Fresno on June 14, 1939. At one point, the rain was coming down at a record-setting rate of 0.65 inch in 10 minutes (a rate of 3.9 inches per hour).[656] This almost certainly caused flooding in the area.

1939 was also the year that the Boyden Bridge (Bridge #42-24) was completed on the newly constructed Highway 180 in Kings Canyon. The Grant Grove approach to that bridge has been washed out on at least two occasions (1955 and 1997), but the original bridge is still in use.

The original Cedar Grove Bridge is believed to have been erected shortly after the Boyden Bridge was completed, but that is based on supposition.

## 1940 Flood

There was at least one, and possibly as many as three, flood events during 1940:
1. flooding at an unknown time on Dry Creek
2. flooding at an unknown time on the Kern River
3. flooding in October in Bakersfield

Sequoia National Park received much more than average precipitation in 1940, but it was largely in the form of rain. Snowfall was less than half of what had occurred in 1939. No flooding was reported in the national parks.

Torrential rains pounded the hills east of the Visalia Electric mainline, flooding Dry Creek and washing away a 45-foot trestle at Dry Creek.

The Kern River flooded enough to damage Highway 178 through the canyon.

A storm dropped 1.51 inches of rain on Bakersfield on October 25, 1940, making that the wettest October day ever in that city.[657] Such storms are typically intense and result in street flooding.

## 1941 Floods (2)

There were two floods in 1941:
1. February
2. Sometime during the April–July snowmelt runoff period

A very intense storm struck Fresno on February 24, 1941. At the peak of the storm, rain was coming down at a record-setting rate of 0.48 inch in 5 minutes (equivalent to 5.78 inches per hour).[658, 659] This almost certainly resulted in flooding in the area. This storm is sometimes reported as having occurred on February 24, 1951. However, the NWS forecast office in Hanford researched their files and confirmed that 1941 was the correct year.

Statewide, the four wettest water years during historic times were 1890, 1941, 1983 and 1995. The heavy rains of 1941 were confined largely to the Sacramento Valley and a narrow strip of land on the South Coast from Santa Barbara to Orange Counties. Both Willows (near Chico) and Santa Ynez (near Santa Barbara) had rain totals for the water year with recurrence intervals in excess of 5,000 years.[660]

In the Tulare Lake Basin, runoff in the spring of 1941 was heaviest in the south end of the valley. It was a much bigger flood on the Kern than on the Kings or Kaweah. A photograph on file in the national parks shows the Kern in flood at the historic wooden Bellevue Weir in Bakersfield (just upstream from the present-day Stockdale Highway) in 1941.

Total flow for water year 1941 was 146% of the 1894–2011 average for the Kings, 149% for the Kaweah, 169% for the Tule, and 187% for the Kern. So much water was delivered to Tulare Lake that the lake's elevation rose 12.2 feet (from elevation 184.5 to 196.7). Tulare Lake has not been that high since.

### 1941 Wind Event

The February 24, 1941 storm wasn't the first vigorous weather system to hit the Tulare Lake Basin during that winter. On January 8, a windstorm of almost hurricane velocity struck Garfield Grove at about 7,000 feet elevation. According to the superintendent's annual report, about 1,000 trees were blown down, including some giant sequoias 20 feet in diameter. The storm then moved north into the drainages of the East Fork and Middle Fork of the Kaweah where hundreds more trees were blown down. Cleaning this up (in crosscut saw days) created "unusual trail maintenance problems."

The national parks have experience with extensive tree failures during the winter, primarily from heavy snow loads and avalanches. Winds have also caused small-scale blowdowns. However, the January 1941 event is apparently the only large-scale blowdown to occur in the national parks in historic times.

Broadly speaking, the Southern Sierra could theoretically experience at least five categories of strong winter winds capable of causing forest blowdowns:
1. Winds associated with the passage of a cold front (either with or without embedded thunderstorm cells)
2. Mono winds
3. Mountain waves (caused by either the passage of a cold front or a low-level jet stream that crosses the crest)
4. Jet stream winds protruding from up in the stratosphere and coming down near the surface.

5.  Low-level barrier jet winds hitting the west slope of the Sierra, resulting in strong upslope winds.

The first category, winds associated with the passage of a cold front, includes both local thunderstorm outflow winds and the cold-front generated winds that accompany those fronts. Cold font storm systems can generate gusty winds and downdrafts that result in the blowdown of a few trees here and there. The following two examples illustrate this type of wind event:

*   Frontal winds associated with a storm front (a cold front) on January 1–2, 2006 brought down the Telescope Tree and the second largest limb on the General Sherman Tree. That storm also blew down a number of other trees and power poles throughout the central and southern San Joaquin Valley. Gary Sanger at the NWS forecast office in Hanford said that this event likely was dominated by frontal winds. However, there may have been unreported embedded thunderstorms (and thundersnow) in the cold front's convective band. The National Severe Storms Laboratory has documented a few instances of microbursts associated with heavy showers that did not generate thunder. In those instances, there was apparently drying below the cloud base, and the dry air was caught in the updraft into the cumulus clouds. This caused the rapid cooling of the center of the storm, with the cold, denser air dropping toward the ground (same as a collapsing thunderstorm core). So the winds on January 1–2, 2006, may have had an isolated microburst embedded in the general wind field, but there was not enough evidence to conclusively state this. For more information on this event, see the section of this document that describes the 2005–06 floods.

*   The Southern Sierra experienced four days of wind as three storms moved through Central California from January 20–23, 2012. These wind events were a combination of (1) local thunderstorm outflow winds and (2) cold-front generated winds that accompanied the fronts. The second of the three storms moved through Central California on January 21. The passage of this cold front triggered pre-dawn thunderstorms over the region, including Yosemite. There were numerous witnesses to the pre-dawn Yosemite thunderstorm on January 21. Strong winds associated with the passage of that storm caused the failure of four live trees in a small portion of Yosemite Valley shortly thereafter. Brian Mattos, Yosemite's forester, reported that the tree failures all seemed to radiate from a point near the east end of Stoneman Meadow. One of the trees that failed was a large green ponderosa pine (dominant) which fell on a tent cabin in Curry Village, killing the concession employee inside. That employee was living there while waiting for the Badger Pass Ski Area to open so that he could start his winter job. He had previously served as an NPS ranger at Yosemite and at Devils Postpile NM. According to Gary Sanger, the winds that caused the blowdown were probably outflow from a collapsing thunderstorm cell. An extreme case of a collapsing cell would be a wet microburst in which the falling core acts like a piston to force the surface winds rapidly away from the collapsing cell. Gary thought that there wasn't enough information to determine whether the Stoneman Meadow event was a microburst or a more general, less extreme, thunderstorm collapse. According to Rhett Milne, warning coordination meteorologist at the NWS in Reno, it is very rare for downburst/outflow winds of any sort to cause blowdowns, especially in the middle of winter. The cold front that caused the Stoneman Meadow downburst outflow winds produced strong winds as it moved south. At least two funnel clouds were observed over Fresno County. Winds gusted to 90 mph as measured by the RAWS automated weather station at the Isabella Dam. (The River Kern RAWS near Kernville often reports strong gusts that funnel through that part of the Kern River Canyon into the Lake Isabella area. In Gary Sanger's opinion, the 90 mph gust was a freak event in terms of its strength.) Steve Bumgardner recalled that the wind was very strong in Lodgepole on January 21, and a few trees were blown down in that area. The Lodgepole weather station reported a

thunderstorm in the distance to the east on that day. Gary Sanger thought that the winds at Lodgepole were likely frontal in nature, generated as the cold airmass behind the front pushed under the warmer air ahead of the front. The warmer airmass lifts along the frontal boundary, enhancing instability. Often cold fronts bring moisture and even flooding. However, this was a dry cold front, so it did nothing to break the dry spell that began on November 20, 2011.[661]

The second category of strong winter winds, Mono winds, is capable of much bigger blowdowns. A Mono wind is a type of katabatic wind like the Santa Ana wind in Southern California. A katabatic wind is the technical name for a drainage wind, a wind that carries high-density air from a higher elevation down a slope under the force of gravity. Katabatic winds can rush down elevated slopes at hurricane speeds, although most are not that intense. Not all downslope winds are katabatic. The term does not include rain shadow winds where air is driven upslope on the windward side of a mountain range, drop their moisture, and descend leeward, drier and warmer (e.g., the Chinook wind that occurs along the Front Range of the Rockies).[662]

The Mono wind originates from cold, dry air over the Great Basin. The most pronounced winds cut through a relatively low portion of the Sierra in Mono County, California, hence the name. The wind then spills out of high mountain valleys and streams down the canyons on the west slopes of the Sierra.

As the wind descends, the pressure on the air mass increases. This pressure increase causes the temperature of the air to increase. People in the path of the wind often experience a dramatic temperature increase.

Areas downwind of Mono County are subject to winds of gale force speeds (32–63 mph). Mono winds have knocked down 100-foot-tall trees and have been clocked at 100 mph in Yosemite Valley.[663]

The classic Mono wind pattern is northeast to southwest.[664] Research by Michael Fosberg determined that Mono winds are responsible for most of the trees blown down on the Kings River Ranger District in the Sierra National Forest.[665] That district is located in the North Fork Kings River Basin, a drainage that trends generally northeast to southwest. Mono County is northeast of the North Fork, so that drainage is perfectly aligned for the winds that come from the Mono County area.

There are no significant breaks in the Sierra mountain wall south of Bishop, California. The high mountain crests of the Sierra and Great Western Divide generally prevent Mono winds from reaching the surface within Sequoia and Kings Canyon National Parks.[666]

It's instructive to look at what a powerful Mono wind event looks like in the Tulare Lake Basin. The east side of the Sierra experienced very strong winds from the NNE from November 30 – December 2, 2011. They were particularly high on the night of November 30. An automated station at the summit of Mammoth Mountain recorded 14 hours of sustained winds over 120 mph with gusts in excess of 150 mph (the limit of the anemometer). (For comparison, a sustained wind of 150 mph is equivalent to an EF3 Tornado or a Category 4 hurricane.)

Rhett Milne analyzed the event. Nothing remotely like this wind event had been recorded on Mammoth Mountain in the past 12 years; the sustained winds in this event were much higher and lasted for a much longer period. Rhett estimated that the top gusts may have been roughly 180 mph (30% greater than the sustained speed of 140 mph).

The strong winds in this event were caused by a huge high pressure system off the West Coast coupled with a huge low pressure in the Desert Southwest. This resulted in incredible pressure gradients, especially east-west across the Sierra. Much of Southern California was windy in this event, but the area around Reds Meadow and Devils Postpile National Monument had the perfect topography and NE/SW alignment to be severely impacted by this downslope windstorm. This wind event affected large sections of the San Joaquin River Basin from Tuolumne Meadows to Mt. Whitney. An estimated 5,000 trees were knocked down just in Devils Postpile National Monument alone.[667, 668]

On the west side of the Sierra, this was perceived as a Mono wind event. Winds gusted to 45 mph at Fresno on December 1, just shy of the record gust for December of 48 mph set on December 28, 1991. Winds gusted to 60 mph at Tioga Pass. Trees were blown down at several locations, including Clovis and Mariposa. Brian Mattos recalled that Yosemite experienced multiple tree failures from elevation 4,000 feet up to over 9,000 feet during this event. Yet despite the strong winds in the Fresno / Yosemite area, the southeastern part of the San Joaquin Valley remained wind-sheltered, allowing areas of dense fog to develop during the night of November 30 – December 1.[669] Steve Bumgardner recalled that the wind was quiet in Lodgepole during this event.

Mono winds such as this occur periodically in Yosemite National Park. However, Mono winds are a generally unremarkable phenomenon in Sequoia and Kings Canyon National Parks except along the ridges and peaks. Mono winds strong enough to cause forest blowdowns don't appear to occur south of the North Fork Kings River.

The third category of strong winter winds is mountain waves. A mountain wave is an atmospheric standing wave formed on the lee side of a mountain range when wind blows over that mountain.

The lee slope of mountains may experience strong downslope winds or many eddies of various sizes which roll down the slope. Within each wave downstream from the mountain range, a large roll eddy may be found with its axis parallel to the mountain range. Roll eddies tend to be smaller in each succeeding wave downstream. The waves downwind of the mountains are referred to as lee waves or standing waves. If sufficient moisture is present, cap clouds will form over the crest of the mountains, roll clouds will be found in the tops of the roll eddies downstream, and wave clouds will be located in the tops of the waves.

Mountain waves occur only on the lee side of mountains. They are often caused by the passage of a cold front. So in our area, mountain waves that are caused by the passage of a cold front occur only on the east side (the lee side) of the Sierra.

But Gary Sanger says that mountain waves often occur when there is a low-level jet (around 700 mb or lower) that crosses the crest. The location of the wave is dependent on the orientation of the jet, and can be on either side of the crest, whichever side is downwind (the lee side).

The Indian Wells Canyon area (15 miles northwest of Ridgecrest) is particularly prone to mountain waves when east-flowing winds funnel through Walker Pass into the Inyokern area. The Bureau of Land Management has a RAWS automated weather station located in the hills there at about 4,000 feet in elevation, downslope of Walker Pass. That station often records very high winds. The mountain waves at Indian Wells Canyon can be indicative of either a low-level jet or a cold front. All that is required is that the winds at 5,000-6,000 feet (around 700 mb or a bit lower) line up perpendicular to the crest in that location to generate standing waves downstream of the crest.

Forest blowdowns due to mountain waves touching down are an occasional occurrence in the Rockies. The largest such forest blowdown ever recorded in the Rockies was the October 25, 1997 event that blew down 20,000 acres on the west side of the Park Range northeast of Steamboat Springs, Colorado.[670]

We have not found a forest blowdown in the Sierra that was attributed to a mountain wave, but it is a theoretical possibility. A key feature of such blowdowns is that they occur in the lee of mountains, downwind of the crest.

Gary Sanger researched the Kaweah blowdown event of January 8, 1941. Bakersfield experienced a gust to 47 mph from the east-southeast on that day. (Bakersfield is roughly 100 miles SSW of where the Kaweah blowdown event occurred.) The Giant Forest weather station reported strong winds out of the southwest on January 8. From this we know that the winds were blowing upslope. This eliminates the possibility of the blowdown being caused by a mountain wave touching down. Mountain waves form only on the lee side of a mountain range, so the wind would have had to be blowing out of the east.

The fourth category of strong winter winds capable of causing forest blowdowns is when the jet stream protrudes from up in the stratosphere and comes down near the surface (aka a tropopause fold). This lowering of the jet stream does not occur very often and is most likely to be experienced in a high mountain range like the Sierra. The President's Day Cyclone of 1979 in the Northeastern U.S. was partially attributed to the lowering of the jet stream. The huge flare-up of the 1988 Canyon Creek Fire in Montana was also attributed to the lowering of the jet stream. We have not found a forest blowdown anywhere in the U.S. that has been attributed to a lowering of the jet stream, but it is at least a theoretical possibility.

The fifth category is low-level barrier jets. A barrier jet is a jet-like wind current that forms when a low-level airflow approaches a mountain barrier and turns to the left to blow roughly parallel to the axis of that barrier. In the Sierra, that results in barrier jets that blow from the south or southwest. Despite the similarity in name, low-level barrier jets aren't directly related to the jet stream. The airflow is upslope, barrier jets occur on the windward side of the mountain. The strongest winds tend to be elevated off the surface, but top wind speeds can reach up to 100 mph.

Gary Sanger speculated that a low-level barrier jet (around 800 mb) might have been the culprit in the 1941 wind event. Although we aren't aware of any similar situation, Gary thinks that a southwesterly jet at around 6,000 feet elevation hitting the west slope of the Sierra might have triggered the type of strong upslope winds that occurred in the 1941 event.

This seems plausible, and most of the other explanations have been eliminated. The most likely alternative explanations would be:
- the jet stream protruding from the stratosphere and coming down near the surface
- the passage of a very powerful cold front

We don't know enough to eliminate the jet stream from consideration. Cold fronts typically don't produce winds nearly strong enough to cause the damage observed in the 1941 event. In any case, the records from the Giant Forest weather station don't suggest that a particularly strong cold front passed through during the 1941 event. Looking at the temperatures, the cold front on January 4 can be seen not only in the wind shift, but also in the cold airmass behind the front. (The high on the 4th was only 36, down 13 degrees from the 3rd). For January 8, the temperatures are consistent with warm-sector precipitation, but there was neither a subsequent shift in prevailing wind direction or evidence of a cold (post-frontal) air mass.

Therefore, we're left with a low-level barrier jet as seemingly the most likely cause. But since there are a lot of unknowns, we really cannot draw any firm conclusions. Regardless of the cause, this was a most unusual event. If the 1941 event were caused by a low-level barrier jet (or by a lowering of the jet stream), this might be the only situation of a forest blowdown being caused by such a wind in the Sierra or anywhere else in the U.S.

## 1942 Flood

Flooding in 1942 occurred in January.

The winter of 1941–42 was an El Niño event. One source said that it was a strong El Niño, but that could not be confirmed. The NOAA index of El Niño / La Niña events only goes back through 1950.)[671]

Flooding occurred on the Kaweah and Kern Rivers and possibly other rivers within the Tulare Lake Basin. The national parks' records make no mention of any flooding that year.

The Kaweah had a peak average daily flow at McKay's Point of about 11,000 cfs on February 2.[672] Visalia was flooded, though not nearly as badly as it would be in the 1955–56 flood. There was a small break in the south levee on the St. Johns River near Miller's Bridge (Fourth Ave East), northeast of Visalia.

Bakersfield was flooded in 1941, so perhaps this was a December 1941 – January 1942 flood. In any case, Highway 178 through the Kern Canyon was damaged in the 1942 flood.

Total flow for water year 1942 was 117% of the 1894–2011 average for the Kings, 114% for the Kaweah, 97% for the Tule, and 111% for the Kern.

The 1937 flood had brought Tulare Lake back to life on February 7, 1937. Subsequent floods kept the lake generally at an elevation of 190 feet or above through 1944, a level that hadn't been seen in decades. American white pelicans thrived during this period. In 1942, Frances von Glahn made a color movie of them nesting at the lake (video on file in the national parks). So much water was delivered to Tulare Lake in water year 1942 that the lake's elevation was raised 10.6 feet (from elevation 183.3 to 193.9 feet).

## 1943 Floods (3)

There were three floods in 1943:
1. January
2. March
3. April

The dry season of the previous year lasted until January 20. There wasn't enough snow to ski in Giant Forest until January 31. A severe storm occurred from January 20–23, and it rained almost continuously through at least the end of the month. The storm dropped a record 20 inches of rain in Giant Forest and 8 inches at Ash Mountain.

Central and Southern California received a widespread series of storms during the last half of January. Hoegees Camp near Mt. Wilson in the San Gabriels received 26.12 inches of rain in 24 hours on January 22, setting a state record. The recurrence interval for that storm event was 11,000 years.[673]

There was much storm damage in the national parks. Roofs were blown off, and there were heavy landslides and washouts on the road and trail systems. The giant sequoia at Puzzle Corner fell during the first couple of days of the storm.

The Kings River peaked at Piedra on January 21 at 46,900 cfs. The Kaweah peaked near Three Rivers on January 22 at 17,000 cfs.[674]

A second round of flooding occurred during March. The floods were concentrated in the Sierra south of the Feather River. During February, there had been occasional periods of light rain as well as warm weather conducive to melting of the mountain snowpack. By early March, the ground was moist and the river stages moderately high. There were a series of light rainstorms from March 4–8. Then heavy rains fell in the mountains and foothills on March 9–10 and March 17–18. The rains on the night of March 9 were particularly heavy and widespread in the foothills. Ten inches of rain fell in Giant Forest on top of a snow base of 29 inches. That cloudburst occurred just as the rivers were nearing crests from the earlier rains.

The Kings River passed flood stage on March 9–10 with only minor damage.

The Kaweah's peak natural flow occurred at McKay's Point on March 9: 17,765 cfs. (That was the peak hourly flow; the peak average daily flow, as reflected in Table 15, was 9,714 cfs.) Based on the flood exceedence rates in Table 16, this had a recurrence interval of 7 years for the Kaweah.

Flooding on the Kaweah, Tule, and Kern Rivers caused considerable damage. The Kaweah peaked at McKay's Point at 14,300 cfs on March 9.[675]

The Tule River peaked near Porterville on March 9 at 15,500 cfs. This was the biggest flood on that river since record-keeping began in 1901. It would remain the flood-of-record until the 1950 flood.

The White River had a major flood: 2,300 cfs. This remains the flood-of-record for that river.[676]

The Kern River peaked near Bakersfield on March 9 at 21,700 cfs.[677] This was the biggest flood on that river since record-keeping began in 1896. It would remain the flood-of-record until the 1950 flood. The Kern River was so high in this flood that it overtopped the old Olcese's Ranch Bridge, a mile downstream from the mouth of the Kern River Canyon.

There was a major flood on Caliente Creek in April 1943, causing extensive flood damage to the Lamont/Arvin area. Presumably this was caused by an intense storm.

The overflow from these streams raised the level of Tulare Lake to near the top of the lakebed levees. Wave action caused levee breaks and the flooding of 28,000 acres. These levee breaks increased the size of Tulare Lake from 46,000 acres to 74,000 acres. By summer, 100,000 acres would be flooded.

Total flow for water year 1943 was 117% of the 1894–2011 average for the Kings, 156% for the Kaweah, 261% for the Tule, and 166% for the Kern. This was one of the very rare years when flows on Tule River exceeded what could be used for beneficial use by the holders of water rights. Enough water was delivered to Tulare Lake in water year 1942 to raise the lake's elevation 5.8 feet (from elevation 189.9 to 195.7 feet).

## 1943–51 Drought

This drought affected parts of the state from 1943–51. It affected the entire state from 1947–49.

The drought was most severe in central and southern coastal areas, where accumulated deficiencies in runoff approached, and in some instances exceeded, those of the drought of 1929–34. Water year 1951 ranks as the driest of record at several gaging stations in Southern Coastal California.

Recurrence intervals for the drought of 1943–51 were about 20 years in the Central and Northern Sierra because of the short duration there. They were about 20–80 years in the rest of the state, where this drought is exceeded in duration and severity only by the drought of 1929–34.

The historical record of the Sacramento Valley Water Year Index indicates that the drought of 1943–51 (recurrence interval of 55 years) ranks second only to the drought of 1929–34.

Table 31 shows how the San Joaquin Valley Water Year Index categorized the drought years in that basin.

Table 31. Rating of drought severity during the 1943–51 drought.

| Water Year | San Joaquin Basin Water Year Classification |
|---|---|
| 1943 | Wet |
| 1944 | Below normal |
| 1945 | Above normal |
| 1946 | Above normal |
| 1947 | Dry |
| 1948 | Below normal |
| 1949 | Below normal |
| 1950 | Below normal |
| 1951 | Above normal |

1947 was the driest year ever in Paso Robles, with only four inches of rain.

After the 1937 flood, the elevation of Tulare Lake was relatively stable for the next 10 years. Then, thanks in large part to the 1943–51 drought, the lake was dry from July 17, 1946, to November 19, 1950.

At a state level, the drought is recognized as spanning the years 1943–51. Rivers within the Tulare Lake Basin had about average yearly flows for the first four years of the drought (1943–46). Then the drought moved into the basin. From 1947–50, flows in the Kings and Kaweah Rivers were generally 55%–75% of average. Total flows in the Tule and Kern were generally 35%–60% during this same period.

In the national parks, the drought was viewed as beginning in the early winter of 1946–47. Drought conditions of record-breaking proportions prevailed over the parks through February 1948. Livestock by the thousands were shipped from the parched ranges of Central California. Demands arose to open the national parks to commercial grazing as one method of relief for stock growers. Fortunately, above-average rain and snow fell during March, April, and May, 1948.

Sequoia National Park's district ranger Clarence Fry took measurements on July 10, 1949, at three locations along the Middle Fork Kaweah between Potwisha and Ash Mountain. He found only seepage between pools, not enough for fish to swim through or pass over.

By 1951, all the rivers of the Tulare Lake Basin — except the Kern — had returned to approximately average flows.

## 1944 Flood

Flooding in 1944 occurred in March. This flood occurred during the 1943–51 drought.

There was a major flood on Caliente Creek in March 1944, causing extensive flood damage to the Lamont/Arvin area. Presumably this was caused by an intense storm.

Judging from historic photographs, Southern California Edison's (SCE) Borel hydroelectric facility had a flood canal cut sometime shortly before March 6, 1944 (photograph on file in the national parks). It's tempting to think that this was due to the same storm that caused flooding on Caliente Creek, but that isn't known.

## 1945 Floods (3)

Flooding occurred three times in 1945:
1. January–February
2. October (twice)

These floods occurred during the 1943–51 drought.

The storm of January 30 – February 3 dropped a total of over 13 inches of precipitation at Giant Forest. Precipitation in the national parks consisted of more rain than snow below about 7,500 feet. At times it was apparently raining as high as 8,500 to 9,000 feet, but rain at the higher elevations was absorbed by the already good snowpack.

239

Visalia received 3 inches of rain during February 2–3. Ash Mountain received 6 inches during the storm. Giant Forest received 12 inches during the storm, of which 8 inches fell on the night of February 2.

The American and Sacramento Rivers flooded, as did presumably most of the rivers in the Sacramento River Basin.

The Kings River peaked at Piedra on February 2 at 49,300 cfs.[678] Parts of Centerville were inundated when the Kings flooded. There was extensive flooding farther downstream, both north and south of Hanford.

The flood was written up in a special edition of the *New York Times*.[679] According to that account, many houses had to be evacuated in the San Joaquin Valley and farms were inundated.

The Kaweah's peak natural flow occurred at McKay's Point on February 2: 18,554 cfs. (That was the peak hourly flow; the peak average daily flow, as reflected in Table 15, was 9,890 cfs.) Based on the flood exceedence rates in Table 16, this had a recurrence interval of 8 years for the Kaweah.

The national parks' annual report made no mention of any flooding in 1945.

The North Fork Kaweah washed out the Airport Bridge, leaving that portion of the Three Rivers community cut off from the highway.[680]

Visalia flooded on February 2–3. This was described at the time as the most severe flooding ever to hit the town. The flooding in Visalia was big enough news that troops in the South Pacific heard about it on the radio.

The causes of the flooding — and the reasons that it was so severe — were described in a series of newspaper articles.[681, 682, 683, 684, 685, 686] The 1906 and December 1937 floods were much bigger events, but Visalia withstood those earlier floods better than it did the 1945 flood. There appear to have been three reasons for this:
1. Visalia was now a bigger and more modern town. Unlike in 1906, it was no longer built for periodic inundations.
2. The primary difference from the earlier floods was that the St. Johns River on the north side of town was no longer able to carry big floodflows. The levee along that river's south bank wasn't being maintained. The channel wasn't being kept clean of debris and trees. Islands and flood deposits had built up in the floodway. A bridge had been built across the channel at the narrowest portion, acting rather like a dam. Most important, a levee had been built on the north side of the St. Johns so that the floodwaters couldn't spread out; the river was effectively restricted to a narrow and very inadequate channel.
3. The 1945 flood came on as a sudden rolling wave instead of a gradual rise. This may have been due in large part to the St. Johns being channelized between two levees rather than being allowed to spread out across its floodplain.

Flooding of Visalia in 1945 resulted from four levee breaks on the St. Johns, all in the vicinity of Miller's Bridge (Fourth Ave East), northeast of the city (photograph on file in the national parks). The breaks occurred about 10 p.m. on February 1, and the water reached Visalia about three hours later. The Kaweah peaked at McKay's Point at 10:30 a.m. on February 2, and the depth of water in Visalia peaked about 6 p.m. that afternoon. Downtown Visalia was heavily damaged.

Water was 3–4 feet deep in the northeastern part of the city and more than a foot deep on some of the downtown streets (multiple photographs on file in the national parks). The current coming down Center Street was particularly strong. For the first time since 1906, a rowboat appeared on Visalia's city streets. It was seen going down Center Street between Court and Church on the morning of February 2, bobbing along the turbulent, muddy current. It then turned up Church Street and continued on to Main.

Main Street was closed to vehicular traffic by 10:00 that morning to stop the wakes that were being thrown into adjacent businesses and homes. Similar problems were happening on nearby streets. Mrs. C.C. Bennett lived at 301 East Mineral King Ave. When she stepped outside to sweep away the water from a wave caused by a passing car, there on her porch was a small, golden-colored river fish which had come to town on the flood.

Over two-thirds of Visalia was flooded. It was a common sight to see a man pick up a lady and carry her across a flooded area. Among the many flooded areas were the Fox Theater, the Tulare County Courthouse (located on Court Street between Oak and Center), and homes on Bridge Street near where the Visalia Convention Center now stands. The flooding was so extensive that it closed TAD's Drive-In (later renamed Mearle's) located in what was then considered the far south side of town.[687]

The city water supply remained safe to drink; the floodwaters only reached one well, and that well was isolated from the rest of the system.

The State Division of Forestry declared Visalia an area of emergency and sent pumps to help in the city as well as bulldozers to assist with repairing the levee.

The Tule River peaked near Porterville on February 1 at 12,600 cfs.[688]

The Warthan Canyon Highway near Coalinga was closed on February 2–3, indicating that there was flooding on the west side of the Tulare Lake Basin.

1945 was the first year of use of the new works, built by the USACE, to keep the Kings River out of Tulare Lake. They did not work quite as designed. A break in the bypass occurred on February 3, about 20 miles south of Hanford at the height of the flood. Some ranchers were driven from their homes on the east side of the bypass and considerable grain was flooded on the west side. Nearly 1,000 people were forced to evacuate their homes.[689]

The J.G. Boswell Co. bought the Cousins Ranch in 1946. At that time, the ranch had been under the waters of Tulare Lake for eight years, ever since the 1938 flood.

On October 6, a cloudburst dropped 2.75 inches of rain on the town of Tehachapi in 1½ hours.[690] Rainfall intensity in the nearby mountains was evidently greater. A wall of water estimated to be eight feet high swept down Tehachapi Canyon, killing three people and causing property damage estimated to be $62,000. About half of this damage was to property in Tehachapi. Several hundred acres of crop land around Tehachapi were heavily damaged. Several hundred feet of railroad track at Keene and near Caliente were washed out. Transportation (both rail and highway) and communication lines were shut down for 24 hours. This was presumably the same storm that caused a major flood on Caliente Creek, causing extensive flood damage to the Lamont/Arvin area.

Apparently there was a cloudburst somewhere in the Kings River Basin on October 29 or 30.[691] The Kings at Piedra was slightly above flood stage on October 30, but the floodwater was diverted into canals and no damage resulted. The archives at the NWS forecast office in Hanford have nothing to explain where the storm was located, so apparently it was in the mountains east of any of the reporting stations. This is presumably the same event described in the national parks' monthly report as a late October storm.

Table 32 gives the precipitation totals for the reporting stations in the national parks.

Table 32. Precipitation during October 1945 storm.

| Reporting Station | Storm Total (inches of rain) |
| --- | --- |
| Grant Grove | 4.66 |
| Giant Forest | 4.42 |
| Ash Mountain | 1.37 |

### *St. Johns Levee — Condition in 1945*

At the time that Visalia was founded in 1852, the flow of the Kaweah River was distributed largely through four channels that flowed along the south side of the Kaweah Delta. That all changed thanks to the huge 1861–62 flood and the even bigger 1867–68 flood. One of the legacies of those floods was the creation of the St. Johns River which routed the majority of the Kaweah River floodwaters along the north side of the delta, to the north of Visalia.

It didn't take Visalia long to erect a levee along the south bank of the newly formed river. The responsibility for maintenance and repair of the south levee on the St. Johns has generally remained with the local levee district since its construction. Only in emergency situations have the state and federal governments stepped in to assist with levee repairs.

The south levee was built using material pulled up out of the river channel. From an engineering standpoint, that levee had a number of significant structural shortcomings. Repairs have generally been made using the same material and engineering.

It didn't take long for the south levee to fail in a flood. The first record that we have of failure of that levee was in the 1877 flood. It would not be the last. However, flooding in the early years (roughly 1875–1940) put only moderate pressure on the levee and caused only moderate flooding problems. That was partly because Visalia was built for periodic inundations. But it was also because the St. Johns had a wide floodplain north of town. Since the floodwaters spread out in a shallow sheet across this floodplain, it put relatively low pressure on the south levee.

But by 1945, the situation had changed dramatically. In the 1945 flood, the south levee failed in four places, and Visalia flooded like never before in its history.

On February 7, 1945, the Visalia Chamber of Commerce hosted a meeting of all the concerned parties to assess the condition of the levees on the St. Johns and determine what could be done to bring them back up to minimum standard.[692] The meeting amounted to a fairly thorough after-action review. It was quickly realized that the problem was bigger than a simple levee repair. It was determined that the channel of the St. Johns in the vicinity of Miller's Bridge (Fourth Ave East), northeast of the city, was now so deteriorated and constrained that it could only pass 3,000–4,000 cfs. The St. Johns had peaked at 14,900 cfs on February 2.

It wasn't immediately obvious that channel capacity could be restored or that the south levee could be upgraded sufficiently to protect Visalia in the event of another flood of similar magnitude. The St. Johns River is subject to floods that are much larger than the 1945 flood. The city and county faced four challenges:
1. Remove Miller's Bridge so that it didn't act as a constriction on the St. Johns.
2. Restore the natural floodplain and channel of the St. Johns so that it could handle projected floodflows.
3. Strengthen the south levee of the St. Johns, correcting its obvious deficiencies.
4. After restoring the channel and levee, find some way to keep them adequately maintained.

### St. Johns Levee — Condition in Recent Years

Over six decades have passed since that after-action review, but there are still major concerns about the ability of the St. Johns channel to safely pass floodwaters. The south levee was partially rebuilt after the 1945 flood, but a portion of it failed again in the December 1955 flood. Using federal emergency funding, the USACE quickly rebuilt the damaged levee. However, federal law required that agency to restore the levee to the same condition that it was in prior to the flood, not fix its obvious deficiencies.

The 2005–06 Tulare County Civil Grand Jury investigated the St. Johns levee and found that it was not constructed to USACE certification standards, it was not being adequately maintained, and there was no adequate source of funds for its maintenance. The grand jury found that after passage of Proposition 13, incoming taxes were insufficient to maintain the levee.[693]

The Tulare County Resource Management Agency surveyed property owners in the levee district in 2002, but those owners were generally uninterested in levee maintenance and did not want to put more of their tax dollars into maintenance. Because the south levee was in such bad shape, $17 million was then needed to bring it up to USACE certification standards. However, no source for those funds has yet been found, and the levee is in approximately the same shape now that it was in when the grand jury assessed the situation.

In June 2009, the Federal Emergency Management Agency (FEMA) found that the levee was in such bad repair that it provided essentially no reliable flood protection for Visalia. About 8,900 additional homes and other properties were designated as being at high risk for flooding. This resulted in some of those homeowners having to purchase flood insurance at relatively expensive rates. No one disputes the finding by the grand jury and FEMA that the St. Johns levee is poorly maintained and that the levee falls far short of USACE certification standards. The problem is that local agencies and districts don't have the funds to maintain the levee, let alone bring it up to standard.

On February 3, 2011, 27 U.S. senators sent a letter to FEMA requesting that the agency recognize and try to quantify the degree of protection provided by poorly built and poorly maintained levees such as the St. Johns' when determining flood risk. The reason given by the senators was that FEMA's current policy puts American jobs at risk.

## 1949 Flood

Flooding in 1949 occurred in March.

On March 7, two thunderstorms hit Bakersfield in the same day. They unleashed heavy downpours that flooded first floors of offices and damaged house foundations, as well as inundating landscaping, streets and storm drains.[694]

## 1950 Flood

Flooding in 1950 occurred in November –December.

The winter of 1950–51 was a moderate La Niña event.

The flood affected the Sacramento and San Joaquin Valleys. It had a recurrence interval of 80 years on some rivers. Flooding in the Tulare Lakebed continued for a few months after river flooding subsided.

October 1950 was the wettest October since 1899, with a precipitation total 300% of average. The storms of November and December were general over all of California north of the Tehachapis. Heavy rain fell on both the Coast Ranges and Sierra. However, the flooding was limited to the Central Valley.

The flooding was caused by a series of storms which brought exceptionally warm, moisture-laden air against the Sierra, resulting in intense rainfall instead of snowfall at unusually high altitudes. There were four distinct storms:[695]
1. November 13–15
2. November 16–21 (this was the big one)
3. December 2–4
4. December 6–8

In the Tulare Lake Basin, most of the precipitation fell as rain. Because it was early in the season, there was relatively little snowpack in place. Rain was relatively light in the valley but heavy in the foothills and Sierra.

The first storm began in the national parks on November 13.

A high-elevation storm passed through Central California on November 18–19, 1950. The rainfall distribution in this storm was quite similar to the January 30 – February 1, 1963 storm. The rainfall in both of these storms was heavy in the coastal mountains as well as in the Sierra. The 1963 storm affected areas south of the wetter zone of the November 18, 1950 storm. The greatest daily total rain for the 1950 storm was 13.16 inches at Giant Forest. Highest-ever daily rainfalls were reported at 30 stations. Nine stations reported rainfall totals in excess of a storm with a recurrence interval of 1,000 years. Seven of these were in the Stanislaus, Merced and San Joaquin River Basins. Highway 140 into Yosemite was washed out near El Portal. Extensive flooding was reported on the lower San Joaquin River. Calaveras Dam in Alameda County received 7.17 inches in one day, which was 33% of its annual average rainfall. The recurrence interval for that event was 23,000 years.[696]

244

In Three Rivers, the rain was continuous for 20 hours. Long-time residents of that community could not recall such a heavy downpour.[697]

Three days of heavy rain from November 17–19 in the Sierra brought more than 15 inches of rain to some areas as high as 5,500 feet elevation and heavy rain as high as 10,000 feet, which melted the small snowpack. Although the rain was heavy and continuous, the greatest recorded intensity was 0.9 inches per hour at Giant Forest on November 18.[698, 699]

Table 33 gives the precipitation totals for the reporting stations in the Kaweah River Basin.[700]

Table 33. Precipitation during the November 18–19, 1950 storm event.

| Reporting Station | Storm Total (inches of rain) |
|---|---|
| Giant Forest | 15.22 |
| Three Rivers | 7.24 |

As shown in Table 34, this storm broke precipitation records that were set in the 1920s and 1940s.[701]

Table 34. November precipitation comparisons.

| Reporting Station | November 1950 | | All Novembers during period of record prior to 1950 | | | |
|---|---|---|---|---|---|---|
| | Total for month (inches) | Greatest daily total (inches) | Greatest monthly amount (inches) | Year | Greatest daily amount (inches) | Year |
| Grant Gove | 14.51 | 6.12 | 11.85 | 1944 | 5.70 | 1949 |
| Giant Forest | 18.87 | 9.55 | 18.73 | 1926 | 5.97 | 1924 |
| Ash Mountain | 6.13 | 3.19 | 7.62 | 1926 | 2.91 | 1946 |
| Three Rivers | 7.86 | 4.10 | 6.56 | 1926 | 2.31 | 1926 |
| Springville | 20.50 | 10.27 | 9.19 | 1946 | 5.16 | 1924 |

The 1950 flood was so newsworthy that it was written up in at least two issues of the *New York Times*. The first story described the flooding that was occurring from Sacramento to Bakersfield, including on the Kings and Kern Rivers.[702] The second article also included the flooding on the Tule.[703] Flooding in Three Rivers, Woodlake, and Visalia was featured in a special issue of the *Exeter Sun* (now *The Foothills Sun-Gazette*).

The 1950 flood caused major damage in the Central Valley. Damage was estimated to be 33 million dollars, 669,400 acres were flooded, and two people died. The $33 million included $1.2 million in damage to Highway 140 (the All-Year Highway to Yosemite Valley) near El Portal in the Merced River Canyon and $509,000 in damages to park infrastructure within Yosemite. The hardest hit areas in the San Joaquin Valley were Merced, Chowchilla, Centerville, Visalia, Porterville, Oildale, Isabella, and Kernville. Approximately 25,000 people were evacuated from their homes during the entire flood period. Governor Earl Warren proclaimed a state of emergency on November 21.[704]

Flood crests on the Kings, Kaweah, Tule, and Kern Rivers exceeded all previous records. At the time, this was the biggest flood to occur on those rivers since the 1867–68 flood, an event that occurred before the onset of formal record-keeping. The 1867–68 flood remains the biggest flood to have occurred in historic times in the Tulare Lake Basin.

The Kings River peaked at Piedra at 3:00 a.m. on November 19 at 91,000 cfs. That was the highest flow since record-keeping began in 1895.

A recording stream gage was installed in the South Fork Kings from 1950–1957. (See the section of this document that describes the Stream Gages on the South Fork Kings.) Either because of good planning or incredible good luck, that gage began operation on November 16, 1950 just as the second and biggest of the 1950 storms struck.

Thanks to that stream gage, we know that the South Fork Kings peaked at 2:00 a.m. on November 19, 1950 at 10,000 cfs.[705]

That puts it in a category with other Cedar Grove floods of the past 70 years (1937, 1950, 1955, 1966, 1969, 1978, 1982, 1983, and 1997) that rise to the level of the modeled 50-year flood: that is, a flood event that occurs about every eight years on average. See the section of this document that describes Cedar Grove Flooding.

Work had begun on the Pine Flat Dam in 1947. The dam would be partially operational in 1952, but it wouldn't be completed until 1954. On November 19, 1950 the Kings washed out the newly completed weir, the cofferdam, and foundation work of the dam. It also destroyed a 300-person motion photograph theater that was closed on the day of the flood. Damage at the dam site totaled $900,000; that was the most costly single item of property damaged by the 1950 flood.[706]

Thirty-five people living on an island at Piedra were marooned by the sudden rise of the river, but were saved by rescue workers.

From Piedra to Highway 99 (immediately south of Kingsburg), about 17,000 acres of agricultural land was flooded; most of that land was in Centerville Bottoms. The lower Kings River Bridge in Reedley was washed out. About 500 families were forced to evacuate their homes. Loss of livestock in the area was especially severe. About 30,000 turkeys, valued at $500,000, were lost.[707]

Downstream from Highway 99, the Kings River inundated the following areas during the November 19–21 flood:[708]
- 36,100 acres between Highway 99 and the Crescent Weir
- 13,100 acres between the Crescent Weir and the San Joaquin River along the north distributaries (Fresno Slough and James Bypass)
- 3,000 acres along the south distributaries on the way to Tulare Lake

As the result of breaks in the river levees at numerous points near Laton, that community was virtually surrounded by the floodwaters. Farther downstream, the floods encroached upon Riverdale and fringe communities near Hanford.

A second, somewhat smaller, flood came down the Kings on December 4–6.

This was the last major uncontrolled Kings River flood event.

The flood caused significant damage to roads and trails in Sequoia, Kings Canyon, and Yosemite National Parks.

Flooding occurred in Sequoia and Kings Canyon National Parks from November 18 – December 8. According to the parks' monthly report, it was not possible to record the maximum high-water level on the Kaweah, as the Potwisha gaging station was under several feet of water. Presumably this was referring to the Marble Fork of the Kaweah.

Rain continued to fall until December 8, causing additional damage by slides and washouts. Slide and washout conditions prevented the use of heavy equipment to keep drainage channels open. In some instances it was necessary to use dynamite to dislodge jams of drifted material endangering structures.

In the 1930s, the CCCs had dug a swimming pool in the Marble Fork of the Kaweah for recreational use by the Lodgepole campers. The pool had gradually silted in, but may have survived in some form until 1950. It was located in the riverbed adjacent to the old campers' market (the now defunct Walter Fry Nature Center building). There was no significant dam as far as we know, just a deep hole in the river. The 1950 flood was likely the end of the Lodgepole swimming pool. You can still see the rows of rock on the south bank of the river that led to the diving board.

Flooding in the national parks was so extensive that personnel were sent from the NPS Washington office, the regional office, and from the Bureau of Public Roads to assist with the damage assessment. There were numerous washouts, including one on the Generals Highway 1½ miles east of the Wye (apparently at Mill Creek). Over 5,000 cubic yards of slide material had to be removed from roads.

The 1950 flood did extensive damage to the national parks' trails. It appears that the damage to trail bridges — and therefore the extent of flooding — was much worse in the Kaweah River Basin than in the Kings or the Kern River Basins. Some of the major trail bridges that were washed out were Castle Creek, Middle Fork Kaweah in River Valley (on Route 70, the trail from Bearpaw to Redwood Meadow), Paradise Creek (near Buckeye Flat Campground), East Fork Kaweah (below Atwell Campground), and Clough Cave (near South Fork Campground).

Bob Meadow's research found that the 1950 flood also destroyed the following bridges in the Bearpaw area: Buck Canyon, Lone Pine Creek, upper and lower trail crossings of Granite Creek, Middle Fork Trail below Redwood Meadow, Lower Buck Canyon, and across the Middle Fork on the Castle Creek Trail.

One source said that this was the flood that washed out the Hospital Rock Bridge. The 1950 flood did damage this bridge and wash out the abutments, but the bridge survived. It is not clear what flood finally destroyed the bridge; perhaps it was the 1955 or 1966 flood. The piers for this bridge were beautiful, as was the bridge itself (photograph on file in the national parks). Those piers were largely demolished by the national parks in about 1974.

The 1949 trails inventory still showed the Board Camp Dome Trail (Route 98, originally constructed as the Hockett Trail) running up the South Fork of the Kaweah past present-day Ladybug Camp to the Hockett Meadows. (*Possibly* that section of trail was later called the Stakecamp Dome Trail.) The bridge on that trail crossed the South Fork Kaweah just north of Garfield Creek and was one of the few trail bridges in the Kaweah River Basin that survived the 1950 flood. Bill Tweed recalled crossing it in 1960. That bridge would eventually be washed out in the January 1969 flood and was never replaced.

The national parks' Ash Mountain headquarters development (originally known as Alder Creek) began operation in 1921. The development originally obtained its water solely from Alder Creek.

247

Starting in 1939, there was an extended period of study and debate about how to obtain a better water supply for the Ash Mountain development. Finally, in fiscal year 1950, 4,900 feet of 3-inch pipe was laid to tap into Southern California Edison's Kaweah #3 flume at Milk Creek.

Shortly after the flume-tapping project was completed, the pipe and pump were damaged during the November 1950 flood. The system was repaired and returned to operation in 1951. The system seems vulnerable to floods because it had to cross the Middle Fork Kaweah River. It's tempting to think that it might have been damaged again in the big floods of 1955 and 1966. In any case, it was eventually replaced with a system that pumped water directly out of the river. That river pump would survive until the 1997 flood.

The peak of the 1950 flood began coming through Three Rivers late on the night of November 18 and continued rising into the early morning hours of the 19th. (One source incorrectly said that it occurred on the night of November 20.)

The North Fork Kaweah peaked at 1:00 a.m. on November 19 at 9,150 cfs.[709] That was the highest flow on that river since record-keeping began in 1910. (In the December 1955 and December 1966 floods, the North Fork would experience flows over twice this great.)

The scene was particularly dramatic at Archie and Mary McDowall's chicken ranch up the North Fork, just above the Kaweah Post Office. (Mary was the principal, superintendent, and teacher at the Three Rivers school for 40 years. McDowall Auditorium is named for her.) Their daughter Bobbie was a junior in high school at the time of the flood.

When the downpour hit Three Rivers on November 18, Archie realized that it was also hitting the mountains up in the national parks. Even though there wasn't much of a snowpack, that meant there would be floodwaters arriving at their ranch that night. He thought that the North Fork wouldn't be affected as much as the mainstem of the Kaweah. All the same, Archie knew that they would be having flooding on their property. So he and Bobbie went out into the night to move chickens in a low part of their property to a chicken house on higher ground. Sixty years later, Bobbie still has a vivid memory of that night.

The McDowall ranch was ½ mile downstream from the Upper North Fork Bridge (about three miles up the North Fork from Three Rivers). Unknown to Archie and Bobbie, debris (trees, brush, etc.) was building up against the upper side of that bridge. The bridge gave way from the power of the rushing floodwaters against that debris. Archie and Bobbie found themselves in the path of that onrushing wall of water. They were carried downstream to the far side of their property until a barbed wire fence stopped them.

Bobbie had on cape-type raingear that kept pulling her under. Her dad yelled for her to get that garment off because it kept pulling her under and keeping her from getting firm hold of the fence. Her dad never cussed. But his colorful choice of words on that occasion — words that were unusual for him — sank in. Bobbie was able to get her garment over her head and reach for his hand. He was hanging onto a tree. Archie was able to get to that tree just as he went under for the second time. (Seeing her dad go down was the most frightening part of the whole experience for Bobbie.) How he got back to that tree has always been a miracle to her.

After the two of them got their adrenalin in another gear, they were able to get to a slower portion of the floodwaters and gradually work their way over to the North Fork Road. That took a long time.

But they still hadn't reached dry land. They found the water running so deep on the road by the Kaweah Post Office that they could hardly walk even there.

The next morning they went back to check on the damage. The fence that they had held onto the previous night was gone; much of the property including a chicken house full of chickens was gone. Archie's pickup truck and much more had also disappeared.

Specific damage in Three Rivers included:[710]

- Most of the bridges in town suffered major structural damage or had their approaches washed away. Only the Pumpkin Hollow Bridge, the Dinely Bridge, and the North Fork Bridge (the one next to the Three Rivers Market that had been built in 1938) remained passable.
- The Upper North Fork Bridge washed out. The Airport Bridge survived, but the approaches were washed out. The North Fork roadbed was badly eroded down to bedrock.
- At least the first three bridges on the South Fork Road (Conley Creek and the first two bridges over the South Fork of the Kaweah) were washed out. Parts of the roadbed were badly eroded when the river rerouted. Huge slabs of asphalt, some as long as 25 feet, were ripped and thrown up by the force of water.
- The Kaweah Park (presumably located at the junction of the South Fork and the mainstem of the Kaweah River) was extensively damaged.
- On the morning of November 19, crowds gathered at the South Fork crossing of Highway 198 (now Cherokee Oaks Drive) to watch that raging river pound and crumble the one highway access for the entire community. By the end of the day, the bridge was knocked out and Three Rivers was cut off.
- At least five homes washed away and many others were damaged or undermined.
- Southern California Edison's (SCE) #2 flume was heavily damaged: 200 feet of the canal bank was washed away and the upper end was filled with mud. That affected not only power production but left many families without domestic water.
- The November 1950 flood did even more damage to the Kaweah Hatchery than the 1937 flood had. The hatchery was shifted off its foundation. Equipment in the interior was greatly disarranged. Pumps, motors, and the entire grounds were covered with tons of sand and debris. This time, the movable property was repaired and transferred to other installations, and the hatchery was permanently closed.
- So many water systems were contaminated that the county health department set up a program to inoculate all Three River residents against typhoid.

The mainstem of the Kaweah peaked in Three Rivers in the pre-dawn hours of November 19, 1950. The Three Rivers gaging station (USGS gage #11-2105 Kaweah R. nr Three Rivers) recorded 45,700 cfs at 1:30 a.m. before that gage was destroyed. The 1950 flood was the largest flood on that stretch of the river with respect to both peak and volume since stream gaging began in 1903.[711, 712] It would remain the flood-of-record until the December 1955 flood. The Three Rivers gaging station was rebuilt after the 1950 flood. It was located just above the junction of the Kaweah with Horse Creek. The Three Rivers gaging station would continue operation until 1961 when this site was submerged under Lake Kaweah.[713]

The Kaweah's peak natural flow occurred at McKay's Point at 3:30 a.m. on November 19: 54,332 cfs. (That was the peak hourly flow; the peak average daily flow, as reflected in Table 15, was 16,640 cfs.) The 54,332 cfs figure was apparently determined by sloped area measurement: the flood destroyed the gaging station.

249

Based on the flood exceedence rates in Table 16, this had a recurrence interval of 13 years for the Kaweah. It would have had a recurrence interval of 30 years if calculated using the 54,332 cfs peak flow. (One source reportedly calculated this as having a recurrence interval of 125 years. That result could not be reproduced with either the peak or the daily flows, even when using the now-outdated 1971 flood frequency curves.)

By November 21, the Kaweah River flows were dropping and the danger of flooding in Three Rivers had ended. One source said that eight bridges in the town had been so badly damaged that they remained impassable.

In addition to losing the only highway bridge in and out of town, families on the North Fork were isolated from the main part of town because the Upper North Fork Bridge was gone. More than a dozen families on the South Fork Road were also isolated from the main part of town until the three bridges on that road could be replaced.

One of the bigger challenges was how to get feed to Bob Lewis's turkey ranch up on the South Fork Road. The 10,000 birds in his flock required bringing in 28 bags of feed every day. By November 20, crews of volunteers had made a plank crossing of Conley Creek, the first washed-out bridge. Men carried the sacks of feed on their backs over the plank crossing, and were met by a private weapons carrier which took the feed over the second washed-out bridge and across the next section of partially washed-out road. At the third washed-out bridge, neighbors had erected a pulley system and the feed bags were taken across in pack animal panniers. People were also being transported in those same panniers.[714]

Upstream from the foothill line, the Kaweah and its tributaries destroyed a total of 7 highway bridges, damaged an extensive stretch of highway, and destroyed or damaged 25 homes.[715]

Terminus Beach was severely damaged, as were two gravel plants and a number of homes in the area.

On November 19, debris carried by the Kaweah lodged against the Visalia Electric trestle near McKay's Point. This created a jetty, diverting the floodwaters toward Woodlake. A total of about 50 homes in that community were flooded. Six homes were destroyed, and others were extensively damaged. The trestle eventually collapsed, resulting in the destruction of several thousand feet of railroad track and embankment.[716] (A similar event would happen at this trestle in the December 1955 flood.)

From Woodlake to Visalia, the flooded area was from 2–4 miles in width.

Mill Creek caused serious flooding in Visalia, but not as bad as in the 1945 flood. A lake formed on E. Main St. extending east from Santa Fe St. The water averaged 6–12 inches in depth, although in several places it was 18 inches deep.[717]

The St. Johns River flooded extensive tracts of agricultural land on the Kaweah Delta north of Visalia. Wide areas of agricultural lands were flooded south and east of Visalia along the St. Johns River and Cross Creek.

The total area flooded by the Kaweah River was about 48,000 acres. Approximately 200 people were evacuated in Woodlake, and 2,000 people were evacuated in Visalia.[718]

Damage in the valley reached $5 million, including $500,000 in damage to bridges.

The Tule River near Springville (see Station 2032 in Table 49) peaked at 22,400 cfs. This gage had just been installed. This flood would remain the flood-of-record until the December 1966 flood.[719]

The South Fork Tule River near Success (see Station 2045 in Table 49) peaked at 7,000 cfs. This was the greatest flow at this site since record-keeping began in 1930. It would remain the flood-of-record until the December 1966 flood.[720]

One source said that the Tule River peaked near Porterville at 4:30 a.m. on November 19 at 25,500 cfs.[721] Another source that the Tule peaked at 32,000 cfs.[722] In either case, that would be the highest flow on that river since record-keeping began in 1901. It would remain the flood-of-record until the December 1966 flood.[723]

The flooding in Porterville was shallow and was largely confined to a small portion of the residential area. Between Porterville and Highway 99, the Tule spread over agricultural areas to a width of 3–4 miles. The United Concrete Corporation plant on Highway 99 was heavily damaged and was closed for two weeks. Roads and bridges suffered severe damage throughout the Tule River Basin. The total area flooded by the Tule was about 32,000 acres.[724]

The North Fork of the Kern River (see Station 1860 in Table 49) peaked at Kernville in November 1950 at 27,400 cfs. This was the greatest discharge on that river since record-keeping began in 1912. (This record would be matched in the December 1955 flood. The December 1966 flood would have a discharge more than twice as great (60,000 cfs.)[725]

Floodwaters covered portions of the town of Kernville and most of the town of Isabella and forced a mass evacuation of about 1,000 inhabitants.

The mainstem of the Kern at the site of the future Isabella Dam (see Station 1910 in Table 49) peaked at 39,000 cfs. This would remain the flood-of-record until the December 1966 flood. That flood would have a computed discharge 2.5 times as great (96,900 cfs) as the 1950 flood.[726]

Upstream from the head of the lower canyon near the Isabella Dam site, the river flooded SCE's Kern No. 3 power plant. It almost completely destroyed the State of California's fish hatchery.[727] It also inundated summer homes, commercial recreation developments, and USFS recreational developments.

Isabella Dam was under construction when the flood occurred and equipment being used to build the dam was flooded or buried, including a big power shovel. Most or all of the damage happened at the lower tunnel.

In the lower canyon, floodwaters damaged several commercial recreational establishments, two hydroelectric facilities (SCE's Kern River #1 and #3), USFS recreational facilities, and the state highway.

Five bridges were washed out in the Bakersfield area, including the Kernville Bridge. The flood also washed out the old Olcese's Ranch Bridge, a mile downstream from the mouth of the Kern River Canyon.

The Kern peaked near Bakersfield at 4:30 p.m. on November 19 at 36,000 cfs. This was the highest flow recorded at that gage since record-keeping began in 1896. The south bank levees protecting Bakersfield almost failed. Heroic efforts by 500 volunteers supported by heavy equipment saved the city from inundation.[728] A similar emergency effort had been required on this levee in the 1937 flood.

The south bank of the Kern wasn't the only trouble spot during the November flood. Floodwaters behind the historic wooden Bellevue Weir just upstream from the present-day Stockdale Highway Bridge were threatening to overflow on the north bank into the Goose Lake / Jerry Slough system. Historically, floodwaters from the Kern would regularly flow through that slough on their way to Goose Lake (see Figure 9), but that had apparently not happened for a while. If that were to happen now, farms through Rosedale and even 20 miles farther west would be inundated. Given the time available, there was apparently no feasible way to contain the Kern along its north bank. Instead, private interests used heavy equipment to hurriedly throw up a levee to minimize the flooding farther downstream along the slough in the vicinity of the Lerdo Highway. They finished the levee just before the floodwaters arrived. That levee would be tested again in the 1952 flood.[729]

A large portion of the Kern floodwaters entered the Goose Lake / Jerry Slough system, and residents of the Rosedale and Stockdale areas had to be evacuated. Flow was continuous in the Goose Lake Slough until December 7 or 8. The Goose Slough Bridge on Highway 139 was washed away. A total of about 18,500 acres of the Goose Lake / Jerry Slough system was flooded. None of the floodwaters extended past Goose Lake.[730]

The rest of the Kern River floodwaters continued in the main channel to Buena Vista Lake. The total inundated area on the Kern River (including the 18,500 acres of the Goose Lake / Jerry Slough system) was about 37,300 acres. None of the Kern floodwaters made it to the Tulare Lakebed in 1950.[731]

The rivers in the Tulare Lake Basin crested on November 19. But as of mid-December, 12 Tulare County bridges still remained impassable.

Tulare Lake had dried up on July 17, 1946, thanks in large part to the extended drought of 1943–51. The flood brought the lake back to life on November 19, 1950, if only for a few months. Table 35 details inflows to Tulare Lake during the 1950 flood.[732]

Table 35. Inflow to the Tulare Lakebed during water year 1950.

| Stream | Total Lakebed Inflow (acre-feet) | Percent Contribution |
|---|---|---|
| Kings River | 39,000 | 57% |
| Kaweah River | 14,000 | 21% |
| Tule River | 15,000 | 22% |
| Kern River | 0 | |
| Total | 68,000 | |

Emergency levees were constructed as soon as it became apparent that the floodwaters would enter the lake. This initially confined inflow to the lakebed to 12 sections (7,680 acres) of agricultural land that had recently been planted to barley. Maximum depth of water in the lake was 7.4 feet on December 10, 1950. A high wind on that day caused waves that broke the last of the emergency levees. This inundated 4½ additional sections of land. Total flooded area was 10,600 acres.[733]

The lake was dry again by March 10, 1951, and remained dry until the 1952 flood.

## 1951 Flood

Flooding in 1951 occurred in July.

Despite several false reports, there was only one flood in 1951 that we are sure of: a localized event in July in Kings Canyon.

According to one account, there was a major flood on the Kings River in the spring of 1951. That was almost certainly a mistake, and the flood being referred to actually happened in the spring of 1952.

The national parks' 1952 annual report suggested that 1951 was a heavy snow year. Judging from the runoff in the Kaweah, the winter of 1950–51 was an average year. However, water year 1952 (which reflects the winter of 1951–52) saw nearly twice the average runoff. There is no reliable record of a major flood occurring on either the Kings or the Kaweah River in the spring of 1951.

According to a published government report, a very intense storm struck Fresno on February 24, 1951, resulting in flooding. That was an error; the storm really occurred on February 24, 1941.

There was a record flood on the American River in 1951 (marking this as a record required ignoring the 1861–62 flood). There is no evidence that this flooding extended into the Tulare Lake Basin.

In July 1951, a cloudburst in Kings Canyon caused a significant debris flow to come down across the highway (photograph on file in the national parks). Judging from the photograph, that debris flow may have occurred near Deer Cove.

## 1952 Floods (3)

There were apparently three flooding events in 1952:
1. January
2. March
3. April–July snowmelt runoff period

The winter of 1951–52 was truly ferocious in California; it was a moderate El Niño event. There were two sets of particularly severe storms: one in January and another in March.

Precipitation in both the San Joaquin River and Tulare Lake Basins was consistently greater than normal throughout the winter. Widespread storms began in October and occurred intermittently until the end of March. Most of the storms brought abnormally cold air and produced snow down to and below an altitude of 1,000 feet. Very little of this snow melted, and a very large snowpack accumulated over the entire mountain area.[734]

By New Year's Day, substantial snow had begun to accumulate from one end of the Sierra to the other. January began with several relatively light storms, but on January 12–13, the first in a series of powerful and cold storms began to move through the state.

Human activity in the Sierra came to a halt as snow fell faster than it could be removed. East of Sacramento, U.S. Highways 40 (now Interstate 80) and 50 closed. Shortly after noon on Sunday, January 13, a huge snowslide west of Donner Pass stalled the westbound *City of San Francisco*, the Southern Pacific Railroad's transcontinental passenger train, in the snow. On the train were 226 people.

Highway 40 was nearby, but it was buried in snow, lost in the blizzard.

With heavy snow falling and the wind blowing up to 100 miles per hour, the Southern Pacific set out to free the train from the drifting snow. Within a few hours, the railroad had plowed to the train and attached another locomotive, but the train was already frozen into the rapidly accumulating snowdrifts and could not be moved. The railroad sent out more equipment, but getting to the train proved terribly slow and difficult. Numerous avalanches had to be cut through and still the snow fell. As the afternoon faded, things got worse as the snow deepened and equipment, pushed to the limit, failed. Finally, another avalanche hit the tracks, flipping a huge steam-powered rotary snowplow on its side, blocking the tracks completely and ending any immediate hope of rescue for the snowbound train.

Meanwhile, in the Southern Sierra, things were not much better. Facing the same huge storm, national park and state crews attempting to maintain access to Giant Forest and Grant Grove found their efforts completely frustrated. Highway 180 was closed entirely above 5,000 feet, as was the Generals Highway into Giant Forest from Three Rivers. The storm continued for the next two days.

By January 16, rescuing the stalled train and its passengers on Donner Pass had become a national priority. A wide variety of rescuers were trying to get to it. The Sixth Army tried unsuccessfully to reach the train using their over-the-snow Studebaker Weasels. Pacific Gas and Electric (PG&E) did succeed in bringing in relief supplies with their double-trucked Sno-Cat. A doctor was brought in from Reno, making the last leg of the trek by dog sled. The Coast Guard sent a helicopter, one of the few at that time on the West Coast.

The Southern Pacific continued working their huge rotary snowplows around the clock to reopen the track. But the storm was unrelenting; avalanches and equipment failures gradually knocked the plows out of commission. The railroad called in all of its reserves, reaching far afield. Finally a rotary was brought down from Oregon — the oldest and last one running in the fleet. On January 16, it succeeded in plowing to the Highway 40 overpass west of Emigrant Gap, near the Nyack Lodge. A special rescue train was then brought in to that point.

At Donner Pass, the snow finally stopped at dawn on January 16. By afternoon, the state highway department had managed to open the section of Highway 40 from Emigrant Gap to the stalled train. The passengers aboard the train were led to the highway and driven by a small fleet of private automobiles five miles to Nyack Lodge where they were fed and their needs attended to. By evening, they were on their way to Sacramento and Oakland. The train was finally freed on January 20, seven days after it was stranded.

The Central Sierra Snow Laboratory monitoring site near Donner Pass received 12.8 feet of snow between January 10–17.

The first flood in 1952 occurred in January; it was apparently a rain-on-snow event. The Kaweah's peak natural flow occurred at McKay's Point on January 25: 8,851 cfs. (That was the peak hourly flow; the peak average daily flow, as reflected in Table 15, was 5,918 cfs.) Based on the flood exceedence rates in Table 16, this had a recurrence interval of 4 years for the Kaweah.

By the end of January, the weather station in Giant Forest had recorded more than 300% of average January precipitation and more than seven feet of snow covered area meadows. At Lodgepole, several national park buildings were damaged and a major equipment shed collapsed under the snow load. Park highways sustained considerable damage. Between Deer Ridge and Eleven Range Overlook, a major section of the Generals Highway slid away into the canyon. The road was closed for three days and travel was restricted for several more weeks as a result. (We assume that the damaged area was eventually repaired with a wood bin wall, but this has not been confirmed.)

In March, another set of similar storms swept the Sierra, again disrupting travel and halting most human activity in the mountains. By the end of March, almost 30 feet of snow had fallen at Giant Forest, one of the wettest seasons ever recorded in the Southern Sierra.

On March 15, a big, late-season snowfall struck the Sierra. Grant Grove received 37 inches of snow in a 24-hour period. This was the second time that month that 30 inches or more of snow was recorded in 24 hours. Grant Grove received a total of 168 inches (14 feet) of snow during March, making it the snowiest month ever at that location.[735] By March 19, the Central Sierra Snow Laboratory monitoring site near Donner Pass had accumulated a snowpack of 20.57 feet, the most ever recorded at that facility. That was reminiscent of the amount that Walter Fry reported at Giant Forest during the record-setting winter of 1905–06.

The second flood of the year occurred in March. It is poorly documented. It was apparently a rain-on-snow event. Evidently so much water was delivered to the Tulare Lakebed that it caused a levee failure within the lakebed, the first of many that year.

On June 27, 1952, Lodgepole received 0.2 inches of snow. This brought the seasonal snowfall to 449.5 inches (37½ feet) (sometimes incorrectly reported as 522.9 inches), making 1951–52 the snowiest winter on record at that location.[736] This record would eventually be broken by the winter of 2010–2011. Snowfall in the winter of 1905–06 was even bigger than this, but that was before a weather station had been established at Lodgepole.

Table 36 provides a monthly record of snowfall at Lodgepole for the winter of 1951–52.

Table 36. Lodgepole snowfall during winter 1951–52.

| Month | Snowfall (inches of snow) |
|---|---|
| September 1951 | 0.0 |
| October 1951 | 4.5 |
| November 1951 | 36.9 |
| December 1951 | 84.1 |
| January 1952 | 119.5 |
| February 1952 | 30.3 |
| March 1952 | 158.2 |
| April 1952 | 15.8 |
| May 1952 | 0.0 |
| June 1952 | 0.2 |
| Total | 449.5 |

(The national parks began systematically collecting weather data in 1920. The park has operated two cooperating weather stations in the Lodgepole / Giant Forest area in the period since then. One was operated at the Giant Forest Museum in Round Meadow from June 6, 1921 – November 7, 1968. The other was operated at the old Lodgepole Ranger Station from February 22, 1951 through December 31, 1955. Collection of weather data at this station may well have begun much earlier, perhaps since the mid-1930s, but no data from that period have yet been found. After being deactivated for 13 years, the Lodgepole cooperating weather station was reactivated at the new Lodgepole ranger station on November 8, 1968. Lodgepole is only four miles away and 300 feet higher than Round Meadow, but it has significantly different weather. Bill Tweed lived in Giant Forest and Lodgepole from 1978–88 and often marveled at how the weather at Lodgepole differed from that in Giant Forest. Lodgepole sits at the bottom of the Marble Fork Basin and is closer to the Sierra. It is significantly colder and receives about 10% more moisture than the western side of Giant Forest. As a result, it receives much more snow. It often snows in Lodgepole when it is raining or slushing in Giant Forest. Bill thinks that the difference in the snowpack is upwards of 20–40%.)

By April 1, 1952, a huge snowpack had accumulated in both the San Joaquin River and Tulare Lake Basins. That snowpack, in all the sub-basins, exceeded that existing on the same date in 1938, which had been the greatest snowpack on record since the beginning of the California Cooperative Snow Surveys record in 1930.

After the heavy winter of 1951–52, the USGS reexamined available records of snowfall at stations in the San Joaquin River and Tulare Lake Basins from the winter of 1905–06.[737] Based on the limited data available, the USGS study tentatively concluded that the 1952 snowpack appeared to equal or exceed the snowpack that caused the great snowmelt floods of 1906.

Table 37 compares the snowpack for the winters of 1938 and 1952.

Table 37. Comparison of April 1 snowpack for 1938 and 1952.

| | Average snowpack in basin | |
|---|---|---|
| **Drainage Basin** | **1938** | **1952** |
| Upper San Joaquin River | 170% | 190% |
| Kings River | 155% | 190% |
| Kaweah River | 155% | 220% |
| Tule River | 180% | 265% |
| Kern River | 205% | 260% |

Thus, in April of 1952, enough snow had accumulated to cause the greatest snowmelt flood on record, with the possible exception of 1906. (The huge 1850 snowmelt flood would have been well before the period of record.) That such a flood did not occur was largely due to the temperature pattern during the snowmelt period. Weather continued to be cold; temperatures from April through July were generally below normal — in June about 5 degrees below normal. The occasional intervals of hot weather that usually cause the peak flows during the period of the snowmelt runoff were short and not as hot as usual.

Based on the data available, the 1952 snowpack appeared to equal or exceed the 1906 snowpack. However, the only way to be certain was to wait until the snowpack melted and ran off. The results turned out to be quite clear, and rather surprising. As the melt occurred, the volume of the April–July runoff approached closely that of 1938, and on the Mokelumne, Stanislaus, Kaweah, and Kern Rivers, exceeded it. However, in no case where the period of record included the year 1906, did the 1952 snowmelt runoff exceed that of 1906 on any river.[738] That confirmed that all those watersheds had a bigger snowpack in 1906 than in 1952.

The difference was particularly remarkable in the Kaweah River Basin. The winter of 1951–52 set a modern-day snowfall record at Lodgepole, one that would last until the winter of 2010–11. However, as shown in Table 38, the snowmelt runoff in the Kaweah River Basin in 1906 was 38% greater than in 1952.

Table 38. Comparison of snowmelt runoff for 1906, 1938, and 1952.

| | Total Runoff April 1 – July 31 (thousand acre-feet) | | | Maximum daily flow (cfs) | | |
|---|---|---|---|---|---|---|
| **River** | **1906** | **1938** | **1952** | **1906** | **1938** | **1952** |
| Kings River | 2,980 | 2,320 | 2,200 | 24,900 | 22,800 | 15,500 |
| Kaweah River | 814 | 562 | 588 | 7,260 | 5,130 | 5,170 |
| Tule River | 192 | 129 | 115 | | 2,820 | 860 |
| Kern River | 1,390 | 962 | 1,120 | 9,500 | 7,300 | 8,360 |

The maximum daily discharge during the April–July 1952 runoff period was greater than the corresponding flow in 1938 on the Mokelumne, Kaweah, and Kern Rivers. However, on those streams where the period of record included the year 1906, the maximum daily flow during the snowmelt period did not exceed that of 1906.

The 1952 snowmelt flood affected all the rivers on the east side of the San Joaquin River and Tulare Lake Basins. We're fortunate that this flood was so well documented by USGS.[739]

Maximum releases from Friant Dam in combination with a maximum inflow of about 4,600 cfs of Kings River water via the Fresno Slough produced a peak of about 8,800 cfs near Mendota on May 29.

One source said that the flood on the Kings River in the spring of 1952 was of a magnitude similar to the big flood of 1914. That seems like a stretch.

Because there was a recording stream gage on the South Fork Kings, we know that stretch of the river peaked on June 5, 1952, at 7,460 cfs. That compares with a peak flow of 10,000 cfs in the 1950 flow.[740]

The 1952 flood, which was a combination of rain and snowmelt runoff, produced much more runoff than the rain flood of November/December 1950.[741]

In May 1952, the Kern River overflowed its north bank west of Bakersfield and entered the natural flood channel known as Goose Slough and Jerry Slough. The floodwaters put pressure on the levee downstream that had been hurriedly thrown up in the 1950 flood. A 40-foot-wide break formed in that levee just north of the Lerdo Highway. Private interests worked into the night, using heavy equipment to repair that break.[742] Larry Frey recalled that the last real Kern River floodwaters coming down Jerry Slough into Goose Lake were in the fall of 1951 and the spring of 1952. That's probably in large part because Isabella Dam began operation in 1954.

Flood crests on the various rivers within the Tulare Lake Basin were not particularly high in 1952. However, moderately heavy flows over a long period caused considerable damage in the valley, particularly in the Tulare Lakebed. The USACE conducted a three-month-long survey to determine the extent of the damages incurred. The results are shown in Table 39.[743]

Table 39. Damages incurred during 1952 snowmelt flood.

| Water Body | Flooded Area (acres) | Public Institutions and Utility Damage (million dollars) | Agricultural Damage (million dollars) |
|---|---|---|---|
| Kings River | 5,200 | $ .063 | $ .136 |
| Tulare Lake | 72,700 | 1.418 | 6.877 |
| Tule River | 300 | .017 | .04 |
| Kern River | 30,700* | .005 | 1.21 |
| Total | 108,900 | $ 1.50 | $ 8.26 |

*The 30,700 acres flooded by the Kern River included 23,500 acres in the Buena Vista Lakebed and 7,000 acres south of Sand Ridge.

The Tulare Lakebed had been dry from March 10, 1951 through January 19, 1952. During the 1952 flood, the lake rose by 15.5 feet to a maximum elevation of 194.6 feet (elevation 194.6 - 179.1 feet). The lake has not been this high since, although the 1969 flood came close.

Presumably that was high enough to threaten Corcoran. The last time that the lake had been this big was during the years 1937–44. The *New York Times* reported on the return of Tulare Lake and that a dike broke, flooding cotton fields.[744] Levee failures in the lakebed occurred from March until June 2, 1952.

As shown in Table 40, all four major rivers contributed water to the Tulare Lakebed during 1952.[745]

Table 40. Inflow to the Tulare Lakebed during water year 1952.

| Stream | Total Lakebed Inflow (acre-feet) | Percent Contribution |
|---|---|---|
| Kings River | 258,000 | 44% |
| Kaweah River | 175,000 | 30% |
| Tule River | 50,000 | 9% |
| Kern River | 100,000 | 17% |
| Total | 583,000 | |

Inflows to Tulare Lake would have been even larger, but 44% of the floodwaters were stored elsewhere before they reached the lake:

- The USACE operated the partially completed Pine Flat Dam, storing about 130,000 acre-feet of Kings River floodwater there.
- A local water storage district stored 232,000 acre-feet of Kern River floodwaters in Buena Vista Lake. That flooded 23,500 acres of that lakebed.
- Local interests dammed the Kern River channel where it flowed through Sand Ridge. This created a 7,000 acre lake south of that structure, which stored about 100,000 acre-feet of water. That lake was where the southern extension of Tulare Lake had formerly been, the lake which had been known to Native Americans as Ton Taché.

The flood peaked in the Tulare Lakebed about June 20, 1952. Agricultural land in the lakebed stayed flooded from March 1952 through at least September 1953. The last of the J.G. Boswell Co. land was pumped dry on September 22, 1953.[746]

Flooding also occurred on the west side of the San Joaquin Valley near Coalinga sometime during 1952.

Total flow for water year 1952 was 163% of the 1894–2011 average for the Kings, 191% for the Kaweah, 229% for the Tule, and 202% for the Kern. This was one of the very rare years when flows on the Tule River exceeded what could be used for beneficial use by the holders of water rights.

## 1955–56 Floods (2)

There were two flooding events in 1955–56:
1. River flooding due to storms that occurred from mid-November through December, 1955.
2. River flooding due to a storm that occurred on January 25, 1956.

The winter of 1955–56 was a strong La Niña event.

During December 17–27, 1955, a warm rainstorm melted accumulated snowfall up to an elevation of 10,000 feet. This storm was heaviest in the Central Sierra, the Feather, Yuba, Bear, American, Cosumnes, and Calaveras Rivers, as well as the Russian and Napa Rivers, and the streams of the South Bay Area.

The rainfalls of the higher elevations of both the Coastal Mountains and the Sierra were affected by this storm sequence. A total of 20 stations reported storm intensities in excess of a storm with a recurrence interval of 1,000 years. The Santa Clara Valley was the hardest hit in terms of rainfall events with large recurrence intervals.

This was a relatively high-elevation storm in the Central Valley. Over half the stations reporting a storm with a recurrence interval of 100 years were located at an elevation over 1,000 feet. The highest for the Sierra stations was 36.57 inches at Strawberry Valley at an elevation of 3,800 feet in the Yuba River Basin. Lake McKenzie, located southwest of San Jose at an elevation of 1,800 feet, received 42.27 inches. Honeydew in the Mattole River Basin received 49.20 inches in 8 days.

A total of 19 stations reported daily rainfall in excess of 10 inches in one day during the December storm. These were located in the Upper Sacramento, Feather, San Joaquin Rivers, and in the Clear Lake area. Lakeshore in Shasta County received 15.34 inches in 24 hours on December 20. This was the heaviest 24-hour rain event ever reported for the Central Valley up to that time. (Hockett Meadows would break this record in December 1966.) This storm did not produce heavy, short bursts of rain, but rather rain continued all week with few breaks. It saturated the soil and filled the surface reservoirs. It resulted in extensive flooding which devastated Yuba City and forced the evacuation of 20,000 people.[747]

Although the river flooding occurred in December 1955 and January 1956, flooding continued in the Tulare Lakebed for about four months thereafter. That was very much how the flooding had played out in 1916.

The December 1955 flood brought large flows to many locations in the Sacramento River Basin. A levee break on the Feather River caused severe flooding in the Yuba City area. The flow in the American River at Fair Oaks was controlled to 70,000 cfs because Folsom Reservoir was nearly empty at the beginning of the event. Had Folsom been up to allowable storage capacity, the project would have exceeded its design outflow and the flow at Fair Oaks probably would have been more than 115,000 cfs. At the Sacramento Weir, 30 gates were opened, and the peak flow reached 48,800 cfs. The peak flow in the Sacramento River at I Street was about 95,000 cfs. Total flow at the latitude of Sacramento, including the Yolo Bypass, was about 380,000 cfs.[748]

Floods in the San Joaquin River Basin reflected those in the Sacramento River Basin. Flows on the San Joaquin River were completely controlled by Friant Dam. Prior to the December 1955 flood, Millerton Reservoir was well below flood-control pool. If storage had been at allowable flood management levels, uncontrolled flows would have exceeded 37,100 cfs and resulted in extensive damage between Friant Dam and the mouth of the Merced River. The peak flow of 62,500 cfs was a record on the Stanislaus River at Ripon, while the Middle Fork of the Tuolumne River at Oakland Recreation Camp reached a record flow of 4,920 cfs. During the 1955 floods, two of the three forks of the Tuolumne River reached record flows.[749]

Flooding occurred from late November through December 1955, but the worst of the flooding occurred around December 23–24. The flooding affected the northern and central parts of the state as far south as the Tehachapis. It resulted from a family of cyclones originating in the mid-Pacific Ocean. The flood had a recurrence interval of up to 100 years, depending on the river. It caused 76 deaths and $166 million in property damage and was one of the five costliest floods in California's history.

In the national parks, the first fall storm arrived on November 13. Two more storms came during the next 10 days, resulting in considerable snow at the higher elevations. There was significantly more rain and snow during December. Torrential rains fell in the Kaweah River Basin on December 23–24. Giant Forest recorded 11 inches on the 23rd and 4 inches on the 24th. It kept raining at a lesser rate through December 27.

The valley and much of the Sierra experienced heavy rainfall from December 22–25. As shown in Table 41, the precipitation was particularly intense on December 23.

Table 41. Precipitation during the December 23, 1955 storm event.

| Reporting Station | Storm Total (inches of moisture) |
|---|---|
| Fresno | 1.72* |
| Grant Grove | 9.22 |
| Giant Forest | 11.04 |
| Ash Mountain | 7.24 |
| Woodlake | 3.3 |

*This set the record for the wettest December day ever in that city.[750]

Heavy rainfall in the San Joaquin Valley and much of the Sierra from December 22–25 led to flooding across the area. Some rivers and streams reached their highest level on record at the time. It was an unusually warm Christmas in Visalia, even by valley standards.

December 1955 is still the wettest December of record, at least in the Northern Sierra and parts of the Tulare Lake Basin. Fresno's total precipitation during December was 6.73 inches, making it the fourth wettest month ever, and the wettest December on record.[751] Visalia received 6.06 inches of rain in December, the most since record-keeping began in 1898.

The flood merited three front-page stories in the *New York Times*. The story in late November talked about flooding from Sacramento to Bakersfield.[752] There were two stories at Christmastime. One talked about the coast floods starting to recede.[753] The other story talked about renewed flooding in the Sacramento–San Joaquin River Delta.[754] All three stories apparently addressed conditions in Visalia.

The floods of December 1955 were memorable not only for the magnitude of peak discharge, but also for the duration of rain and the extent of the area affected. Rain fell in coastal areas on 39 of the 44 days between December 15, 1955 – January 28, 1956 as several storms crossed the northern two-thirds of the state. In most areas, the storm of December 21–24 caused the most damage.

Warm, moist air from the southwest released rains that drenched the mountains and melted much of the snow that had accumulated in the Sierra. During December 15–27, extremes of up to 40 inches of rain fell at several locations, and quantities greater than 20 inches were common in the coastal mountains and the Sierra.

The floods of December 1955 produced peak discharges in much of the area that were in excess of any previously recorded. Flooding was particularly notable on the Klamath River on the North Coast, Alameda Creek in the San Francisco Bay area, the San Lorenzo River at Santa Cruz, the Feather River near Yuba City, the Kaweah River at Visalia, and the Carson River east of the Sierra. Peak discharges at these widely separated rivers were generally 1½–2 times the discharge of the previously

261

recorded peak flows. On many streams, the floods ranked among the greatest since the 1861–62 flood.

The Merced River at Happy Isle peaked on December 23 at 9,860 cfs. One source said that this flood had a recurrence interval that exceeded 100 years. However, Mary Donahue calculated the recurrence interval as 45 years.[755]

Several rivers, including the Eel, experienced their greatest discharge-of-record during this flood. Overflow of the Klamath River resulted in almost complete destruction of the town of Klamath.

About 382,000 acres of the Sacramento River Basin were flooded. Unusually high tides aggravated the situation by impeding the passage of floodwater through the Sacramento–San Joaquin River Delta.

The American River experienced a record flood, the second record flood in seven years.

On December 24, a levee failure on the Feather River flooded more than 3,000 homes in Yuba City, killing 38 people and forcing the evacuation of 12,000 people. The city was inundated with floodwater as much as 12 feet deep.

Marysville was surrounded by the merged floodwaters of the Yuba and Feather Rivers. The entire city, all 12,500 residents, was ordered to evacuate. A photograph (on file in the national parks) shows the city awash even inside the moat-like ring levee and evacuating cars streaming out the only exit.

The Friant Dam contained the entire runoff of the flood on the mainstem of the San Joaquin River and prevented widespread flooding of the agricultural lands from Mendota north to the vicinity of Los Banos. (However, the northeastern part of the town of Los Banos was flooded, forcing the evacuation of 65 people.) Other reservoirs effectively reduced the flooding on the Merced County Stream Group (Burns, Bear, Owens and Mariposa Creeks). The Exchequer Reservoir regulated the flood on the Merced River to a safe outflow of 10,800 cfs.[756]

However, the Fresno and Chowchilla Rivers were not regulated, and the Melones Reservoir on the Stanislaus River and the Don Pedro Reservoir on the Tuolumne River both overflowed; the uncontrolled spill at the peak being 59,400 cfs and 41,700 cfs respectively.[757]

This resulted in a huge lake that extended along the San Joaquin River and several of its tributaries from the Fresno-Madera county line north through Merced and Stanislaus Counties and into San Joaquin County, flooding an estimated 300,000–400,000 acres.[758] Many streams and rivers in this area flooded. Water covered the Chowchilla business district and inundated Highway 99 up to 5 feet deep.[759] This forced the closure of Highway 99 at Chowchilla at the height of the Christmas travel period. The State Division of Highways and the Highway Patrol worked together to effect a side-road detour around this large water barrier.

This would be one of two great lakes to form during the December 1955 flood. The other would develop around Visalia and be fed solely by runoff from the Kaweah River.

Panoche/Silver Creek west of Mendota flooded at about Christmastime 1955.[760]

Because there was a recording stream gage on the South Fork Kings, we know that river peaked on December 23, 1955 at 13,900 cfs. That broke the previous record of 10,000 cfs set for this stretch of river on November 19, 1950.[761] This remains the highest flow recorded on the South Fork Kings since record-keeping began in 1950.

The peak day natural flow at Pine Flat Dam occurred on December 23. That is the largest peak day of the year at Pine Flat since the dam was built in 1954. Based on the flood exceedence rates in Table 16, this had a recurrence interval of 100 years for the Kings River at Pine Flat. That puts it in a category with other Cedar Grove floods of the past 70 years (1937, 1950, 1955, 1966, 1969, 1978, 1982, 1983, and 1997) that rise to the level of the modeled 50-year flood: that is, a flood event that occurs about every eight years on average. See the section of this document that describes Cedar Grove Flooding.

Specific damage in Kings Canyon included:
- The Highway 180 Boyden Bridge survived the 1955 flood, but the Grant Grove approach to the bridge was washed out (photograph on file in the national parks). The Grant Grove approach would be washed out again in 1997. The bridge is still standing and in use. It was constructed in 1939 and has withstood many floods.
- The South Fork Kings River caused serious damage to Highway 180 in the Kings River Canyon. Many miles of Highway 180 were washed out or undermined (multiple photographs on file in the national parks). In the vicinity of Boulder Creek, where the highway and the Kings River occupy a steep, narrow canyon, approximately 4,000 feet of road was completely washed out. There were approximately seven or eight other washouts of a serious nature. Even by foot, the road was only passable to a point ½ mile above Boulder Creek.
- The left embankment of the Cedar Grove Bridge washed out. Significant quantities of fill and riprap were required to repair this washout. The damage done to the Cedar Grove Bridge in this flood was similar to the damage that would later occur in the 1997 flood.
- One of the piers on the Roaring River Bridge was reportedly undermined, causing the bridge to tilt.[762]

The various branches of the Kaweah River took out many bridges during this flood, one of which was in the national parks. The Visalia City Council hosted what amounted to a miniature after-action review on January 31, 1956. Fred Walker represented the national parks. R.C. Sorenson represented the USACE. Joe Garcia, Jr. represented the Tulare County Road Commission. One of the discussion items at that meeting was the desirability of replacing the washed out bridges with suspension bridges.

Specific damage in the Ash Mountain area included:
- The Marble Fork Bridge near Potwisha was washed out (photograph on file in the national parks). A temporary footbridge was constructed over the bridge piers.
- A washout occurred ½ mile below Potwisha, undermining the Generals Highway (photograph on file in the national parks).
- There was a slide on the Generals Highway ½ mile above the national parks headquarters (photograph on file in the national parks).
- The national parks' approach to the Pumpkin Hollow Bridge completely washed away (multiple photographs on file in the park). A very tall ladder was placed in the resulting hole (or new riverbed) so that people could climb up to the bridge and then continue on to Three Rivers. Both approaches to this bridge had washed out in the 1937 flood. The 1966 flood would largely wash

out the parks' approach to this bridge. But the bridge itself has withstood all these floods; it is still standing today.

- About 400 feet of Highway 198 washed out in the lower portion of Pumpkin Hollow (photographs on file in the national parks). Now there is a large curving concrete wall at this point. From the photographs, it is evident that the dirt required to repair the damage was trucked down from the national parks. Perhaps it was brought up Dinely Dr. and through the Riverway Ranch.

It isn't clear when the section of the Generals Highway between Ash Mountain and Giant Forest was reopened to visitor traffic. Colony Mill Road had been considered unsafe for visitor traffic for many years; that route may not have been used as a detour since 1937. At the least, visitors were probably kept off this section of the Generals Highway until a new bridge could be constructed across the Marble Fork Kaweah at Potwisha. That presumably took at least a year. This may have been the first time since 1937 or 1938 that the Generals Highway had been closed for more than a couple of weeks.

Jim Harvey recalled that the national parks' bridge over Yucca Creek washed out in this flood. This was on the road known as the West Boundary Truck Trail.

In Three Rivers, the various branches of the Kaweah rose steadily all day on December 22 due to the heavy downpour. There was no cause for alarm until shortly after midnight, when the rivers surged up with a thunderous roar and swept everything from their path. Many of the town residents were caught up in the battle to save lives that night. One sample from among many in the first post-flood issue of the *Three Rivers Current*:[763]

> *Willie Clay was another near-flood victim. He started down the road just before 3 a.m. to lend a helping hand when just past Kath's, a wall of water spun his car around like a toy. He climbed out and headed for shore about 30 feet away when a second wave hit and tumbled him over and over. He managed to hang on to the first solid thing he felt which was a tree and he climbed up above the water. He spent a terrifying 5 hours wavering above the black rush of water and crashing debris. Shortly after dawn, Fred Walker, John Wollenman, and Leroy Maloy got out to him from the upper side of the road.*[764]

In December 2005 or thereabouts, the Kaweah Commonwealth published a 50th anniversary edition featuring the 1955 flood. It had a photograph of the mainstem of the Kaweah at flood stage going around the south side of the Three Rivers Market.

Specific damage in the Three Rivers area included:[765]
- In addition to the damage that occurred in the Pumpkin Hollow area (described above); there were several washouts on Highway 198 between Pumpkin Hollow and Slick Rock.
- The Dinely Bridge (the one that had been built after being washed out in the December 1937 flood) was overtopped and swept away at 3:15 a.m. When that bridge went out, it reduced the height of the floodwaters on the nearby Buchholz' house. A new bridge was built in 1957.
- The Upper North Fork Bridge washed out. The Airport Bridge survived, but both approaches were washed out. Some of the homes on the North Fork were surrounded by floodwaters, trapping the occupants. Many of the homes and buildings in that area were flooded and badly damaged. The McDowall chicken ranch suffered severe damage. The North Fork Road was badly eroded, isolating residents from the rest of town.

- The mainstem of the Kaweah overtopped and washed away the North Fork Bridge, the one near the Three Rivers Market that had been built to replace the one washed out in the 1937 flood.
- Water and sand flooded all three of SCE's powerhouses, and all were knocked out of operation. The Kaweah #3 hydroelectric complex that is located just inside Sequoia National Park was particularly hard hit (multiple photographs on file in the national parks). One transformer at Kaweah #3 was tipped over, and the other was undermined. The generator floor at Kaweah #3 was flooded to a depth of three inches. There were a number of washouts on the flumes. One of the damaged areas was a major break in the Middle Fork Kaweah concrete flume near Station 60 that was repaired with redwood. Some records associated with SCE's photographs documenting the damage incorrectly identified the flood as occurring on January 18, 1956. The damage was really sustained in the December 23, 1955 flood.
- The first two bridges on the South Fork Road (Conley Creek and South Fork of the Kaweah) were washed out when these streams merged and created what was described as "a scene of devastation" (photograph on file in the national parks).
- The Three Rivers Motel and Trailer Court was demolished.
- The Sequoia Hardware Store was undermined and badly damaged (multiple photographs on file in the national parks). That building was located just east of the present Hummingbird Restaurant.
- Many homes were washed away and many others were severely damaged.
- Much livestock was lost.

At the time of the flood, there were two ways to access Three Rivers from the west, bridging the South Fork of the Kaweah. The primary access was on Highway 198, passing through what is now the Kaweah Park Resort. Farther to the south, the county road known as Old Three Rivers Road followed the former alignment of Highway 198. (The westernmost portion of that road is known today as Cherokee Oaks Drive). The flood caused serious damage to both of those roads, cutting off all access to Three Rivers and to Sequoia National Park.

The Highway 198 bridge over the South Fork of the Kaweah River withstood the flood. (It appears in a photograph from the 1966 flood and is presumably still there today.) However, immediately east of that bridge, the mainstem of the Kaweah River overtopped and washed away 1,600 feet of the highway, and occupied an area approximately 1,000 feet south of the previous location of the route (multiple photographs on file in the national parks). This washed away five houses and the Dunlap Motel (one source incorrectly said that it was the Noisy River Lodge). Because of the magnitude of this washout, the state would decide to abandon that alignment rather than rebuild it.

The South Fork of the Kaweah undermined the center of the county's South Fork Bridge and caused it to collapse (photograph on file in the national parks). That was the old highway bridge that connected what is now Cherokee Oaks Drive with Old Three Rivers Road. That was the route that had formerly been the alignment of Highway 198. That bridge had been reconstructed by the state after it was destroyed in the 1937 flood.

The mainstem of the Kaweah peaked in Three Rivers at 80,700 cfs on December 23. This was the flow of record for this gaging station (USGS gage #11-2105 Kaweah R. nr Three Rivers) during its period of operation from 1903–1961. (The December 1966 flood would have only a 2% greater flow through this stretch: 82,700 cfs.[766]) The Three Rivers gaging station was located just above the junction of the river with Horse Creek. This gage location is now submerged under Lake Kaweah.

As soon as the flood passed, the national parks sent equipment and employees to help Three Rivers. The assistance they provided included:[767]

- Worked for days to keep the Buckeye route passable for at least truck traffic through the Riverway Ranch to the North Fork Road.
- Repaired the washed-out approach to the Pumpkin Hollow Bridge.
- Assisted with repairs to the lower portion of Pumpkin Hollow.
- Dozed a route over the Shepherd's Saddle Road to the North Fork Road.
- Cleared mud and debris from many roads in the Three Rivers area.

This may have been the last time that federal accounting oversight controls allowed the national parks to send such assistance to the Three Rivers community.

The county worked to repair the many bridges that had been damaged or destroyed within Three Rivers. One of the high priorities was to reestablish access across the mainstem of the Kaweah River. Almost immediately after the flood, a cable trolley was rigged at the site of the washed-out North Fork Bridge so that people could be pulled back and forth (multiple photographs on file in the national parks). (A cable trolley had previously been used at this location after the North Fork Bridge had been destroyed during the December 1937 flood.) On December 31, a temporary log bridge was constructed for vehicles.

But a much more permanent temporary bridge was needed. The road commissioner pointed out that it had taken 1½ years to replace the North Fork Bridge after it had been destroyed in the 1937 flood. So the county authorized the rental of a Bailey Bridge at about $550 per month, to be installed just above where the Chevron Station is today. That may be where the temporary log bridge had been constructed. It was in place by January 18, 1956 (photograph on file in the national parks).

The state worked to restore service to Three Rivers and the national park as soon as possible. This was done by using Old Three Rivers Road (the former state highway) as a detour. This required shoring up and re-decking the county bridge that had collapsed during the flood.

The state then built a new Highway 198 alignment midway between the two bridges that had failed. The new bridge over the South Fork of the Kaweah (Bridge #46-29) was completed in 1957.

The flood caused several washouts on the Mineral King Road, with the major damage above Faculty Flat.

During the height of the flood, the mainstem of the Kaweah River was very wide at Slick Rock and immediately above. A photograph in the *Exeter Sun* (on file in the national parks) seems to indicate that the river extended to roughly where the new USACE toilets and parking lot are located.

The *Visalia Times-Delta* published a special flood edition on January 20, 1956. That paper presumably contains interesting material about the details of this flood, but we haven't seen it.

Dry Creek peaked below present-day Terminus Dam at 6,070 cfs.[768] Those floodwaters merged with the Kaweah, adding to the destruction.

The area from Terminus Beach to Highway 99 was hard hit by flooding on December 23. When the Kaweah reached the Lemon Cove–Woodlake Road (Highway 216), it was about a mile wide, stretching from the intersection with Dry Creek Drive almost to the intersection with Highway 198 (photograph on file in the national parks). Both the Terminus Beach and McKay's Point Resorts were inundated.

The Kaweah overtopped and broke the Friant-Kern Canal south of Woodlake, causing significant damage to that canal (multiple photographs on file in the national parks).

The Kaweah swept away 350 feet of the Visalia Electric mainline trestle and 1,800 feet of track near McKay's Point, about a mile below the Lemon Cove–Woodlake Road (photograph on file in the national parks). A similar event had happened at this trestle in November 1950.

The flood damage below Terminus Beach had two separate components:
- Damage to agricultural lands because floodwaters were greater than levees were designed to contain.
- Damage to Visalia due to failure of the diversion structure at McKay's Point.

The Kaweah's peak flow occurred at McKay's Point on December 23. The gage at McKay's Point was swept away that morning, before the peak of the flood. There have been a variety of estimates presented for how big the flood was at McKay's Point when it peaked.

In a 1956 report, the California Disaster Office reported that the Kaweah River set a record when it peaked at 74,400 cfs during the December 1955 flood.[769]

In 1970, USBR (with assistance from USACE) reported that the Kaweah was flowing at 80,700 cfs when it peaked on December 23.[770] The 80,700 cfs figure was just a reflection of the flow upstream at the Three Rivers gage (USGS gage #11-2105).[771]) It did not include the flow that was added from Dry Creek.

In recent years, the USACE at the Lake Kaweah Visitor's Center reports that the Kaweah peaked at 87,400 cfs at McKay's Point.

A December 2010 issue of the *Kaweah Commonwealth* reported that the USACE has modeled the December 1955 flood and that it had a recurrence interval of 100 years. The model estimated that at the peak of the flood, the Kaweah was flowing at an estimated 110,000 cfs in the vicinity of Three Rivers.[772] We haven't been able to locate this USACE study, and those model results do not appear to have been used when the flood frequency curves were revised for the Kaweah in 2005.

This document uses the estimates used by the USACE when the flood frequency curves were revised for the Kaweah in 2005.

The Kaweah's peak natural flow occurred at McKay's Point on December 23: 84,332 cfs. (That was the peak hourly flow; the peak average daily flow, as reflected in Table 15, was 44,512 cfs.)

Based on the flood exceedence rates in Table 16, this had a recurrence interval of 85 years for the Kaweah. It would have had a recurrence interval of 55 years if calculated using the 84,332 cfs peak flow. (One source reportedly calculated this as having had a recurrence interval of 232 years. That result could not be reproduced with either the peak or the daily flows, even when using the now-outdated 1971 flood frequency curves.)

Since 1870, there had been a weir at McKay's Point to divert most of the Kaweah River floodwaters into the St. Johns River channel and north around Visalia. Judging from the photographs (on file in the national parks), the 1955 flood cut a channel around the north end of the weir. In addition, the flood plugged the mouth of St. Johns channel with a huge amount of sediment.

This failure of the McKay's Point diversion resulted in most of the floodwaters going down the Lower Kaweah River channel toward Visalia. The Lower Kaweah River channel ends east of Visalia (just north of the Ivanhoe turnoff (Road 156/158) on Highway 198). The floodwaters rushing down the Kaweah on December 22 and 23 found an outlet to the southwest through Cameron Creek, Packwood Creek, and Mill Creek, none of which had nearly the capacity to handle the volume of water coming their way. The combined capacity of those three creeks was well under 5,000 cfs.

A very big lake was about to form on the east side of Visalia. At its peak, that lake would have a length of 10–15 miles.[773] It flooded an estimated area of 183,000–300,000 acres.[774] The Kaweah Delta was accustomed to flooding, but it had not seen anything like this in historic times (multiple photographs on file in the national parks).

Cameron Creek (a distributary of Deep Creek, southwest of Kaweah Oaks Preserve) caused widespread flooding, especially in the area from the Cameron Creek Colony to Farmersville. Worman (Saw-)Mill was severely damaged (multiple photographs on file in the national parks). Its entire inventory of 140,000 board feet of dried lumber was swept away as well as 50,000 board feet of logs. The mill office (essentially a small house) was swept off its foundation and was later found over a half-mile downstream where it had come to rest on the Farmersville Road. The owner of the mill gave the office away, saying that he had no further use for it; the flood had wiped him out completely.

Packwood Creek flooded a large area, including several sections of Highway 198 near the Ivanhoe turnoff (Road 156/158). Packwood Creek caused severe erosion in numerous areas, such as the farmland around the intersection of Lovers Lane and Caldwell (multiple photographs on file in the national parks). Nearly all of the bridges over Packwood and other creeks south of the Mineral King Highway were destroyed during the flood.

The town of Exeter was just outside the flood zone, but the land around Exeter, stretching out toward the communities of Woodlake, Farmersville and Lindcove, was so widely inundated that about 500 families had to leave their homes. Some 50 of those families were from Woodlake. Woodlake and the surrounding area were thoroughly inundated (multiple photographs on file in the national parks). Eight motorboats were initially rushed to Farmersville to help with evacuation of that area, and the sheriff's office was looking for more. Many of the families in the Woodlake area had to be evacuated by helicopter.[775]

The network of roads in the agricultural area around Visalia effectively became a network of canals. One migrant labor camp containing some 200 cottages was entirely inundated, and the workers there escaped with little personal property.[776]

The flood swept through Visalia on Christmas Eve, reaching the town just as darkness closed in. Damage was particularly heavy on the southeast side of town. Many residents had to be evacuated from their homes by boat during the night. The National Guard assisted the sheriff's office and the police department with the evacuations. At 11 p.m. on Christmas Eve, an area covering 21 blocks was sealed off due to the flooding. The National Guard and a detachment of U.S. Marines from Tulare assisted the police with patrolling the town to prevent looting.

The depth of the water in Visalia was not great. While it reached five feet in some places, most areas were only about a foot deep. Because flooding was historically a relatively common occurrence in Visalia, very few buildings had basements. Even so, a lot of silt and mud was left behind in the flooded buildings.

Water was generally flowing in a shallow sheet along the Kaweah Delta, bound for Tulare Lake. South of Mineral King, Mooney Blvd. worked like a levee. The water pooled up when it reached Mooney, flooding the houses on the east side of that road. Houses on the west side of Mooney were grateful for the flood protection.

Pumps were brought in to provide relief for the houses on the east side, and began pumping the floodwaters across Mooney. A particularly large pump was placed at the northeast corner of Meadow Lane and Mooney, where the parking lot for Carrows Restaurant is currently located. The round-the-clock pumping achieved the goal of moving the floodwaters across Mooney, but the homeowners on the west side were not happy at receiving all that water.

Most of the campus of the College of the Sequoias was flooded (multiple photographs on file in the national parks). The damage was most severe in the building that housed the college gym because it had a basement. (The building had been designed under the assumption that floodwaters could never reach this part of Visalia.) There was widespread flooding in the neighborhoods around the COS campus.

Farther downstream, Mill Creek broke across the Visalia Airport and flooded the depressed portion of the Visalia Plaza interchange (aka Highway 99 and 198 interchange, since reconfigured but in the same general location). The water came in faster than the pumps could bail the water, and finally the pumps quit working. Traffic had to be detoured around that interchange for the better part of the week of December 25–31.

Largely isolated by the flooding of Highway 99, Visalia and the neighboring communities had to call for food and medical supplies from the outside. The State Office of Civil Defense made those arrangements, with food for the city coming from Holloman Air Force Base near Alamogordo, New Mexico, and medicine being sent from McClellan Air Force Base near Sacramento.[777]

The danger from contaminated drinking water was so great that the Tulare County Health Department embarked on an emergency program to inoculate 35,000 persons.[778] This apparently represented the majority of the county since the combined population of Visalia and Tulare was less than 24,000.

In Visalia, 332 families had to be evacuated from their homes, chiefly in the southeast section of town. Combined with the 500 families from the rural areas around Exeter and some scattered evacuees, an estimated 880 families were driven out by floods in Tulare County. Three mass care centers had to be opened in the Visalia area, and it was eight days before the last of the evacuees could return to their homes.[779]

Immediately after the flood crest passed, emergency work began to reopen the mouth of the St. Johns River at McKay's Point. Among the first requests made to the State Office of Civil Defense from

269

Visalia was one for five draglines to dredge out the sandbar in the St. Johns River. Tulare County Civil Defense officials were able to locate the necessary draglines, and the California Highway Patrol made arrangements to get the bulky caravan through to its destination. By December 28, 25 pieces of heavy equipment were involved in the effort to reopen the St. Johns channel (multiple photographs on file in the national parks).[780]

This emergency work was overseen by the USACE as was levee repair on the St. Johns more than a mile downstream. The repair work was done using federal funds. Under the terms of the federal law governing such emergency repair work, the USACE could only restore the St. Johns levee to the same condition that it was in prior to the flood, not fix its obvious deficiencies. When the January 1956 flood hit, a portion of some of the newly rebuilt levees failed again. It was a frustrating situation for all parties concerned.[781]

Approximately 183,000 acres of land in and around Visalia was inundated by the Kaweah during the flood. (This had initially been estimated to be more than 300,000 acres.) There was no loss of life in Tulare County, but 100 injuries were attributed to the flood, and there were 24 traffic accidents under flood conditions that involved personal injury.

Private residences in Visalia sustained nearly $1 million in damage, and businesses suffered an additional loss of $600,000. There was $200,000 damage to public property, not including state highways. In the rural areas, there was $200,000 damage to farm homes. Tulare County sustained approximately $10 million in loss to the current agricultural crop plus an additional $10 million in permanent damage to land, orchards and vineyards.[782]

According to the California Disaster Office, the Kaweah River Basin sustained the greatest damage of any area in the San Joaquin River Basin during this flood.[783] The damage suffered in the Visalia area in the 1955 flood was the impetus for Terminus Dam to be constructed. Flooding in Visalia has been much less of a problem since that dam was completed in 1962.

Tulare was largely outside the flood zone and had only a few blocks flooded. However, roads to that city were closed off in most directions.

Tagus Ranch is located about four miles north of Tulare. Highway 99 was shut down at Tagus Ranch when a large lake covered the roadway there, leaving 25–50 cars stranded (photograph on file in the national parks). Residents of the area around Tagus Ranch had to be evacuated. Trains were also stalled as a result of the area flooding.

Springville is located just downstream from the confluence of the North Fork and Middle Fork of the Tule River. Springville had its water system flooded and sustained some other damage during the 1955 flood. Twenty families had to be evacuated.[784]

The Tule River flowing past Porterville flooded, but its peak flow of 13,900 cfs was considerably less than the 25,500 cfs recorded in the 1950 flood. Still, it was a river to be reckoned with. The Santa Fe Railway trestle bent under the onslaught of the floodwaters, but survived. The highway system was not so fortunate. Traffic coming into Porterville from the west had to be detoured when the east approach to the West Olive Street Bridge caved in, revealing a cavity the width of the highway and 12 feet deep. Homes along the riverbank in Porterville were threatened, and many were evacuated. Thousands of acres were flooded west of Porterville. The highway to Visalia was cut off north of Lindsay.

Damage caused by flooding on the Tule was minimal compared to that caused by the Kaweah. However, Harlan Hagen, the local congressman, used the 1955 flood as an opportunity to push through funding for both Terminus and Success Dams. Success Dam would be completed in 1961.

Over 15 inches of rain over a two-day period caused some flooding along the Kern River. Homes and roads were flooded during this time. The state fish hatchery, rebuilt after being destroyed by flooding in 1950, was washed away again. They lost 683,000 fish this time. The areas near Kernville were evacuated.[785]

The North Fork of the Kern River at Kernville peaked at 27,600 cfs. (That was the peak hourly flow; the peak average daily flow was 12,787 cfs.) Based on the flood exceedence rates in Table 16, this had a recurrence interval of 13 years for the Kern River.

Isabella Dam had been completed in 1953. That reservoir held virtually all the runoff from the 1955 flood, protecting Bakersfield from flooding. It impounded 55,000 acre-feet during the month of December.[786]

On December 23, President Eisenhower declared Northern California a major disaster area. The following day, he added two Nevada counties to the disaster list.

The North Fork of the Kern River was the southernmost stream listed by the California Department of Water Resources as setting any record in the December 1955 flood.[787] During the period that northern communities were being deluged, the southern part of the state was experiencing a drought.

About one month after the December flood, Central and Southern California experienced flooding as a result of an intense rainstorm on January 25, 1956. The degree of flooding from the January storm varied dramatically by area.

The Kings River flooded on January 25, apparently brought on by a sudden downpour. It was not a major flood. Peak flow at Pine Flat Dam was less than 20% of flows experienced in the 1955 flood.

There was no report of damage in the national parks from the January flood.

Heavy rain fell in the lower Sierra on January 26, 1956, causing rivers to swell. Numerous farms were flooded near Chowchilla and Madera. Flooding was most significant near Visalia.[788] The flows in January 1956 were apparently much lower than they had been in December 1955. With a little luck, the flooding in Visalia should have been relatively minor.

However, luck was not with Visalia. There was apparently no grating at the head of Mill Creek to keep debris from flowing into the aqueduct/conduit that goes under the town. A large tree with a big root wad floated in with the floodwaters. It was some 3 feet in diameter and about 12–14 feet long. The story soon started that it was a sequoia log, but that seems to be just an urban legend.

When the log got into the downtown area (under Garden Street, between Main and Center), it jammed. A major (unseen) debris jam quickly built up behind it. All that water pressure was not to be denied. Geysers started erupting at various places in the vicinity of the debris jam. One of the major clusters of those eruptions was inside the Harvey Hotel (aka Harvey House). Another major cluster of geysers erupted directly out of Garden Street. A sandbag enclosure was constructed in an attempt

271

to contain where the water was erupting out of the street, but that failed. There was simply too much water. Mill Creek was essentially forced to the surface and started flowing through town on the city streets (multiple photographs on file in the national parks).

Some 40,000 sandbags were used to convert streets and alleys into canals to handle the runaway floodwaters. More than 50% of Visalia was underwater at some point during the flood. Floodwaters inundated up to 72 city blocks for five days. A "glory hole" was cut through the pavement on Church Street, allowing the water to flow back into the aqueduct/conduit.

The USACE brought in a truck-mounted clamshell and started excavating, searching for the log that had to be under the city streets somewhere. Before it was all done, there were a total of four gaping holes in Visalia's city streets (two on Garden, one on Church, and one on Center). Eventually the huge log was found and extracted. It was rather like a root canal, done on a huge scale.

Along with the key log, at least eight truckloads of logs and timbers were removed from the hole where the debris jam had formed. On January 30, Mill Creek resumed flowing in its underground aqueduct/conduit. The city council decided to leave the sandbags in place because the levee on the south bank of the St. Johns was deemed to be in serious condition, and it was feared that another flood might occur before the season ended. The USACE estimated that damage in Visalia would be on the order of $1.5 million. The Harvey Hotel was so severely damaged by the flood that the building was condemned.

A screen of welded steel pipe to catch floating debris was installed across Mill Creek at Burke Street (near E. Center St.) where the underground conduit begins. This was apparently the first such screen ever installed to prevent blockage in that conduit.

In addition to flooding Visalia, the January flood caused damage elsewhere in Tulare County. The Kaweah River washed out a levee where the People's Ditch takes water out of the Kaweah. This levee had just been rebuilt by the USACE, but they had rebuilt it out of sand because they were required by law to rebuild it out of the same material as before. In the January 1956 flood, the St. Johns washed out a 150-foot section of the south levee northwest of Visalia. The Tule River washed away much of the repair work that had just been done on the Oettle Bridge.

The flooding was so serious in Southern California that it merited an article in the *New York Times*.[789] Los Angeles received seven inches of rain in what was described as one of the worst rainstorms in Southern California history. Some 1,500 people had to abandon their homes as a result.

Total flow for water year 1956 was 149% of the 1894–2011 average for the Kings, 168% for the Kaweah, 150% for the Tule, and 123% for the Kern. Flooding occurred in the Tulare Lakebed in water year 1956. It resulted from runoff from both the December 1955 flood and the January 1956 flood. Tulare Lake had been dry from about July 1, 1953, through December 23, 1955.

The December 1955 rain-type storm on the Kings River Basin was outstanding in both peak and volume of flow, exceeding the previous record runoff of December 1950. However, the entire runoff above Pine Flat Dam was controlled by the reservoir, and no floodflows from the Kings reached Tulare Lake.

The runoff from the Kaweah River was also a record for rain-type storms, exceeding that of December 1950 in both peak and volume.

Tule River flows were not as great as the 1950 quantities.

Runoff on the Kern River was also very large. The North Fork of the Kern River (see Station 1860 in Table 49) peaked at Kernville in December 1955 at 27,400 cfs. That tied the record set in the 1950 flood. (The December 1966 flood would have a discharge more than twice as great (60,000 cfs.)[790] Isabella Reservoir was able to completely control the December 1955 flood.

Inflows to the Tulare Lakebed, in large part from the Kaweah River, are shown in Table 42.[791]

Table 42. Inflow to the Tulare Lakebed during water year 1956.

| Month | Lakebed Inflow (acre-feet) |
|---|---|
| December, 1955 | 19,000 |
| January, 1956 | 36,800 |
| February, 1956 | 10,000 |
| Total | 65,000 |

Three of the lakebed sumps were flooded. It is unclear whether these portions of the lakebed were included in the calculation of the 183,000 acres of lake flooding that occurred around Visalia in the 1955–56 flood.

The 1955–56 inflow to Tulare Lake was eventually distributed over various portions of the lakebed and absorbed into the ground. The lake was dry by about April 21, 1956 and would remain dry until the 1958 flood.

The principal damages in the lakebed in the 1955–56 flood were the loss of a crop of barley growing on the flooded land, the loss of irrigation equipment, and the erosion of levees and land. Total losses in the lakebed were about $575,000.

## 1957 Flood

Flooding in 1957 occurred in June. It was a snowmelt flood.

The winter of 1956–57 was a moderate to strong La Niña event.

We know about this flood only from the stream gage record.

The peak day natural flow at Pine Flat on the Kings River occurred on June 4: 13,077 cfs. The Kings remained high from June 2–8. Thanks to the presence of a recording stream gage, we know that the South Fork Kings peaked on June 4 at 7,220 cfs.[792] That was nearly as big as the much more famous June 1952 flood (7,460 cfs).

The peak day on the Kern occurred on June 5: 3375 cfs. The Kern remained high from June 3–9.

## 1958 Floods (3)

There were three periods of flooding during 1958:
1. March (west of Mendota)
2. April (near Coalinga and west of Mendota)
3. Flooding occurred in the Tulare Lakebed during February–June as the result of a combination rain and snowmelt flood event.

The winter of 1957–58 was a moderate El Niño event.

According to the national parks' monthly report, March was particularly wet in the parks. There were 22 days of storms during the month. As shown in Table 43, this resulted in more than twice the average precipitation at all three of the parks' reporting stations.

Table 43. Total precipitation during March 1958.

| Reporting Station | 1958 Precipitation (inches of rain) | Average Precipitation (inches of rain) |
|---|---|---|
| Grant Gove | 16.49 | 6.84 |
| Giant Forest | 17.90 | 6.51 |
| Ash Mountain | 9.72 | 4.38 |

Most of the precipitation at higher elevations fell as snow during March. The biggest storm of the month occurred on March 11–17, dropping 38 inches on Giant Forest and 52 inches on Grant Grove. By the end of March 1958, Grant Grove had received a total of 252 inches (21 feet) of snowfall, compared with 96 inches at the same time the previous year. Giant Forest had received 218 inches by the end of March 1958.

The storm system seemed to break at the end of March. It felt like spring had arrived, giving the national parks time to prepare facilities for normal spring opening. Then April 1–8 brought an almost unparalleled late storm, dropping 72 inches at Grant Grove and 57 inches at Giant Forest. Precipitation and snowfall were the greatest since 1952 and came close to all-time records for a single storm. The weather for the remainder of April was generally clear with below-average temperatures, which greatly reduced the flooding in the valley below.

On March 16, heavy rain triggered debris flows that caused a bridge to wash out 21 miles west of Mendota. A car drove into the raging water, resulting in one boy being killed.[793]

A series of storms off the coast with an associated series of fast-moving fronts swept over California during late March and early April, 1958. The San Joaquin Valley experienced several small tornadoes. Thunderstorms were widespread. We know about two of these: one in Stanislaus County and one near Coalinga. We have much better information on the one that occurred in Stanislaus County.

Woodward Dam is located seven miles northwest of Oakdale in Stanislaus County. On April 3, Woodward Dam received 5.72 inches of rain, an amount equal to 45% of its average annual rainfall. That is 8.55 standard deviations above the average maximum daily rainfall with a recurrence interval of almost 300,000 years.[794]

Sometime in April, there was a major flood event near Coalinga (presumably from Los Gatos and/or Warthan Creeks). It mainly affected agricultural lands and public facilities such as roads and bridges. This was one of the three biggest flood events to occur in the Coalinga area during historic times. Panoche/Silver Creek west of Mendota also flooded in April 1958.[795]

Flood releases of 8,000 cfs or greater occurred at Friant Dam on the San Joaquin River during the April–July snowmelt period. Total flow for water year 1958 was 146% of the 1894–2011 average for the Kings, 148% for the Kaweah, 160% for the Tule, and 149% for the Kern.

The flood of February-June 1958 was a combination rain and snowmelt flood. The rains began in February and continued into April. On the Kings River, much of the precipitation above Pine Flat Dam fell as snow, although some intense rain also occurred at low elevations. The rain-flood runoff which occurred in April was well below the record of December 1955. The snowmelt runoff, which began in late May, was well below the 1906 record runoff.

Flooding occurred in Tulare Lake in 1958. The lake had been dry from about April 21, 1956 until March 31, 1958.

Pine Flat Dam had been completed in 1954. It contained most of the runoff from the Kings River. However, a small amount of Kings water did reach Tulare Lake, largely in June.

Runoff from the Kaweah and Tule River Basins in the 1958 flood was like that from the Kings River in that the runoff from rain was well below the December 1950 record runoff and the snowmelt runoff was less than the 1952 record runoff. However, considerable water did reach Tulare Lake.

Inflows to the Tulare Lakebed are detailed in Table 44.[796]

Table 44. Inflow to the Tulare Lakebed during water year 1958.

| Stream | Total Inflow (acre-feet) | Percent Contribution |
|---|---|---|
| Kings River | 24,000 | 14% |
| Kaweah River | 75,000 | 44% |
| Tule River | 72,000 | 42% |
| Kern River | 0 | 0% |
| Total | 171,000 | |

No Kern River water reached Tulare Lake during the 1958 flood. Flows in the Kern were completely controlled by the operation of Isabella Reservoir, which stored about 350,000 acre-feet of runoff between March 1 and June 25.

Maximum depth of water in Tulare Lake was about 9.7 feet on April 20, 1958. However, by August 14, the lake was dry. Presumably much of the water had been used to irrigate lands which had not been flooded in April and May. The lake then remained dry until December 6, 1966.

At some point in 1958 or 1959, a short section of the Colony Mill Road slid off the hillside. That section of the road was immediately downhill of the junction with the Admiration Point Trail. It's tempting to think this slide was a result of the very wet winter of 1958. In any case, the slide resulted in the Colony Mill Road remaining closed until June 1960 when it had to be hurriedly reopened to support the Tunnel Rock Fire. Rather than constructing a short bypass around the section of road that had failed, a mile or so of new road was constructed (the Over-the-Hump Road) on the other side of the ridge. Bill Tweed recalled that the reason for doing this was to provide a road that was not as exposed to the fire, which was on the south side of the Ash Peaks Ridge.

## 1959–62 Drought

This drought affected the entire state, but was most extreme in the Sierra and Central Coast. Recurrence intervals were greatest along the Central Coast, in the Sierra, and in the Southern California desert (30–75 years).

Table 45 shows how the San Joaquin Valley Water Year Index categorized the drought years in that basin.

Table 45. Rating of drought severity during the 1959–62 drought.

| Water Year | San Joaquin Basin Water Year Classification |
|---|---|
| 1959 | Dry |
| 1960 | Critically dry |
| 1961 | Critically dry |
| 1962 | Below normal |

1959 was the driest year on record in Kern County; total precipitation during the year was less than two inches.

Total annual flows for the rivers within the Tulare Lake Basin were generally less than 50% of average for water years 1959–61. In the southern half of the state, 1961 was the driest year of the drought, ranking among the driest years of record at many sites. Total flow for water year 1961 was 33% of the 1894–2011 average for the Kings, 27% for the Kaweah, 18% for the Tule, and 25% for the Kern.

On January 22, 1962, Fresno experienced its biggest snow in 32 years when 2.2 inches fell. The snow closed schools and caused a rush of people to stores seeking to buy film to photograph this unusual event. Many roads were slippery and some were closed altogether. Five people died on valley roads due to the slick conditions. Other amounts in the valley included 4.0 inches at Madera, 3.0 inches at Wasco, 2.0 inches at Hanford, Avenal, Buttonwillow, and 1.5 inches at Los Banos. The higher elevations were buried in snow, 33 inches was reported at Badger Pass in Yosemite.[797] But this was just an interesting interlude, the drought would continue for another year.

In 1963, a record number of mid-winter foggy and rainless days were recorded at Sacramento associated with high barometric pressures and stagnant winds. This was one of the worst mid-winter droughts of record in Central California.[798]

National park records indicate that this was a severe and extended drought. In the last week of January 1963, there was so little snow in the parks that three people were able to complete a trip to East Lake, Reflection Lake, and over Langley Pass to South Guard Lake. They made it just in time. The drought ended abruptly with a major storm that began on January 29, 1963.

The similarity of meteorological conditions of the 1860–61 and the 1959–62 droughts are notable. Both were severe droughts that ended with severe flooding.

Tulare Lake was dry throughout the 1959–62 drought. The lake went dry in August 1958, and would stay dry until the 1966 flood brought it back to life, if only for a few brief months.

## 1962 Floods (2)

There were two floods in 1962:
1. a small rain flood in February
2. a small snowmelt flood in May

We know about these floods only from the stream gage record.

The peak day natural flow at Pine Flat on the Kings River occurred on February 10: 10,236 cfs.

The Kaweah's peak natural flow occurred at Terminus Dam on February 10: 8,000 cfs. (That was the peak hourly flow; the peak average daily flow at the dam was 3,707 cfs).

The Tule peaked on February 10: 1,337 cfs. The Kern peaked on February 11: 2,438 cfs.

The second flood of the year was a snowmelt flood. In that flood, the peak day natural flow at Pine Flat on the Kings River occurred on May 6: 12,724 cfs. The Kings remained high from May 4–9. Thanks to the presence of a crest-stage gage, we know that the South Fork Kings peaked at 5,600 cfs on about May 6.[799]

The Kaweah's peak average daily flow at Terminus Dam during the runoff occurred on May 5: 2,652 cfs. The Kaweah remained high from about May 3–9.

The runoff came early on the Tule in 1962. The peak day occurred on April 9: 573 cfs. The runoff on the Tule in 1962 really didn't amount to a flood in any conventional sense.

The peak day natural flow on the Kern occurred on May 6: 3,574 cfs. The Kern remained high from about May 4–10.

## 1963 Flood

Flooding in 1963 occurred in February. It was caused by a storm that came out of the North Pacific.

Most of January was very dry but extremely cold in the national parks. There was little snow on the ground even at the high elevations. The combination of extreme cold and lack of snow caused damage to the national parks' water systems. This was considered the worst mid-winter drought (or extended dry season, depending on your point of view) in the state's history.

This extended dry season came on the end of four years of severe drought (1959–62). The drought was finally broken by a storm that lasted from January 29 – February 2, 1963. This is often treated as a three-day storm, lasting from January 30 – February 1. The rainfall distribution in this storm was quite similar to the November 18–19, 1950 storm. The rainfall of both these storms was heavy in the coastal mountains as well as in the Sierra.

This 1963 storm resulted in the heaviest-ever three-day rainfalls at 45 stations. These extreme rainfalls were generally at high elevations in the Southern Sierra. The heaviest rainfalls were centered south of Yosemite. Florence Lake received 64% of its average annual precipitation in this storm, which represented a recurrence interval of 33,000 years. Other Sierra stations with a recurrence interval greater than 1,000 years were the South Entrance of Yosemite National Park and Tollhouse.[800]

Table 46 gives the total precipitation during the January 29 – February 2, 1963 storm event for selected reporting stations.

Table 46. Precipitation during the January 29 – February 2, 1963 storm event.

| Reporting Station | Storm Total (inches of moisture) |
|---|---|
| South Entrance of Yosemite | 22.99 |
| Wishon Dam (near Shaver Lake) | 23.25 |
| Grant Gove | 17 |
| Giant Forest | 21 |
| Ash Mountain | 12 |

The snowline associated with this storm was generally over 8,000 feet and at times as high as 11,000 feet. Snowmelt was a major factor in the flooding associated with this storm. Many streams reported record-high flows during this storm.[801] The snowline on the west side of the Great Western Divide was 7,000–9,000 feet. Snow depths were progressively greater to the east and north.

Major flooding occurred to the north of the Tulare Lake Basin, including the cities of Napa, Marysville, and Reno.

Within the national parks, mudslides (or debris flows) occurred in the drainage above Simpson Meadow, suggesting considerable rain before the snow began.

The peak day natural flow at Pine Flat on the Kings occurred on February 1. That is the sixth largest peak day of the year at Pine Flat since the dam was built in 1954. This was a bigger flow than occurred in the much more famous 1983 flood. It seems likely that this was a very high-flow period in Cedar Grove as well.

Thanks to a pair of crest-stage gages, we know that Grizzly Creek peaked at 293 cfs on about February 1.[802] This was the highest flow recorded on that creek between 1960–1973.

The Kaweah's peak natural flow occurred at Terminus Dam on February 1: 30,900 cfs. (That was the peak hourly flow; the peak average daily flow, as reflected in Table 15, was 18,405 cfs.) Based on the flood exceedence rates in Table 16, this had a recurrence interval of 16 years for the Kaweah.

Runoff from this storm was very rapid and the Kaweah River reached flood stage before dawn on February 1. Lake Kaweah was able to catch much of this runoff, preventing flooding downstream. Despite the intensity of the storm and the resulting flooding, the national parks received less damage than expected.

In the Kern River Basin, over 14 inches of rain fell in February over a short duration. Forty people were evacuated from their homes in the Kernville area as floodwaters from the North Fork of the

Kern River threatened their homes. Once again the state fish hatchery sustained damages, and all the fish were lost — some 225,000 rainbow trout that were about to be released. Everyone staying in low-lying areas was evacuated.[803]

The South Fork Kern River near Onyx (see Station 1895 in Table 49) peaked at 3,460 cfs in February 1963. This was the greatest discharge on that river since record-keeping began in 1911. However, the December 1966 flood would be eight times greater.[804]

There was also flooding on the west side of the valley. Damage was sustained largely by agricultural lands and public facilities. The area around Coalinga was one of the areas that was damaged. We know this happened sometime in 1963, but we don't know that it was during the February storm.

## 1964 Flood

Flooding in 1964 occurred in December.

The winter of 1964–65 was a moderate La Niña event.

The flooding resulted from meteorological conditions similar to those of the December 1955 flood. An arctic air mass moved into Northern California on December 14, and precipitation on December 18–20 produced large quantities of snow. Beginning on December 20, a storm track 500 miles wide extended from Hawaii to Oregon and Northern California. Warm, moist air collided with the arctic air and resulted in turbulent storms that produced unprecedented rainfall on Northern California and melted much of the snow from the previous storms. In the Mattole River Basin, nearly 50 inches of rain was reported during December 19–23, with 15 inches observed in 24 hours.

The six-day period from December 19–24 was the wettest ever recorded at 78 Northern California stations. The North Coast had the worst flooding ever experienced in that region. Every major stream in the North Coast produced new high values of extreme peak flows. A total of 34 California counties were declared disaster areas. This storm had three major centers of activity: the Eel River, the Upper Klamath and the Yuba River in the Central Sierra.[805]

Branscomb in the Eel Basin received 31.71 inches during the storm event. Most stations in the Eel River Basin reported their highest-ever rainfall during this storm. Gazelle in the Klamath River Basin reported 8.09 inches. That was 7.78 standard deviations above the average with a recurrence interval of over 300,000 years. A total of 35 stations reported daily rainfalls of 10 inches or more on December 22. These stations were located in the North Coast streams as well as in the Central Sierra. The highest-ever rainfalls occurred in the Yuba and Bear River Basins, where Lake Spaulding (east of Grass Valley) received 32.60 inches of rain during the storm event.[806]

Floods were widespread across the northern half of the state. The main center of precipitation was in the Feather, Yuba, and American River basins. Runoff from streams of the Coast Ranges, almost without exception, produced peak stages and peak flows that exceeded previous records. Runoff from the Sierra into the Feather, Yuba, and American rivers surpassed all previous records.[807]

Bridges on every major stream were destroyed. Several towns along the Eel and Klamath Rivers were totally destroyed. The floods caused $239 million in property damage and 24 deaths statewide. The property damage in north coastal California was about 50% greater than had occurred in the

December 1955 floods. The December 1964 flood remains the greatest known flood in the history of Northern California.

Exceptionally large flood peaks were recorded on rivers in north coastal California. Peak discharges of the Eel, Klamath, and Smith Rivers were 30–40% greater than the 1955 peaks and exceeded flood stages of the 1861–62 floods.

The American River experienced a record flood, the third record flood in less than 15 years. Several rivers, including the Salmon and Klamath, experienced a flood event with a recurrence interval of greater than 100 years. Botanic and geomorphic evidence indicated that floods exceeding the magnitude of the 1964 flood may not have occurred since about 1600.

(During the 1600–1610 time period, a dramatic climatic change was happening across North America. Major precipitation events occurred in the areas near the Sacramento River Basin, Mono Lake, Southern California, Mexico, and elsewhere. See the section of this document that describes the California mega-floods for more about that event.)

The Klamath is the second largest river in California: more than twice as large as the third largest river. It drains a 12,000 square mile watershed that extends more than 260 miles inland to the Klamath Basin in Oregon. The December 1964 flood was the most devastating flood of the Klamath ever. It swept away all of downtown Klamath, destroyed the Highway 101 bridge, and washed away a great many homes. The river peaked at over 550,000 cfs, 17% greater than the average flow of the Mississippi River.

At the time of the flood, an 800 pound Angus bull named Bahamas was living in the valley of the Klamath. On December 22, the Klamath topped out at 52 feet, covering its valley in a maelstrom of churning logs, brush, lumber and debris. Bahamas was swept up in all the debris and carried downriver and out to sea. He apparently survived the ride by climbing on top of a raft of flotsam.

Bahamas rode there on the open ocean on a constantly disintegrating raft of logs and brush, through huge storm waves. Eventually he and his raft arrived at the Crescent City Harbor, 16 miles up the coast from the mouth of the Klamath River. Somehow he had stayed aboard his accidental raft of slippery, tossing logs and brush.

Bahamas was discovered the next day, 200 feet offshore in the 10 acre mass of floating, churned debris that plugged Crescent City Harbor. He was helped to go from log to log until he reached shore. He was more or less adopted by the town. A novel was written about him: *Beloved was Bahamas*. He lived out his life in his own grassy paddock in Klamath, his feat commemorated by a large sign on the fence. He was visited by many people over the years who regarded him as a living touchstone of courage and will. He died in the spring of 1983 and was buried in his green pasture.

The peak flow in the American River at Fair Oaks, controlled by Folsom Dam, reached 115,000 cfs. In the remaining watersheds of the Sacramento Valley, peak stages and flows tended to equal those experienced in 1955. At the Sacramento Weir, all 48 gates were opened, and the peak flow reached 84,000 cfs. The peak flow in the Sacramento River at I Street was about 100,000 cfs. Total peak flow at the latitude of Sacramento, including the Sacramento River and the Yolo Bypass, was about 475,000 cfs.[808]

The December 1964 floods did not extend as far south as those of December 1955. In the Sacramento River Basin, many streams had peak discharges that were greater than during December 1955. However, peak discharges in the San Joaquin River Basin were substantially less than during 1955. In both basins, flood-control operations generally were able to confine downstream flows within flood-control channels. As a result, loss of life was avoided, and damage was less than half that caused by the 1955 flooding.

Although the worst of the storm was in the north, it still brought significant precipitation to the Tulare Lake Basin. The peak day natural flow at Pine Flat on the Kings occurred on December 24, 1964. It was obviously a flood, but it was only half as high as the peak day natural flow for 1963. Thanks to a crest-stage gage on the South Fork Kings, we know that stretch of the river peaked at 5,200 cfs on about December 24.[809]

As detailed in Table 47, the national parks experienced intense storm activity between December 19–28. This consisted of three separate storms, the most intense of which occurred on December 27–28.

Table 47. Precipitation during the December 19–28, 1964 storm event.

| Reporting Station | Storm Total (inches of rain) |
|---|---|
| Grant Gove | 7.45 |
| Giant Forest | 12.61 |
| Ash Mountain | 6.17 |

The national parks did not report any significant flooding. Most of the precipitation in Giant Forest consisted of rain. The storm lowered the snow on the Wolverton ski area to an unsafe level.

The heavy rains caused some damage to the Generals Highway. On December 27, a portion of the wooden cribbing one mile above Ash Mountain failed, leaving a one-lane roadway for about 60 feet. (That wood cribbing was replaced with a galvanized metal bin wall, constructed by the national parks, in April 1965.) A section of dry rubble retaining wall at the 5,000 foot elevation also failed because of the heavy rains.

The winter 1964 flood is remembered because it did so much damage in Northern California. However, the winter 1963 flood was a more memorable event in the Southern Sierra.

## 1965 Floods (2)

There were two floods in 1965:
1. March
2. August

The winter of 1964–65 was a moderate La Niña event.

Fresno received 1.55 inches of rain on March 12, setting a daily rainfall record. Most of the rain fell in a five-hour window from 4 p.m. to 9 p.m., inundating streets and poor drainage areas with water described as up to hip deep. A number of transformers in the city shorted out, plunging many homes and businesses into darkness.[810]

Daily computed natural flow at Pine Flat Dam suggests that the Sierra experienced a series of storms during the week of August 11–17.

From August 11–19, Cedar Grove had daily rains with downpours that caused debris flows which blocked the road to traffic until cleared away. Cedar Grove had 2.81 inches of rain for the month of August.

Heavy rain occurred in the national parks during the week of August 11–17. Park records document that this storm affected the area of the South Fork Kings. Thanks to a pair of crest-stage gages, we know that Grizzly Creek peaked at 247 cfs on about August 17. [811]

The Marble Fork of the Kaweah flooded through Lodgepole Campground on August 17 (photograph on file in the national parks).

## 1966 Flood

Flooding in 1966 occurred in December.

A very large storm brought a strong inflow of warm moist Pacific air across Central California from December 3–7. In the Tulare Lake Basin, the most severe effects of the storm were felt south of the Kings River. The storm penetrated deeply inland, bringing significant moisture into the Owens Valley. The heaviest rain was in a narrow band that ranged from SCE's Kern River Intake #3 in the south to the White Mountains over 100 miles to the northeast.

December 1966 was the wettest five days ever at 58 California stations in an area stretching from the Kern River to the White Mountains, and into Tulare and San Bernardino Counties. A total of 19 stations reported 10 inches or more of rain on December 6. These stations were located mainly in Tulare and San Bernardino Counties.

The heaviest 24-hour rainfall ever recorded in the Central Valley, 17.0 inches, occurred on December 6 at Hockett Meadows. This record would last for 20 years. It would eventually be exceeded by the 17.6 inches recorded at Four Trees in the Feather River Basin on February 17, 1986.

The record downpour that occurred on Hockett Meadows and the surrounding plateau on December 6, 1966, generated unprecedented runoff. Cahoon Meadow was predisposed for erosion by years of heavy grazing. It is possible that this was the event that caused much of the gullying that we see today in Cahoon Meadow.

A total of 42 stations recorded their highest-ever 5-day rainfalls during this storm event. A total of 11 stations reported rainfall totals in excess of a storm with a recurrence interval of 1,000 years. The highest rainfall was reported at Johnsondale with a 5-day total of 30.45 inches.[812]

Table 48 shows the elevation and precipitation for the December 2–7 storm event. Springville received its greatest precipitation of the storm event on December 5. The other reporting stations received their greatest amount on December 6.[813]

Table 48. Precipitation during the December 2–7, 1966 storm event.

| Reporting Station | Elevation (approximate) | Storm Total (inches of rain) | Drainage Basin |
|---|---|---|---|
| General Grant | 6,600 | 23.04 | Kings |
| Giant Forest | 6,358 | 27.75 | Kaweah |
| Ash Mountain | 1,730 | 15.52 | Kaweah |
| Three Rivers Powerhouse #2 | 950 | 11.85 | Kaweah |
| Springville Ranger Station | 1,050 | 10.78 | Kaweah |
| Johnsondale | 4,680 | 30.45 | Kern |
| Glennville | 3,140 | 8.62 | Poso |
| Wofford Heights | 2,700 | 11.00 | Kern |
| Kern River Powerhouse #1 | 970 | 3.46 | Kern |

Rain fell as high as 9,000 feet. The rain apparently melted all the snow on the ground at Grant Grove, Giant Forest, and the Wolverton Ski Bowl. However, there may not have been a particularly heavy snowpack to melt.

Grant Grove experienced an exceptionally severe rain and wind storm on the night of December 5. That event brought down a 100-foot forked-top sugar pine, demolishing a visitor cabin that was fortunately closed for the season.

On December 7, the weather turned cold, reducing the amount of flooding. A second storm had been feared, but stayed north of the Tulare Lake Basin.

Mountain Home received 23 inches of rain during the storm, and Camp Wishon (northwest of Camp Nelson) reported 36 inches of rain in 69 hours.[814]

One source said that the December 2–7 event was the most severe rainstorm on record in the southern San Joaquin Valley.

Paso Robles received 5.25 inches of rain in 24 hours on December 6, setting a record for that city.

Some rivers had a recurrence interval greater than 100 years. It was the first significant storm of the winter. Flooding occurred throughout Northern California, including on the Russian and Eel Rivers. However, this was not to be a replay of the 1964 flood. This time, flooding was most severe in the Kaweah, Tule, and Kern River Basins.[815] San Bernardino and Riverside Counties also sustained serious flooding.

Continuously above-average precipitation from December 1966 through March 1967 resulted in the flooding of 35,000 acres of the northern San Joaquin River Basin. The San Joaquin River above Millerton Lake experienced high runoff during early December. A maximum mean daily inflow of 18,450 cfs was recorded at Friant Dam. However, releases of only 52 cfs were made to the San Joaquin River.[816]

Significant amounts of flooding occurred in both mountain areas and on the valley floor. A total of 141,800 acres flooded, including 122,400 acres of valley floor and 19,400 acres in mountain and foothill areas. These record-breaking floods inundated parts of the towns of Kernville, Springville, Three Rivers, Lindsay, and Lamont.[817]

Panoche/Silver Creek west of Mendota also flooded in December 1966.[818]

The peak day natural flow at Pine Flat on the Kings occurred on December 6, 1966. That is the second largest peak day of the year at Pine Flat since the dam was built in 1954. (The flow of record occurred on December 23, 1955.)

Thanks to a crest-stage gage on the South Fork Kings, we know that reach of the river peaked at 11,800 cfs on about December 6.[819] That put this flood about midway in size between the 1950 flood (10,000 cfs) and the 1955 flood (13,900 cfs). There was major damage to the state's portion of Highway 180 in Kings Canyon.

Based on the flood exceedence rates in Table 16, this had a recurrence interval of 70 years for the Kings River downstream at Pine Flat. That puts it in a category with other Cedar Grove floods of the past 70 years (1937, 1950, 1955, 1966, 1969, 1978, 1982, 1983, and 1997) that rise to the level of the modeled 50-year flood: that is, a flood event that occurs about every eight years on average. See the section of this document that describes Cedar Grove Flooding.

Pine Flat Reservoir appears to have caught (or at least diverted) the entire flood on the Kings. No floodwaters from the Kings made it into the Tulare Lakebed during 1966.

In the Tulare Lake Basin, the December 1966 flood was generally the biggest flood-of-record on most major streams south of the Kings since the 1867–68 flood. These include Sand Creek draining the area west of the North Fork Kaweah River and Deer Creek draining the area west of the Kern River and south of the Tule River, and Poso Creek draining the area west of the lower Kern River Basin.[820]

Past records of peak flow and 3-day storm-runoff volume in the Kaweah, Tule, and Kern River Basins were greatly exceeded by the floods of December 1966. Extremely high peak discharges occurred at most gaging stations between 11:00 p.m. December 5 and 6:00 p.m. December 6. Snowmelt was not a major cause of the floods, although some snow that had accumulated during minor November and early December storms was melted.[821]

Thanks to a USGS report, the December 1966 flood was particularly well documented.[822] Table 49 summarizes the flood discharge data for selected streams in the Tulare Lake Basin.

Table 49. Summary of peak flood discharges for the December 1966 storm event.

| River | Station ID | Maximum previously known | | | Maximum December 1966 | |
|---|---|---|---|---|---|---|
| | | Period of record | Year | Discharge (cfs) | Day | Discharge (cfs) |
| Kern River near Quaking Aspen | 1853.5 | 1960–66 | 1963 | 4,060 | 6 | 9,360 |
| Little Kern River near Quaking Aspen | 1854 | 1955, 1957–66 | 1955 | 12,200 | 6 | 13,100 |
| Kern River near Kernville | 1860 | 1912–66 | 1950, 1955 | 27,400 | 6 | 60,000 |
| Kern River at Kernville | 1870 | 1905–66 | 1950 | 38,700 | 6 | 74,000 |
| South Fork Kern near Olancha | 1882 | 1956–66 | 1958 | 1,280 | 6 | 1,010 |
| South Fork Kern near Onyx | 1895 | 1911–14, 1919–42, 1947–66 | 1963 | 3,460 | 6 | 28,700 |
| Kelso Creek near Weldon | 1897 | 1958–66 | 1965 | 1,340 | 6 | 5,800 |
| Kern River below Isabella Dam | 1910 | 1945–66 | 1950 | 39,000 | 30 | 2,160* |
| Kern River near Democrat Springs | 1925 | 1950–66 | 1950 | 40,000 | 6 | 10,100* |
| Kern River near Bakersfield | 1940 | 1893–66 | 1950 | 36,000 | 7 | 9,290* |
| Poso Creek near Oildale | 1978 | 1958–66 | 1958 | 2,750 | 6 | 4,300 |
| White River near Ducor | 1995 | 1942–53, 1958–66 | 1943 | 2,300 | 6 | 1,080 |
| Deer Creek near Terra Bella | N/A | N/A | N/A | N/A | 6 | 10,000 |
| North Fork of Middle Fork Tule River near Springville | 2020 | 1939–66 | 1955 | 12,400 | 6 | 16,900 |
| North Fork Tule River at Springville | 2031 | 1957–66 | 1963 | 4,600 | 5 | 24,200 |
| Tule River near Springville | 2032 | 1950–66 | 1950 | 22,400 | 6 | 49,600 |
| South Fork Tule River near Springville | 2045 | 1930–54, 1956–66 | 1950 | 7,100 | 6 | 14,300 |
| Middle Fork Kaweah near Potwisha | 2065 | 1949–66 | 1955 | 46,800 | 6 | 23,300 |
| Marble Fork Kaweah at Potwisha | 2080 | 1950–66 | 1955 | 12,500 | 6 | 6,400 |
| East Fork Kaweah near Three Rivers | 2087.3 | 1952–55 1957–66 | 1963 | 2,850 | 6 | 13,000 |
| Dorst Creek near Kaweah Camp | 2090 | 1959–66 | 1963 | 1,540 | 6 | 2,010 |
| North Fork Kaweah River at Kaweah | 2095 | 1910–66 | 1955 | 21,500 | 6 | 23,900 |
| Kaweah River at Three Rivers | 2099 | 1955, 1958–66 | 1963 | 30,900 | 6 | 73,000 |
| South Fork Kaweah River at Three Rivers | 2101 | 1955, 1958–66 | 1955 | 10,000 | 6 | 11,600 |
| Kaweah River at McKay's Point | N/A | 1905–66 | 1955 | 84,332 | 6 | 105,000* |
| Dry Creek near Lemon Cove | 2113 | 1959–66 | 1963 | 1,600 | 6 | 14,500 |

*Some of the gages in the above table were below dams. Unless noted otherwise in the text, the associated discharges for the 1966 storm generally don't reflect full natural flow of those rivers because the flows were affected by storage and/or diversion upstream. That is not the case with the discharges for those gages that have been included in Table 15 on page 133. In order for the latter table to show full natural flow, those discharges have been adjusted to remove the effects of dams upstream of the gages.

Severe flooding extended over the Kaweah, Tule, and Kern River Basins in a 60- by 100-mile area in the Sierra northeast of Bakersfield. Moderate flooding occurred in the Kings River Basin and other basins to the north and in streams draining from the Coast Ranges to the west. Flood peaks were the greatest of record at many gaging stations in the Kaweah, Tule, and Kern River Basins. Damage was severe in all headwater areas. Culverts were overflowed or plugged with debris, or usually a combination of both. Most highway bridges were destroyed or severely damaged.[823]

The flood caused major damage to roads, trails, and other facilities in the national parks.

Highway 180 was closed at Snowline Lodge by a rock slide, isolating Grant Grove. According to the parks' monthly report, there was major damage to the state's portion of Highway 180 in Kings Canyon. However, there was only minor damage to the section of the road that was within the park boundaries.

As shown in Table 49, Dorst Creek's peak natural flow occurred on December 6: 2,010 cfs. A crest-stage gage (USGS 11209000) was located on Dorst Creek near where the Generals Highway crosses that creek. That gage was operated from May 1960 through May 1973. The December 1966 flood is the biggest flow recorded on Dorst Creek during the period of record. It was 30% greater than the 1963 flood, the previous high flood in the area.

Giant Forest was isolated for a week by slides on the Generals Highway between the two national parks; access between Grant Grove and Giant Forest wasn't restored until December 14.

According to the parks' monthly report, the section of the Generals Highway between Giant Forest and Ash Mountain was badly damaged:
* The Marble Fork Bridge near Potwisha was washed away (photograph on file in the national parks). This bridge had been badly damaged in the 1938 flood and washed out altogether in the 1955 flood. This time it would be rebuilt in a new location, a hundred feet or so upstream. Manuel Andrade recalled that, as an interim measure, a national park crew constructed a temporary highway bridge at the old location using brace-and-bits and crosscut saws. That bridge was open for administrative traffic in less than two weeks.
* Numerous major rock falls and landslides (multiple photographs on file in the national parks), some of which contained several thousand cubic yards each. From the photographs, one of those appears to have been just below Amphitheater Point.
* Slides / washouts in 10 locations, each of which required cribbing to restore the road to two-lane width. One of those was at Alder Creek (photograph on file in the national parks). Leroy Maloy recalled that one of the 10 sections was a big failure above Deer Ridge that was replaced with a metal bin wall in the summer of 1967. Howard Mancha (oiler on the crane that placed that bin wall) recalled that the Deer Ridge bin wall reconstruction happened after the April 1965 construction of the Ash Mountain bin wall.

According to the parks' monthly report, park crews had pushed a one-lane road through the slides by December 30. This finally permitted administrative access between Ash Mountain and Giant Forest. All other travel to Giant Forest had to be by way of the Big Stump entrance.

It isn't clear when this section of the Generals Highway was reopened to visitor traffic. At the least, visitors were probably kept off this section of the Generals Highway until a modern bridge was built across the Marble Fork Kaweah at Potwisha. That presumably took at least a year. This section of highway had probably been closed for the summer of 1956 when the previous bridge had to be replaced.

There are suggestions that the general practice in those years was to keep the Generals Highway open to single-lane visitor traffic even when bin walls were being constructed in the Deer Ridge area. It appears that no concrete barricades were used for such construction; only cones and lane delimiters were used.

As shown in Table 49, the East Fork Kaweah's peak natural flow occurred on December 6: 13,000 cfs. A stream gage (USGS 11208730) was located on the East Fork at the diversion dam for SCE's flume, about ¼ mile above where the Mineral King Road crosses the river.[824] It was a complex gage (a water-stage recorder coupled with an acoustic velocity meter) designed to give accurate measurements in variable and low velocity flow situations. The gage was operated from June 1952 through October 2010. It was knocked out of operation during the 1955 flood. The gage apparently wasn't read from October 1979 – September 1993. The December 1966 flood is the largest flow recorded on the East Fork during the period of record. It was 15% greater than the January 1997 flood.

As shown in Table 49, the North Fork Kaweah's peak natural flow occurred on December 6: 23,900 cfs. A stream gage (USGS 11209500) (a water-stage recorder) was located on the North Fork a mile above the Upper North Fork Bridge (the Bailey Bridge). The gage was operated from October 1910 through September 1981. The December 1966 flood is the largest flow recorded on the North Fork during the period of record. It was 11% greater than the December 1955 flood.

As shown in Table 49, the South Fork Kaweah's peak natural flow also occurred on December 6: 11,600 cfs. A stream gage (USGS 11210100) (a water-stage recorder) was located on the South Fork about ½ mile upstream from where Highway 198 crosses that river, 200 feet upstream from an unnamed tributary. The gage was operated from October 1958 through September 1990, but partial data is available for a longer period extending from December 1955 through January 1997. The December 1966 flood is the largest flow recorded on the South Fork during the period of record. It was 16% greater than the 1955 flood and 36% greater than the 1997 flood.

Specific damage in Three Rivers:
- Ash Mountain was isolated for two days because the national parks' approach to the Pumpkin Hollow Bridge was so badly eroded that only pedestrian traffic was allowed across.[825] The parks' approach to this bridge had completely washed out in the 1937 and 1955 floods.
- Both approaches to the Dinely Bridge were badly eroded, but the bridge survived.[826]
- The Upper North Fork Bridge (the Bailey Bridge) washed out. Residents past that point were isolated from the rest of the community.
- The North Fork Road washed out a short distance past the Upper North Fork Bridge and was apparently badly damaged in one or more other locations.
- The Airport Bridge on the North Fork Kaweah was either badly damaged or washed out.
- Many properties on the North Fork were damaged. The River Isle Trailer Park was heavily damaged (see below for details). The C&D Trailer Park was also damaged. Archie McDowall lost 2,500 chickens and a new pickup. Luther Smeltzer lost his home. B.F. McKinley, C.E. Fairman, and Leroy Maloy received extensive damage to their homes.
- Heavy damage totaling $1.75 million was reported to homes along the South Fork Kaweah. The South Fork flooded as severely as the North Fork, but there were relatively few homes on the South Fork.
- Huffaker's Candy Shop and Calloway's Drive-In were flooded. These were the only two Three Rivers businesses that were damaged in this flood. Other businesses were threatened, but these were the only businesses that were actually flooded.[827]
- The Three Rivers Golf Course was heavily damaged. It was cut in two by a re-channeling of the mainstem of the Kaweah. The river's new course isolated the main clubhouse, destroyed

fairways, and washed away green #2 and part of green #6. Many logs were washed onto the fairways.[828]
- The county park on the lower side of town was washed out. The river was up to the edge of Highway 198 at this point.
- SCE suffered $40,000 in damage to its transmission lines and distribution systems in the Three Rivers area.

The River Isle Trailer Park was located several miles up North Fork Drive. It is reported to still be functioning as a trailer park of some sort in 2012. The trailer park was surrounded and overtopped by the 1966 flood, scattering trailers everywhere. Mobile homes stood on end, upside down, and sideways, completely ruined by the water. The situation was so dire that 19 people had to be evacuated by helicopters from Lemoore Naval Air Station. On one flight, a Lemoore helicopter ran out of fuel and had to make a forced landing with civilians on board. A commentary on the odd things people do in an emergency was the comment from a helicopter crew member that one girl evacuated from the trailer park took with her only hair curlers, hairspray, a comb, and a change of clothing.[829, 830]

Bobbie McDowall Harris was living on the North Fork Kaweah near the River Isle Trailer Park at the time of the flood. She has a vivid memory of the floodwaters separating their family (they had five little kids at the time) from the road. As the river started rising, Bobbie and her husband decided to save their cars by moving them up onto the North Fork Road. Her husband went first in the Ford, and she followed him in a VW bus. He got stalled halfway through the water. Knowing that she had an engine in the rear, Bobbie yelled for him to get back in the car and she'd push him. It worked, and they ended up on the road. They rushed back to their children before the water got too high. It was very scary. They were stranded in their home for many days, cut off by the flooding river. They were offered the opportunity to get their family out by helicopter. But they chose to stay put, feeling that getting kids that young in a hovering helicopter was more dangerous than staying.

The Upper North Fork Bridge was replaced in 1967. The new bridge has low-profile guardrails that can be removed in anticipation of an oncoming flood. The bridge hasn't washed out since.

Sometime after the 1966 flood, the USACE constructed a levee from the Upper North Fork Bridge down past the McDowall property all the way to the River Isle Trailer Park.

Movie maker Harold Schloss was filming his movie *One on Beetle Rock* at the time that the flood hit. Floodwaters covered the runway of the Three Rivers Airport, keeping employees from flying in food for his menagerie of characters: mountain lions, a wolf, a hawk, a raven, a badger, and two coyotes. All were stranded on location at the Roping Arena. Shooting of the movie was delayed for several days. The upside was that there was a storm scene in the movie and Schloss got some excellent storm footage.

Fifteen-foot waves were reported to have been common on the mainstem of the Kaweah in Three Rivers at the peak of the flood.[831] Floodwaters apparently came up several feet on the river side of the Three Rivers Market. Shortly before noon on December 6, it was announced that some businesses in Three Rivers would have to be evacuated.

The December 1955 flood had been the biggest flood on the Kaweah since record-keeping began. Comparing the size of the 1966 flood in Three Rivers with the 1955 flood is not straightforward.

Floodmarks at two gaging stations in Three Rivers were observed to be slightly higher in the 1966 flood than in the 1955 flood. One of those gages was on the South Fork Kaweah and one was on the mainstem of the Kaweah, upstream from Lake Kaweah.

There was a gage near Three Rivers on the mainstem of the Kaweah (USGS gage #11-2105 Kaweah River near Three Rivers) located just above the junction with Horse Creek (latitude: 36:24:24N longitude: 118:57:12W). The total drainage area upstream of that gage was 519 square miles. That gage was operated from 1903–1961. But by 1966, it was submerged under Lake Kaweah.

Lake Kaweah is roughly comparable to the former Three Rivers gage. It has a total drainage area of 561 square miles. Lake Kaweah is located slightly farther downstream, but it measures essentially the same tributary streams as the former Three Rivers gage. The difference is that the Lake Kaweah gage also measures the inflow from Horse Creek.

The computed maximum bihourly inflow to Lake Kaweah on December 6, 1966 was 82,700 cfs. That was only 2% greater than the peak flow at the former Three Rivers gage on December 23, 1955: 80,700 cfs.[832]

Realizing how big the 1966 flood was predicted to be, a sack-concrete barrier was placed on the spillway of Terminus Dam to provide additional storage.[833] That proved to be good planning. About 10,250 acre-feet of floodwater was surcharged against that spillway barrier, bringing the reservoir to almost 5½ feet above full-pool level.

Lake Kaweah reached a peak storage of 147,200 acre-feet just before 2 a.m. on December 8.[834] Runoff from the upper basin was completely controlled by the reservoir, and no downstream releases were made until after the first of the year when Dry Creek and other downstream tributaries had subsided.[835] Thanks to Terminus Dam (and a break in the weather), Visalia escaped with no flooding other than from surface water. An evacuation center had been set up in the city but wasn't needed.

The Lake Kaweah marina was ripped loose from its moorings and set afloat in the middle of the lake. Many acres of debris clogged the upper end of the lake following the flood. This included logs, trees, shoes, chickens, dead fish, boats, and general flotsam.[836] Most of the heavier woody material was removed somehow the following summer, presumably by burning.

As reflected in Table 15, the peak average daily flow occurred at Terminus Dam on December 6: 53,280 cfs.

Dry Creek below Terminus Dam peaked at 14,500 cfs, the highest flow since record-keeping began on that stream. (For comparison, this was 44% greater than Merced River in Yosemite Valley in the much more famous January 1997 flood.) This flow on Dry Creek had a recurrence interval of approximately 85 years.[837] Dry Creek Road was closed from Lemon Cove to Badger during the flood. Culverts were washed out on the Eshom Valley Road. About 48,600 acre-feet of Kaweah River floodwater reached Tulare Lake in December. Dry Creek was the principal source of most of that water, with lesser amounts contributed by Yokohl, Cottonwood, Sand, Lewis and Cameron Creeks.

The Kaweah's peak natural flow occurred at McKay's Point on December 6: 105,000 cfs. It was the largest peak day of the year since record-keeping began in 1905; it remains the flow of record for this stretch of the river for floods that have occurred since 1905. However, recorded history on the Kaweah Delta began in about 1850. During recorded history, the 1867–68 flood is considered the flood-of-record for the Kaweah; that flood just occurred before any stream gages were installed on this river.

Above Terminus Dam, the 1955 and 1966 floods were roughly equal in size. But because Dry Creek added so much water, the flood below Terminus Dam was a much more impressive event. The peak average daily flow at McKay's Point in 1966 was 20% higher than the more famous December 1955 flood (peak average daily flow of 53,280 cfs in 1966 versus 44,512 cfs in 1955).

Keep in mind that the above flows for 1966 (105,000 cfs and 53,280 cfs) reflect the natural flow of the river, after adjusting to remove the effects of the dam upstream of the gage. It is how big the 1966 flood would have been if Terminus Dam had not been there.

Based on the flood exceedence rates in Table 16, this flood had a recurrence interval of 170 years for the Kaweah. It would have had a recurrence interval of 100 years if calculated using the 105,000 cfs peak flow. (One source reportedly calculated this as having had a recurrence interval of 140 years. Presumably that was done using the peak flow and the now outdated 1971 flood frequency curves. That result could not be reproduced.)

Yokohl Creek flooded dramatically, undermining the Visalia Electric track adjacent to Highway 198. It also put four to five inches of water into the Yokohl Store and damaged at least one house in the area. Bridges were damaged along Yokohl Creek and at Rocky Hill. Yokohl Creek eventually flows into the Consolidated Peoples Ditch; this is near the Lower Kaweah River but well below McKay's Point.

Flooding occurred in Lindsay, East Woodlake, Terra Bella, and some isolated areas near Cutler, Orosi, Yettem, and Seville. Flooding was up to three feet deep in Toneyville near Lindsay; 150 people were evacuated from that community.[838]

The Tule was flowing 18,000 cfs at Globe at 10 p.m. on December 5 when the gage was swept away. That was probably well before the peak of the flood.[839]

Many bridges were washed out in the Springville area. Some local residents recall that Highway 190 remained passable throughout the flood. But the Visalia Times-Delta printed an aerial photo at the time showing that this bridge had washed out.[840] Will Wood and George Costa both confirm newspaper reports from the time that the lower and upper bridges on the Globe Drive loop were washed out.

There was damage to the bridges on the North Fork of the Tule. The Bear Creek Bridge on the road leading to the SCICON school was destroyed. The North Fork Bridge (located about 7 miles up the North Fork of the Tule on the Balch Park Road) was overtopped; the approaches may have been damaged, but the bridge survived.[841]

Damage was extensive in Springville with 9–12 homes washed away, 35 homes extensively damaged, and 75–100 people left homeless. Springville's domestic water supply system was knocked out of commission early in the flood, and it was many days before it could be brought back on line.[842, 843]

Five cabins washed away at Camp Wishon northwest of Camp Nelson.[844]

The Tule Indian Reservation was hit hard. Most of the Native Americans live along an eight-mile stretch of the Tule. The floodwaters washed out roads, destroyed all bridges, and swept away telephone and power lines leading into the area. Bulldozers were flown into the reservation to begin construction of a makeshift road. Other equipment went to work from the Porterville end, expecting to meet within four days.

Johnsondale in the southeast part of Tulare County lost power. It also lost all road access, being cut off by the floodwaters of both the Tule and Kern Rivers. Helicopters from Lemoore Naval Air Station were used to fly in food, water, and emergency generators. Later, a temporary access road — rough, but passable — was opened over Parker Pass to California Hot Springs.[845]

The December 1966 maximum flow of 49,600 cfs at the gaging station on the Tule River near Springville (see Station 2032 in Table 49) was the greatest flood-of-record and more than double the previous record flow of 22,400 cfs in November 1950. Records at a former gage site inundated by Lake Success in 1961 show that the 1950 flood was the greatest during the period of record (1901–1960). The December 1955 peak discharge was slightly less than that in 1950 and thus was the third-highest recorded flood on the Tule River.[846]

The December 1966 peak discharge of 14,300 cfs on the South Fork Tule River near Success (see Station 2045 in Table 49) was also more than double the previous record flow of 7,000 cfs in November 1950.[847]

The computed maximum bihourly inflow to Lake Success near Success of 52,800 cfs on December 6 similarly was 1.7 times the peak flow of 32,000 cfs in November 1950 at a former gaging station near the damsite.[848] That was the peak bihourly flow. The reservoir had an instantaneous inflow on December 6 of 76,900 cfs and a daily inflow on that date of 40,000 cfs; both of these are flows of record.[849]

(The same qualifier applies here as on the Kaweah. This is the greatest flow measured since stream gages were installed on the Tule. However, during recorded history, the 1867–68 flood is considered the flood-of-record for the Tule; that flood just occurred before any stream gages were installed on this river.)

The peak average daily flow at Success Dam, as reflected in Table 15, was 40,085 cfs. It was the largest peak day of the year since that dam was built in 1961; it remains the flow of record. Based on the flood exceedence rates in Table 16, this had a recurrence interval of 200 years for the Tule.

Terminus Dam had been able (just barely) to capture the floodwaters of the Kaweah and prevent flooding downstream. Success had a stated capacity in 1966 of 85,400 acre-feet. The reservoir filled, but the floodwaters kept coming. Water began spilling. At one point, the reservoir was holding 101,400 acre-feet (85,400 acre-feet of nominal storage + 16,000 of surcharge storage).[850]

About 250 people were evacuated in the Porterville area, mostly from the Doyle Colony. The Tule River broke through levees in the Porterville area. Flooding in the lower Tule continued for several days. Heavy flooding was reported at the Pixley National Wildlife Refuge. But for all that, Success Dam prevented the bulk of the flooding that would have otherwise occurred. At Porterville, an official put it succinctly: *Without Success Dam there would be no Porterville today.*[851]

Some flooding occurred in agricultural areas downstream from Lake Success during sustained release of floodwater December 6–11.[852] Approximately 1,000 sheep were discovered marooned on a levee next to the Tule near Ave 184 and Rd 152 in the Woodville area. They were too heavy to swim because of their full coats of wool. County crews evacuated them by constructing a temporary bridge. Many other cattle, sheep, and other animals weren't so lucky and drowned in the flood, contributing to the health problems caused by flooded wells.[853]

As shown in Table 50, the Tule sent nearly as much floodwater into the Tulare Lakebed in 1966 as the Kaweah did. That rarely happens.

Deer Creek peaked near Terra Bella on December 6: 10,000 cfs. This is the flood-of-record for that creek. It was a particularly destructive flood. The main road to California Hot Springs follows Deer Creek and crosses it at several locations. All bridges were destroyed or badly damaged. Downstream irrigation diversion structures were washed out and were further damaged by deposition of coarse sediment.[854]

White River flooded on December 6, but it was only half as big a discharge as the 1943 flood.[855]

There was a major flood on Caliente Creek in December, causing extensive flood damage to the Lamont/Arvin area.

There was a severe rainstorm over the Kern River Basin on December 2–7. Almost 21 inches of rain fell in the area in two days.[856] The flooding was most severe in the Kernville area. Flooding there isolated an area of 150 square miles and forced the evacuation of 200 persons. All roads in that area were under water.

Peak flows of the December 1966 flood exceeded previous maximums at most gaging stations in the Kern River Basin except:[857]
1. High altitude stream channels. Little storm runoff occurred above 9,000 feet elevation where the precipitation fell during part of the storm as snow and during the remainder as rain, as the freezing level changed during the storm. When inspected on December 9, the channel of Golden Trout Creek was nearly full of ice and frozen saturated snow. There was no evidence of substantial high flow during the storm. Snow already on the ground apparently absorbed and held most of the rain that fell upstream from this gaging station.
2. Stream channels below the Isabella Dam.

The flood damaged many stream gages. It destroyed the water-stage recorder structures and the measuring cableways at two sites:[858]
- North Fork Kern River at Kernville
- South Fork Kern River near Onyx

The South Fork Kern River near Onyx (see Station 1895 in Table 49) had a peak discharge of 28,700 cfs. This was eight times greater than the previous maximum discharge of 3,460 cfs recorded in the February 1963 flood. That makes the 1966 flood by far the highest flow on the South Fork since record-keeping began in 1911.[859]

292

The North Fork Kern River near Kernville (see Station 1860 in Table 49) had a peak discharge of 60,000 cfs on December 6, 1966. That was more than twice the previous maximum discharge of 27,400 cfs recorded in the floods of November 1950 and December 1955. That makes the 1966 flood by far the highest flow on the North Fork since record-keeping began in 1912.[860]

Isabella Reservoir had a computed maximum bihourly inflow of 96,900 cfs. That was 2.5 times the previous maximum flow of 39,000 cfs which was recorded at the damsite in November 1950 prior to dam construction.[861] The reservoir had an instantaneous inflow of 118,600 cfs and a daily inflow of 72,782 cfs; both of these are flows of record.[862]

The road from Johnsondale downstream to Kernville (Mountain 99, aka Kern County SM99) is close to the North Fork Kern at many locations. This road was obliterated at the outside bank of many river bends, and the pavement was scoured away in other locations.[863]

A hydroelectric plant upstream from the Kernville Bridge washed out, hitting the bridge and collapsing it.[864] (Presumably this was SCE's Kern No. #3 power plant.) Bridge debris, including one 3-foot by 40-foot steel girder, was moved several hundred feet downstream.[865]

The state fish hatchery, to the surprise of none, washed away.[866] The hatchery had been destroyed in the 1950 flood and washed away again in the 1955 flood.

Sequoia National Forest reported that an aerial survey showed considerable amounts of timber felled by the storm and erosion damage. The Kernville Road (State Highway 155) was badly damaged. Large portions of that road, some sections up to 2½ miles long, were washed away.[867]

A trailer court and other buildings along the river at Kernville were badly damaged.[868] Many people were evacuated from the Kernville area. Prisoners at a work camp were evacuated and sent to facilities in Bakersfield. A section of the golf course at Kernville was washed away. All of the mountain roads were either washed out or closed by landslides. All trailer parks, motels, lodges and cabins were swept away by floodwaters. The fire station at Lake Isabella was flooded. Highway 178 was closed due to flooding and debris. The historic wooden Bellevue Weir just west of Bakersfield washed out. Two people lost their lives in Kern County.[869]

The flooding on the Kern River was covered in the *New York Times*.[870]

Lake Isabella was able to fully contain the flood. The only release from the dam for the first 10 days during and after the flood was the 300-500 cfs released to the Borel Canal for power production.[871]

The USACE estimated that if Lake Isabella had not existed, flow on the Kern River six miles upstream of Bakersfield at the First Point of Measurement gage would have been approximately 80,000 cfs, resulting in significant damage in the city. Actual flow was only 9,300 cfs and consisted primarily of inflow from tributary streams entering the river below the dam.[872]

There was also significant flooding on the west side of the San Joaquin Valley near Coalinga. Flooding caused extensive road and bridge damage in the upper reaches of Arroyo Pasajero (aka Los Gatos Creek) and Warthan Creeks. East of Coalinga, sewage-treatment facilities and the levees along Warthan Creek were damaged, the Los Gatos Creek channel was severely eroded, and there was extensive damage to utilities and agricultural land. Damages totaled approximately $570,000, and floodwaters inundated 4,500 acres.

Many people in Tulare and Kern Counties were displaced from their homes or otherwise needed assistance. The American Red Cross launched a major relief operation. Evacuation centers were operated in Farmersville, Woodlake, Lindsay, Three Rivers, and perhaps other areas. The Red Cross provided food, clothing and other relief services throughout the affected area. They provided services to remote areas as soon as those areas were reachable. This included organizing a pack train to get food, water, and other supplies into the Tule Indian Reservation. The only other way to access the reservation was by helicopter.[873]

On December 9, Governor Edmund Brown declared portions of Tulare, Kern, and Riverside Counties to be disaster areas. There were three deaths and $18 million in property damage. Damage to Tulare County roads and bridges was initially estimated to be $2.5 million, but that was soon deemed to be way too low. Damage to roads and bridges in the Kern River area of the county was particularly bad. The Tulare County civil defense chief estimated damage to ranch property as simply "astounding."[874]

Tulare Lake had been dry since about August 14, 1958. It came back to life on December 6, 1966. The river flood occurred in 1966, but the flooding in the lakebed continued into 1967. From the standpoint of the lakebed, it could be thought of as a flood with two phases. The lakebed flooding of December 1966 wasn't fully dissipated when the April 1967 flooding arrived.

As detailed in Table 50, the December 1966 rain flood delivered a total of 87,600 acre-feet to the Tulare Lakebed.[875]

Table 50. Inflow to the Tulare Lakebed during water year 1966.

| Stream | Total Inflow (acre-feet) | Percent Contribution |
|---|---|---|
| Kings River | 0 | |
| Kaweah River | 48,600 | 55% |
| Tule River | 37,900 | 43% |
| Deer Creek | 1,100 | 1% |
| Kern River | 0 | |
| Total | 87,600 | |

This resulted in the flooding of 26,560 acres in the lakebed. However, that area included 14,750 acres flooded by diverted floodwaters for which there was no flood damage. That is, the land was considered of no value. Flooding first occurred in Sumps #1 and 2, after which agricultural land was flooded. After the heavy flows in these two sumps had subsided, much of this water was transferred to lands in the south and southeastern portions of the lakebed. There it would do little harm. This procedure made Sumps #1 and 2 again available for storage of the snowmelt flood runoff which was anticipated to occur later in the flood season.

The upstream federal reservoirs were all full at the end of the flood, and it was just the beginning of the rainy season. It was imperative to empty those reservoirs as soon as possible to restore their capacity to capture the next potential flood. However, the downstream interests in the Tulare Lakebed warned that this would cause a major disaster. Thousands of acres of farm land were already under water as a result of the flood. But there was no choice; the reservoirs had to be emptied.[876]

## 1967 Flood

The 1967 flood was a snowmelt flood, extending from April into early July.

A vast amount of snowmelt from April to July compounded the flood damage already experienced from the 1966 flood. Significant flooding also occurred along the Cosumnes River, in the Morrison Creek and Beach-Stone Lake areas, and in Madera County streams in the lower portions of the Fresno and Chowchilla rivers.[877]

In addition to the snowmelt, it was a wet spring. Fresno experienced its wettest April ever, with over four inches of precipitation.

The valley floor was already flooding from the 1966 flood, but the reservoirs were able to hold only a portion of the 1967 runoff. Table 51 shows the quantities of snowmelt sent downstream in the spring of 1967:[878]

Table 51. Snowmelt flows sent downstream in the spring of 1967.

| Basin | Total Flow (million acre-feet) |
|---|---|
| San Joaquin River Basin | 7.8 |
| Tulare Lake Basin | 3.9 |
| Total | 11.7 |

The federal reservoirs, acting in concert, made every effort to keep floodwaters out of Tulare Lake in the 1967 flood. However, that goal proved to be more than the system was up to. The reservoirs were full at the beginning of the year because of the December 1966 flood. They would have to make releases at some point in order to provide storage space to catch the predicted large snowmelt runoff that was coming in 1967.

There were sustained high flows on the Kings River from mid-May through late July. The peak day natural flow at Pine Flat on the Kings occurred on July 1.

Thanks to a pair of crest-stage gages, we know that Grizzly Creek peaked at 233 cfs, probably sometime during June.[879]

The maximum mean daily inflow to Pine Flat was 19,739 cfs, which was controlled to an outflow of 15,034 cfs. The natural flow of the Kings peaked at 20,500 cfs, but PG&E cooperated in manipulating the storage remaining in their upstream reservoirs, reducing this flow at the critical time to assist in minimizing outflows from Pine Flat.

Pine Flat Reservoir held all the water that it possibly could. It reached a peak stage of 1.3 feet above full pool. Every effort was made to keep Kings River water out of the Tulare Lakebed. However the runoff was simply too large. It couldn't all be held in the reservoirs or diverted through the Fresno Slough Bypass. About 62,000 acre-feet of Kings River floodwater wound up in Tulare Lake.

Lake Kaweah started the year full as a result of the December 1966 flood. Like all the other federal reservoirs, it did everything possible to keep floodwaters out of Tulare Lake. However, in April and May, 23,000 acre-feet of water was sent down the Kaweah River to the lakebed in preparation for the snowmelt runoff season.

The same thing happened on the Tule River. In April and May, 9,300 acre-feet of water that had been stored in Lake Success during the December 1966 flood had to be released in preparation for the snowmelt runoff season. That water was passed through to the Tulare Lakebed. In mid-May, a sack-concrete barrier was placed on the spillway of Success Dam to provide an additional 10,000 acre-feet of storage. This extra freeboard turned out to be a good idea; Success Reservoir would eventually peak that summer at about 1.6 feet above full pool. Due to flood inflow into Success Reservoir, eventual maximum outflow reached 8,300 cfs, exceeding channel capacity downstream.[880]

During May and June, large releases were made from Isabella Reservoir for spreading and irrigation use. This made it possible to keep the reservoir from going out of control (that is, to keep it from spilling). The maximum reservoir storage reached was 539,000 acre-feet on July 20. This was 2.8 feet short of reaching the spillway.

Total flow for water year 1967 was 191% of the 1894–2011 average for the Kings, 238% for the Kaweah, 268% for the Tule, and 223% for the Kern. This was the fourth-highest flow for the Kaweah since record-keeping began in 1894.

As detailed in Table 52, about 94,000 acre-feet of floodwater entered Tulare Lake during the April-July snowmelt period.[881] As a result, about 40,000 acres of the lakebed was flooded, a little more than in the 1966 flood. Two levees within the lakebed were in danger of failing during this period. If that had occurred, the acreage flooded would have been significantly larger.

Table 52. Inflow to the Tulare Lakebed during water year 1967.

| Stream | Total Inflow (acre-feet) | Percent Contribution |
|---|---|---|
| Kings River | 62,000 | 66% |
| Kaweah River | 23,000 | 24% |
| Tule River | 9,300 | 10% |
| Kern River | 0 | |
| Total | 94,300 | |

## 1969 Floods (3)

There were three floods in 1969:
1. January (rain flood)
2. February (rain flood)
3. April through July (unusually large snowmelt)

The winter of 1969 was very wet in Southern California and the Southern Sierra. Southern California experienced some of the severest flooding since 1938.

Over 200 stations, mainly in Southern California, reported their highest-ever rainfalls for 60 consecutive days. Mount Baldy Notch in the San Gabriel Mountains received 88.50 inches in 60 days from January 13 – March 13. Stations reporting extremely high rainfalls for the 60 days ranged from Cottonwood Creek at 10,600 feet in the Southern Sierra to Death Valley at 194 feet below sea level. A total of 13 stations reported rainfall totals in excess of a storm with a recurrence interval of 1,000 years. The valley floor portion of the San Joaquin Valley also had heavy rainfalls with high recurrence intervals.[882]

The 1969 flood was a major flood in the San Joaquin River Basin, especially in January. It was one of the most damaging natural catastrophes in California's history. Property damage was about $400 million, and 60 lives were lost. Governor Reagan declared a state of emergency for the January storm; a total of 40 counties were declared disaster areas. President Nixon also declared the State of California a disaster area for the January storm.

In the Sacramento Valley, floodwaters produced by the January storms were largely controlled by major reservoirs, flood channels, and the bypass system. As a result, flows in the mainstem of the Sacramento River and its major tributaries remained well below project design flows. Peak flow at the latitude of Sacramento was approximately 250,000 cfs.[883]

During January 18–27, a series of storms, drawing on a strong flow of warm, moist air from the southwest, moved across Central and Southern California. Massive quantities of precipitation fell on the coastal mountains from Monterey Bay to Los Angeles and in the Southern Sierra. Lytle Creek Powerhouse in the San Gabriel Mountains northwest of San Bernardino received 24.92 inches of rain in a 24-hour period on January 24. The peak discharge on the Santa Ynez River near Lompoc was 78% greater than during the flood of March 1938.

Fresno recorded 8.56 inches of rain in January 1969, making that the wettest month ever for this city. In all, 22 days of the month recorded precipitation.[884] As of January 27, Dinuba had received 21 inches of rain for the season, an all-time record.

As the heavy rains continued in the valley, a snowpack of unprecedented depth and water content accumulated in the mid- to higher elevations of the Sierra. Record after record was broken during the winter of 1968–69:

- The Central Sierra Snow Laboratory monitoring site near Donner Pass received 13.7 feet of snow between January 20–31, the second biggest snowstorm ever recorded at that site. (An even bigger snowstorm would come in March–April 1982.)
- In late February, a series of northwestern cold-front storms moved south along a low-pressure trough that had formed over the California coast. Incredible all-time 24-hour snowfall records were set in parts of the Sierra on February 24–25 with 46.0 inches of snow measured at Lodgepole and 36.0 inches of snow at Grant Grove.[885]
- Lodgepole received 187 inches (15.6 feet) of snowfall during the month of February. This is the greatest amount of snowfall ever recorded in one month at that location.
- On February 26, 1969, the snowpack at Lodgepole reached 197 inches (16.4 feet). This is the greatest snowpack ever recorded at that site.[886] Despite the similarity of numbers (187 and 197), this is a different record from the preceding bullet. "Total snowfall" refers to new snow falling during a storm event. It is computed by summing the 24-hour snowfalls measured daily during the time period of interest. "Snowpack" refers to the total amount of snow on the ground, including existing snow from previous storms. The snowpack may actually be less than the total snowfall as the snow at the lower depths may be compacted by the weight of the overlying snow.
- On March 13, Grant Grove measured a snowpack of 179 inches (14.9 feet), the greatest ever recorded at that site.[887] Much more was to come.
- The Montecito-Sequoia Lodge was damaged from a 20-foot snow-dump.
- As detailed in Table 53, Lodgepole received a total snowfall of at least 440.5 inches (36.7 feet) during the winter of 1968–69. (This only reflects snowfall after November 8, 1968, the date that the cooperating weather station was reactivated. Therefore, the total snowfall for the winter may

297

have been somewhat under-measured.) This is the fourth biggest winter at that location. The winters of 1905–06, 1951–52, and 2010–2011 were all larger.

Table 53. Lodgepole snowfall during winter 1968–69.

| Month | Snowfall (inches of snow) |
|---|---|
| September, 1968 | no data |
| October, 1968 | no data |
| November 1968 | 13.0 |
| December 1968 | 67.0 |
| January 1969 | 93.5 |
| February 1969 | 187.0 |
| March 1969 | 50.0 |
| April 1969 | 27.0 |
| May 1969 | 3.0 |
| Total | 440.5 |

Almost the same areas were flooded in February as in January. Peak discharges in Southern California were slightly less than in January, but on February 26 the Salinas River at Spreckels had a new peak discharge-of-record that exceeded the March 1938 peak by 11%.

The flood peak discharges were the largest in 30 years in Central and Southern California and in many places equaled or exceeded those of the March 1938 floods. In the Santa Clara, Santa Ynez, and Salinas River Basins, flood levels may have approached those of 1861–62.

Flood releases of 12,000 cfs or greater occurred at Friant Dam on the San Joaquin River during the April–July snowmelt period.

Panoche/Silver Creek west of Mendota flooded in January 1969.[888]

During the period of the January and February floods, storage in Pine Flat Reservoir increased from about 420,000 acre-feet on January 1 to 820,000 acre-feet on March 1. In that time about 223,000 acre-feet of Kings River water was passed through the dam and routed to the San Joaquin River via the Kings River North Channel and Fresno Slough Bypass.

During the preceding dry season, the USACE had fortuitously increased the capacity of the Kings River North Channel from 3,500 cfs to 5,500 cfs. This project had been completed just prior to the onset of the January 1969 flood.

During March, April, May, and June, increasingly large quantities of water were released from Pine Flat for diversion to the San Joaquin River. The total diversion during those four months was about 1,185,000 acre-feet.

No floodwaters from the basin above Pine Flat Dam reached the Tulare Lakebed before June. However, some uncontrolled flows, largely from Mill Creek (a southside tributary of the Kings below Pine Flat Dam) did reach the lake.[889]

Runoff on the Kings during water year 1969 was the second highest since record-keeping began in 1894 (1983 would be even higher). The huge snowpack in the Kings River Basin resulted in the largest-ever releases from Pine Flat: 17,000 cfs.

Runoff for the Kings River at Pine Flat during water year 1969 was 4.2 million acre-feet. This was 251% of the 118-year average (1894–2011) for that river.

One source said that the January 1969 flood on the Kings River was in the same class with the 1914 and 1952 floods. It seems probable that there were other floods (particularly 1861–62, 1867, and 1937) that also belong in this category.

In any case, the peak day natural flow at Pine Flat on the Kings occurred on January 25, 1969. That is the fifth largest peak day of the year at Pine Flat since the dam was built in 1954. Based on the flood exceedence rates in Table 16, this had a recurrence interval of 25 years for the Kings River at Pine Flat.

That puts it in a category with other Cedar Grove floods of the past 70 years (1937, 1950, 1955, 1966, 1969, 1978, 1982, 1983, and 1997) that rise to the level of the modeled 50-year flood: that is, a flood event that occurs about every eight years on average. See the section of this document that describes Cedar Grove Flooding.

Jim Harvey recalled that the flood on the South Fork Kings was very impressive in Cedar Grove. The trail bridge across the South Fork Kings in Upper Paradise Valley was swept away.

No one imagined that Dinuba was susceptible to a major flood. The Kings River was way to the north, the Kaweah way to the south. Flooding was something that Reedley and Visalia worried about, not Dinuba. The January 1969 flood had a big surprise in store for that town.

The Alta Irrigation District gets its water from the Kings River with irrigation releases from Pine Flat Dam. In January, the East Branch of the Alta Canal was running full thanks to the flood flows on the Kings, when an intensive localized rainstorm caused it to overflow. Overflow events had happened in the past on this canal (1937, 1950, 1955, and 1966) and would happen again in the future (1993). However, this overflow would prove particularly memorable. As a result of this overflow, the canal suffered three ruptures near Smith Mountain, one of which was massive, some 60–80 feet long. The floodwaters poured out and flowed cross-country.

Dinuba was eight miles away, but there was nothing to divert the flood before it got there. The downtown area was flooded, as was much of the surrounding ranch land. Flooding was heaviest on the night of January 21. China Town was particularly hard hit. An evacuation center was set up, and a police car with loudspeaker went through China Town urging the residents to evacuate. A spokesman for the American Red Cross said that the China Town residents simply did not want to leave their homes even though they were underwater.

Apparently there was no headgate on the East Bank Canal; the philosophy being to take whatever irrigation releases were available, more is better. But that meant that there was now no way to shut off flow into the canal. As a result, the canal continued hemorrhaging floodwaters into the Dinuba area. Crews from the district and the town worked for a week, struggling to plug the leak. The final leak couldn't be plugged until the Kings River went down. The canal wall was finally repaired on January 27.

Farther to the east, the normally dry Sand Creek flooded Cutler and East Orosi on January 25. Cutler was inundated under a solid sheet of water, several feet deep in places. At least 350 people had to be evacuated, and it was several days before many of those could return to their homes.

Flooding on the Kaweah River washed away the Kaweah Public Beach (aka River Park; the exact name of that county park is unclear) on the lower side of Three Rivers.

As on the Kings, runoff on the Kaweah during water year 1969 was the second highest since record-keeping began in 1894 (1983 would be even higher).

The Kaweah's peak natural flow for the first rain-flood period occurred at Terminus Dam on January 25: 35,200 cfs. (That was the peak hourly flow; the peak average daily flow, as reflected in Table 15, was 22,437 cfs.)

Based on the flood exceedence rates in Table 16, this had a recurrence interval of 25 years for the Kaweah. It would have had a recurrence interval of 20 years if calculated using the 35,200 cfs peak flow. (One source reportedly calculated this as having had a recurrence interval of 12 years. Presumably that was done using the peak flow and the now outdated 1971 flood frequency curves. That result could not be reproduced.)

The trail bridge over the South Fork of the Kaweah above Ladybug Camp was apparently one of the very few trail bridges in the Kaweah River Basin to survive the 1950 flood. However, one source said that the January 25, 1969 flood washed out this bridge and many others.[890] The loss of this bridge closed the section of the Hockett Trail above this point. See the section of this document that describes the 1950 flood for a description of this section of the trail. Since losing this section of the Hockett Trail, the route has gone through Garfield Grove to get up onto the Hockett Plateau.

Inflow to Lake Kaweah was 153,000 acre-feet between January 19–27. Storage rose from 8,300 acre-feet on January 18 to 139,800 acre-feet on January 27, an increase of over 131,000 acre-feet. Most of the subsequent flow, which was released because of flood operating criteria, found its way into the Tulare Lakebed.

Panoche/Silver Creek west of Mendota flooded in February 1969. This may have been a separate flood, or it may have been a continuation of the January flooding.[891]

Peak discharges for the second, and somewhat smaller, rain-flood period on the Kaweah occurred on February 24–28.

Lake Kaweah was drawn down to 82,000 acre-feet on February 23, just prior to the second heavy rain flood. Storage rose to 117,300 acre-feet by February 28, and eventually reached a maximum of 158,800 acre-feet on June 26.

Runoff for the Kaweah below Terminus Dam during water year 1969 was 1,272,000 acre-feet. This was 295% of the 118-year average (1894–2011) for that river.

The USACE, with cooperation from local organizations and parties, used every possible means to reduce Kaweah River flows into the Tulare Lakebed. Throughout the months of January to July, efforts were made to apply maximum quantities of water in the Kaweah service area, drawing down the reservoir. Sandbags were again placed on the spillway of the dam, increasing the storage capacity of the reservoir by 10,000 acre-feet.[892] Even with all these efforts, 430,000 acre-feet of Kaweah floodwaters made it into the Tulare Lakebed in 1969.[893]

Dry Creek peaked below Terminus Dam at 5,710 cfs. This was less than half the flow observed in the 1966 flood.

Joe Childress was the manager of the Wutchumna Water Company for many years. He recalled that Antelope Creek produced a large amount of floodwater in the area west of Woodlake's Presbyterian Church during the 1969 flood. That was particularly remarkable since the Antelope Creek Basin seldom has any flow of note.

As on the Kings and Kaweah Rivers, runoff during water year 1969 on the Tule River was the second highest since record-keeping began in 1894 (1983 would be even higher). Runoff for the Tule below Success Dam during water year 1969 was 504,000 acre-feet. This was 361% of the 118-year average (1894–2011) for that river.

The 1969 storm pattern on the Tule River was similar to that on the Kaweah River. Based on the flood exceedence rates in Table 16, this had a recurrence interval of 67 years for the Tule River. Storage in Lake Success rose from 11,100 acre-feet on January 18 to 77,230 acre-feet on January 28. Gross storage in Lake Success was considered at the time to be 85,400 acre-feet. Available storage was decreased after the January rain flood to about 63,000 acre-feet. The February rain flood caused storage to rise to 83,800 acre-feet on February 25. As at Lake Kaweah, USACE and local water-using agencies worked together, making every effort to reduce the amount of water that had to be spilled into the Tulare Lakebed.

Sandbags were again placed on the Lake Success spillway, increasing the storage pool above the designed 85,400 acre-feet. During the snowmelt flood runoff season, storage in Lake Success rose to a maximum of 95,300 acre-feet on June 20 and was above 85,400 acre-feet from May 19 through July 15.

The lower Tule flooded in both the January and February floods, and 215,000 acre-feet were passed downstream to the Tulare Lakebed.

Even normally dry Deer Creek was flowing into the Tulare Lakebed in the spring.[894]

The Kern's peak natural flow was 22,359 cfs. (That was the peak average daily flow.) Based on the flood exceedence rates in Table 16, this had a recurrence interval of 56 years for that river. The inflows to Isabella Reservoir during the peak of the rain flood in January and February were not nearly as great as those during the even more impressive December 1966 flood.
- Maximum mean daily inflow to Isabella Reservoir on January 25, 1969, was 22,200 cfs. The comparable value for the 1966 flood was over three times as great (72,800 cfs on December 6, 1966).
- Maximum 10-day rain-flood inflow to Isabella Reservoir was 132,000 acre-feet from January 19–28, 1969. The comparable value for the 1966 flood was nearly twice as great (254,000 acre-feet for the period from December 5–14, 1966).

There was a major flood on Caliente Creek in February, causing extensive flood damage to the Lamont/Arvin area.

The third flood of 1969 was a snowmelt flood. A great snowpack had accumulated in the Southern Sierra by the beginning of April. It contained over 200% of the average water content.[895] That set the stage for the flooding that was to occur during the April–July runoff period.

The Kings River is controlled by Pine Flat Dam, but the reservoir holds only 59% of the 118-year average runoff (1894–2011) of that river.

See Figure 13 on page 109 to understand why 1969 was not an average year. Pine Flat Dam recorded the largest snowmelt of record during April through July of that year.[896] The 1969 snowmelt exceeded all previous years since record-keeping began in 1895. Pine Flat Dam was operated to control outflow to a maximum of 17,100 cfs.[897]

Flood control releases from Pine Flat Dam in 1969 totaled 1,017,000 acre-feet.[898, 899] Floodwaters were routed both to the San Joaquin River and to the Tulare Lakebed. The James Bypass experienced a maximum daily discharge of 5,570 cfs on June 7, 1969; that remains the flow of record for this channel.[900] During 1969, the USACE had to pass a total of 4,197,901 acre-feet of Kings River water through to the Tulare Lakebed.

Runoff for the Kern near Bakersfield during water year 1969 was 2,406,500 acre-feet. This was 330% of the 118-year average (1894–2011) for that river. This was the third highest runoff for the Kern since record-keeping began in 1894 (1916 and 1983 were both higher: 2,463,790 and 2,442,500 respectively).

Isabella Reservoir was operated with great care during the 1969 flood. Because of the heavy snowpack, it was known that Isabella would fill for the first time since operation began in 1954. Efforts were made to hold the storage down through March and April so that space would be available for the snowmelt runoff which was expected to be heavy during May and June. Much of the water released from Isabella Reservoir was used in the service areas below the reservoir for irrigation or spreading. Some water was stored in Buena Vista Lake.

Historically, the Kern would fill Buena Vista Lake before spilling over into Tulare Lake. However, in 1969, a giant dike protected two-thirds of Buena Vista Lake from being filled. When the other third of the lake filled, the Kern then spilled or passed through to Tulare Lake. The decision to keep the remainder of Buena Vista Lake dry was not appreciated by those downstream in the Tulare Lakebed.

In 1952, the Tulare Lake Basin Water Storage District had stored Kern River floodwaters in Buena Vista Lake. That was presumably possible because the J.G. Boswell Co., which had a long-term agricultural lease for the Buena Vista Lakebed, was willing to have its land flooded. In any case, no such water storage was allowed in 1969. That created hard feelings among some who were being impacted by the flooding that was occurring in the Tulare Lakebed in 1969. Emotions ran high as did financial losses.

The decision to pass through the Kern River floodwaters was challenged in court. But in the meantime, the floodwaters continued to come.[901] As a result, about 222,000 acre-feet of Kern River water flowed into the Tulare Lakebed. The majority of the Buena Vista Lakebed remained dry, safe behind its giant levee.

The primary reason that this water storage didn't happen in Buena Vista Lake in 1969 was because the J.G. Boswell Co. stood in the way. They controlled the Tulare Lake Basin Water Storage District and didn't want their cropland in the Buena Vista Lakebed to be flooded. The resulting lawsuit made it all the way to the U.S. Supreme Court. In a 6–3 split decision, the court eventually found that it was permissible under the U.S. Constitution for the water district to be controlled by the J.G. Boswell Co. to the exclusion of all the other landowners and residents of the Tulare Lakebed.[902]

On May 8, 1969, the USACE received approval for a half-million-dollar project to throw up levees to connect the separated segments of Sand Ridge, south of the current Tulare Lake, creating a gigantic holding pond to contain Kern River floodwaters.[903]

There was significant flooding on the west side of the San Joaquin Valley near Coalinga. Arroyo Pasajero (aka Los Gatos Creek) and Warthan Creek experienced extremely high flows in February. The resulting floodwaters covered 16,600 acres and caused approximately $4.5 million in damage. Flooding extended from the foothills west of Coalinga to the valley east of the city. Bridges and roads were washed out, agricultural land was eroded, farm and ranch improvements and petroleum installations were damaged and destroyed, areas were isolated, traffic was disrupted, and residential and commercial areas in the northwest and southeast portions of the city were damaged. This was one of the three largest and most damaging flood events to occur in the Coalinga area during historic times.

Tulare Lake reappeared on January 20, 1969; it had been completely dry since August 9, 1967. By the end of March, 125 square miles (80,000 acres) of farmland had been inundated. The total lakebed inflow in 1969 was about 1.155 million acre-feet. This is the second biggest lakebed flood (both by volume and by area flooded) since the federal reservoirs were completed; only the 1983 flood would be bigger.

In 1969, 88,700 acres were inundated, significantly more than the 72,700 acres flooded in 1952. The J.G. Boswell Co. had more land flooded in the Tulare Lakebed than any other landowner (almost 50,000 of the total 88,700 acres).

During May, the USACE closed the channel where the Kern River flowed through Sand Ridge into the Tulare Lakebed. This caused a lake to form south of Sand Ridge. As a result, about 235,000 acre-feet of Kern River water was prevented from entering Tulare Lake.[904]

This essentially recreated the southern extension of Tulare Lake, which had been known to Native Americans as Ton Taché. Along with the South Wilbur Flood Area (located north of Sand Ridge), that is the area known today by Tulare Lake water storage districts and irrigators as the South Flood Area.

As detailed in Table 54, the combined runoff of the four rivers in the Tulare Lake Basin during 1969 was 8,379,585 acre-feet, the second highest since record-keeping began in 1894; only 1983 would be larger (see Table 66 and Figure 13).

Table 54. Runoff in the Tulare Lake Basin during water year 1969.

| Watershed | Total Runoff (acre-feet) | % of average (1894–2011) |
|---|---|---|
| Kings | 4,197,901 | 249% |
| Kaweah | 1,271,328 | 295% |
| Tule | 503,856 | 361% |
| Kern | 2,406,500 | 330% |
| Total | 8,379,585 | 281% |

On June 24, 1969, Tulare Lake reached a peak height of 192.5 feet elevation. (This was the highest the lake had been since 1952, when the lake reached 194.6 feet.)

In 1969, two middle-age fathers and their sons took advantage of the high water to boat from Bakersfield through Buena Vista Lake and Tulare Lake to San Francisco Bay. They traveled in two small motorized fishing boats. The boats had to be light enough to carry around weirs and head gates. Their wives met them at various prearranged stops along the trip route, as they would overnight with friends or motel it up for the night. Ed Nelson recalled that their trip was written up in some newspaper such as the *Fresno Bee*. He remembers thinking how neat a trip that would be no matter what your age. It sort of brought out the Tom Sawyer / Huckleberry Finn in you. This was the fourth documented trip between the lake and the bay. (The other trips were in 1852, 1868, 1938, and 1982–83.)

The 1969 flooding in Tulare Lake caused the complete inundation of sumps #1, 2, 3, and 4, as well as some additional lands outside of those sumps. (That is presumably the area that we now know as the South Wilbur and Hacienda flood units.) Maximum storage in Tulare Lake reached 960,000 acre-feet in June. The difference between total inflow to the lake and maximum storage, 195,000 acre-feet, was lost by evaporation and absorption into the lakebed. A small amount of water was also diverted from the lake for irrigation of lands around the perimeter of the flooded area. The lakebed would remain at least partially flooded through 1971 (see Figure 11). The lakebed was finally dry in calendar year 1972.

The lake threatened the west side of the town of Corcoran during the 1969 flood (multiple photographs on file in the national parks, also see the back cover photographs). An emergency levee was hurriedly built just west of the Corcoran Airport.

The J.G. Boswell Co. took the lead on the levee building with much assistance from Salyer and the smaller farmers in the area (Boyett and Gilkey). Boswell also took the lead on the purchase and movement of junk cars to the levees. These were used as riprap to protect the levee from erosion.

In 1969 Mo Basham was a 12-year-old girl living in Corcoran, so she remembers much of what happened. Tulare Lake was deep enough to cause significant erosion to the emergency levee, even though it was faced with the junk cars. The chop on the water during windy/stormy weather was pretty significant, so the levees were constantly monitored.

Mo recalled going out on rodent patrol to spot/kill ground squirrels and gophers that would dig into the levees. Mo also recalled going on crawdad (crayfish) hunts. The kids would bring back hundreds at a time, and their families would eat them just like lobster. Mo later recalled the adventure:

> *My dad, whose ulterior motive was having crawdads to feast on, was the one that got all the neighborhood kids together and loaded us in the back of his pickup (back in the days when it was still legal to do that), and out to the lake's edge we went. He gave us little or no instruction as I recall but did give us each a rake and a burlap sack. We were to walk along the lake levee, and whenever we saw a crawdad try and sweep it out of the water and onto the levee where it couldn't get back into the water. And that is exactly what we did. It didn't take long before we all got the real hang of sneaking up on a crawdad and sweeping it up onto the levee. And some of those crawdads went flying several feet before hitting the ground. We spent several hours each time we went out and with a half-dozen or so neighborhood kids, you can accumulate quite a few crawdads. Now we thought we were having fun and often came home with mud up to our hips but enjoyed the whole process, and after that would always ask Dad when we would be going out again to catch them crawdads.*

The crawdad hunts continued for about two years before the lake receded back behind the larger levees that the J.G. Boswell Co. had built like the El Rico, North Central and South Central levees which were miles out of town instead of just on the other side of the emergency levee by the Corcoran Airport.

The emergency levee was taken down once the waters receded enough to eliminate the threat to Corcoran. As best Mo can recall, that was sometime in early 1971.

Inflows to the Tulare Lake, in large part from the Kaweah River, are shown in Table 55.[905]

Table 55. Inflow to the Tulare Lakebed during water year 1969.

| Stream | Total Lakebed Inflow (acre-feet) | Percent Contribution |
|---|---|---|
| Kings River | 195,000 | 17% |
| Kaweah River | 430,000 | 37% |
| Tule River | 215,000 | 19% |
| Deer Creek & other sources | 93,000 | 8% |
| Kern River | 222,000 | 19% |
| Total | 1,155,000 | |

On August 9, 1969, a mature giant sequoia failed in Hazelwood Picnic Area in Giant Forest, triggering the failure of three other mature giant sequoias. That event killed a park visitor, resulting in the permanent closure of this picnic area. The failure likely had multiple causes, one of which may well have been the heavy precipitation of the preceding winter.

The January-February 1969 flood was unusual in being a major event in both the Tulare Lake Basin and in Southern California.[906] The heavy January storm saturated the ground so that the February storm produced particularly high levels of runoff.

San Luis Obispo County experienced flooding in both the January and February storms. Flood damage occurred in places such as San Luis Obispo, Morro Bay, and in state and USFS campgrounds. Considerable debris was washed up on beaches. The most severe damage occurred in the January storm. Total damage for the county was $5 million.

Santa Barbara County experienced flooding in both the January and February storms. Flood flows were of unprecedented magnitude. The January flows are generally the flood-of-record in the area. Many areas of the county had severe flooding, notably the Solvang and Lompoc areas. The Santa Ynez River peaked near Lompoc at an estimated flow of 100,000 cfs. The spillways of Gibraltar and Cachuma Dams spilled flows exceeding their design flow. Total damage for the county was $4.5 million with 5 deaths.

Ventura County experienced flooding in both the January and February storms. All rivers in the county flooded, and highway damage was heavy. The entire city of Santa Paula was evacuated during both the January and February storms. The February event was the largest flood-of-record in the Simi Valley and Moor Park areas. Total damage for the county was $43 million with 12 deaths.

Los Angeles County experienced severe flooding in both the January and February storms. People who lived in mountain areas such as Topanga Canyon, Mandeville Canyon, and Big Tujunga Canyon were especially hard hit. Landslides, debris flows and overflowing debris basins were a major problem. Total damage for the county was estimated to be $68 million plus $16 million to remove debris; 73 lives were lost.

Orange County experienced flooding from January 18–28 and again in February. The January flood damaged and destroyed bridges, roads, rail lines, and homes. Had Prado Dam not been in place, it was estimated that the Santa Ana River would have flooded in the January flood at 75,000 cfs, resulting in $440 million damage. The county experienced even worse flooding from the storm of February 18–27. The Santa Ana River threatened to breach its levees and emergency work (assisted by the U.S. Marine Corps) was required to prevent disastrous flooding. Total damage for the county from the two storms was $22 million with 7 deaths and 15 serious injuries.

San Bernardino County experienced flooding from January 18–28 and again in February. The January storm was an intense event, Etiwanda (part of present-day Rancho Cucamonga) received 16 inches of rain during this 9-day period. Many people had to be evacuated, and many homes were damaged or destroyed. Along the Santa Ana River, many highway bridges were lost or damaged. Many transportation routes between Bernardino and surrounding areas were closed and impassable until April or later. The February storm generated greater runoff and consequential damage to infrastructure. Places that were damaged in the January event were damaged again in February. Total damage for the county for January and February was more than $54 million with 13 deaths.

Riverside County was struck by flooding in both the January and February storms. The February flooding was even worse than the January flooding. Roads, railroads, and homes were heavily damaged in both floods. The mainline of the Southern Pacific Railroad was washed out. Floodwaters covered the Corona Airport up to 10 feet deep. Total damage for the county was $32 million with 4 deaths.

## 1970–71 Floods (2)

Flooding occurred in the Tulare Lakebed in both 1970 and 1971.

This lakebed flooding occurred despite the fact that there was no storm event of note in either year. Bakersfield did get hit by a powerful storm on May 26–27, 1971, setting a 24-hour precipitation record for the month. But that was just a local event during a dry year; none of that moisture made it to the lakebed.

Runoff was below average in both 1970 and 1971, bordering on drought conditions. Runoff during water year 1970 was 81% of the 118-year average (1894–2011) for the four major rivers (Kings, Kaweah, Tule, and Kern) combined. In 1971, it was only 66%.

The flooding in the lakebed in 1970 and 1971 was all left over from the big 1969 flood. As illustrated in Figure 11, the lakebed would remain at least partially flooded through 1971.

Lakebed flooding is a social construct; it is counted based on the number of growing seasons that are missed. The lakebed was flooded for three growing seasons: 1969, 1970, and 1971. So this is counted as three floods from the perspective of the lakebed farmers, even though the flood event occurred only once.

Something similar happened in the lakebed in 1982–84 and 1997–99. In each of those cases, lakebed flooding continued into a non-flood year.

## 1972 Floods (2)

There were two floods in 1972:
1. June
2. August

On June 7, an intense thunderstorm centered over the north Bakersfield area caused flooding and damages in its wake.[907, 908] The storm dropped 1.09 inches at Meadows Field in Bakersfield in 45 minutes, making it the wettest June day ever in that city. There was one report of 3.50 inches of rain in an hour in one part of the city. Lightning struck six substations, knocking out the power to most of Bakersfield.[909] The storm produced wind gusts to 50 mph, damaging automobiles and buildings.

Domestic water supply lines were washed out, roads severely damaged, and cars lifted and moved during the high runoff period. Houses were flooded with up to 4½ feet of water, and apartments were flooded and destroyed. Two people drowned, one a high school senior who was returning from his class picnic. Debris flows closed some roads and highways. Highway 178 was closed because of mud and landslides. The cross-town freeway in Bakersfield was closed due to flooding. Kern General Hospital had a flooded basement and first floor, which closed the emergency room services. Memorial Hospital was threatened with evacuation if the flood control canal broke, but it held.[910]

Hurricane Gwen formed off the coast of Mexico on August 24. It then spent a few days heading west-northwest and became a major hurricane on August 27. After retaining that intensity for over a day, it rapidly weakened and became a tropical storm on August 29.

On August 30, a cloudburst associated with Gwen off the coast of Southern California dropped 0.99 inches of rain 14 miles southwest of Coalinga in the Bear Canyon Jupiter area resulting in flash flooding.[911]

## 1973 Flood

We don't have a clear understanding of this flood.

The winter of 1972–73 was a strong El Niño event.
The winter of 1973–74 was a strong La Niña event.

Total flow for water year 1973 was 124% of the 1894–2011 average for the Kings, 143% for the Kaweah, 161% for the Tule, and 131% for the Kern. Roughly 25% of years have more runoff than this.

However, flooding still occurred in the Tulare Lakebed. Perhaps this was due to the timing of the runoff rather than to the total quantity of the flow. The lakebed had been flooded for three years, from 1969–71. After three years of below-average runoff, it was dry again in 1972. But as shown in Figure 11, the lakebed flooded again in 1973.

Judging just from Figure 11, this flooding event was relatively small. It doesn't look like an event that would threaten any of the valley towns. However, one report said that the lake threatened Corcoran and Alpaugh and stretched toward Kettleman City and Lemoore. That makes it sound like it was big as the 1969 flood, which seems unlikely. It seems almost certain that person was confusing the two floods.

On the other hand, the national parks' files have a photograph of a valley town (perhaps Stratford or Avenal) with a lot of water in the town. That photograph is identified as having been taken in 1973. Conceivably that photograph was mislabeled and it was actually taken during the height of the 1969–71 flood.

What actually occurred in the Tulare Lakebed in 1973? Was it the small flood documented in Figure 11, or was it the extensive flood described by other reports?

As illustrated in Figure 10, we have an invaluable record of how Tulare Lake changed in elevation for 120 years (1850–1969). If we knew the elevation of the lake in the 1973 flood, we'd have a good measure of how big the flood really was that year. Unfortunately, it has proved impossible to obtain access to the lake gage data for years since 1969. As a result, we really can't be sure just how big the lake was in the 1973 flood.

## 1975 Flood

Flooding in 1975 occurred from September 8–12 in Kern County.

In the Isabella area, a high intensity flash flood caused considerable damage in Kern County. One woman was swept from Highway 14 and drowned. High levels of sediment and debris deposits were a clean-up-chore on highways, roads, and on agricultural lands. Agricultural lands saw some damages, mostly to crops waiting to be picked.[912]

The South Fork of the Kern also flooded on September 8–12.

## 1976–77 Drought

This drought affected the entire state. It was the driest two years in the state's history. It was most severe in the northern two-thirds of the state.

Within California's roughly 100-year period of recorded hydrologic data, the driest single water year for runoff was water year 1977. Statewide runoff (including coastal and Southern California watersheds) was only about 15 million acre-feet. This represents 21% of the statewide average annual 71 million acre-feet.

In water year 1977, 47 of the state's 58 counties declared local drought-related emergencies. Water year 1976 ranks as the second driest at gaging stations in the central part of the Coast Ranges and among the five driest in the Central and Northern Sierra. Water year 1977 was the driest year of record at almost all gaging stations in the affected area. The two-year deficiency in runoff during the drought was unequaled at gaging stations in the affected area. The recurrence interval was more than 100 years.

In terms of recurrence intervals, the droughts of 1929–34 and 1976–77 are similar; both are of unsurpassed severity among droughts of corresponding duration during the period of systematic record collection. The drought of 1929–34 was longer and accumulated a larger deficiency in runoff. The drought of 1976–77 was more intense and had greater annual deficiencies in runoff. Arguments can be made that either was the most severe drought in the history of the state.

Table 56 compares the 1976–77 drought with other severe droughts of the 20th century.

Table 56. Comparison of selected 20th century droughts.

| Drought Period | Sacramento Basin Runoff | | San Joaquin River Basin Runoff | |
|---|---|---|---|---|
| | Average yearly runoff (million acre-feet) | % of average 1901–2009 | Average yearly runoff (million acre-feet) | % of average 1901–2009 |
| 1929–34 | 9.8 | 56% | 3.3 | 56% |
| 1976–77 | 6.6 | 38% | 1.5 | 26% |
| 1987–92 | 10.0 | 57% | 2.8 | 48% |

Table 57 shows how the San Joaquin Valley Water Year Index categorized the drought years in that basin.

Table 57. Rating of drought severity during the 1976–77 drought.

| Water Year | San Joaquin Basin Water Year Classification |
|---|---|
| 1976 | Critically dry |
| 1977 | Critically dry |

George Durkee recalled that the 1976 dry season was very late in ending in the Yosemite area. In early January 1977, he and others drove over Tioga Pass and ice-skated at Ellery Lake on the east side. One of the first winter storms hit just after, finally closing the Tioga Pass Road.

1976–77 were back-to-back critical drought years for the Kings River Basin. DWR estimated that about 125,000 acres of irrigated cropland were fallowed due to water shortages in 1977, mostly in

Fresno and Kern Counties, despite a significant increase in groundwater extraction to compensate for reduced surface water supplies.

1977 is the smallest tree-ring in Sequoia National Park since 1580. It is even smaller than the 1924 tree-ring.

On July 25 and 29, 1977, there was no inflow to Pine Flat Reservoir; all three forks of the Kings River had run completely dry.[913] That is the first time this has occurred on the Kings since record-keeping began at Pine Flat in 1953.

Roy Lee Davis recalled that the spring in the draw below Cactus Point in the national parks' Tunnel Rock Pasture ran dry in 1976–77. However, he said that spring and the associated stream were definitely flowing again in 1978.

The 1976–77 drought appears to have been less severe — or at least the effects less noticeable — in the High Sierra than in the lower elevations. Although we don't have gaging data for that zone, we do have observational data from several individuals. Bob Meadows recalled that none of the lakes that he visited in Yosemite during the summer of 1977 were significantly lower. Based on Bob's research, none of the daily wilderness ranger logs for Sequoia or Kings Canyon National Parks make any mention of lakes drying up in 1977.

George Durkee was the Crabtree wilderness ranger in 1977. It was his first year there, so he had no baseline to compare conditions to. The signs of drought, while no doubt present, were not particularly obvious. He didn't observe any lakes drying up. A large multi-day tropical storm occurred in August 1977, bringing significant rain and snow to the Sierra Crest but little to the mid-elevations. That event not only partially recharged the lakes, but caused lots of campers to scurry out of the high country, abandoning their gear. There was one fatality from the storm: a backpacker who died from hypothermia in Lamarck Col.

The spring near the Crabtree Ranger Station continued to flow throughout the summer of 1977. There was a pool above the station that had a stump that had apparently become rooted in a drought during the 1100s (see the section of this document that describes California's mega-droughts). That pool did not lower significantly during the summer of 1977, and the spring that fed it continued to flow.

On the other hand, Dave Graber, NPS regional chief scientist, recalled that some of the lakes in the High Sierra of the national parks did in fact dry up by the end of 1977; they were nothing but mud.

Total flows for water year 1976 was less than 35% of the 1894–2011 average for each of the four rivers within the Tulare Lake Basin. Flows for water year 1977 was the lowest experienced on the Kings, Kaweah, and Tule since record-keeping began in 1894. (The Kern experienced lower runoff in 1931.)

As shown in Table 58, the combined runoff of the four rivers in the Tulare Lake Basin during 1977 was only 696,572 acre-feet, the lowest since record-keeping began in 1894. For a sense of how widely flows vary in our area, the runoff in 1983 was 8,746,222 acre-feet (see Figure 13 on page 109 for a graph of other years).

Table 58. Runoff in the Tulare Lake Basin during water year 1977.

| Watershed | Total Runoff (acre-feet) | % of average (1894–2011) |
|---|---|---|
| Kings | 386,007 | 23% |
| Kaweah | 93,641 | 22% |
| Tule | 15,884 | 11% |
| Kern | 201,040 | 28% |
| Total | 696,572 | 23% |

## 1976 Floods (4)

There were at least four periods of flooding during 1976:
1. February
2. September/October (3)

The winter of 1975–76 was a strong La Niña event.

February 10 was the seventh consecutive day of measurable rain in Fresno, with 4.01 inches falling from February 4–10. Daily precipitation records were set on both February 5 (0.83 inches) and February 9 (1.50 inches). The Fresno City Works department distributed 1,800 sandbags to Fresno residents as several streets and poor drainage areas in that city flooded.[914] Table 59 gives the total precipitation during the February 4–10 storm event for selected reporting stations.

Table 59. Precipitation during the February 4–10, 1976 storm event.

| Reporting Station | Storm Total (inches of rain) |
|---|---|
| Fresno | 4.01 |
| Sanger | 4.27 |
| Dinuba | 3.86 |
| Batterson | 4.15 |
| Coarsegold | 4.89 |
| North Fork | 5.57 |
| Poison Ridge | 5.60 |

This storm also dropped an inch of snow in San Francisco. That city wouldn't see any significant snow again until February 2011.

The September/October floods in the Tulare Lake Basin occurred near the beginning of the 1976–77 drought.

Hurricane Kathleen formed off the coast of Baja California on September 9. It was a hurricane for only six hours and was a tropical storm when it made landfall on September 10. Kathleen weakened to a depression after it crossed the U.S./Mexico border near El Centro, but its circulation allowed gale-force winds to be recorded in Arizona and California.

The storm wasn't finished yet, as flooding rains continued to plague the Southwest along with gale-force winds. Kathleen's rapid forward speed allowed it to keep its strength for a long time over land. Kathleen is one of only six recorded tropical cyclones in the eastern Pacific Ocean known to have brought gale-force or hurricane-force winds to the continental United States.

Kathleen moved northward through the deserts of California bringing rain to interior Central California from September 9–11. The heaviest one-day totals were on the 10th at most locations. Two people were swept to their deaths when Interstate 8 was washed out.[915] Lodgepole set an all-time 24-hour precipitation record for the month of September with 5.06 inches of rain. Rainfall totals for the 3-day event were between ½ and 1 inch in the valley, averaging 1–2 inches in the foothills and 3–6 inches in the Sierra. Specific event totals are shown in Table 60.[916]

Table 60. Precipitation during the September 9–11, 1976 storm event.

| Reporting Station | Storm Total (inches of rain) |
|---|---|
| Yosemite South Entrance | 2.08 |
| Grant Grove | 3.96 |
| Lodgepole | 6.21 |
| Fresno | 0.90 |
| Bakersfield | 0.53 |

Tropical Depression Kathleen dissipated on September 11 while over southern Nevada as it continued accelerating.

Thunderstorms struck the central and southern San Joaquin Valley on September 29 with up to 2½ inches of rain falling in some areas. Dramatic lightning displays were seen from Fowler to Delano, and marble-size hail fell in Visalia and Porterville. The storm knocked out power to several thousand customers and struck two F106 jets operated by the Fresno Air National Guard, causing burn marks on the planes. The heavy rain also caused a roof to collapse at a building under construction as well as flooding homes, businesses and streets. It also caused additional damage to crops that had been seriously affected by the rain associated with Tropical Depression Kathleen.[917]

Heavy rain drenched parts of the central San Joaquin Valley on October 1. Fresno received 1.46 inches of rain, setting a daily precipitation record. Several roads were heavily flooded in that city, temporarily stranding some motorists. Los Banos received ½ inch of rain in just 30 minutes. Many roads and fields in Mendota were flooded.[918]

Flooding occurred on the west side of the San Joaquin Valley near Coalinga in 1976. Presumably this resulted either from Kathleen on September 10 or from the storm system that came through the valley on September 29 – October 1.

## 1977 Flood

Flooding in 1977 occurred in December. This flood occurred near the end of the 1976–77 drought.

Soaking rains fall in Kern County from December 27–28. Storm totals were 2.05 inches in Lost Hills and 1.11 inches in Bakersfield. Water was two feet deep at some intersections in Bakersfield, stranding some motorists.[919]

## 1978 Floods (3)

There were three periods of flooding in 1978:
1. February
2. Early summer
3. September

The winter of 1977–78 was a moderate El Niño event.

Flooding occurred on the west side of the valley in February. Flooding occurred on the east side of the valley and in the Tulare Lakebed in early summer due to high runoff. River flooding occurred in September due to a tropical downpour throughout the Sierra.

Former state climatologist Jim Goodridge rated the February 1978 flood as one of the 10 most damaging floods in the state's history. Ventura County received over 13 inches of rain in one day, resulting in floods and landslides.

A vigorous winter cyclic storm with widespread flooding and mudslides developed on the windward slopes of the South Coastal Basin on February 10. There was $120 million in storm-related property damage and 18 deaths. This storm was still quite robust as it moved northeasterly into the rain shadow zone of the comparatively dry Buena Vista Lake Basin.

The storm of February 10, based on recurrence interval, was centered in the area around Buena Vista Lake. Blackwells Corner (intersection of Highways 33 and 46) received 3.90 inches of rain on February 10, which was 74% of its average annual precipitation. This was 7.41 standard deviations above the average maximum day with a recurrence interval of 28,000 years. A total of 32 stations reported recurrence intervals in excess of 100 years and 16 stations reported recurrence intervals in excess of 1,000 years.[920]

Bakersfield received 2.29 inches of rain on February 9. That was the wettest day ever in that city.[921] Bakersfield received a total of 5.36 inches of rain during February, making it the wettest month ever in that city.[922] That record would eventually be broken in December 2010.

This was a heavy rain event combined with snowmelt runoff in some areas. Over 6,000 acres were flooded in Kern County, causing extensive damage to agricultural lands. The Lamont/Arvin area was flooded. Mudslides, landslides and debris flows were common. One woman died in Kern County when her car was swept off Interstate 5 by a mudslide. Transportation routes, including rail traffic, were suspended for as long as three days. A total of 91 county roads were closed. The California Aqueduct was damaged. Bridges, culverts and other flood control works were badly damaged. Domestic water supply and sewer lines were washed out. Oil field facilities were also damaged. Total damage in Kern County was approximately $25 million. President Ford declared Kern County a disaster area on February 15.[923]

Within the Tulare Lake Basin, the storm's effect was felt primarily on the west side of the valley. Panoche/Silver Creek west of Mendota flooded in February.[924] Hanford received 2.4 inches of rain on February 10, the most that city has ever received in any 24-hour period. Flooding occurred along Los Gatos Creek from the foothills to the valley floor and damaged agricultural lands, roads and bridges, and utilities. An estimated 4,500 acres were flooded, and damage totaled $160,000.

313

There is a strong resemblance between this storm and the remnants of a hurricane which came onshore near Monterey Bay on September 11, 1918. Both were robust cyclic storms which vigorously entered the rain shadow areas to the northeast, resulting in a deluge in normally dry areas.

The February 1978 Buena Vista Lake storm also resembled the March 1995 storm. That storm produced devastating rainfalls on the windward slopes of the Coast Ranges. It was still quite energetic as it moved into the rain shadow area to create further devastating floods. That was the storm that washed out the Interstate 5 bridges near Coalinga.

One of the side-effects of the February 1978 storm was the grand display of wildflowers seen in the vicinity of the Tulare Lakebed by mid-March of that year. That was a result of the thorough soaking of the ground at an optimum time of the year.[925]

The February 1978 storm was spectacular south of the Tehachapis. In contrast, the September 1978 storm would be spectacular in the Kings River Basin and to the north. In the Kaweah River Basin, the February storm was a much bigger flood event than the September storm. It produced a peak average daily flow at Terminus Dam of 8,135 cfs while the September storm produced a flow of only 3,890 cfs. The Kaweah's peak natural flow occurred at McKay's Point on February 9: 14,700 cfs. (That was the peak hourly flow; the peak average daily flow was 8,135 cfs.)

Flood releases of 8,000 cfs or greater occurred at Friant Dam on the San Joaquin River during the April–July snowmelt period.

The peak daily flow at Pine Flat on the Kings occurred on June 9 during snowmelt.

Hurricane Norman was a powerful Category 4 hurricane with a 40-mile-wide eye and sustained winds of 140 mph. It developed in early September well off the coast of Acapulco, and then slowly weakened as it moved over cooler waters west of Baja California. By early on the morning of September 4, moisture from that hurricane had spread north, initiating rains in California. Norman then recurved, turning north toward Southern California. It made landfall as a tropical depression on September 5–6.

When a Pacific hurricane degrades, it usually makes landfall in Southern California or in Mexico. Norman came ashore in the LA area, but its track was aimed straight for the southern end of the Sierra. As the remnants of Norman plowed inland, heavy rains fell across the Sierra, with a maximum amount of 7.01 inches reported at Lodgepole. Rivers rose so quickly that roads closed, and campers found themselves marooned all over the Southern Sierra.

George Durkee (national park wilderness ranger) recalled that the storm was known in the national parks as "Stormin' Norman." Tighe Geogehagan was the parks' fire dispatcher, and one day she reported "rain, rain and more rain" on the weather report. Many backpackers got drenched and sought refuge in the wilderness ranger stations. Two backpackers died of hypothermia at Trail Camp on the east side of Mt. Whitney. They had apparently gotten soaked in the rain down by Crabtree and then encountered cold temperatures and probably sleet at Trail Crest.

Approximately 4 inches of rain fell at Cedar Grove. At Pine Flat, the natural daily flow on the Kings River on September 5 was over 10 times greater than it had been the previous day. It was not quite as high as the natural daily flow that had occurred on June 9 during snowmelt, but it was still a very impressive event.

Labor Day occurred on September 4. Jerry Torres, Kings Canyon National Park's trails supervisor at the time, recalled that the flood over the Labor Day Weekend covered the North Side Road and spilled over onto Highway 180 in two spots: west of Grizzly Falls and near the Boyden Bridge. Based on the flood exceedence rates in Table 16, this had a recurrence interval of only 8 years for the Kings River downstream at Pine Flat. Still, that puts it in a category with other Cedar Grove floods of the past 70 years (1937, 1950, 1955, 1966, 1969, 1978, 1982, 1983, and 1997) that rise to the level of the modeled 50-year flood: that is, a flood event that occurs about every eight years on average. See the section of this document that describes Cedar Grove Flooding.

This was a negligible flood on the Kaweah and the rivers to the south.

Norman also caused flooding on the east side of the Sierra. At Lake Sabrina, a 10-hour duration rainfall of 1.02 inches was recorded on September 5, producing a peak flow of 940 cfs at Power Plant No. 6 on Bishop Creek. This was the second largest flood-of-record on that creek and was estimated to have a recurrence interval of 50 years. To prevent flood damage to Bishop and surrounding areas, the Los Angeles Department of Water and Power activated the Owens River Canal Bypass.

Total flow for water year 1978 was 200% of the 1894–2011 average for the Kings, 193% for the Kaweah, 195% for the Tule, and 220% for the Kern.

By some measures, 1978 was the wettest water year in Kern County since record-keeping began in 1889. Bakersfield received 12 inches of rain.

Flooding occurred in the Tulare Lakebed; this was the first significant flooding since 1973 (see Figure 11). In order to minimize flooding, 9,000 acre-feet of river floodwater was pumped into the Friant-Kern Canal and routed to the Los Angeles area.

## 1980 Flood

There were two floods in 1980:
1. January
2. February

The national parks' records make no mention of any flooding in 1980.

Flood releases of 8,000 cfs or greater occurred at Friant Dam on the San Joaquin River during the winter (prior to snowmelt).

The peak day natural flow at Pine Flat on the Kings occurred on January 13. The flow on that date was 37 times greater than the flow just four days earlier. It was approximately as large as the peak day flow in the 1963 flood. It seems likely that this was a high-flow period in Cedar Grove as well.

The Kaweah's peak natural flow occurred at Terminus Dam (or possibly this was McKay's Point) on January 13: 34,000 cfs. (That was the peak hourly flow; the peak average daily flow, as reflected in Table 15, was 16,933 cfs.) Based on the flood exceedence rates in Table 16, this had a recurrence interval of 13 years for the Kaweah and a recurrence interval of 10 years on the Tule.

In the Tulare Lake Basin, the second flood of the year occurred on February 18. The Kern had a much bigger relative response to this storm than the rivers farther to the north. This was the same pattern as in the 1916 flood when the storm was to the south of the Tulare Lake Basin.

The February 14–21, 1980 flooding was most severe in Central and Southern Coastal California. It had a recurrence interval of up to a 50 years on some rivers. Disastrous and record-breaking rainfalls in the South Coastal Basin resulted in the highest-ever rainfall totals over a broad area. Record-high eight-day rainfalls occurred at 133 stations. Recurrence intervals in excess of 100 years were reported at 70 stations. Over 1,500 homes were damaged or destroyed; there was a total of $270 million in property damage, and there were 18 storm-related deaths. Seven counties were declared disaster areas.[926]

Total flow for water year 1980 was 177% of the 1894–2011 average for the Kings, 205% for the Kaweah, 236% for the Tule, and 221% for the Kern. Flooding occurred in the Tulare Lakebed; this was the first significant flooding since 1978 (see Figure 11). In order to minimize flooding in the Tulare Lakebed, 5,000 acre-feet of river floodwater was pumped into the Friant-Kern Canal and routed to the Los Angeles area.

## 1982–83 Floods (10)

There were at least 10 floods in 1982–83:
1. April 1982 (rain-on-snow event)
2. June 1982 (severe storm)
3. September 1982 (due to remains of Hurricane Olivia)
4. December 1982 (rain flood)
5. March 1983 (severe storm)
6. Memorial Day Weekend, 1983 (four debris flows)
7. May–July, 1983 (runoff)
8. August 1983 (two severe storms caused by monsoonal moisture)
9. September 1983 (severe storm)

The 1982–83 El Niño was the strongest such event recorded over the past 50 years. The winter of 1981–82 experienced heavy snowfall in the Sierra. Record after record was broken:
- Echo Summit received 67 inches (5.6 feet) of snow in 24 hours on January 4–5, 1982, breaking the state record that had been set by Giant Forest in January 1933.
- On January 5, 1982, 29.5 inches of snow fell in Yosemite Valley, setting the record for the biggest 24-hour snowfall ever at that location.[927]
- The Central Sierra Snow Laboratory monitoring site near Donner Pass received 15.5 feet of snow between March 27 – April 8, 1982, the biggest snowstorm ever recorded at that site.

That heavy snowpack set the stage for a big spring runoff on all the rivers.

Flood releases of 8,000 cfs or greater occurred at Friant Dam on the San Joaquin River during the April–July, 1982 snowmelt period.

On April 11–12, 1982, a combination of snowmelt and rainfall caused the Merced River to overflow its banks and flood parts of Yosemite Valley. The national park headquarters building was damaged, and parts of a road were washed out.[928] The Merced River peaked at Happy Isle at 4,880 cfs. By Merced River standards, that's a fairly modest flood, having a recurrence interval of 8 years.

Easter Sunday fell on April 11, so this flood is sometimes referred to as the Easter 1982 flood. Table 61 gives the precipitation totals for the reporting stations in Sequoia and Kings Canyon National Parks.

Table 61. Precipitation during the April 11–12, 1982 storm event.

| Reporting Station | Storm Total (inches of rain) |
|---|---|
| Grant Grove | 7.44 |
| Lodgepole | 9.33* |
| Ash Mountain | 4.57 |

*6.83 inches of this total fell on April 11

Jim Harvey recalled that Elk Creek had a big flood in the early 1980s; this seems like the probable time when that event would have occurred. Jim said the flood damaged the Generals Highway, much as it had in the 1935 flood.

The peak day natural flow at Pine Flat on the Kings occurred on April 11, 1982. The flow that day was 10 times larger than the flow of the previous day. That is the fourth largest peak day of the year at Pine Flat since the dam was built in 1954. That would suggest that there was very high water in Cedar Grove as well, but we have no national park records to substantiate that.

The Kaweah's peak natural flow occurred at Terminus Dam (or possibly McKay's Point) on April 11, 1982: 28,800 cfs. (That was the peak hourly flow; the peak average daily flow, as reflected in Table 15, was 18,514 cfs.) Based on the flood exceedence rates in Table 16, this had a recurrence interval of 17 years for the Kaweah.

On June 18, 1982, an intense thunderstorm occurred at Forni Ridge in the Eldorado National Forest. This storm and the resulting debris flow were outside the Tulare Lake Basin, but the story merits inclusion in this document as an example of how intense a summer storm can be. The storm lasted only a short time, but is notable because 4.02 inches of rain was measured in 30 minutes (a record-setting rate of 8.04 inches per hour). The storm was centered over a recently burned steep mountain slope adjacent to U.S. Highway 50. The storm was followed by a debris flow that closed the highway.[929]

On June 30, 1982, there were numerous reports of funnel clouds over Clovis, and one touched down near Fresno State University. Thunderstorms caused street flooding in Farmersville and also flooded homes in other parts of the valley. Dinuba received particularly heavy rain.[930]

The June 30 thunderstorm system extended into the national parks. Table 62 gives the precipitation totals for some of the reporting stations in the area.

Table 62. Precipitation during the June 30, 1982 storm event.

| Reporting Station | Storm Total (inches of rain) |
|---|---|
| Grant Grove | 1.15 |
| Lodgepole | 0.65 |
| Ash Mountain | 1.30 |
| Dinuba | 1.62 |

Thanks to a photograph taken by Bill Tweed, we know that the Marble Fork Kaweah flooded through Lodgepole in June (photograph on file in the national parks). This suggests that there was a strong thunderstorm cell in the Tablelands area.

Hurricane Olivia formed about 400 miles south of Acapulco on September 19, 1982. It developed winds of 130 mph, becoming the strongest storm of the season. It gradually weakened as it passed over cooler waters west of Baja California. Olivia then recurved and came ashore as a tropical depression. When a Pacific hurricane degrades, it usually makes landfall in Southern California or in Mexico. Olivia came ashore near the U.S./Mexico border, but its track was aimed for Utah. Olivia's storm track, as illustrated in Figure 19,[931] followed a fairly typical pattern for Pacific hurricanes that degrade and then make landfall in Southern California.

Figure 2. Hurricane Olivia.

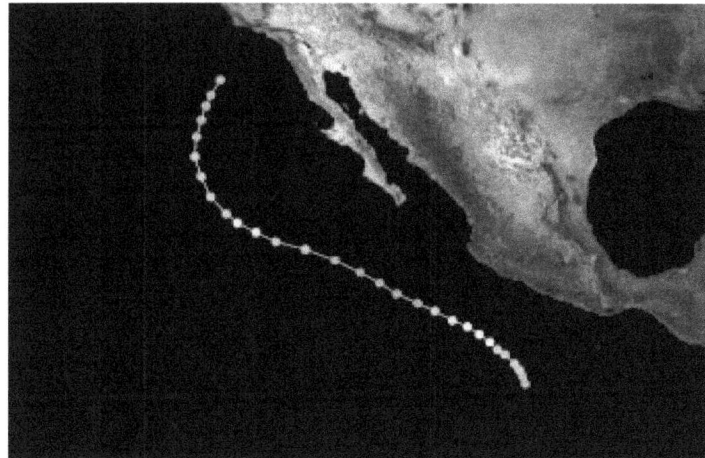

Figure 3. Hurricane Olivia's storm track.

As the remnants of Olivia plowed inland, heavy rains fell across the San Joaquin Valley and the Sierra from September 23–27. Measurable rain fell from September 24–26 in both Fresno and Bakersfield. Storm totals included 0.70 inches at Bakersfield and 1.10 inches at Fresno, although locally heavier amounts were reported. Far heavier amounts fell in the Sierra with a maximum amount of 7.19 inches reported at Grant Grove.[932]

The heavy rain wiped out half of California's raisin crop, a quarter of the wine crop, a tenth of the tomato crop, and also damaged the almond crop. The rain caused power outages to over 10,000 customers in Fresno County.

Jerry Torres, Kings Canyon National Park's trails supervisor at the time, recalled that it rained nonstop for at least two days in Cedar Grove, resulting in a major flooding event in the Kings River Basin. Pine Flat experienced a large and abrupt increase in flows on September 25.

The Middle Fork Kings Bridge at Dougherty Creek was constructed in the summer of 1979. Jerry was part of that construction. That bridge was washed out in the September 1982 flood.

Flooding also occurred on the South Fork Kings. In Cedar Grove, the western ¼ mile section of the North Side Road had water over it. The flood damaged a section of Highway 180, 100 yards west of Grizzly Falls. But the biggest damage occurred two miles west of Grizzly Falls where an entire hillside was washed away, including a section of Highway 180.

Access into Cedar Grove was closed for four days until Caltrans used a dozer to literally cut a road out of the mountain. The September 1982 flood is sometimes remembered in the park as the "Great Trails End Flood" because it occurred during the annual "Trails End" end-of-year celebration in Cedar Grove.

Based on the flood exceedence rates in Table 16, this had a recurrence interval of 35 years for the Kings River downstream at Pine Flat. That puts it in a category with other Cedar Grove floods of the past 70 years (1937, 1950, 1955, 1966, 1969, 1978, 1982, 1983, and 1997) that rise to the level of the modeled 50-year flood: that is, a flood event that occurs about every eight years on average. See the section of this document that describes Cedar Grove Flooding.

The Kaweah's peak average daily natural flow, as reflected in Table 15, occurred on September 26, 1982: 6,308 cfs. Based on the flood exceedence rates in Table 16, this had a recurrence interval of 4 years for the Kaweah.

All that we know about the December 1982 flood comes from the gaging stations at the various dams. The gages all jumped sharply on December 22. Evidently it was a rain-on-snow event. The rise in the rivers was least noticeable in the Kern River Basin, so apparently the storm was located more to the north.

The Kaweah's peak natural flow occurred at Terminus Dam (or possibly McKay's Point) on December 22, 1982: 11,100 cfs. (That was the peak hourly flow; the peak average daily flow, as reflected in Table 15, was 8,325 cfs.) Based on the flood exceedence rates in Table 16, this had a recurrence interval of 6 years for the Kaweah.

One of the 1982 flooding events, perhaps the September one, was a major flood on the Tule.

Total flow for water year 1982 was 181% of the 1894–2011 average for the Kings, 179% for the Kaweah, 165% for the Tule, and 157% for the Kern.

The heavy spring runoff in 1982 resulted in flooding in the Tulare Lakebed. This was the first significant flooding in the lakebed since 1980 (see Figure 11). In order to minimize flooding in the lakebed, 33,000 acre-feet of river floodwater was pumped into the Friant-Kern Canal in 1982 and routed to the Los Angeles area.

319

New record-high total annual rainfalls were reported from stations located over a broad range of California during water year 1983 (October 1982 – September 1983). California received a long sequence of storms which left poorly drained areas soaked for many months. This soaking resulted in unusually extensive flooding in several parts of the state. In all regions, the high rainfall totals were associated with a quite noticeably increased numbers of rainy days, rather than with large individual rainfalls.

It had been 93 years since California had as much rain as in water year 1983. The last year with rainfalls as high was 1890. One of the factors which make the 1983 year even more unusual was that 1982 was also one of the wettest years of record. A total of 58 stations reported 100 or more inches for water year 1983. A total of 511 stations reported their wettest year ever during 1983.

During water year 1983, half of the state's land area had rainfalls in excess of a recurrence interval of 100 years. During that same period, 45 stations reported yearly rainfall totals that were in excess of the 1,000-year amounts. These were distributed from the Klamath River Basin in the north to the Borrego Desert in the south.[933]

The winter of 1982–83 was a potent El Niño event. The impact of this El Niño on California's weather in 1982-83 was complex. The high-pressure ridge between 10 and 20 degrees north latitude was magnified by the heat from the warmer than normal ocean water. Simultaneously, extremely low air pressures developed over the Gulf of Alaska. These contrasting pressure extremes caused the westerly airflow across the Pacific to double. The jet stream that directs storms into California was intensified and displaced to the south so that storms hit the Central California coast fiercely and more often.[934]

Northern and Central California experienced flooding incidents from November 1982 through March 1983 due to numerous storms. The melting of the record snowpack then created a second episode of flooding from May through July of 1983.

Statewide, the two wettest water years during historic times were 1890 and 1983.[935] The statewide precipitation for water year 1983 averaged 190% of normal, with many areas well over 220%. New precipitation records were set at 49 locations in the state.[936]

Table 63 summaries the increased precipitation for the three drainage basins from the Upper San Joaquin to the Kern River.[937]

Table 63. Precipitation totals during winter 1982–83.

For the three drainage basins from the Upper San Joaquin to the Kern River.

| Season | Percent of Average |
|---|---|
| Fall 1982 (September, October, November) | 318% |
| Winter 1983 (December, January, February) | 183% |
| Spring 1983 (March, April, May) | 199% |

Yosemite recorded 66.39 inches of precipitation during 1983, breaking the record of 61.09 inches set in 1938. The mean yearly precipitation for Yosemite is 35.26 inches.[938]

Fresno received a total of 23.57 inches of precipitation during water year 1983, making that the wettest water year on record for that city.[939]

The stage for a disastrous year of flooding had been set in the fall of 1982. In some parts of California, September 1982 was one of the wettest Septembers on record, thanks to subtropical moisture from the remains of Hurricane Olivia. Soils were saturated, and there was less than normal flood control space in many reservoirs. The 1982 and 1983 water years are the wettest pair of years on record.[940]

As shown in Table 64, the snowfall total in Lodgepole in the winter of 1982–83 was a rather awe-inspiring 429.8 inches (36 feet).

Table 64. Lodgepole snowfall during winter 1982–83.

| Month | Snowfall (inches of snow) |
|---|---|
| September, 1982 | 2.0 |
| October, 1982 | trace |
| November 1982 | 46.0 |
| December 1982 | 51.3 |
| January 1983 | 84.5 |
| February 1983 | 74.0 |
| March 1983 | 113.0 |
| April 1983 | 40.0 |
| May 1983 | 19.0 |
| Total | 429.8 |

Up to four feet of snow fell in the Sierra in less than 24 hours on December 22, 1982. Lodgepole received 27 inches of snow with a storm-total liquid water equivalent of 10.09 inches. The snowpack on some of the higher peaks from this storm was raised to nearly 100 inches.[941]

(We tend to think of the incredible winter of 1982–83 as being record-setting at Lodgepole, but there were at least four bigger snowpacks that we know of at that location. Bill Tweed recalled that having been through all that snow, the people who spent the winter at Lodgepole were disappointed that they had not set a new record. The winter of 1968–69 received at least 440.5 inches. In the winter of 1951–52, Lodgepole received 449.5 inches (37½ feet), setting the record for this weather station. That record would eventually be broken in the winter of 2010–2011. The most impressive winter in this area that we know anything about was the winter of 1905–06. In that winter, the snowpack reached a maximum depth of 29 feet on the level in Giant Forest. Even by June 25, 1906, the snowpack in Giant Forest had only melted down to about 12 feet on the level.)

In the winter of 1982–83, snowpack records were set at three-fourths of the Sierra snow courses. On May 3, 1983, snow water content in the Sierra exceeded 230% of normal; the ensuing runoff resulted in four times the average volume for Central Valley streams. [942]

The trans-Sierra highway over Tioga Pass in Yosemite National Park and many Sierra wilderness trails that normally open by early summer remained blocked by snow. Snow survey measurements found the snowpack persisting later into the year leading to an unprecedented July snow survey in the Kings River Basin.[943]

Jack Vance recalled being at the Hockett Ranger Station over the Fourth of July Weekend, 1983. Hockett Meadow was one big lake with miniature icebergs floating on it. The high watermark from that lake can still be seen inside the tackshed.

An intense storm struck the Tehachapi Mountains on March 1–2, 1983. Heavy rainfall of 2–7 inches fell during that two-day period, including 6.50 inches at Frazier Park. This triggered flash flooding on several creeks; Caliente Creek peaked at 15,000 cfs as it flowed into the southeast end of the San Joaquin Valley. Most severely impacted was Lamont, where 1,973 homes were damaged or destroyed — over half of the town. Over 33 roads were washed out in Kern County, and two 100-car trains had to be abandoned after water washed out parts of tracks. The town of Caliente was also flooded, resulting in 77 people having to be rescued by helicopter. Agricultural lands and irrigation works were also damaged and destroyed. Irrigation works were washed out. Total damage from the flood was an estimated $58.7 million. A series of storms resulted in continued flooding through March 13.[944, 945]

During the 1983 Memorial Day Weekend, Kings Canyon experienced three debris flows. Another debris flow occurred in the Redwood Creek drainage in the Mineral King area at apparently the same time. These four debris flows are described at the end of this section along with other mass wasting events that occurred in 1983.

All of the major reservoirs in the Sacramento River and San Joaquin River Basins reached or nearly reached design capacity during the June and July runoff. At least two levees failed in the Sacramento River Basin. Levee breaks caused flooding at four locations along the San Joaquin River. In the Sacramento–San Joaquin River Delta, four levees failed, resulting in partial or total flooding of some islands. Damage exceeded $91 million in the Sacramento River Basin and $324 million in the San Joaquin River Basin.

Flood releases of 12,000 cfs or greater occurred at Friant Dam on the San Joaquin River during the April–July snowmelt period.

The peak day natural flow at Pine Flat on the Kings occurred on May 29, 1983. It was a significant flood, but only half of what the peak day natural flow had been during the much less famous 1982 flood. It seems likely that this was a very high-flow period in Cedar Grove as well. That puts it in a category with other Cedar Grove floods of the past 70 years (1937, 1950, 1955, 1966, 1969, 1978, 1982, 1983, and 1997) that rise to the level of the modeled 50-year flood: that is, a flood event that occurs about every eight years on average. See the section of this document that describes Cedar Grove Flooding.

Two severe storms occurred in August 1983: one in the southeastern part of the Tulare Lake Basin and the other farther north in the Kings Canyon high country. Both were caused by monsoonal moisture which typically originates in the vicinity of the Four Corners Area.

The first storm occurred in the southeastern part of the Tulare Lake Basin. On August 16, more than 1½ inches of rain fell in the Tehachapi Mountains in an hour, washing out portions of Highway 58.[946] On August 17, portions of California City were flooded after heavy rain fell in the Tehachapi Mountains and caused Cache Creek to swell. Water was the height of car windows and some houses flooded.[947]

Apparently the monsoonal surge continued moving northwest along the crest of the Sierra. Lodgepole received 1.48 inches between August 15–18. Huntington Lake received 1.00 inches between August 15–17. The moisture never reached Yosemite. There may have been multiple severe storms or cloudbursts over the national parks between August 15–18, but only the following account survives.

The second severe storm of August 1983 occurred when a very large black cloud brought intense rain to a section of the Kings Canyon high country. George Durkee witnessed that event while he was the wilderness ranger at the McClure Ranger Station.

Ralph Kumano, who was a wilderness ranger on patrol on the Monarch Divide, first spotted the black cloud when it was over Enchanted Gorge, northeast of Tehipite Dome. From there, the cloud moved north to Mt. Darwin near the Evolution Valley.

The cloud settled over Darwin Canyon. It apparently dumped onto the Lamarck and Darwin Glaciers. It resulted in major flooding of Darwin Creek. A lot of debris plugged the creek where it crosses the Pacific Crest Trail (UTM 347825E 4115568N NAD83 Zone 11). It was a very localized event; there was no rain at all at McClure Meadow just two miles down the trail.

At the Pacific Crest Trail, there is a log crossing over Darwin Creek where the water flows under the log. There's a pool just upstream from the trail crossing. The flood filled that pool with sand and rocks, forcing the creek to flow over the log. (The idea is that the creek is supposed to flow *under* the log; hikers should walk on top of the log.) Both 1982 and 1983 were El Niño years with high runoff, but the flows in those years hadn't blocked up the log crossing, so it shows what a single event can do. George got out into the pool and dug out under the log until he could clear it a little. He had to do that a couple of times over a week or two because debris kept clogging the opening under the log.

Lower Darwin Lake drains Lamarck Glacier. After the flood, George observed that this lake had turned the milky blue that is associated with suspended glacial silt. That suggests that the flood may have breached an ice dam on Lamarck Glacier. Alternatively the change in the lake could have been caused by the intense rain coming down on the glacier, which drains through a hole in front of the Little Ice Age moraine. In either case, the storm had washed down a lot of fine glacial silt off of the glacier.

After the storm, George also checked out Enchanted Gorge. That gorge is very narrow and had been mostly filled with avalanche snow, much of which had melted. Toward the bottom of the gorge, there was recent scarring on trees about four feet up from the base and there was a lot of sand at the confluence with Goddard Creek. There was no evidence of flooding in Goddard Creek above its confluence with the gorge.

Large-scale debris flows usually occur in small, steep stream channels and are often mistaken for floods. The Darwin Canyon event may have been a flash flood that carried a large amount of debris, or it may have been a debris flow. We just don't have enough data to clearly classify it. The Enchanted Gorge event was probably a flash flood.

The 10th flood of the 1982–83 period occurred on September 30, 1983. We know relatively little about it. Apparently it was caused by a thunderstorm system that spread along the Sierra. Table 65 gives the precipitation totals for the reporting stations in the national parks.

Table 65. Precipitation during the September 30, 1983 storm event.

| Reporting Station | Storm Total (inches of rain) |
|---|---|
| Grant Grove | 1.14 |
| Lodgepole | 1.10 |
| Ash Mountain | 1.22 |

Thanks to a photograph taken by Bill Tweed, we know that the Marble Fork Kaweah flooded through Lodgepole in September (photograph on file in the national parks). This suggests that there was a strong thunderstorm cell in the Tablelands area.

As shown in Table 66, the combined runoff of the four rivers in the Tulare Lake Basin during water year 1983 was 8,746,222 acre-feet, the largest runoff since record-keeping began in 1894. For comparison, that is 5.4 times the combined current capacity of the federal reservoirs on those four rivers.

Table 66. Runoff in the Tulare Lake Basin during water year 1983.

| Watershed | Total Runoff (acre-feet) | % of average (1894–2011) |
|---|---|---|
| Kings | 4,286,703 | 254% |
| Kaweah | 1,402,005 | 325% |
| Tule | 615,014 | 440% |
| Kern | 2,442,500 | 334% |
| Total | 8,746,222 | 293% |

This was the highest flow year for the Kings, Kaweah, and Tule Rivers since 1894. It was the second highest for the Kern River, only 1916 was higher.

This was the last year that the Kaweah River (via the Consolidated People's Ditch) flooded a significant portion of Kaweah Oaks Preserve.

The 1983 flood had a greater total runoff than the 1969 flood. Unfortunately, some parts of the runoff were not actually measured and can only be estimated. An unknown amount of the Kern River flowed into the Tulare Lakebed, but 759,000 acre-feet of the Kern was diverted into the California Aqueduct and routed to the Los Angeles area. The total estimated lakebed inflow of the other three rivers (Kings, Kaweah, and Tule) was about 1.069 million acre-feet. That was 27% more than the 0.840 million acre-feet from those same three rivers in 1969. The total inflow to the Tulare Lakebed in water year 1969 from all streams was 1.155 million acre-feet. If the 1983 runoff were 27% greater than that, then it would have been on the order of 1.467 million acre-feet if there had been no diversion into the California Aqueduct. That would have been almost as large as the total 1.530 million acre-feet inflow that occurred in the 1906 flood.

The 1983 flood brought the lake to a peak elevation of 191.44 feet, slightly lower than the modern 192.5 foot record set in 1969. In order to protect Corcoran, the USACE spent $2.7 million to construct emergency flood protection levees along Cross Creek and the Tule River. Unfortunately those levees were not strong enough and were breached.

Tulare Lakebed inundation began in January and peaked in July. By July 13, 82,000 acres of prime agricultural land were flooded.[948] Based on a comparison of maps, the area flooded in 1983 was slightly greater than the area flooded in 1969.

Mo Basham recalled that the eastern edge of Tulare Lake came to about Avenue 10½. This is about 3 miles west of where the emergency levee was built near the Corcoran Airport during the 1969 flood. Bill Tweed recalled that the lake was so big in the summer of 1983 that you could see it from the High Sierra, shining through the valley haze. To see it was like seeing a ghost, a relic of another time (photograph on file in the national parks, see back cover).

In order to minimize the flooding in the lakebed, an unknown amount of river floodwater was pumped into the Friant-Kern Canal in 1983 and routed to the Los Angeles area.

Table 67 summarizes the damages incurred in the southern end of the San Joaquin Valley during the 1983 floods.[949]

Table 67. Damages incurred during 1983 floods.

| County | Private Damage (thousand dollars) | Public Damage (thousand dollars) | Road Damage (thousand dollars) | Agricultural Damage (thousand dollars) | Total Damage (thousand dollars) |
|---|---|---|---|---|---|
| Merced | $200 | $414 | | | $614 |
| Madera | $200 | $0 | $100 | $40,000 | $40,300 |
| Fresno | $100 | $7,060 | $616 | $5,648 | $13,424 |
| Kings | $420 | $1,998 | $550 | $95,000 | $97,968 |
| Tulare | $100 | $844 | $37 | $23,750 | $24,731 |
| Kern | $2,750 | $356 | $1,328 | $7,500 | $11,934 |
| Total | $3,770 | $10,672 | $2,631 | $171,898 | $188,971 |

In 1982–83, two young naturalists from Bakersfield took advantage of the high water to kayak from Tulare Lake to San Francisco Bay. They had to do a good bit of portaging, but they made it. Dave Graber, NPS regional chief scientist, recalled that their trip was written up in the *Fresno Bee*. This was the fifth documented trip between the lake and the bay. (The other four were in 1852, 1868, 1938, and 1969.)

After two years of flooding (1982 and 1983), cotton growers decided to drain their lands. The Tulare Lake Irrigation District applied for a permit to pump the excess water over the top of the Tulare Lake sill. It appears that there was considerable opposition to granting this permit. Under an emergency proclamation issued by the USACE during the spring of 1983, reclamation districts and land companies remade the channel along some 29 miles of the lower Kings River (see Figure 12) to dewater the lake and drain the water north into the Sacramento–San Joaquin River Delta region.

A series of pumps were installed with a total lift of 43 feet. The project was designed to remove approximately 2,000 acre-feet of water per day from the lakebed. Pumping began on October 7, 1983 and continued intermittently until the program was terminated on January 19, 1984. Only about 90,000 acre-feet was pumped northward under this program. Pumping was stopped earlier than scheduled due to concern that white bass might be transferred from Tulare Lake to the San Joaquin River. The lakebed would not be fully drained until water year 1985.

An outstanding feature of the 1982–83 storm event was the number of significant landslides and debris flows that resulted. Some of those were in the Central Sierra, north of the Tulare Lake Basin. Those events merit inclusion in this document because they were well studied, and they can inform risk management planning in our area.

### Landslide: South Fork American River

This event occurred near White Hall on U.S. Highway 50 about 26 miles east of Placerville and 34 miles west of South Lake Tahoe. At this point, the highway is squeezed in a narrow canyon between the South Fork American River and a steep cliff.

This landslide occurred in the El Dorado National Forest. It is described in several secondary sources.[950, 951]

The soil was derived from weathered granitic material. It had lots of voids that could hold water. In addition, the landslide included rock with some boulders that measured more than 16 feet (5 meters) in diameter.

The highway and the river had undercut the base of the slope, reducing the overall stability of the hillside. The long period of heavy precipitation during the 1983 water year had raised groundwater levels and increased pore-water pressures within the hillside. That, in combination with the removal of the base of the hill, acted together to trigger the landslide.

At 5:10 a.m. on April 9, 1983, a large section of the hillside gave way. The landslide moved rapidly downhill, across the highway, and dammed the river.

Maximum depth of the lake was 50 feet (15 meters). The river began breaching the landslide dam at 11:30 a.m. the next morning. There were enough large boulders in the dam to prevent rapid breaching and downstream flooding. During the following months, the river gradually eroded the dam down to the original riverbed. By June 1983, the area of the lake had decreased to roughly one-third of its original size.

The landslide had an estimated total mass of about 1,000,000 cubic yards (765,000 cubic meters). It took Caltrans 75 days to reopen the highway.

The Mill Creek Landslide would happen just 0.6 miles west of this location on January 24, 1997.

### Landslide: Slide Mountain, Nevada

This event occurred just northeast of Lake Tahoe in the Toiyabe National Forest. It is described in a secondary source.[952]

The winter of 1982–83 was unusually wet and built a record snowpack. A sudden sustained warm period beginning in late May greatly reduced the snowpack and promoted infiltration of water into the subsurface. That increase of moisture content increased local pore pressure in discontinuities and in the unconsolidated surficial deposits covering the bedrock.

At about noon on May 30, 1983, a large section of the hillside gave way. Several types of mass wasting processes were involved, including a rock slump, a rockfall avalanche, and a debris avalanche. The rock slump composed the largest part of the slide and was up to 100 feet (30 meters) thick.

Along the northeastern margin of the landslide, a rapidly moving rockfall avalanche of large boulders and a debris avalanche of gravelly sand entered Upper Price Lake, displacing most of the water in the lake, which breached a low dam. The water then breached the dam of Lower Price Lake and sent a torrent down the gorge of Ophir Creek. This created a debris flow which damaged and destroyed homes, overtopped old U.S. Highway 395, and caused one death.

Total volume of the slide was estimated to be up to 940,000 cubic yards (720,000 cubic meters).

### *Debris Flow: Camp Creek*

The April 10–11, 1982 rain-on-snow storm event was responsible for triggering numerous landslides and debris flows in the Sierra. One of those was a debris flow in Camp Creek, a tributary of the San Joaquin River. That event was analyzed by Jerry DeGraff, a geologist for the USFS.[953] The debris flow began on a 50% slope in a soil composed largely of fine gravel and sand. It flowed about 1½ miles to Mendota Pool reservoir. The Camp Creek debris flow had one of the fastest peak velocities ever recorded for a Sierra debris flow: 16 mph (26 km/hr).

### *Debris Flow: Garnet Dike*

The Garnet Dike debris flow also happened in 1982.[954] It occurred on a tributary of the South Fork Kings River in the Sierra National Forest. It was several miles from Big Creek. That is part of the Kings River Special Management Area. The event was analyzed by Jerry DeGraff.[955]

It was a relatively slow debris flow for the Sierra, only 11 mph (18 km/hr). Despite being a big debris flow, it was able to pass around trees without doing much damage or abrasion to them. It did form a boulder levee on its flanks.

### *Debris Flow: Calvin Crest*

On July 5, 1983, a debris flow occurred on the Sierra National Forest adjacent to the Calvin Crest Conference Center near Oakhurst, California. The event was thoroughly analyzed by Jerry DeGraff.[956]

The debris flow originated at an elevation of 2,500 feet (762 m). It was located on a 30% slope at the broad head of a small drainage basin. The slope had an open stand of mixed oak and Jeffrey pine with an understory of herbaceous vegetation. Where the 30% slope flattened to 10%, there was evidence of seasonal groundwater seepage at a number of points in the vicinity of the debris flow. The area where the debris flow occurred was underlain by granitic bedrock. A few thousand feet upslope was the contact with remnant meta-sedimentary bedrock capping the hill top. Near this contact, a number of seeps and wet meadows were present.

By July 7, the debris flow was about 600 feet (200 m) long and 72 feet (22 m) wide. The deposit had the consistency of very wet cement. Water discharged from the end of the deposit. Groundwater flowed from the upper scarp and other points along the debris flow track. This made the bottom of the track too soft and muddy to be examined more closely. It remained saturated until the following winter. In succeeding years, grass grew over the debris flow scar and flow path. Despite revegetation, it remains fairly wet to damp throughout most of the year.

Usually debris flows are triggered either by a storm event or by melting of a snowpack. But that wasn't the case with Calvin Crest; it was apparently triggered by groundwater conditions resulting from above-average recharge. No precipitation was received during the previous 24 days at the South

327

Entrance of Yosemite National Park, the nearest station to the debris flow. Precipitation totaled only 0.6 inches (15.5 mm) for the 35 days prior to the debris flow. The pre-movement observation of groundwater flow from a depression at the base of the 30% slope suggested high pore-water pressures were present in the slope materials.

### *Debris Flow Complex: Kings Canyon National Park*
During the 1983 Memorial Day Weekend, Kings Canyon experienced three debris flows that we know of:
1. Bubbs Creek
2. Unnamed tributary of Lewis Creek
3. Castle Dome Meadow

These debris flows could be thought of as one event that occurred in multiple locations. The event was triggered by a heavy snowpack and an extreme change in temperatures in a 48-hour period.

We know about these events primarily because of the outstanding memory of Jerry Torres, Kings Canyon National Park's trails supervisor at the time.

The Bubbs Creek debris flow began in two unnamed tributaries of Bubbs Creek, high on the side of Glacier Monument (UTM 365530E 4072600N NAD83 Zone 11, elevation 9,600). From there it flowed south down the steep hillside 1 mile to Bubbs Creek, elevation 6,700. That point was approximately 1½ miles (2½ km) east of the Sphinx Creek Bridge. Aerial photography shows significant scouring all along the flow path. At that point, the debris flow had dropped 2,900 feet in elevation.

That hillside had been burned seven years earlier in the 1976 Sphinx Fire. However, the primary triggering event was probably the large amount of moisture infiltrating into loose soils from the melting snowpack.

Once the debris flow got to the bottom of the hill, it turned and followed the course of Bubbs Creek east. It scoured Bubbs Creek for approximately the first ⅓ mile (0.5 km), taking vegetation including large trees, rock, tons of soil, and a good swath of the trail downstream. That section of Bubbs Creek was not any steeper than the sections farther downstream. However, the debris flow had just come off a very steep hill, so possibly that was the cause of the scouring.

The debris flow created a large earthen dam on its eastern flank. This restricted the downstream flow of Bubbs Creek, creating a very large pool which lasted several years. Today a medium-size pool or slow-water area still exists there. Although the Sphinx Creek Bridge was unscathed by the event, it changed the Bubbs Creek bridge channels, drying two of the channels and increasing the flow in one twofold. There were some large tree jams at and above these bridges as well as areas of the forest above the fourth and third bridges that were toppled by the event.

The Bailey Bridge crossing on the South Fork Kings (elevation 5,000 feet) two miles east of Roads End marked the western extent of the effects from the Bubbs Creek debris flow. The debris flow had a total length of about 3.6 miles (5.8 km) with a drop of 4,600 feet. The average gradient was about 1,300 feet/mile, but that's misleading. The gradient on the mountain section (2,900 feet in 1 mile) was 2,900 feet/mile while the gradient on the Bubbs Creek section (1,700 feet in 2.6 miles) was only about 650 feet/mile.

The second debris flow occurred in the Lewis Creek Basin. The first tributary of Lewis Creek above its confluence with the South Fork Kings experienced a high-energy debris flow. This debris flow began on a steep, sparsely vegetated slope in a remote trailless area of the park (UTM 352609E 4076728N NAD83 Zone 11). This tributary (now informally referred to as Tsunami Creek) had a massive wall of water and debris come barreling down its channel late on Friday morning (May 27, 1983) prior to the Memorial Day Weekend. The wall scoured the channel, depositing mud fifty feet up the trunks of those trees that survived the onslaught of the "tsunami." Bill Tweed recalled that huge logs came crashing down in the debris flow, and that there was a lot of silt in the debris. The presence of large logs in the debris suggests that this event has a long recurrence interval.

It has been speculated that the severity of this debris flow might have been attributed in part to the 1980 Lewis Creek Fire. However, this appears to have been just the coincidence of association. The other debris flows that occurred during this storm event were not associated with previous fires. Moreover, there has never been a debris flow in the national parks which was clearly linked to the effects of fire.[957]

There is a more reasonable explanation for what probably triggered this debris flow. The unusually wet spring presumably soaked the ground to depth. The extreme change in temperatures in a 48-hour period then melted much of the heavy snowpack, promoting further infiltration of the soil. That raised groundwater levels and increased pore-water pressures within the hillside. That reduced the soil's frictional strength, causing the soil mass to begin moving downslope as a flowing mass. That is what triggered the landslide and debris flow at Slide Mountain, Nevada on May 30, 1983.

The Lewis Creek Basin has several areas of loose granitic sand on steep slopes that are sparsely vegetated. A similar high-energy debris flow event would occur in this drainage (but in the main Lewis Creek channel) in the July 15, 2008 flood.

The national parks used to have the main pump house for the Kings Canyon development near Lewis Creek. Maintenance worker Ron Cook was checking the water intake on the bank of the creek on the morning of May 27, 1983, when the debris flood struck. It came upon him so fast that it caught him by surprise, knocking him off his feet. As Ron told the story, he was almost caught up in the maelstrom and killed. He wrapped one arm around a small tree to hold on while he radioed for help with the other.

The debris flood nearly took out the Lewis Creek pump house. (That pump house has since been removed and the Kings Canyon development now gets all of its water from the facility at Sheep Creek.)

The debris flow closed Highway 180 below the Lewis Creek Bridge for most of the day. A bulldozer was used to push Lewis Creek back into its former channel. Remnants of the flood channel are still visible today, just west of the Lewis Creek Bridge on the north side of the road. Once the creek was pushed back, crews removed rocks and several feet of mud and other debris from the roadway.

The third Kings Canyon debris flow of the Memorial Day Weekend occurred at Castle Dome Meadow. It covered a 200-yard section of trail and meadow with decomposed granite and sand.

### *Debris Flow: Redwood Creek*

René Ardesch discovered this debris flow in July 1983 when he was on a cross-country backpacking trip. He later recalled the discovery and the impression that it made on his group:

> *Back in the late spring of 1983 four of us decided to go on a backpack trip to the Castle Rocks in Sequoia National Park. As we were aspiring wanderers we opted to go on a cross-country course with the destination of Pine Top Mountain for the first night. We started our adventure at the road up to the old Camp Conifer below Atwell Mill, which is where we parked and assembled our gear. One of our group's families had a cabin in Camp Conifer long ago and we wanted to see how it looked since all the cabins were removed.*
>
> *We walked with our heavy packs through the grove and checked out some acorn grinding holes onsite. We then walked on into the forest towards our bivy spot for the night. After sometime of up and down hiking we noticed in the distance an odd scene that looked like a large opening in the heavy forest we were in. As we finally came to the edge we all stood together with a sense of awe, just blown away with what we were looking at. A huge path of destruction up to 100 feet wide with whole trees reduced to logs in big piles gathered along the edges and rocks of all sizes everywhere up and down what appeared to be a creekbed (photograph on file in the national parks). We assumed it to be Redwood Creek as we were headed in a westerly direction and the flow was southerly. The land was fairly level here, and we were at an elevation of around 7,000 feet.*
>
> *We put our packs down and tried to gather our thoughts about what it all meant and how it happened. Still in the giant sequoias (in the Redwood Creek Grove), the smell in the air was of fresh, moist soil and the ground was damp all around. In one sandy area we moved to later we saw large cat tracks that we compared to our own foot size. We had to circumvent this area for quite a ways as it was in our direct path and the whole time we were talking about what might have caused this catastrophic event. This one sighting was in our minds for many years to come.*

Redwood Creek is a tributary of the East Fork Kaweah. The area where they encountered the debris flow was relatively flat, so René inferred that they were near the bottom of the run, and that it had started far above them. Circumstantial evidence suggested that the debris had occurred within the previous couple months, perhaps over the Memorial Day Weekend when the other three debris flows occurred in the Kings Canyon area.

## 1984 Floods (5)

There were four periods of flooding in 1984:
1. July (twice)
2. August
3. September
4. Lakebed flooding

On July 15-16, a high-intensity, short-duration thunderstorm produced flood conditions in the Goat Ranch Canyon and Long Canyon areas. This storm followed the 26,000 acre lightning-caused Bodfish Fire that began on July 7. Debris flows and debris blocked Highway 178 and many other roads. Uffert Park was covered by debris flows that were about 6 inches deep. Three houses in the Long Canyon area became completely uninhabitable when debris flows inundated them. A small levee in this location was breached and eliminated by the flood. Debris flows threatened homes in the Bodfish Creek area.[958]

On July 30, an intense thunderstorm occurred in Scodie Canyon, causing flooding in the community of Onyx. The floodwaters overflowed channels and eroded new channels. Thirty mobile homes were washed away, and nine of these were completely destroyed. Stranded residents had to be airlifted out. Damage was estimated to be $3 million. One man was killed by lightning. A state disaster was declared for Kern County on July 31.[959, 960]

An intense storm occurred in the hills east of Lake Isabella on August 20. Two-thirds of an inch of rain fell in just 40 minutes. Scodie Creek (Sometimes incorrectly listed as Sodie Creek) overflowed its banks, flooding the community of Onyx. Four homes were damaged by mud and one home was washed away.[961] This is a different event from the July 30 flood.

Gary Sanger at the NWS forecast office in Hanford researched the two storms in Scodie Canyon. There was abundant monsoonal moisture in Southern California during July 27-30, as reflected by reports of scattered rainfall. However the thunderstorm that occurred in Scodie Canyon on July 30 was considerably more intense than any other storm that was reported during the July 27-30 period. It was a relatively isolated event.

There was also some monsoonal moisture in Southern California during August 19-20, although less than during the July 27-30 period. The storm that occurred in Scodie Canyon on August 20 was considerably more intense than any other storm that was reported during the August 19-20 period. Once again, it was a relatively isolated event. It seems to have been just a bizarre coincidence that Scodie Canyon got hit with two back-to-back huge thunderstorms in this three-week period. You have to wonder what terms area residents used to describe these two events that had been visited upon them.

The July 30 event was reported to be a thunderstorm; we don't know about the August 20 storm event. Gary observed that storms such as this don't necessarily have to be thunderstorms. A nearly stationary storm could also produce very heavy rainfall over a small area, resulting in flash flooding. Conversely, a fast moving, intense storm might spread rain over a much larger area, reducing the impact of the flash flooding.

An intense storm occurred in Lake Isabella on September 19. Over an inch of rain fell in just 45 minutes, washing out ¼ mile of one road, covering others with mud, and destroying two mobile homes.[962]

The Tulare Lakebed flooded in 1984.

This lakebed flooding that occurred in 1984 was completely unrelated to the floods that occurred that year; none of those flood events contributed significantly to the flooding that occurred in the lakebed. Nor did any of the rivers contribute runoff to the lakebed in 1984.

The Tulare Lakebed flooded in the spring of 1982. The floods of 1983 greatly expanded the size of the lakebed flooding. As illustrated in Figure 11, the lakebed was still extensively flooded at the beginning of 1984. The lakebed would not be fully drained until water year 1985.

Lakebed flooding is a social construct; it is counted based on the number of growing seasons that are missed. The lakebed was flooded for three growing seasons: 1982, 1983, and 1984. So this is counted as three floods from the perspective of the lakebed farmers, even though flood events occurred in only two of those years.

Something similar happened in the lakebed in 1969–1971 and 1997–99. In each of those cases, lakebed flooding continued into a non-flood year.

## 1986 Floods (4)

There were three periods of flooding in 1986:
1. February
2. March (twice)
3. April–July snowmelt period

The first storm event lasted from February 11–24. The actual transport mechanism was an atmospheric river that brought phenomenal amounts of precipitation to a large portion of Northern and Central California and western Nevada.[963] *Rivers of Fear: The Great California Flood of 1986* is supposed to be the most comprehensive source of data available for this flood.

A series of four tropical storms pounded the state between February 11–20. Rains from the first three storms saturated the ground and produced moderate to heavy runoff before the arrival of the fourth storm. The heaviest precipitation from those storms was in a band 200 miles north to 100 miles south of a line from San Francisco to Sacramento to Lake Tahoe.[964]

A total of 200 stations reported their highest-ever rainfalls for 10 consecutive days. Half of the average annual rain fell in the 10 days between February 11–20 at 150 stations in the state. Mono Lake had 95% of its annual average rainfall occur during those 10 days. Bucks Lake in the Feather River Basin had 49.44 inches, which was 71% of its average annual rainfall.

Rains from the first storm started the evening of February 11 and peaked the next day. This storm originated in the Pacific just north of Hawaii and brought up to 6 inches of precipitation to the upper Feather River Basin. On February 13, a second storm developed northeast of Hawaii. A strong cold front generated by this storm moved across Northern California on February 14. Gusty winds and heavy rains hit the entire state. Behind this front, a pattern of overrunning (warm moist air flowing over cold air) produced additional rainfall through much of the following day.[965]

On February 15, a strong, deep flow of warm moist air from Hawaii advanced south of California. On February 16, weather satellites showed enormous development along the jet stream between Hawaii and California. Southwest winds of 210 mph were reported in the jet stream. This storm (the third storm), which entered south of California, began moving slowly north as a warm front. North of the warm front, strong overrunning by a deep moist southwest flow began producing heavy rainfall from the North Bay counties to the Sierra. In many areas, this heavy rainfall continued with only brief breaks through February 17. Rainfall of ½ to ¾ inch per hour was common.[966]

Another Pacific weather system (the fourth storm) approached Northern and Central California on February 18. This storm originated well north of Hawaii, and thus was a much colder front in comparison to the previous three storms. The snow level dropped to 5,000 feet for this storm; during the previous storms, the level was about 7,000 feet.[967]

In the Sierra, the storms affected mainly the area from the Feather River Basin in the north to Yosemite on the south. The Sierra stations that received rainfalls in excess of the 1,000-year recurrence interval ranged in a band from Clarks Peak north of Sierra Valley in the Feather River Basin to Calaveras Big Trees in the Cosumnes River Basin in the south.

The heaviest 24-hour rainfall ever recorded in the Central Valley, 17.6 inches, occurred on February 17 at Four Trees in the Feather River Basin, 30 some miles north of Oroville. This broke (just barely) the old record that had been set at Hockett Meadows on December 6, 1966. Four Trees received a total of 56 inches of rain for the month, the greatest February total recorded for any station in the state during 1986.[968]

Calistoga, in the Napa River Basin, had 29.61 inches in 10 days. This represented a recurrence interval of 2,600 years.[969]

Due to the storms' tropical nature, snow levels fluctuated between 7,000 and 8,000 feet. Between February 11–20, more than 34 inches of rain fell at Blue Canyon on the American River east of Grass Valley. Above 8,000 feet, storm-total estimates ranged from 15–20 feet of new snow with 20–30 inches of water content.

The widespread drenching rains led to extensive flooding and mudslides. The floodwaters destroyed many bridges and punched through several levees. This was the flood that caused the big levee failure on the Yuba River at Linda, south of Marysville.

Statewide, more than 50,000 people fled their homes, and 13,000 homes and businesses were either damaged or destroyed. Damage was estimated to be $500 million, 13 flood-related deaths occurred, and 96 were injured.

Flooding was widespread with 23% of streamflow gaging stations in California reporting significant discharges. Flooding was most severe in the northern half of the state. It had a recurrence interval of 100 years on some rivers.

Over much of the area, the precipitation ranged from 100 to 200% of normal February precipitation for the 9-day period from February 11–19. In many rivers and streams, those storms produced either record or near-record flows. At 16 stream gages, the peak flow recorded either equaled or exceeded the previous maximum. A record flow of 640,000 cfs was estimated at the latitude of Sacramento.[970]

The 1986 flood was a record flood on the American River, the fourth record flood in 36 years. The American River dumped more water into Folsom Lake than it was designed to handle. After two days of releases at the maximum design release level of 115,000 cfs, officials were forced to boost releases to 134,000 cfs.

Peak discharge-of-record occurred in the Napa River and upper Feather River Basins. Inflow of the Feather River into Lake Oroville reached a high of 266,540 cfs. Record flood management releases of 150,000 cfs made room for this unexpected volume of water.

The Napa River crested near Napa with a peak discharge of 37,100 cfs; it had a recurrence interval of 75–100 years.

The bypasses for the Sacramento River Basin provided much needed storage and flow capacity during the peak of the flood. Before the mid-February storm systems, overflow at each of the weirs had been minor or nonexistent. By February 17, however, all weirs were flowing and all but one of the weirs continued flowing until the last week of March. The peak flow exceeded the project design flow at three of the weirs.

System breaks in the Sacramento River Basin included two disastrous levee breaks on the Feather River. Levee breaks along the Mokelumne River caused flooding in the community of Thornton and the inundation of four Delta islands. Damages exceeded $172 million and $15 million in the Sacramento River and San Joaquin River Basins respectively.

Much of the San Joaquin River Basin was spared the full impact of the 1986 storms. The major projects for the San Joaquin River Basin did not encroach on their flood-control pool as did their counterparts in the Sacramento River Basin. The exception in the San Joaquin River Basin was Millerton Lake where only 16% of the flood-control pool remained at the end of the February event. [971]

A major frontal storm system crossed the Central Sierra in mid-February 1986. The southern edge of this storm triggered three debris flows on the north-facing slopes of Shingle Hill, near Greeley Hill, California, within the Merced River Basin. These three debris flows are described at the end of this section along with two other debris flows that occurred in 1986.

The Kings River experienced a flood from February 13–19. The peak day natural flow at Pine Flat on the Kings occurred on February 18, 1986: 25,060 cfs. This was a slightly bigger flow than occurred in the much more famous 1983 flood. It seems likely that this was a very high-flow period in Cedar Grove as well.

Damage was much greater in Fresno County than in Tulare.

Jerry Torres and David Karplus (Kings Canyon National Park trails supervisors) recalled that there were a large number of avalanches throughout both Sequoia and Kings Canyon National Parks that spring, causing considerable damage. One of those avalanches pushed the Palisade Creek Bridge off its footings. One of the largest and best known avalanches was the Paradise Valley (in Kings Canyon National Park) "logalanche" which scoured a large swath of Middle Paradise Valley below the Kidd Creek headwaters.

In addition to the avalanches, the floods created several large log jams along the Kings River as well as other water courses in the national parks.

The Kaweah's peak natural flow occurred at Terminus Dam (or possibly McKay's Point) on February 13: 9,852 cfs. (That was the peak hourly flow; the peak average daily flow, as reflected in Table 15, was 9,428 cfs.) Based on the flood exceedence rates in Table 16, this had a recurrence interval of 7 years for the Kaweah.

**Floods and Droughts in the Tulare Lake Basin**
Specific Floods and Droughts

Table 68 summarizes the damage incurred in the southern end of the San Joaquin Valley during the 1986 flood.[972]

Table 68. Damages incurred during February 1986 flood.

| County | Private Damage (thousand dollars) | Public Damage (thousand dollars) | Total Damage (thousand dollars) |
| --- | --- | --- | --- |
| Merced | $70 | | $70 |
| Madera | $210 | $38 | $248 |
| Fresno | $840 | $450 | $1,290 |
| Kings | | | 0 |
| Tulare | $20 | | $20 |
| Kern | | | 0 |
| Total | $1,140 | $488 | $1,628 |

An F0 tornado touched down in Kingsburg on March 7. That tornado was spawned by a strong thunderstorm complex that produced heavy rain over the Southern Sierra, causing flash flooding in Mariposa and Madera counties.[973]

On March 10, an intense thunderstorm struck Fresno during the height of the evening commute. About an inch of rain fell in downtown Fresno resulting in widespread flooding, stranding dozens of cars, some with water up to the rooftops. The deluge flooded basements in a number of buildings in downtown Fresno and caused part of the roof to collapse on a store. Hailstones as large as mothballs fell in nearby farm areas and accumulated up to 4 inches deep. In parts of Biola (west of Fresno), up to 3 inches of hail was still on the ground at noon the next day. Locally heavy rain fell farther south, causing the White River to surge over its banks and flood Highway 98 between Earlimart and Delano.[974]

Flood releases of over 15,000 cfs occurred at Friant Dam on the San Joaquin River during the April–July snowmelt period. This was the second biggest release since the dam was completed in 1942.

Pine Flat Dam on the Kings River recorded the largest 30-day inflow of record during late May and early June of 1986.[975]

Spring snowmelt was heavy enough to cause flooding in the Tulare Lakebed. In order to minimize flooding in the lakebed, 94,000 acre-feet of river floodwater was pumped into the Friant-Kern Canal and routed to the Los Angeles area. For comparison, that is half as much water as the total capacity of the newly expanded Lake Kaweah.

Total flow for water year 1986 was 189% of the 1894–2011 average for the Kings, 189% for the Kaweah, 177% for the Tule, and 197% for the Kern.

335

## *Debris Flow: Shingle Hill*

A major frontal storm system crossed the Central Sierra in mid-February 1986. The southern edge of this storm triggered three debris flows on the north-facing slopes of Shingle Hill, near Greeley Hill, California, within the Merced River Basin. The event was analyzed by Jerry DeGraff, a geologist for the USFS.[976]

The storm precipitation fell in the form of rain. No snow accumulation was present on Shingle Hill which ranges from about 2,099 feet to over 3,100 feet. A rural county road along the base of the hill was open the evening of February 17. On the morning of February 18, deposits were blocking the road at two locations. A third deposit was found in an ephemeral channel a few tens of feet up-gradient from the road. The debris flows occurred the night of February 17 or early on the morning of February 18, a time which coincided with a high intensity rainfall period. The slopes ranged from 50%–65%.

Two other significant debris flows occurred in 1986:[977]
- Wolfin debris flow (Tuolumne River Basin). This debris flow began on a 60% slope adjacent to an existing intermittent channel. The mass entered perpendicular to the direction of the channel and immediately began moving down-channel in a clear indication that remolding into a flowing mass took place. It had one of the fastest peak velocities ever recorded for a Sierra debris flow: 15.5 mph (25 km/hr).
- Minarets Highway debris flow (San Joaquin River Basin). It had one of the slowest peak velocities ever recorded for a Sierra debris flow: 6 mph (9 km/hr).

## 1987–92 Drought

This drought began over most of the state in 1987. However, parts of the state were in drought from 1984–93.

The recurrence interval of this drought in the Sacramento River Basin was approximately 70 years based on the 1906–92 record. On the San Joaquin River, where the drought was more severe, the recurrence interval was approximately 300 years.[978] These statistics reflect both the six-year length and the severity of the drought.

A significant portion of the country experienced drought conditions during the general period that California was in drought: 1987–92. By June 1988, 54% of the contiguous U.S. was in drought condition.[979] The drought of 1988 became the worst drought in the U.S. since the Dust Bowl 50 years earlier. Not until 2012 would the U.S. see a drought this extensive. The drought of 1988 remains the costliest U.S. natural disaster ever. Hurricane Katrina ranks second and Hurricane Andrew third.[980]

Table 69 compares the 1987–92 drought with the other severe droughts of the 20th century. Both the 1929–34 and the 1976–77 droughts had a recurrence interval of more than 100 years, at least by some measures.

Table 69. Comparison of selected 20th century droughts.

| Drought Period | Sacramento Basin Runoff | | San Joaquin River Basin Runoff | |
|---|---|---|---|---|
| | Average yearly runoff (million acre-feet) | % of average 1901–2009 | Average yearly runoff (million acre-feet) | % of average 1901–2009 |
| 1929–34 | 9.8 | 56% | 3.3 | 56% |
| 1976–77 | 6.6 | 38% | 1.5 | 26% |
| 1987–92 | 10.0 | 57% | 2.8 | 48% |

The 1987–92 drought was notable for its six-year duration and the statewide nature of its impacts. In 1991, the single driest year of the drought, the State Water Project terminated deliveries to agricultural contractors and provided only 30% of requested urban deliveries. The federal Central Valley Project provided 25–50% supplies to urban contractors and 25% to agricultural contractors.

At that time (1991), 23 of the state's 58 counties had declared local drought-related emergencies. Many of the declarations were prompted by economic impacts associated with loss of dryland cattle range, damage to timber resources and associated wildfire damage, and diminution of water-based recreational and tourism activities, rather than by shortages of developed water supplies.

Extremely dry periods frequently last more than one year. However, based on the historical record, long droughts exceeding three years are relatively rare in our area. The only exception for the Tulare Lake Basin since 1850 was the 1929–34 drought.

Table 70 shows a comparison of multi-year droughts for the Sacramento River Basin since 1560.[981]

Table 70. Sacramento River multi-year droughts.

| Period | Length (in years) | Average Runoff (million acre-feet) |
|---|---|---|
| 1579–82 | 4 | 12.4 |
| 1593–95 | 3 | 9.3 |
| 1618–20 | 3 | 13.2 |
| 1651–55 | 5 | 12.3 |
| 1719–24 | 6 | 12.6 |
| 1735–37 | 3 | 12.2 |
| 1755–61 | 6 | 13.3 |
| 1776–78 | 3 | 12.1 |
| 1793–95 | 3 | 10.7 |
| 1839–41 | 3 | 12.9 |
| 1843–46 | 4 | 12.3 |
| 1918–20 | 3 | 12.0 |
| 1929–34 | 6 | 9.8 |
| 1959–62 | 4 | 13.0 |
| 1987–92 | 6 | 10.0 |

Table 71 shows how the San Joaquin Valley Water Year Index categorizes the drought years in that basin during the 1987–92 drought.

Table 71. Rating of drought severity during the 1987–92 drought.

| Water Year | San Joaquin Basin Water Year Classification |
|---|---|
| 1987 | Critically dry |
| 1988 | Critically dry |
| 1989 | Critically dry |
| 1990 | Critically dry |
| 1991 | Critically dry |
| 1992 | Critically dry |

The 1987–92 drought was the worst extended sequential critical drought during recorded history in the Kings River Basin. However, based on tree-ring analysis, it was not as severe as the 1580 drought.

Bob Meadows was the wilderness ranger at Ranger Lake in 1989 and 1990 and recalled the effect of the 1987–92 drought on that lake. Ranger Lake has an extremely small watershed and almost no regular inflow. However, during the two years that Bob was there, the lake dropped no more than about a foot or so.

Total flows for water years 1987–92 generally ranged from 30%–60% of the 1894–2011 average for each of the four rivers within the Tulare Lake Basin.

As a result of the 1987–92 drought, the Tulare Lakebed was largely dry from 1987 through 1994. This is the longest period that the lakebed has remained dry since the drought years of 1922–34. During that series of droughts, the lakebed was largely dry from 1924 until February 7, 1937.

In 1991, the "March Miracle" brought abundant snow to the middle and upper elevations of the Sierra. Lodgepole received a total of 147 inches of snow, the third greatest monthly snowfall at that location since record-keeping began.

On February 15, 1992, a strong winter storm lowered the snow level to about 2,000 feet in the Tehachapi Mountains. That was one in a series of storms that dumped a total of 17.32 inches of precipitation on Frazier Park during the month. The cause was an inflow of subtropical moisture that moved over the mountains from the south.[982]

This brought relief to portions of Kern County. However, the drought was generally considered to continue through most of calendar year 1992. A series of major Pacific storms brought abundant moisture to the state between December 1992 and February 1993. As a result, Governor Pete Wilson declared the drought to be officially over on February 24, 1993.

### Ash Mountain Pasture

The national parks' Ash Mountain Pasture has experienced six multi-year drought since it began being used in 1921:
1. 1922–34
2. 1943–51
3. 1959–62
4. 1976–77
5. 1987–92
6. 2007–09

The pasture was badly damaged by drought and overuse during the 1922–34 drought. Conditions were so bad by 1934 that the park was seriously considering killing some of its livestock.

The parks' stock are used to support wilderness operations during the summer. But in the winter, the stock have to be brought back to lower elevation pasture. In the early years, the parks' stock were kept on the Ash Mountain Pasture during the winter. But beginning in about 1970, the national parks began sending most or all of their stock outside the parks during the winter whenever they could.

From about 1975 to the present, most of that winter pasture has been on the Horse Pasture Unit at the Pixley National Wildlife Refuge. Pixley manages their Horse Pasture Unit for a particular conservation objective: maintaining average residual dry matter of 800 pounds per acre at the beginning of summer. This is done for the benefit of two threatened and endangered species that live on this pasture: the blunt-nosed lizard and the Tipton kangaroo rat.[983]

This partnership between the parks' stock and the refuge's conservation objective worked reasonably well until the 1987–92 drought. After the first two dry winters (1987–88 and 1988–89), Pixley's managers informed the national parks that they would have to discontinue putting their stock on the refuge for a while; the refuge Horse Pasture Unit could meet its residual dry matter objective without any grazing. That was because Mediterranean grasses don't grow when the rains don't come.

The parks weren't allowed to put any stock on the refuge during the winters of 1989–90, 1990–91, or 1991-92. As a result, the parks apparently kept most or all of their stock on the Ash Mountain Pasture during those three winters. By the second year (1990–91), the pasture was so depleted that the parks had to purchase a very large amount of supplemental feed to get the stock through the winter.

The drought finally broke in December 1992, and Pixley allowed the parks to bring most of its stock back onto the refuge that winter.

### 1988 Flood

This flood occurred during the 1987–92 drought.

The winter of 1988–89 was a strong La Niña event.

Flood releases of 8,000 cfs or greater occurred at Friant Dam on the San Joaquin River, supposedly during the April–July snowmelt period in 1988. That is difficult to explain. Total flow for the water year in the Tulare Lake Basin was less than 50% of the 1894–2011 average, and there was no report of flooding in the basin. *Possibly* the flood release was related to an unusual storm event that occurred in January.

A very intense storm passed through California on January 17, associated with high winds and surf. Several deaths occurred when people became snow-bound in the mountains of Southern California. A 7-foot tide combined with a 15–20 foot surf caused an estimated $50 million in damage to coastal Southern California. Tornadoes were reported in Orange County.

This storm moved out of the Gulf of Alaska and developed into a violent cyclone when it came ashore near Aleva Beach at 1 p.m. on January 17. All-time low barometric pressure was recorded at several Southern California weather stations as the storm moved onshore about 20 miles north of Santa Barbara. Table 72 gives the total precipitation received during the January 17 storm event for selected reporting stations.

Table 72. Total precipitation during the January 17, 1988 storm event.[984]

| Reporting Station | Total Precipitation (inches) |
| --- | --- |
| Turlock | 1.10 |
| Modesto | 1.69 |
| Newman (northwest of Los Banos) | 4.10 |

Newman's 92-year average annual rainfall is 10.27 inches. The 4.10 inches that Newman received on January 17 was 6.67 standard deviations above the average extreme annual storm. The associated recurrence interval is about 20,000 years. It appears that a local thunderstorm was embedded in the larger statewide storm that hit Newman on January 17.[985]

It isn't clear what effect this storm had on the Tulare Lake Basin. Storms such as this often result in localized flooding. The flow of the Kings and Kaweah Rivers roughly doubled on January 18, so there was apparently a strong rain in the northern part of the basin on that day. We haven't found any records to indicate whether this storm caused any localized flooding in that part of the Tulare Lake Basin. But even if it didn't, the story merits inclusion in this document as an example of how intense a storm can be in Central California.

## 1991 Flood

Flooding in 1991 occurred in early March. This flood occurred during the fifth year of the 1987–92 drought.

The peak day natural flow at Pine Flat on the Kings occurred on March 4, 1991: 13,078 cfs. This was a sharp peak, 13 times bigger than the flow of the previous day. It seems likely that this was a very high-flow period in Cedar Grove as well.

The March 1991 storm event caused a debris flow at El Portal. The event was analyzed by Jerry DeGraff, a geologist for the USFS.[986] Although El Portal is north of the Tulare Lake Basin, it's worth including this event as an example of how a community can prepare for anticipated debris flows.

This small community was originally established in the Merced River canyon to serve the needs of workers on the local railroad and those working at timber harvest and mining. More recently, it provides residences for employees of Yosemite National Park and its concessionaires. In August 1990, a major wildfire burned parts of the national park and the adjacent Stanislaus National Forest. The burned area included the small, steep watersheds which empty into El Portal. A routine

340

assessment of possible landslide hazard resulting from loss of vegetation was carried out. Previous landslide mapping indicated a large number of past debris flow scars present on the slopes of the burned watersheds. Several closed debris basins were constructed on two drainages where the severity of vegetation loss from the fire, indications of past debris flow activity, and presence of houses at the mouths of these ephemeral channels represented a high risk for debris flow damage.

A major storm occurred in the Sierra from February 27 – March 4, 1991 which triggered debris flows from the drainages above El Portal. No precipitation was received in the El Portal area for 21 days prior to this storm. After several days of significant rainfall, the greatest amount of daily precipitation was received on March 3. Interviews with residents of El Portal disclosed that a period of intense rainfall occurred shortly after 11:00 p.m. on Sunday, March 3. The time was firmly established by a number of residents who watch the late television news. At that time water and debris was seen spilling from the closed basin at Chapel Lane. A total of 100 cubic yards of debris was trapped in that basin.

Smaller debris flows occurred on two other drainages leading into El Portal at the same time that the basin at Chapel Lane trapped its debris flow. At these drainages, the site conditions had not permitted construction of debris basins. At one location, water, mud, and occasional cobbles flowed against a house and passed between the house and an outbuilding. Fifteen minutes earlier, only water was seen flowing through this location. The debris flow was described by the residents as swift enough to "carry away a small child." A few tens of meters to the east, residents at another house felt vibrations that seemed greater than the impact of the intense rainfall. Turning on their outside lights, they saw water and debris issuing from a channel and flowing across their backyard. A deposit about one foot thick was formed against the back of their house.

The debris flows at El Portal illustrate the potential for damage to property and threat to life which exists in the Southern Sierra. The Chapel Lane debris basin built as part of burned area emergency rehabilitation, and the small volume of the debris flows from the other two drainages kept costs to a minimum.

## 1993 Floods (2)

Two floods occurred in 1993, both in January.

January was a very wet month in the Southern Sierra, at least in parts of it. A foot of snow fell in Yosemite Valley on January 28, bringing the monthly total for that site to 175 inches (14.6 feet), a record for any month.[987]

An energetic series of storms swept through Southern California during January 5–19, 1993. High wind and tornadoes were associated with this storm sequence. Extensive flooding occurred in Southern California during this period. The greatest 15-day rainfall totals of record occurred at 132 stations during this storm. Ten stations reported rainfall totals in excess of a storm with a recurrence interval of 1,000 years. Total property damage was $600 million and 20 lives were lost due to flooding.[988] While the great majority of those storms were to our south, some appear to have affected the Tulare Lake Basin.

Horse Creek Dam in Sand Canyon in the Tehachapi Mountains failed on January 9, causing localized flooding.[989]

341

On January 13, a series of winter storms brought between 1 and 2 inches of rain, flooding numerous farm fields in Fresno County. Several houses between Fresno and Madera were flooded with water up to 3 feet deep and numerous roads were flooded. A debris flow occurred on Highway 33 near Coalinga. A levee collapsed north of Orosi.

Total flow for water year 1993 was 147% of the 1894–2011 average for the Kings, 127% for the Kaweah, 100% for the Tule, and 115% for the Kern.

## 1995 Floods (4)

There were four floods in 1995:
1. January (3)
2. March

The winter of 1994–95 was a moderate El Niño event.

Statewide, water year 1995 had the third-highest rainfall total in historic times. It was exceeded only by rainfall totals for the years 1890 and 1983. A total of 100 stations reported their highest-ever water year total precipitation. Thirty stations reported over 100 inches for the year; most of those stations were located in the Feather and Yuba River Basins.[990]

Table 73 shows the total precipitation at two mountain stations in the San Joaquin Valley for water year 1995.[991]

Table 73. Total precipitation in water year 1995.

| Reporting Station | Total Precipitation (inches) | Recurrence Interval (years) |
|---|---|---|
| Panoche 2 W near the crest of the Coast Range | 21.36 | 1,900 |
| Florence Lake in the Sierra | 50.29 | 1,300 |

Historically, years of large rainfall totals were not necessarily years of heavy flood-producing rainfalls. However 1995 was somewhat of an exception as there were numerous periods of robust rainfall activity throughout the state.

The storms of January 1995 extended from Humboldt County in the north to Riverside County in the south. They caused a total of 740 million dollars in damage along with 17 deaths. Extensive debris flows occurred in Santa Barbara County.[992]

The flooding in early January was attributed to a series of two storms originating 500 miles north of Hawaii. The first storm front arrived on January 6. That two-day storm produced moderate precipitation totals in Northern California. The second, and more severe, storm front arrived on January 8 and remained over Northern California through January 10. The evening of January 9–10 brought record rainfall to the already saturated floor of the Central Valley. Sacramento set a new rainfall record, receiving 4.45 inches within a 24-hour period.[993]

Rainfall in December 1994 was just slightly below average, and early January 1995 was well above average for most of the Sacramento River and San Joaquin River Basins. By January 7, Sacramento had received 10 inches of rainfall compared to the average 8 inches.

Record-breaking rainfalls occurred during the 6 days from January 7–12 on the west side of the Sacramento Valley. A total of 50 stations reported their greatest-ever six-day total rainfall. Cobb in the Clear Lake Basin received 35.18 inches in 6 days. Greenville in the Feather River Basin received 30.50 inches in 6 days, which is a recurrence interval of 2,400 years. The main precipitation for this storm series was located in a band extending from Clearlake northeast to the Lake Almanor Region. Another band of high rainfall extended from Whiskeytown north to the McCloud region in the Upper Sacramento River Basin.[994]

The January 10 storm events were embedded in the January 7–12 storm. They occurred almost simultaneously in Sacramento and in Kern County. Needless to say, there were a lot more gages to record the Sacramento event.

On January 10, a major storm event occurred northeast of Sacramento. The peak 24-hour rainfall for this storm was 7.57 inches at the Granite Bay Country Club rain gage. That peak 24-hour storm consisted of three separate rainfall sequences: the first from about 7–11 p.m. on the 9th, the second and heaviest from 4–8 a.m. on the 10th, and another burst of rain from about 1–5 p.m. Twelve Sacramento area stations reported over 5 inches of rain in one day. Based on the 28-year rainfall record available for Rancho Cordova, the recurrence interval for this storm was 4,000 years. The January 10 storm fell on saturated ground; it was preceded by 8 days of rain. High antecedent rains preceding record rainfalls resulted in devastating flooding in the Sacramento area centered on Linda Creek which flows through Roseville and Rio Linda.[995]

On January 10, heavy rain of up to 4 inches caused creeks to swell and washed out several roads in Kern County near Frazier Park and Highway 66 near Maricopa.[996]

On January 24, strong thunderstorms moved through the Central California interior, causing flooding in Lamont.[997]

On January 25, Kern County was drenched by heavy rain. Up to 5 feet of water surged out of Caliente Creek, washing out roads. Parts of Interstate 5 flooded. Numerous crops were damaged in Arvin, and up to 30 chickens drowned in Loraine.[998]

Panoche/Silver Creek west of Mendota flooded sometime in January 1995.[999]

A much stronger than normal Pacific jet stream was displaced well south of its normal position during much of the winter and early spring of 1995 due to El Niño conditions in the Pacific. This forced major moisture-laden storm systems directly into California, 15 to 20 degrees south of their normal locations. During January and March, the state was struck repeatedly by very strong storm systems laden with Pacific moisture.[1000] Flood damages exceeded $498 million for the Central Valley.

Both January and March showed much above-average precipitation over most of the state. Since most of the storms occurred within relatively cool, unstable air masses, much of the precipitation above elevation 5,000 feet accumulated as snow. Water content of snowpack exceeded 150% of average in much of the Sacramento River Basin and Sierra at the end of March.[1001]

As of January 7, most of the major reservoirs in the Sacramento River and San Joaquin River Basins were less than half full and only 75% of normal after the 1987–92 drought and the relatively dry 1994 water year.[1002]

None of the major reservoirs in the Sacramento River Basin greatly infringed on their flood-control pool during the January 1995 floods. The major reservoirs in the San Joaquin River Basin experienced similar operations with over 70% of the flood-control pool remaining in all the reservoirs after the January event.[1003]

Runoff from major Sierra rivers during the January flood was mostly stored by reservoirs. Most of the flooding occurred on small streams.

The January storms more severely affected Northern California, while the March storms concentrated more of their impact on Central and Southern California. During March, most locations in the southern San Joaquin River and Tulare Lake Basins received several times their average March precipitation, as illustrated in Table 74.[1004]

Table 74. Total precipitation during March 1995.

| City | Percent of Average |
|------|--------------------|
| Bakersfield | 326% |
| Coalinga | 603% |
| Five Points | 474% |
| Fresno | 311% |
| Hanford | 356% |
| Visalia | 397% |

The heaviest March rainfalls occurred mainly between March 9–10. There was 1.1 billion dollars in property damage attributed to this storm sequence and 16 deaths.[1005]

Three days of soaking rain from March 9–11 resulted in $146.8 million in damage to crops across interior Central California. Mendota was hard hit where many roads and poor drainage areas flooded and gusty winds toppled trees and knocked out power. Highways 140 and 41 to Yosemite National Park were closed due to water, rocks and debris on the roads.[1006]

The major Central Valley reservoirs had less flood control space to handle the March flood than they had for the January flood. However, runoff from major Sierra rivers was still mostly stored by the reservoirs. Millerton Reservoir in the San Joaquin River Basin had less than 5% of its flood-control pool remaining during the peak of the March event.[1007]

The March storm brought considerable precipitation to the Coast Ranges that borders the west side of the Tulare Lake Basin. It was a major event in the Coast Ranges. Highway 1 was closed by a landslide for approximately a week.[1008]

The highest-ever flood stages were reported on the Salinas River at the Spreckles Highway Bridge. Upstream on the Salinas River, four stations recorded their highest-ever 24-hour rainfall. Paso Robles had a total of 7.40 inches. This event had a recurrence interval of about 1,100 years at Paso Robles.

The March storm on the upper Salinas River spilled over the Coast Ranges into the San Joaquin Valley near Coalinga. Coalinga received 3.74 inches of rain in 24 hours on March 10, breaking that city's previous record of 2.53 inches set in 1914. Since the average annual rainfall for Coalinga is only 7.85 inches, the city received nearly 50% of its average annual precipitation in a 24-hour period.[1009] The recurrence interval for the Coalinga rain in this storm was 2,400 years.

344

Kettleman Station and Westhaven also recorded their highest-ever 24-hour rainfalls during this storm event.

Fresno tied October 5, 1904 for its wettest calendar day when 2.38 inches of rain fell on March 10.[1010]

High flows occurred on some of the Tulare Lake Basin west side tributaries. Panoche/Silver Creek west of Mendota flooded sometime in March 1995.[1011] Arroyo Pasajero (aka Los Gatos Creek) flows from Coalinga east toward Lemoore. On the evening of March 9 (sometimes reported as March 10), extremely high flows in Arroyo Pasajero collapsed the two Interstate 5 bridges near Coalinga, killing seven people. The peak flow in the arroyo was 33,000 cfs, delivering a flood volume of 33,500 acre-feet. That was an even larger flood than had occurred at this location in the 1969 flood event. This was one of the three largest flood events to occur in the Coalinga area during historic times.

Huron is located about 15 miles east of Coalinga on Highway 269 (Lassen Avenue). (Trivia question: In the 2000 Census, Huron had the highest proportion of Hispanics of any city in the United States.) Prior to 1995, Highway 269 was closed an average of 26 days a year due to flooding caused by Arroyo Pasajero. Each time that highway closes, residents of Huron have to drive an additional 20 miles each way when they travel to Fresno. In 1995, Highway 269 was closed for 72 days.

The peak day natural flow at Pine Flat on the Kings occurred on March 11, 1995. The flow on March 11 was nearly five times higher than the flow on the previous day. It was a slightly bigger flow than occurred in the much more famous 1983 flood. It seems likely that this was a very high-flow period in Cedar Grove as well.

The Kaweah's peak natural flow occurred at Terminus Dam (or possibly McKay's Point) on March 11: 12,714 cfs. (That was the peak hourly flow; the peak average daily flow, as reflected in Table 15, was 8,369 cfs.) Based on the flood exceedence rates in Table 16, this had a recurrence interval of 6 years for the Kaweah.

There is a remarkable similarity in rainfall distribution between the March 1995 storm and the February 1978 cyclic storm which dumped record rainfalls in an area to the south of the area affected by this storm. The 1978 storm produced large rainfalls on the windward slopes of Ventura County and then continued over into the rain shadow area in the Buena Vista Lake region.

The March 1995 storm behaved in a similar manner. It appeared to be a cyclic storm since it produced devastating rainfalls on the windward slopes of the Coast Ranges. It was still quite energetic as it moved into the rain shadow area to create further devastating floods.

A similar cyclic storm came ashore near Monterey Bay on September 11, 1918, resulting in extreme rainfalls at Antioch, again in the rain shadow zone. The 1918 storm was caused by the remnants of a tropical hurricane which originated off the southwest coast of Mexico.[1012]

As a result of the March 1995 flood, President Clinton declared 39 California counties disaster areas.

1995 was one of the two wettest years ever at Paso Robles, the other being 1941.

Flood releases of 12,000 cfs or greater occurred at Friant Dam on the San Joaquin River during the April–July snowmelt period.

Table 75 summarizes the damage incurred in the southern end of the San Joaquin Valley during the 1995 flood.[1013]

Table 75. Damages incurred during 1995 flood.

| County | Private Damage (thousand dollars) | Public Damage (thousand dollars) | Business Damage (thousand dollars) | Agricultural Damage (thousand dollars) | Total Damage (thousand dollars) |
|---|---|---|---|---|---|
| Merced | | | | $38,854 | $38,854 |
| Madera | $160 | $1,300 | $10 | $829 | $2,299 |
| Fresno | $80 | $300 | $10 | $20,846 | $21,236 |
| Kings | | | | $2,484 | $2,484 |
| Tulare | | | | $48,515 | $48,515 |
| Kern | $10 | $1,900 | $10 | $21,046 | $22,966 |
| Total | $250 | $3,500 | $30 | $132,574 | $136,354 |

Total flow for water year 1995 was 200% of the 1894–2011 average for the Kings, 201% for the Kaweah, 181% for the Tule, and 182% for the Kern.

Flooding occurred in the Tulare Lakebed; this was the first significant flooding since 1986 (see Figure 11). Katrina Young recalled that parts of Manning Road were under nearly two feet of water where it crossed the normally dry lakebed.

In order to minimize flooding in the lakebed, 13,000 acre-feet of river floodwater was pumped into the Friant-Kern Canal and routed to the Los Angeles area.

## 1996 Floods (2)

There were two floods in 1996:
1. April
2. May

An F0 tornado touched down 9 miles west of Fresno on April 16. Small hail that fell in association with this thunderstorm caused $600,000 in crop damage, mainly to grapes. Heavy rain fell in Fresno, stranding motorists in cars. One report had as much as 0.73 inches of rain falling in just 25 minutes.[1014]

An intense storm struck Yosemite on May 16. The resulting heavy rain-on-snow event caused the Merced River to flood in Yosemite Valley. Over $2 million in flood damages occurred.[1015] On May 26, the Merced peaked at Happy Isle at 5,900 cfs. By Merced River standards, that's relatively big, having a recurrence interval of 15 years. Only the floods of 1937, 1950, 1955, 1964, and 1997 have been bigger.[1016]

The Kings River also flooded in May. Peak flow at Pine Flat was 28,705 cfs on May 17. This was more than twice the flow of the day before, suggesting that it was caused by a heavy rain event. This was a relatively minor flood by Kings River standards. There have been 29 bigger floods since record-keeping began.

346

## 1997 Floods (4)

Flooding occurred three times in 1997:
1. January (twice)
2. July
3. September

For a month that was neither an El Niño nor a La Niña, January 1997 saw a lot of damage. The first January flood is sometimes referred to as the New Year's Day Flood of 1997. By some accounts, this flood was the largest and most extensive flood disaster in the state's history. It was the second costliest in California's history.

Statewide, the impacts were:
- 2 deaths, 50 injuries
- 120,000 people displaced by flooding
- $1.6 billion in damages
- 20,000 homes and 1,500 businesses destroyed or damaged
- Disaster areas were declared in 43 counties

Damage to urban and agricultural lands and the cost to replace, restore, and rehabilitate flood damage reached $524 million in the Central Valley.

As a result of severe storms and flooding, a major federal disaster (DR-1155) was declared on January 4, 1997, for the period December 28, 1996 – April 1, 1997. It covered 49 counties including Fresno, Kings, and Tulare.[1017]

The flood resulted from a relatively short-duration, high-intensity storm. It was derived from a very warm area of ocean just west of Hawaii. The convection and atmospheric steering resulted in a convergence of cold arctic air and vast tropical moisture. The entire average water year's precipitation was received by the end of January. The storm lasted from December 29, 1996 – January 4, 1997. The actual transport mechanism was an atmospheric river.[1018]

Early winter rainfall was well above average throughout the Sacramento River and San Joaquin River Basins. In the Northern Sierra, total December precipitation exceeded 28 inches, making it the second wettest December of record, exceeded only by the 30.8 inches in December 1955.[1019]

The heaviest rainfall fell along the coastal mountains and the Northern Sierra. Over 50 recording stations recorded their historical one-day precipitation totals during this storm. Precipitation was heavy throughout Northern California, with many stations reporting 15–30 inches of precipitation during the nine-day period between December 26 –January 3. Some stations in the Feather River Basins received over 40 inches during that period. Over 14 inches of rain fell in a 24-hour period at Four Trees, north of Oroville.

The flooding resulted from three subtropical storms. Over a three-day period, warm moist winds from the southwest blowing over the Sierra poured more than 30 inches of rain onto the already saturated watersheds. The first of the storms hit Northern California on December 29, 1996. The second storm arrived on December 30. The third and most severe storm hit late on December 31 and lasted through January 2.[1020]

Precipitation totals at lower elevations in the Central Valley were not unusually high, in contrast to the extreme rainfall in the upper watersheds. For example, Sacramento received 3.7 inches of rain while Blue Canyon (at elevation 5,000 feet on the American River east of Grass Valley) received over 30 inches of rainfall, thus providing for an orographic ratio of 8 to 1. A typical storm for this region would yield an orographic ratio of about 3.5 to 1 between those two locations.[1021] "Orographic ratio" is the contrast between mountain and lowland precipitation, which can be expressed as the ratio of the precipitation at those locations.

In addition to these three subtropical storms, snowmelt also contributed to the already large runoff volumes. Several days before Christmas 1996, a cold storm from the Gulf of Alaska brought snow to low elevations in the Sierra foothills. The low-elevation snowpack that formed had a high water content (five inches at Blue Canyon) and that portion below about 6,000 feet in elevation melted when the three warmer storms hit. The effect of the snowmelt contributed approximately 15% to runoff totals.[1022]

George Durkee (national park wilderness ranger) recalled that the rain did not melt the snow very high in the Kings River Basin in the January 1997 event. There was not much evidence of flooding above about 7,000 feet (Cartridge Creek or so). He was skiing in Dusy Basin on really bad sun cups into very early July. (Sun cups are depressions made as snow melts under intense sunshine in a dry atmosphere.)

At the beginning of December 1996, 100% of the flood control space was available. But by Christmas, much of that space was already in use.

Record flows were recorded on rivers throughout California, but particularly on rivers in the Sacramento River and San Joaquin River Basins.[1023] Record peak river flows were recorded at 106 gaging stations. Multiple levees on the Sacramento and San Joaquin Rivers broke due to extremely high runoff from melting snow and heavy rainfall.

The North Fork Feather River crested at Grizzly Creek with a peak discharge of 115,000 cfs; it had a recurrence interval that was greater than 100 years. The Feather River fish hatchery was virtually destroyed.

This was a record flood on the American River, the fifth record flood in 46 years. Folsom Lake on the American River experienced a peak inflow of 255,000 cfs.

The Sacramento River crested at Delta with a peak discharge of 62,300 cfs; it had a recurrence interval of 50–75 years.

The Cosumnes River crested at Michigan Bar with a peak discharge of 93,000 cfs; it had a recurrence interval that was greater than 100 years.

The San Joaquin River crested near Auberry with a peak discharge of 99,200 cfs; it had a recurrence interval of 75 years. The San Joaquin River fish hatchery was virtually destroyed.

This was a major flood on the Tuolumne River.

On January 2, a drowning fatality occurred when a vehicle was swept from a roadway through the Chowchilla River.

On January 1, a series of thunderstorms moved into Yosemite National Park, resulting in major flooding in Yosemite Valley from the combination of heavy rain and melting snowpack.[1024] On January 3, the greatest flood on record occurred in Yosemite Valley. Extensive damage occurred to national park trails, roads, sewer and water systems and housing totaling 178 million dollars in damage there alone.[1025] On January 2, the Merced peaked at Happy Isle at 10,100 cfs. It had a recurrence interval of 89 years; the Merced has not experienced a 100-year flood during historic times.[1026]

As of December 1, 1996, most of the major reservoirs in both the Sacramento River and San Joaquin River Basins were at normal flood control levels (100% of the flood control space was available). Despite this, the San Joaquin River flood management system was pushed beyond its limits during the 1997 flood. Millerton Lake and Don Pedro Reservoir, two of the major projects in the San Joaquin River Basin, exceeded their design capacity.[1027]

The January flood caused significant flooding in the San Joaquin Valley as well as the adjacent foothills. Numerous houses adjacent to the San Joaquin River flooded, while agricultural lands near the Merced River were inundated. Flooding also impacted areas in the South Valley, especially Earlimart and Porterville.[1028]

The record flows stressed the flood management system to capacity in the Sacramento River Basin and overwhelmed the system in the San Joaquin River Basin. Flood storage behind dams reduced flood flows by half or more. However, levees were overwhelmed in some areas. Levees on Sacramento River tributaries sustained three major breaks. The San Joaquin River levee system failed in 36 places and was extensively damaged throughout its length, resulting in widespread flooding.

The San Joaquin River had peak flows upwards of 90,000 cfs. As a result, flood releases of 60,000 cfs or greater occurred at Friant Dam. This was five times bigger than any other flood release since the dam was completed in 1942. Runoff exceeded the flood control capacity of the Don Pedro Reservoir on the Tuolumne River and Millerton Lake on the San Joaquin River.

In the Tulare Lake Basin, the crest of the flood passed through the mountains and upper foothills very late on the night of January 2. The deltas and valley floor felt the brunt of the flood impact on the following day. The peak flow on the valley floor for many rivers occurred late on the night of January 3. The impacts of the flood were documented by the NWS forecast office in Hanford.[1029]

George Durkee and David Karplus (Kings Canyon National Park' wilderness ranger and trails supervisor, respectively) think that it was the January 1997 flood that altered the streambed of the Middle Fork Kings so markedly. It blew out the many logjams from Grouse Meadow down to the confluence with the South Fork, allowing kayakers to successfully attempt that stretch of the river. It also altered the section of the Kings River below Tehipite Valley, changing the riverbed to more boulders when it was hiked from then on.

Flows were very high on the South Fork Kings in Cedar Grove as well. Significant quantities of fill and riprap were required to repair damage done to the left embankment of the Cedar Grove Bridge during this flood. That damage was similar to what occurred during the 1955 flood. A seven-mile section of Highway 180 from Boyden Cave to the national parks' boundary was also badly eroded.

The Highway 180 Boyden Bridge survived the flood, but about 100 yards of the Grant Grove approach was washed out. This was the same thing that happened in the 1955 flood. There was also bridge damage near Big Creek. Total highway damage in the canyon was estimated to be $1.8 million.

The Kings' peak natural flow occurred on January 3: 112,000 cfs. (That was the peak hourly flow; the peak average daily flow, as reflected in Table 15, was 50,217 cfs.) According to the Kings River Handbook, that peak hourly flow of 112,000 cfs was the peak flow of record on the river.[1030] Flows during the 1867–68 flood would have been greater, but no estimate of those flows has ever been calculated.

To give a sense of just how big 112,000 cfs is, the Kings River divides in its lower reaches into distributaries: the South and North Forks. The total channel capacity of those two rivers, measured at Highway 41 north of Lemoore, is about 8,600 cfs.

Based on the flood exceedence rates in Table 16, the 1997 flood had a recurrence interval of 40 years for the Kings River at Pine Flat. That puts this flood in a category with other Cedar Grove floods of the past 70 years (1937, 1950, 1955, 1966, 1969, 1978, 1982, 1983, and 1997) that rise to the level of the modeled 50-year flood: that is, a flood event that occurs about every eight years on average. See the section of this document that describes Cedar Grove Flooding.

Large scale flooding in the Tenmile Creek area in January damaged flood facilities at Hume Lake Christian Camp in Sequoia National Forest.

Heavy rains contributed to high runoff and flooding throughout Sequoia and Kings Canyon National Parks, resulting in significant road, bridge, and trail damage. Eleven inches of rain fell at Hockett Meadows at the 8,500 foot elevation in a 24-hour period.

Kirk Stiltz, the national parks' roads foreman, recalled January 2 very well. He was the only operator on duty that day at Red Fir. It had been raining steadily through the day, and he was quite busy keeping drains open above snowline. The situation got so bad by early afternoon that two additional operators had to be called in to work. There were many rock, mud, and debris slides on the Generals Highway below Giant Forest. These temporarily blocked the road but did not take it out. The only washout that Kirk recalled on the Generals Highway occurred at Halstead Meadow.

Jim Harvey recalled that the 1997 flood caused a big slide on the old Middle Fork Trail, just west of Elk Creek. That now marks the east end of Tunnel Rock unit of the Ash Mountain Pasture. The parks' stock come to that slide and can't go any further.

This was probably the flood that took out the river pump at the national parks' Ash Mountain headquarters. However, Jack Vance recalled that it might have been the smaller February 1998 flood that took out the pump.

This was probably the storm event that caused one of the Ash Mountain sewage ponds to lose its integrity. At the time, the ponds were lined with bentonite. As a result of the storm damage, the parks chose to line the ponds with butyl liners to ensure their integrity during high-water events.

Bill Tweed recalled that the 1997 flood caused a rockslide that damaged the piping feeding the Sycamore Creek stock tanks on the Shepherd's Saddle Road.

The Middle Fork of the Kaweah was high at the Pumpkin Hollow Bridge late on the afternoon of January 2 (multiple photographs on file in the national parks; also see the title page photograph of this document). As shown in Figure 20, the floodwaters kept rising as the last light of day ended.

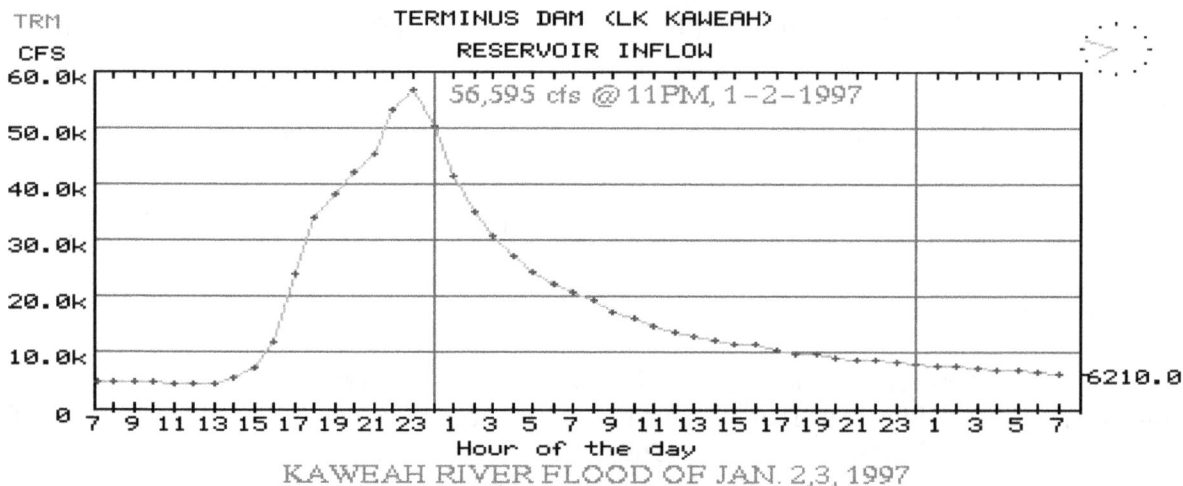

Figure 4. Discharge of the January 1997 flood.

The title page photograph was taken at about 5:00 p.m. on January 2. The flow shown at that time would be plotted as "17" on Figure 20 using a 24-hour clock. Note that the flow was more than three times as great when the Kaweah peaked at 11 p.m. that night.

Harold Werner, the national parks' former wildlife ecologist, recalled that stream flows from this storm were so great that the non-native bullfrogs were flushed out of the North Fork Kaweah River Basin within Sequoia National Park and took several years to recolonize it.

Kirk Stiltz recalled that the Kaweah inundated the low section of the highway between Reimer's candy store and the Three Rivers school, and that there was a report of a propane tank floating down Highway 198 through that area.

Kirk recalled watching, hearing, and feeling the flood from the Dinely Bridge around 11:00 p.m. on the night of January 2. The river put on a kind of other-worldly lightshow as the rocks crashed together underwater.

Richard Fletcher recalled that the flooding South Fork Kaweah backed up in an unnamed tributary deep into the Cherokee Oaks Subdivision. A large number of rainbow trout moved into this relatively quiet backwater to escape the wild waters of the South Fork. At Richard's property near Oakridge Drive, this creek was about 60 feet wide (multiple photographs on file in the national parks). When the floodwaters receded, trout up to 16 inches long were left stranded in yards.

Lake Kaweah took on 40% of its total capacity in a 24-hour period. The lake's elevation went from 620 feet to 670 feet in 36 hours. At that point, the lake was rising nearly five feet per hour.

The Kaweah's peak natural flow occurred at Terminus Dam (or possibly McKay's Point) at 11:00 p.m. on January 2: 56,595 cfs. (That was the peak hourly flow; the peak average daily flow, as reflected in Table 15, was 17,948 cfs.) Based on the flood exceedence rates in Table 16, this had a

351

recurrence interval of 14 years for the Kaweah. (One source reportedly calculated this as having had a recurrence interval of 80 years. However, a USACE Sacramento District hydrologist could not reproduce that result, even using the now-outdated 1971 flood frequency curves.)

Phil Deffenbaugh, park manager at Lake Kaweah, recalled that the flow of Dry Creek exceeded the 5,500 cfs channel capacity of the Lower Kaweah River during the onset of the 1997 flood. Therefore, nothing was released from Lake Kaweah during that part of the flood; the gates in the dam were closed. As the flow in Dry Creek dropped, the gates were opened, allowing the Kaweah River to start flowing through. Lake Kaweah peaked at 115,700 acre-feet on the morning of January 7 with 27,000 acre-feet capacity remaining.

The gates could only be kept shut for a short period, albeit a critical period. Lake Kaweah's flood-control pool is relatively small compared to the size of its watershed. When there is a moderately severe flood, it is necessary to pass much of the flood through; it just isn't feasible to keep the dam entirely closed during such a flood. For example, Lake Kaweah filled and emptied twice during the 1997 flood.[1031]

The flood left considerable sediment and wood debris in the Lake Kaweah lakebed. Annie Esperanza recalled that some of the trees were obviously giant sequoias. The logs and woody debris were piled and disposed of by burning the following summer.

This was a major flood on the Tule River. It peaked in early January, resulting in significant flooding. Success Dam filled and emptied twice during the flood. [1032] A levee broke on the Tule River.

This was also a major flood on the White River; it peaked at the same time in early January as the Tule. State and federal disaster assistance was granted to the town of Earlimart, which suffered millions of dollars of damage to homes and other structures. Highway 99 was closed for over a week due to the flooding. This was the fifth time in 40 years that flooding occurred in the area.

A breach in Poso Creek levees on January 4–5 put water onto the valley floor near Wasco.

The Kern River peaked late on January 2 at about 42,000 cfs near Kernville. (That was the peak hourly flow; the peak average daily flow was 18,780 cfs on January 3.) One mobile home was swept downriver and a couple of others were damaged.

Central San Joaquin Valley agriculture suffered large losses as farm land was inundated from runoff. Uncontrolled small streams and major river flooding caused damage to permanent crops, irrigation equipment, and roads. Agricultural damage was particularly high in Kings County; flooding of the Tulare Lakebed kept acreage from being farmed during the 1997 crop year.

Table 76 summarizes the damage incurred in the southern end of the San Joaquin Valley during the 1997 flood.[1033]

Table 76. Damages incurred during January 2–5, 1997 flood.

| County | Private Damage (thousand dollars) | Public Damage (thousand dollars) | Business Damage (thousand dollars) | Agricultural Damage (thousand dollars) | Total Damage (thousand dollars) |
|---|---|---|---|---|---|
| Merced | $0 | $570 | $0 | $7,610 | $8,180 |
| Madera | $1,400 | $270 | $20 | $2,497 | $4,187 |
| Fresno | $620 | $3,400 | $0 | $1,394 | $5,414 |
| Kings | | | | $38,857 | $38,857 |
| Tulare | $1,500 | $770 | $500 | $6,066 | $8,836 |
| Kern | | | | | |
| Total | $3,520 | $5,010 | $520 | $56,424* | $65,474 |

*Another source put total agricultural damage for these counties at $70.7 million.

Following the heavy rain and snowmelt floods of early January and another storm passage around January 20, another period of heavy rain occurred from the afternoon of January 24 through the evening of January 26.

Storm totals in the Southern Sierra were generally 3–4 inches of rain. Storm totals in the valley were about an inch but rather variable. Table 77 shows the precipitation totals for some of the reporting stations during this storm event.

Table 77. Precipitation during the January 24–26, 1997 storm event.

| Reporting Station | Storm Total (inches of rain) |
|---|---|
| Oakhurst | 4.06 |
| Fresno | 0.83 |
| Visalia | 1.4 |
| Bakersfield | 0.31 |

This heavy precipitation induced a second round of flooding in the San Joaquin Valley and the foothills. In the valley, small streams swelled and poor drainage roads were covered with water.

In Sequoia National Park, the rain brought rock falls and debris flows. Damage to the Generals Highway occurred just below Giant Forest at 4:30 p.m. on January 23.

The USACE had to make large releases from Lake Success in anticipation of the forecasted heavy mountain rains. This resulted in the Tule River running quite high downstream. The event was anticipated, and at-risk structures were closed. A 10-foot section of the Jaye Street Bridge in Porterville was washed away on January 24.[1034]

The second storm of 1997 occurred on July 23. We know this storm from two places. While these were caused by two separate storm events, they could be thought of as one event that occurred in multiple locations:

1. Alder Creek flash flood in the Ash Mountain area. Bill Sullivan recalled that Alder Creek flash flooded on this date. The water was so high and carried so much sediment that the Ash Mountain water plant had to be shut down for a while. The sediment/turbidity situation cleaned itself up by early/mid-evening, and the water plant was producing water again by that time.

2. There was a flash flood on the creek west of Silver City in the Mineral King area. This creek is locally known as Silver Creek or Silver City Creek. This flash flood resulted from an estimated 2 inches of rain that fell within a 45 minute time period over this creek's small watershed. Water washed over the Mineral King Road, but only minor road erosion occurred. The impacts of the storm and resulting flood were documented by the NWS forecast office in Hanford.[1035] This creek would have a much more severe flash flood in 2006.

The third storm of 1997 occurred in September. Monsoonal moisture over the Southwest supported thunderstorm activity over the desert portions of Kern County and the northern Kern County Mountains in early September:[1036]

- On September 2, thunderstorms brought 1 inch diameter hail to Mount Mesa (near Lake Isabella) and dropped 1.11 inches of rain in 30 minutes at Ridgecrest. The heavy rain in Ridgecrest caused numerous intersections in that town to flood and some were covered with up to 6 inches of mud. An automated station just west of Ridgecrest recorded 0.90 inches of rain in just 8 minutes (a rate of 6.8 inches per hour).[1037]

- On the evening of September 3, a particularly large thunderstorm cell produced 4.5 inches of rain in a little over an hour in Red Rock Canyon State Park. The resulting flash flood brought 28,000 cfs down Red Rock Creek, across Highway 14, and on into Koehn Dry Lake. A 12-foot wall of water swept over Highway 14 at 7:10 p.m. Several highway bridges were damaged, and four cars were swept into the floodwaters. The highway had to be closed until repairs and clean-up could be made.[1038] Nearly 100 motorists were stranded by the flooding.[1039] A related thunderstorm on the same evening occurred just west of Mojave. The flash flooding associated with that storm produced flooding four feet deep at the intersection of Highways 14 and 58, floating cars. (The Ridgecrest and Red Rock storms and floods were just outside the Tulare Lake Basin, but the story merits inclusion in this document as an example of how intense a summer storm can be.)

- On September 3, a thunderstorm east of Lake Isabella resulted in flash flooding in Scodie Creek, causing water to flow over Highway 178 at Onyx. Hail was also reported with this thunderstorm up to ½ inch in diameter.

Total flow for water year 1997 was 152% of the 1894–2011 average for the Kings, 177% for the Kaweah, 256% for the Tule, and 171% for the Kern.

About 48,000 acres of agricultural land were submerged in the Tulare Lakebed, returning the lake to 1983 levels. Apparently the western edge of the lake came to about the intersection of 10th and Pueblo on the west side of Corcoran. In order to minimize flooding in the lakebed, 87,000 acre-feet of river floodwater was pumped into the Friant-Kern Canal and routed to the Los Angeles area. For comparison, that's half the total storage capacity of the newly expanded Lake Kaweah. The Tulare Lakebed wouldn't be completely drained until 2000.

An outstanding feature of the 1997 storm event was the number of significant landslides and debris flows that resulted. The large ones that we know about were in the Central Sierra, north of the Tulare Lake Basin. These events merit inclusion in this document because two of them were well studied, and they can inform risk management planning in our area.

### Mill Creek Landslide: South Fork American River

This event occurred near White Hall on U.S. Highway 50, about 25 miles east of Placerville and 35 miles west of South Lake Tahoe. At this point, the highway is squeezed in a narrow canyon between the South Fork American River and a steep cliff. Another major landslide had occurred just 0.6 miles east of here on April 9, 1983.

This landslide occurred in the El Dorado National Forest. The event was thoroughly analyzed by Robert Sydnor, an engineer geologist for the California Division of Mines and Geology.[1040]

The soil was a very wet sandy colluvium, a sandy mud. It contained some silt, but lacked a cohesive clayey matrix to bind it together. It had lots of voids that could hold water. The December 31 – January 1 rain-on-snow event saturated the ground, filling it with water. This was followed two weeks later by a week-long period of sustained heavy rainfall. That was the proverbial straw that broke the camel's back and triggered the landslide.

At 11:20 p.m. on January 24, 1997, a large section of the hillside gave way. The head of the landslide began on a 44% slope. The landslide moved rapidly downhill, across the highway, and dammed the river. The headscarp was about 1,100 vertical feet above the elevation of the river (elevation 3,420 feet).

The landslide dam was breached at 4:30 a.m. the following morning. The river took five hours to erode the dam down to the original riverbed. This occurred slowly enough that no downstream flooding occurred.

The landslide had a total estimated mass of about 2,000,000 cubic yards (1,500,000 cubic meters).[1041] The highway was buried under 75 feet of debris. It took Caltrans 27 days to remove approximately 275,000 cubic yards of debris and reopen the highway.

### Sourgrass Debris Flow: North Fork Stanislaus River

This event occurred on Sourgrass Creek, a small tributary of the North Fork Stanislaus River. The debris flow crossed U.S. Highway 4 six miles east of Dorrington and 14 miles west of Ebbetts Pass. The flow ended in the North Fork Stanislaus, five miles upstream of Calaveras Big Trees State Park.

The debris flow occurred in the Stanislaus National Forest. The event was thoroughly analyzed by Jerry DeGraff, a geologist for the USFS.[1042, 1043, 1044]

This debris flow began in glacial till and eroded into other unconsolidated material.

The event began at about 6:30 p.m. on January 1, 1997. It started as a debris slide on a 40% slope. The headscarp was at an elevation of about 5,960 feet. The debris slide almost immediately disaggregated into a debris flow and continued as such all the way to the North Fork Stanislaus River, elevation 3,960 feet. Total distance from the headscarp to the river was 2.4 miles with an elevation drop of 2,000 feet.

The debris flow had an average gradient of about 830 feet/mile. In the gentler sections, it was moving at only 2–3 mph and not eroding much material. But when the debris flow entered the steeper sections of Sourgrass Ravine, it picked up speed and began scouring the creek channel to bedrock. This would have been a very impressive event to witness. Here and there in the trees, there are large cobbles 12–18-inches in diameter that were hurled out of the channel as the flow passed by. (Imagine a moving catapult, and you get the picture.)

In addition to the scouring, the debris flow cleared everything from its path. Because the flow was cohesive, it left a surprisingly clean swath behind it. About 10% of the flow was deposited on the sides as levees. But except for a few large boulders, scattered rocks, and tree pieces, the debris flow carried everything else to the end.

The debris flow was approximately 500 feet wide and 20 feet high when it mobilized over Highway 4. Because the force of the mass was spread over a wide area, damage to the highway was minimal. No drivers were there that night to witness this phenomenon.

By the time the flow reached the North Fork Stanislaus, it had attained a maximum speed of over 12 miles per hour. This nearly equals the average peak velocity of past debris flows in the Sierra.

The debris flow began with an initial mass of approximately 65,000 cubic yards. The volume of the debris flow increased due to erosion of material along its path. By the time it reached the river, it was about 300 feet wide and 35 feet high and its mass had tripled to about 190,000 cubic yards. There was a campground located at this point, but fortunately it was closed for reconstruction. Otherwise, there could have been a disaster. (Particularly since this happened in the middle of the night.)

The debris flow poured into the North Fork Stanislaus, completely filling its channel and damming the river. The river was experiencing a major flood at the time, flowing at 28,000 cfs compared with a seasonal average of 250 cfs. It took the river's floodwaters about one hour to overtop and erode the dam enough to restore unimpeded flow. About 200 acre-feet of that debris would be washed eight miles downstream and deposited in McKays Reservoir, causing that reservoir to lose 10% of its capacity.

### Other 1997 Debris Flows: Central Sierra
Three USGS geologists conducted an aerial reconnaissance of potential landslide activity in the Central Sierra on January 8, 1997. In addition to the Sourgrass debris slide, their reconnaissance detected the following large debris flows:[1045]

- They observed several large debris flows in the Royal Gorge canyon of the North Fork American River, south of Snow Peak. Those debris flows had fallen from the top of the canyon over a thousand feet down into the river.
- About two miles west of Strawberry Lodge (19 miles west of South Lake Tahoe), they observed a large debris flow that had covered U.S. Highway 50 with mud and granite boulders for a distance of several hundred feet. Several other large debris flows crossed the highway in that vicinity.
- Several large debris flows dropped tons of mud and woody debris into Salt Springs Reservoir on the Mokelumne River. That reservoir is located north of Calaveras Big Trees State Park in the El Dorado National Forest.

## 1998–99 Floods (7)

There were six flooding events in 1998:
1. February, 1998 (twice)
2. March, 1998
3. April, 1998
4. May, 1998
5. September, 1998 (remnants of Hurricane Isis)
6. 1999 Tulare Lakebed flooding

The winter of 1997–98 was a strong El Niño event.

February began with a strong jet stream (170+ mph) oriented perpendicular to the Sierra. The transport mechanism for the moisture was an atmospheric river.[1046] The accompanying storm lasted from February 1–3. The storm's greatest impact was felt on the Central Coast and the Santa Cruz Mountains.

Weather stations reported 14 inches of rain falling in 45 hours over the coastal mountains. Various creeks and rivers reached flood stage throughout the Santa Cruz Mountains:
- Pescadero Creek crested at Pescadero with a peak discharge of 10,600 cfs; it had a recurrence interval of 25–50 years.
- The San Benito River crested at Hollister with a peak discharge of 34,500 cfs; it had a recurrence interval of 50–75-years.
- Tres Pinos Creek crested at Tres Pinos with a peak discharge of 27,200 cfs; it had a recurrence interval of 75–100 years.
- San Lorenzo Creek crested at San Lorenzo with a peak discharge of 10,300 cfs; it had a recurrence interval of 75–100 years.

Although the storm was focused on the coastal mountains, it had a significant impact on the Tulare Lake Basin. The storm began on February 1. By the next day, flooding was being reported from the far south end of the San Joaquin Valley. On February 2, heavy rainfall led to flash flooding and water over Highway 166 southwest of Bakersfield.

The storm peaked on February 3. It resulted in the lowest barometric pressure ever recorded during a February in Fresno, and the lowest barometric pressure ever recorded in any month in Bakersfield. Southerly winds increased throughout the morning of February 3, blowing down trees, power lines, fences, and damaging buildings. Bakersfield experienced near record-setting wind gusts. Many roofs were damaged in Kings, Tulare, and Kern Counties. One woman was injured near Lemoore when a tree fell on her.[1047]

Fresno County was hit by a series of storms that brought heavy rainfall to the Coast Ranges to the west and high wind and heavy rainfall to the San Joaquin Valley floor. Runoff from the Coast Ranges caused flooding in west Fresno County affecting agricultural areas around Mendota, Firebaugh, and Cantua Creek. Approximately 9,300 acres of farmland were flooded.

Some of the worst flooding was about 15 miles southwest of Mendota. The estimated flow in Panoche Creek at Interstate 5 (northwest of Mendota, between Belmont and Nees) was 17,000 cfs on the morning of February 3. Cantua Creek and Arroyo Hondo combined to flood 240 acres of farmland.

Over 100,000 chickens died near Gustine (northwest of Los Banos) when two chicken farms were inundated by the floodwaters of Garzas Creek just before dawn on the morning of February 3. The Los Banos area received 3.16 inches of rain in the previous 24 hours by 10:00 a.m. on February 3. Los Banos Creek peaked at midnight on February 2 with a flow of 14,480 cfs into the Los Banos Creek Reservoir. This set a new record, breaking the 11,500 cfs record flow set in 1955.

That was only the beginning of one of the wettest months on record in the Tulare Lake Basin. Measurable rain fell in Fresno on 21 of February's 28 days, resulting in the third-wettest February on record for that city. With 5.36 inches in rain, Bakersfield experienced the wettest month ever recorded since record-keeping began in 1889. (This record would be broken in December 2010.) The near constant rains kept the ground in the foothills and the San Joaquin Valley floor saturated, and runoff caused persistent problems through the month.

In Bakersfield, significant rain led to ponding water and flooding on many secondary roadways on the evening of February 7.

The same storm system impacted the west side of Fresno County that evening and the following day. Stream flow from Panoche/Silver Creek crested at 13,000 cfs at 10:00 p.m. on February 7. The resulting flooding downstream peaked in Mendota on the evening of February 8. Flooding affected Highway 198 west of Interstate 5 in far western Fresno County.

A second major storm struck on February 23. The impact of that storm was apparently focused on the southeast side of the Tulare Lake Basin from the vicinity of Lindsay south to the Tehachapis.

Lewis Creek near Tonyville and Frazier Creek near Strathmore both overflowed early on February 24, causing an estimated $1.5 million in damage to area homes and businesses. Rainfall in the 24 hours prior to the flooding was estimated to be 1–1½ inches in the lower Tulare County foothills.

The White River had a 700 cfs flow in its shallow channel by midnight on February 23. All the tributaries of the White River (Speas, Chalaney, Coho, and Tyler Gulch Creeks) were flowing heavy. The river breached a levee at 1:30 a.m. on February 24 and flooded the town of Earlimart. Highway 99 had to be closed through the town and remained closed for a week. Up to 250 homes in the town were impacted by the flooding, with 50 homes having 3 feet or more of water in them and 220 people forced to evacuate. Damage in Earlimart was estimated to be $13.7 million.

Poso Creek breached its banks late on the night of February 23 with a peak flow estimated to be 7,000 cfs. The creek flooded 112 homes in the town of McFarland; damage was estimated to be $2.5 million. Poso Creek floodwaters also threatened some rural homes downstream near Wasco later on February 24.

Evacuation of Lamont was begun at 8:00 p.m. on February 23 in anticipation of flooding due to the steady accumulation of rain. Caliente Creek progressed from nuisance flow to flooding before dawn on the morning of February 24. Caliente Creek peaked at 6,000 cfs at Bena (in the Tehachapi foothills upstream of Lamont) by 2:30 a.m. on February 24. Farther downstream of Lamont, water from Caliente Creek flooded and closed the northbound lanes of Highway 99 at Herring Road (8 miles south of Bakersfield) by 8:00 a.m. on February 24.

February precipitation for the Sacramento 8-station index was 265% of average. For the 8 reference stations in that index, average precipitation is 7.9 inches, but February 1998 had 20.9 inches. The statewide snowpack water content by the end of February was running at 160% of average. The impacts of the heavy February rains and resulting flooding were documented by the NWS forecast office in Hanford as shown in Table 78.[1048]

Table 78. Rainfall during February 1998.

| City | 1998 Rainfall (inches of rain) | Average Rainfall (inches of rain) |
|---|---|---|
| Fresno | 5.10 | 1.80 |
| Hanford | 4.26 | 1.45 |
| Bakersfield | 5.36* | 1.03 |

*The 5.36 inches of rain recorded by Bakersfield made that the city's second wettest February on record.[1049]

Seasonal rainfall 10 miles northeast of Springville was 28.86 inches by February 23.

As a result of severe winter storms and flooding, a major federal disaster (DR-1203) was declared on February 9, 1998 for the period February 2, 1998 – April 30, 1998. It covered 41 counties including Fresno, Kern, and Tulare.[1050]

Table 79 summarizes the damages from the various February storms.

Table 79. Total damages incurred from rain and flood during February 1998.

| County | Property Damage (million dollars) | Agricultural Damage (million dollars) |
|---|---|---|
| Merced | $2.0 | 1.4 |
| Fresno | $1.6 | 1.8 |
| Kings | $0.02 | 1.0[1] |
| Tulare | $13.9 | 1.5 |
| Kern | $12.5 | 5.4[2] |
| Total | $30.02 | $11.1 |

[1] In Kings County about $1.0 million in flood protection costs was expended to try to protect agricultural land.
[2] In Kern County, the areas most severely impacted were Arvin-Lamont and McFarland, although flooding also occurred in the Lebec-Frazier Park-Cuddy Valley area and in the Kern River Valley.

Flood releases of 8,000 cfs or greater occurred at Friant Dam on the San Joaquin River during the April–July snowmelt period.

The peak day natural flow at Pine Flat on the Kings occurred on June 17. There were sustained high flows throughout June and July. These were clearly above-average flows, but there had been about a dozen floods greater than this in the previous 45 years. Although the river didn't reach a particularly impressive height, it delivered a tremendous amount of water to the valley floor.

On March 25, a band of quasi-stationary thunderstorms deluged Merced with 3 to 6 inches of rain in a 12- to 18-hour period. One gage in the northern part of the city of Merced had 6.8 inches in a 48-

hour period from late March 23 to the 25th. The Merced Airport recorded 3.25 inches of rain on the 24th alone. Bear Creek reached a crest of 19.3 feet on the morning of March 25, resulting in 1,000 people being evacuated. A total of 65 homes and 19 apartments were flooded. Damages totaled $9.6 million to property with agriculture suffering a $1.5 million loss.[1051]

Panoche/Silver Creek west of Mendota flooded sometime in March 1998.[1052]

On April 1, Lewis and Frazier Creeks swelled due to heavy rains and snowmelt, resulting in flooding in Lindsay, Strathmore and Tonyville, damaging 32 homes.[1053]

On May 2, thunderstorms unleashed locally heavy rain over rugged terrain northeast of Bakersfield. A spotter reported 1.5 inches of rain falling in about an hour. The rapid runoff from those storms resulted in flash flooding on several streets in the Bakersfield area, including Highways 178 and 58. Many vehicles became stalled in the high water, and some residences and apartments were flooded with as much as 3 feet of water.[1054]

On September 4, moisture associated with remnants of Hurricane Isis brought rain to parts of interior Central California. Frazier Park received 1.53 inches of rain and Bakersfield received 0.27 inches of rain, setting a new daily precipitation record. Trace amounts were reported in the valley as far north as Madera.[1055]

Total flow for water year 1998 was 177% of the 1894–2011 average for the Kings, 217% for the Kaweah, 330% for the Tule, and 221% for the Kern. This was the fifth-highest year ever recorded on the Kaweah and the fourth highest on the Tule.

Tulare Lake had reappeared in 1997, and the 1998 flood exacerbated the problems being experienced by landowners in the lakebed. In order to minimize flooding there, 202,000 acre-feet of river floodwater was pumped into the Friant-Kern Canal and routed to the Los Angeles area in 1998. For the same reason, 130,000 acre-feet of Kern River water was also routed to Los Angeles. An additional 984,000 acre-feet of Kings River water was routed through the James Bypass to the San Francisco Bay.

That helped to minimize the lakebed flooding, which was beneficial for the farmers there. However, it meant that the Tulare Lake Basin lost 1,316,000 acre-feet of water that could have been put to productive use or used to recharge our groundwater aquifers. For comparison, that 1,316,000 acre-feet of water is equivalent to 82% of the combined current capacity of all four of the federal reservoirs in the Tulare Lake Basin.

The Tulare Lakebed flooded in 1999. This flooding occurred despite the fact that there was no storm event of note that year.

Runoff was also below average in 1999. Runoff during water year 1999 was only 70% of the 118-year average (1894–2011) for the four major rivers (Kings, Kaweah, Tule, and Kern) combined.

About 32,000 acres of agricultural land were submerged in the Tulare Lakebed in 1998. The flooding in the lakebed in 1998 and 1999 was to some degree left over from the big 1997 flood. The floods of 1998 exacerbated the lakebed flooding. As illustrated in Figure 11, the lakebed would not be fully drained until 2000.

Lakebed flooding is a social construct; it is counted based on the number of growing seasons that are missed. The lakebed was flooded for three growing seasons: 1997, 1998, and 1999. So this is counted as three floods from the perspective of the lakebed farmers, even though flood events occurred in only two of those years.

Something similar happened in the lakebed in 1969–1971 and 1982–84. In each of those cases, lakebed flooding continued into a non-flood year.

## 2000 Floods (2)

Flooding occurred twice in 2000:
1. January
2. October

From January 23–25, a three-day storm brought locally heavy winter rains to the valley, foothills and lower elevations of the Sierra as shown in Table 80.[1056]

Table 80. Precipitation during the January 23–25, 2000 storm event.

| Reporting Station | Storm Total (inches of rain) |
|---|---|
| Bass Lake | 6.78 |
| Shaver Lake | 5.69 |
| Northeast Fresno | 2.29 |
| East Visalia | 2.20 |

Valley urban areas had significant ponding of water and mountain streams exhibited moderately large amounts of flow. There was some flooding along the valley floor and foothill interface.[1057]

An early season storm brought several inches of snow to the Central and Southern Sierra on October 10. Table 81 shows precipitation totals during that storm event.

Table 81. Precipitation during the October 10, 2000 storm event.

| Reporting Station | Storm Total (inches of snow) |
|---|---|
| Lodgepole | 10 |
| Mount Tom | 8 |
| Huntington Lake | 5 |
| Tuolumne Meadows | 4 |

Over an inch of rain fell in some areas in the valley, including Fresno, resulting in the closure of the Fresno Fair for the first time since 1922. The rain caused numerous flooding problems in Fresno and ceilings to collapse in buildings in Tulare.[1058]

## 2001 Floods (2)

Flooding occurred twice in 2001:
1. March
2. August

Although not a flood, an unusual snow event occurred on February 12–13. A low-pressure system came onshore near Point Conception and brought unusually heavy snow to the mountains at the southern end of the San Joaquin Valley. The snowline in the Tulare County foothills dropped to 3,000 feet. Greenhorn Summit received 18 inches of snow, and Frazier Park received 30 inches. Some 500 motorists were stranded on Interstate 5 over the Grapevine.[1059]

Gary Sanger at the NWS forecast office in Hanford speculated that a system such as the above storm might have accounted for the phenomenal snowfall that the Southern Sierra experienced in January and February 1906, especially if a deep atmospheric river were entrained.

Western Fresno County experienced heavy rain from late on March 4–6; Coalinga received 2.99 inches. Several roads in the area were washed out, including Highway 33/198.[1060]

An intense thunderstorm struck a large portion of the Rock Creek Basin on August 9. This caused Rock Creek to quickly flash flood, sending a large quantity of water out onto the Kern Valley floor about two miles north of Kern Hot Spring. Erika Jostad was on patrol and witnessed both the deluge and the flood. (Tony Caprio, the national parks' fire ecologist and his fire effects crew showed up shortly thereafter.) In a place that does not typically flow water, a wall of milky water, gravel, and woody debris flowed across the High Sierra Trail and into the Kern River. It is unclear whether this event was a debris flow or a flash flood that carried large quantities of debris.

Tony recalled that the volume of water coming out of Rock Creek was so great that the Kern River ran milky white for some distance downstream. The flood debris obliterated the High Sierra Trail for a stretch of a few hundred feet (multiple photographs on file in the national parks). In the photographs, tree trunks appear to have a couple feet of gravel piled against them. Some of the images show damage to the standing trees a couple of feet above the level of the gravel bank, giving some indication of the depth of water and debris flowing over the area at peak flow. The High Sierra Trail was not rebuilt but gradually reestablished through use.

A similar flash flood would occur in the Rock Creek Basin in July 2011. Although that flood may have been even larger than the 2001 flood, it did not put any debris onto the High Sierra Trail. We don't know why these two floods differed in this way. Tony Caprio speculated that the 2001 flood might have cleared accumulated material from the stream channel, and there had been insufficient time to accumulate a similar quantity of material before the 2011 flood.

## 2002 Floods (2)

Flooding occurred twice in 2002:
1. May
2. November

Thunderstorms dropped 1.01 inches of rain at the Hanford Airport in just 21 minutes on the afternoon of May 30 (a rate of 2.9 inches per hour), resulting in street flooding in that city. Some 260 lightning strikes were recorded in just an hour in the central and southern San Joaquin Valley.[1061]

Hurricane Huko (the Hawaiian equivalent of the name Hugo) formed in the central Pacific and became a hurricane on October 28. Huko then became a wanderer, impacting all three North Pacific basins (east, central, and west) and eventually morphed into an extratropical cyclone.

Tropical moisture from Huko combined with a major trough from the eastern Pacific to bring copious amounts of precipitation and gusty wind to the Tulare Lake Basin from November 7 until early on the 9th. The impacts of that storm and the resulting flood were documented by the NWS forecast office in Hanford.[1062]

The flood covered a number of locations in the foothills and mountains of Tulare and Kern counties. Flooding problems were most pronounced in the Tulare County mountains and the higher foothills. Table 82 provides precipitation totals for some stations during the storm event. Numerous foothill locations received 5–10 inches of rain during the three-day period.

Table 82. Precipitation during the November 7–9, 2002 storm event.

| Reporting Station | Storm Total (inches of rain) |
|---|---|
| Merced | 1.80 |
| Fresno | 1.76 |
| Lodgepole | 11.60 |
| Ash Mountain | 5.89 |
| Hanford | 1.44 |
| Glennville (NE of Bakersfield) | 6 |
| Bakersfield | 1.29 |
| Johnsondale | 16.38 |
| Tehachapi | 4.67 |

Many valley locations set new 24-hour rainfall records on November 8. For example, Fresno's old record for the 8th was 0.23 inches; the new record set during this storm was 0.98 inches.

Snow levels were relatively high at 9,000 feet. Chagoopa Plateau received 80 inches (6.7 feet) of new snow during the three-day event.

Gusty winds associated with the storm caused 23 pole fires, resulting in 102,000 valley residents losing power.

There was very heavy rainfall in Sequoia National Park on the morning of November 8. After a few hours of such rain, the Kaweah River was rising at a rate comparable to the January 1997 storm. If that rain had continued for a few more hours, it was conceivable that the national parks' approach to the Pumpkin Hollow Bridge would have washed out as it had in 1937, 1955, and 1966. As a precaution, the Ash Mountain Entrance was closed and a partial evacuation of the park was begun. Fortunately the rain stopped before the approaches to the bridge sustained any damage.

According to records provided by the USACE, the Kaweah's peak natural flow occurred at Terminus Dam on November 8: 30,273 cfs. (That was the peak hourly flow; the peak average daily flow, as reflected in Table 15, was 9,436 cfs.

Based on the flood exceedence rates in Table 16, this flood event had a recurrence interval of 7 years for the Kaweah River. It would have a recurrence interval of 14 years if calculated using the 30,273 cfs peak flow. (Another source reportedly calculated this as having a recurrence interval of 18 years. Presumably that was done by using the peak flood flow and the now-outdated 1971 version of the flood frequency curves.)

Bill Sullivan and Call Kessner recalled that Alder Creek flooded with water that was the color of coffee and cream. The water was so high and carried so much sediment that the Ash Mountain water plant had to be shut down. (The floodwaters and debris washed on downstream to where they clogged the culvert on the Generals Highway and caused significant erosion problems there.) After the flood, the sluice gate in the Ash Mountain dam was opened, and much of the sediment was worked downstream. It was two or three days before the water plant could be brought back on line.

During the storm event, a landslide dam apparently formed on a very small stream on the north side of Shepherd's Saddle in the national parks. When that dam failed, a huge wall of water came down this small drainage, washing out the Shepherd's Saddle Road and placing three huge rocks at a surprisingly high elevation (photograph on file in the national parks). Those rocks didn't fall down from a higher elevation. However improbable, the floodwaters from this very small stream lifted the rocks up on high. The road was washed out, but the park was able to repair it the following year for about $90,000. (This was the cost to repair all the damage to the Shepherd's Saddle Road, only about $30,000 of this went to repairing the big washout near the saddle.)

That storm also caused Sycamore Creek, a normally very small stream, to wash away two stock tanks that sit beside the Shepherd's Saddle Road.

With numerous rock falls and debris flows, flooding, and road erosion problems, the Generals Highway and Mineral King roads were closed. Campers were evacuated from Potwisha on the morning of November 9. The Mineral King Road and Generals Highway sustained significant damage and required 1.25 million dollars to repair.

An additional $150,000 was required to repair damage to the Crystal Cave Road in the national parks. A large culvert plugged and allowed a creek to flow down the road washing a deep trench along the uphill side of that road until it finally crossed and washed out the road and fill slope. The road was left impassable due to the sinkhole. The parks used gabion baskets to stabilize the shoulder and used boulders from a slide to fill the sinkhole.

Kirk Stiltz, the national parks' road foreman, recalled that many of the culverts that plugged during the November 2002 storm, plugged as a result of debris flows. The parks experience a lot of small debris flows when there are heavy rain events.

The total cost to repair all the national park roads damaged during this storm event was approximately 1.49 million dollars. That included the Generals Highway, the Mineral King Road, the Crystal Cave Road, and the Shepherd's Saddle Road.

Numerous other roads flooded and debris flows occurred in the foothills of the Southern Sierra. Several roads were flooded in Kern County. Three roads were washed out in southeast Tulare County:
1. The Parker Pass Road
2. The road below the Durwood Resort
3. The road that leads from Johnsondale southward to Kernville along the North Fork Kern (Mountain 99, aka Kern County SM99)

Rock falls and debris flows occurred on Highway 168 and Highway 180 in the Southern Sierra foothills.

The McNally Fire had burned about 150,000 acres of Sequoia National Forest in July and August 2002. Some of that area burned quite hot. When the intense November storm hit that area a few months later, some erosion problems resulted. Debris was spread across many mountain roads in the area as well as contributing to a fish kill in the Kern River.

Peak flow into Lake Isabella from the Kern River of 26,500 cfs occurred on the night of November 8. (That was the peak hourly flow; the peak average daily flow was 10,306 cfs on November 9.) The lake storage increased from 82,000 acre-feet to 109,000 acre-feet, and the lake rose 5 feet in elevation during the two-day period from November 8–9. Flooding and debris flow problems occurred along Highway 178.

## 2003 Floods (3)

Flooding occurred at least three times in 2003:
1. February
2. August (several times)
3. December

An intense storm struck northwest Fresno on February 13, dropping 3.40 inches of rain in just 2 hours. Up to 3 feet of water flooded parts of the area.[1063]

There were monsoonal influences over the Central and Southern Sierra throughout August, resulting in periods of heavy rain, localized flooding, and brief road closures. Multiple flash floods occurred around Kernville, Tehachapi, Johnsondale, and along the Sherman Pass Road. Estimated rainfall rates of 3–4 inches per hour occurred in an area from near Lake Wishon (near Shaver Lake) south to near Lodgepole on the afternoon of August 2. The Mineral King Road flooded in Mineral King from heavy rain.

Heavy thunderstorms drenched parts of the Kern County mountains in the morning hours of August 21. Piute received 1.78 inches of rain.

Monsoonal moisture generated thunderstorms late on August 25 that produced over an inch of rain in some areas. Cottonwood Creek in the Sierra received 1.68 inches, and 1.48 inches fell at Lost Hills in the valley. Roads were closed in parts of Sequoia National Park and in the Kern Plateau.

365

On December 25, locally heavy rain on the Southern San Joaquin Valley floor and adjacent foothills led to flooding at several locations over the South Valley due to runoff, including in Bakersfield.

Fresno received 0.99 inch of rain on December 25, setting a daily precipitation record; the old record of 0.53 inch had been set in 1946. Bakersfield received 0.91 inch that day, also setting a daily precipitation record; the old record of 0.76 inch had been set in 1931. A storm total of 1.54 inches was reported just east of Fresno. Large amounts of snow fell in the neighboring Sierra. There were 4 indirect deaths in vehicles caused by the heavy rainfall southeast of Bakersfield.[1064]

In the fall of 2003, several wildfires raged across the mountains and hills of Southern California. Two of those fires — the Old and Grand Prix — occurred in areas adjacent to Cajon Pass. When winter rains fall on steep Southern California land that has burned, water runoff is greater than on land that has not burned. The possibility of debris flow events also increases.

During a 24-hour period from December 24–25, 2003, more than 4 inches of rain fell on an area that had burned in the Old / Grand Prix complex. This torrential rain resulted in excessive runoff, sending a debris flow through the small town of Devore on December 25. This debris flow was captured in a dramatic video as it came through town (multiple videos on file in the national parks).[1065]

Devore is on the south side of the Tehachapi Mountains. Although it isn't in the Tulare Lake Basin, it merits inclusion in this document because it was nearby and it is one of the few videos that we have to illustrate what a debris flow looks like.

## 2005–06 Floods (8)

Flooding occurred eight times in 2005–06:
1. April 2005
2. May 2005 (twice)
3. August 2005
4. October 2005
5. December 2005 – January 2006
6. April 2006
7. July 2006

Severe thunderstorms struck in the afternoon hours of April 28, 2005, dropping hail as large as 1¼ inches in diameter in Kings County, damaging crops, including 20% of the cherry crop. A number of streets were flooded in Fresno, and 3.57 inches of rain fell in Parlier.[1066] A funnel cloud was spotted and photographed north of Visalia.[1067]

Heavy rainfall from thunderstorms on the afternoon of May 5, 2005, triggered flooding in the Madera area. A roof collapsed on a building in downtown Madera, and several roads were flooded. Parts of Highway 99 and Interstate 5 experienced flooding in Kern County. All the roads in Coalinga were flooded with some described as impassable.[1068]

On May 16, 2005, a rain-on-snow (actually, a rain-through-snow) flood occurred on the Merced River in Yosemite Valley. Although milder in degree, this was similar in nature to the more famous flood that occurred on that river May 16, 1996.

This was the first major flood observed by the new high-country hydroclimatic network in Yosemite. That network was developed by scientists from the USGS, Scripps Institution of Oceanography, California DWR, National Park Service, and other institutions, and consists of streamflow and air-temperature loggers, plus snow-instrumentation sites.

A storm drew warm, wet subtropical air into the Sierra, bringing moderate rain to the Southern Sierra on the night of May 15–16. That resulted in flooding on a number of rivers on May 16. Like so many cool-season floods in the Southern Sierra, the flood was mostly due to unusually warm temperatures and large catchment areas that received (moderate) rainfall rather than snowfall. In Yosemite, temperatures were above freezing up to about 10,000 feet elevation. Rain fell and streams filled up to 10,000 feet compared to typical freezing levels of about 5,000 feet.

With snow levels so high, rainfall amounts averaging 1.75 inches in the mountains combined with a snowmelt runoff contribution of about 1 inch water equivalent caused river flooding on the valley floor in Yosemite. The Merced River rose 3 feet on the morning of May 16, prompting the park to evacuate campers in Yosemite Valley. The Merced River crested at 12.5 feet in Yosemite Valley, forcing the closure of roads into the valley.[1069]

Warm storms — past and future — can unleash floods when rain falls over unusually large catchment areas. (In this flood, the area receiving rainfall may have been as much as five times normal.) Warmer temperatures in the future may increase the frequency and severity of these floods, even as snowpack volumes decline.[1070]

The storm was less intense in the Tulare Lake Basin. The flow on the Kings and Kaweah Rivers on May 16 increased 81% and 74% respectively compared to the previous day. The flow on the Tule and the Kern Rivers only increased 47% and 22%.

A large and severe thunderstorm swept over the southeast part of Kern County on the evening of August 15. The California City Fire Department rain gage measured 5 inches of rain in just one hour from the deluge. This led to flash flooding in that town from Cache Creek and extensive sheet flow through the area. Portions of Highway 14 and Highway 58 flooded.[1071] (This storm and flood was just outside the Tulare Lake Basin, but the story merits inclusion in this document as an example of how intense a summer storm can be.)

Weak low pressure off the Southern California Coast entrained tropical moisture that resulted in an intense storm striking the Tehachapi Mountains from the evening of October 17 into the morning of the 18th. The impacts of the storm and resulting flood were documented by the NWS forecast office in Hanford.[1072]

There were numerous rainfall reports of 2–3 inches of rain from throughout the Tehachapi Mountains during the storm event. Among the various reporting stations, Bear Valley Springs had 2.31 inches and the Piute Forest Service RAWS automated weather station reported 2.31 inches of rain for the event.

Numerous locations from Tehachapi to Taft experienced flash flooding, but the Frazier Park area was especially hard hit. Cuddy Creek overflowed in that community, flooding some areas 4-feet deep and resulting in the evacuation of at least twenty people.

The December 2005 – January 2006 flooding was most severe in Northern California, but did come as far south as the Tulare Lake Basin. Statewide, the storm lasted from December 29, 2005 – January 2, 2006. The transport mechanism for the moisture was an atmospheric river.

Damages totaled $300 million. As a result of severe storms, flooding, debris flows, and landslides, a major federal disaster (DR-1628) was declared on February 3, 2006 for the period December 17, 2005 – January 3, 2006; it covered 30 Northern California counties.[1073] The impacts of the storm and resulting flood on the Tulare Lake Basin were documented by the NWS forecast office in Hanford.[1074]

Rainfall totals for December 24 – January 3 exceeded 20 inches throughout the Northern Sierra. Coastal Range stations in the Russian and Napa River Basins received 18–30 inches. On December 31, there were widespread 24-hour rainfall totals in excess of 5 inches.

As a result of this heavy rain, several rivers throughout Northern California came above flood stage. Recurrence intervals for peak discharges generally ranged from 10–25 years.

Major flooding in the state was concentrated in the Napa and Russian River Basins. The Russian River crested near Guerneville with a peak discharge of 85,800 cfs; it had a recurrence interval of 10–25 years. The Napa River crested near Napa with a peak discharge of 29,600 cfs; it had a recurrence interval of 10–25 years. Approximately 1,000 homes were flooded in Napa.

Sonoma Creek crested at Agua Caliente with a peak discharge of 17,600 cfs; it had a recurrence interval that was greater than 100 years.

The Klamath River crested near Klamath with a peak discharge of 416,000 cfs; it had a recurrence interval of 25–50 years.

Heavy rainfall fell in the San Joaquin Valley from January 1–2. Continuous rain, heavy at times, brought an abnormally high 2.84 inches of rain to Fresno during the January 1-2 storm event; 3.19 inches in Selma, and 2.25 inches at Coalinga in a little over 24 hours. Fresno set a new daily precipitation record of 1.88 inches on January 2. Flooding occurred in the city of Fresno as 15 ponding basins overflowed. Over 150 houses were damaged within the Fresno County.

Rainfall in excess of 2.5 inches in just over 30 hours on January 1–2 led to water-covered roadways in several locations around Kings County. Hanford measured 2.82 inches of rain in a 30-hour period while Lemoore and Corcoran received just over 3 inches. Ponding basins overflowed in Lemoore, and flooding occurred in Huron and Corcoran.

Consistent rains led to more than 3 inches of rain in a 30-hour period from mid-day on January 1 to the evening hours of the 2nd around Visalia and over 3.5 inches of rain in the city of Tulare. Over 2 feet of water flooded portions of West Visalia as well as flooding just east of Tipton. Tulare County had 45 homes damaged to some extent by flooding. Some of the flooding in Visalia was due to detention basins and pumps failing to perform as expected.

Over 7.5 inches of rain was reported during the storm event at the 2,000 foot elevation in the Tulare County foothills. Strong winds in the Tulare County mountains felled several large trees on January 2. Among those were:[1075]

- The second largest limb on the General Sherman Tree
- The Telescope Tree, a large, hollow giant sequoia on the Congress Trail in Giant Forest
- A large tree that fell onto the Runciman cabin in East Mineral King

Strong wind events were reported elsewhere in the San Joaquin Valley during the January 1–2 storm event:[1076]

- During the late morning and early afternoon of January 1, gusty southeast wind commonly hit over 40 mph in the northern portions of the central San Joaquin Valley.
- On the afternoon of January 2, strong wind in the eastern part of Fresno County blew down trees and power lines, leaving over 60,000 customers without power.
- During the mid-afternoon hours of January 2, strong southeast wind in the Taft area in southwest Kern County resulted in downed power poles.
- On the evening of January 2, gusty wind caused damage around the community of Oakhurst and blew down numerous trees.

Gary Sanger at the NWS forecast office in Hanford said that the January 1–2, 2006 events likely were dominated by frontal winds. However, there may have been unreported embedded thunderstorms (and thundersnow) in the cold front's convective band. See the section of this document that describes the 1941 Wind Event for more detail about strong winter winds capable of causing forest blowdowns.

Rainfall totals of almost an inch in less than 24 hours at Bakersfield resulted in significant water flows from nearby mountains onto the south San Joaquin Valley floor. Flooding occurred on Highway 33 north of Highway 46 on the morning of January 2 on the west side of the valley.

The storm dropped significant snow at higher elevations. Lodgepole and Mineral King reported 36 inches of new snow between January 1–2. Charlotte Lake received 41 inches of new snow, and Farwell Gap had 72 inches (6 feet) during the same period.

Table 83 summarizes the damages incurred during the January 1–2 storm event.

Table 83. Damages incurred during January 1–2, 2006 storm event.

| County | Property Damage (million dollars) | Agricultural Damage (million dollars) |
|---|---|---|
| Fresno | $1.5 | not assessed |
| Kings | $0.1 | $1. |
| Tulare | $5.72* | not assessed |
| Kern | $.025 | not assessed |
| Total | 7.345 | $1. |

*$220,000 of the property damage in Tulare County was attributed to falling trees, including one at Mineral King.

In April, there was another round of storms which combined with snowmelt to create more significant flooding over a greater area. As a result of severe storms, flooding, landslides, and debris flows, a major federal disaster (DR-1646) was declared on June 5, 2006, for the period March 29,

2006 – April 16, 2006. It covered 17 Northern California counties.[1077] Minor flooding occurred throughout the San Joaquin River and Tulare River Basins.

The peak day natural flow at Pine Flat on the Kings occurred on April 5. The flood surge lasted three days, but was unremarkable by Kings River standards. This was followed by sustained high flows from May through June. Although the river didn't reach a particularly impressive height, it apparently delivered more than the average amount of water to the valley floor.

A cloudburst occurred above Silver City in the Mineral King area on or about July 20, 2006. It resulted in a flash flood on the creek west of Silver City. That creek is locally known as Silver Creek or Silver City Creek. (This is the same drainage that had a much smaller flash flood in 1997.) The 2006 flood damaged some of the cabins in Cabin Cove and washed out the Mineral King Road (multiple photographs on file in the national parks). Tony Caprio and Joel Despain (the national parks' fire ecologist and geologist, respectively) recalled that the flood eroded these low-gradient stream channels down nearly to bedrock. This level of erosion implies a very high stream discharge far from the statistical norm, an event that would occur only rarely. The effect of the storm was very localized. There was little effect east of Silver City (for example, High Bridge) or west of Deadwood Creek, with the latter having only a moderate increase in flow.

Total flow for water year 2006 was 171% of the 1894–2011 average for the Kings, 164% for the Kaweah, 147% for the Tule, and 150% for the Kern. In order to minimize flooding in the Tulare Lakebed, 29,000 acre-feet of river floodwater was pumped into the Friant-Kern Canal and routed to the Los Angeles area in 2006.

### Debris Flow: Cement Table

The national parks do not know precisely when this event occurred. It may or may not have occurred during the floods of 2005. David Karplus, Kings Canyon National Park's trails supervisor, discovered it in 2008, and at that time it appeared to have been there for a couple of years at least. He knows that it was not there in 1999.

The debris flow began on the east side of Cloud Canyon, and flowed across the trail and into the creek. It started over 1,000 feet up the canyon wall (UTM 365105E 4060830N NAD83 Zone 11). The debris flow was about 100 yards wide and consisted of two channels where it crossed the Colby Pass Trail, each channel about 15 feet deep. It eroded down to bedrock in many places. Rather than fill in the eroded trail, David simply had his crew dig a trail across the new mud slope.

## 2007–09 Drought

California experienced three consecutive dry years during 2007–09. Those years also marked a period of unprecedented restrictions in State Water Project (SWP) and federal Central Valley Project (CVP) diversions from the Sacramento–San Joaquin River Delta to protect listed fish species.

Statewide hydrologic conditions overall were not as severe during 2007–09 as compared to prior droughts of statewide significance. Water years 2007–09 were the 12th driest three-year period in the state's measured hydrologic record, based on the DWR's 8-station precipitation index. That means that the state experienced 11 three-year periods during the 20th century that were more severe than the 2007–09 drought.

Table 84 compares the 2007–09 drought with more severe droughts of the 20th centuries.

Table 84. Comparison of selected 20th century droughts.

| Drought Period | Sacramento Basin Runoff | | San Joaquin River Basin Runoff | |
| --- | --- | --- | --- | --- |
| | Average yearly runoff (million acre-feet) | % of average 1901–2009 | Average yearly runoff (million acre-feet) | % of average 1901–2009 |
| 1929–34 | 9.8 | 56% | 3.3 | 56% |
| 1976–77 | 6.6 | 38% | 1.5 | 26% |
| 1987–92 | 10.0 | 57% | 2.8 | 48% |
| 2007–09 | 11.2 | 64% | 3.7 | 63% |

A system deposited 0.08 inch of rain in Bakersfield on May 27, 2008. That was the only measurable rain to fall in the entire March-May period in that city and tied 1992 for the driest meteorological spring on record.[1078]

A DWR report said that the 2007–09 drought impacts were most severe on the west side of the San Joaquin Valley.

The major difference between the 2007–09 and prior droughts was the severity of SWP and CVP delivery reductions, which began immediately in the first year of the drought.

Table 85 shows how the San Joaquin Valley Water Year Index categorized the drought years in that basin.

Table 85. Rating of drought severity during the 2007–09 drought.

| Water Year | San Joaquin Basin Water Year Classification |
| --- | --- |
| 2007 | Critically dry |
| 2008 | Critically dry |
| 2009 | Below normal |
| 2010 | Above normal |
| 2011 | Wet |

Water years 2007 and 2008 clearly constituted a multi-year drought; that's obvious from looking at Table 85 or Figure 13. The runoff was so low in those years that the state's water year index rated those years as critically dry. From a hydrologic standpoint, the drought lasted only through 2009, the last year of below-normal flows.

Governor Arnold Schwarzenegger proclaimed a state of emergency on June 12, 2008, recognizing the onset of the drought. The 2007–09 drought was California's first drought for which a statewide proclamation of drought emergency was issued. That turned out to be critical. When precipitation returned to near-average or above-average, it was hard politically for the governor to declare an end to the drought. There clearly wasn't enough water to go around.

Other droughts had been declared over when water conditions returned to near-average conditions. That didn't seem possible in this drought. It wasn't until March 30, 2011, after an incredibly wet winter, that Governor Jerry Brown issued a proclamation rescinding the state of emergency. The drought had finally met an official end. That was long after the end of the hydrologic drought.

However, there still wasn't enough water to go around. There was still a significant groundwater overdraft. There were still signs posted along some of our highways expressing the opinion that the drought was caused by Congress. The signs are still there, and the groundwater overdraft persists. For more about these issues, see the section of this document that describes the Groundwater Overdraft.

As explained in the section on Groundwater Overdraft, water users have come to rely on the groundwater aquifer more and more, especially during droughts.

During the four year period between April 2006 and March 2010, water users in the Tulare Lake Basin used a huge amount of groundwater. Not all of those four years were drought years. The San Joaquin Valley Water Year Index categorized water year 2006 as a wet year; the drought didn't move into the valley until 2007.

Apparently there are two separate estimates of the size of that water withdrawal:

- In February 2011, the University of California Center for Hydrologic Modeling at the University of California, Irvine, estimated that based on satellite data, the groundwater loss was more than half the size of Lake Mead (19.5 million acre-feet), the third largest decline in 50 years.[1079]
- In 2012 Bridget Scanlon and her colleagues at the University of Texas published what appears to be a separate analysis of the same four-year period. According to news accounts, Scanlon found that water users in the Tulare Lake Basin had used enough groundwater to fill Lake Mead, the nation's largest man-made reservoir.[1080]

## 2007 Flood

Flooding in 2007 occurred during October. This flood happened during the 2007–09 drought.

The National Weather Service rated this event as one of the largest severe weather outbreaks on record in interior Central California. It occurred during the afternoon and evening hours of October 29. An upper-level low moving inland across Central California interacted with a surge of tropical moisture, triggering thunderstorms that produced hail in many places as large as one inch in diameter and gusty winds as well as locally drenching rains. Hardest hit was the northwest side of Fresno where rainfall totals of one to two inches were reported and a number of streets flooded quickly during the evening rush hour resulting in a good many stalled vehicles. Some streets in northwest Fresno were still covered with several feet of water nearly four hours after the thunderstorms had ended. In addition, hail up to an inch in diameter fell. The combination of the heavy rain and hail resulted in the collapse of the roof on an 80,000 square foot warehouse. Thunderstorm winds also knocked out power to 18,000 customers in Fresno.

Two houses in Visalia had trees fall on them, and about 200 boats were damaged at a boat dock on Lake Kaweah. Downed trees were reported in the valley from Merced County to Tulare County and eastward into the Sierra at Yosemite and Sequoia National Parks.

## 2008 Flood

Flooding occurred in July, primarily from July 12–15. This flood occurred during the 2007–2009 drought.

It was caused by a series of individual storm events, but could be thought of as one event that occurred in multiple locations.

These storm events were caused by the North American Monsoon. From July 7–11, high pressure centered over the Four Corners area dominated the weather over the Southern Sierra. By July 12, a major low pressure area had formed in the Southwest. In addition, moist air influenced by Hurricane Bertha east of Bermuda began to reach the Southwest.[1081]

Conditions were also changing in California. An upper-level high pressure ridge moved inland off the Pacific on July 11, with a low pressure area moving along the coast. This pattern set up a southerly wind pattern over California, drawing up monsoonal moisture from the southeast on July 12. As the air flow from the southeast brought in this moist air, thunderstorms formed over the Southern Sierra. Thunderstorms formed over the Tulare County mountains by early on the afternoon of July 12 and remained in the region through July 15.[1082] Following is a sample of the storm events that occurred during the July 12–15 period.

### *Mud flows: Tioga Pass Road*
On July 14, the Tioga Pass Road in Yosemite National Park was closed due to mud flows across that road. Thunderstorms dropped a lot of hail, resulting in pea-sized hail covering about a two-mile section of the road.[1083]

### *Debris Flow: Oak Creek*
This debris flow was generated within Oak Creek, an east-flowing drainage near Independence. Oak Creek is within the Inyo National Forest. That is outside the Tulare Lake Basin. However this event merits inclusion in this document because of the similarities of this debris flow to the one that would occur in the nearby Lewis Creek Basin just two days later.

This event was analyzed by Jerry DeGraff, Dave Wagner, and others.[1084]

Like Lewis Creek, the bedrock in the Oak Creek Basin is granitic. Perhaps there are some soils in the upper elevations that are composed largely of granitic sands. However, most of the bedrock is apparently overlaid with deposits of alluvium, basaltic lava flows, glacial outwash and moraines, landslide deposits, alluvial fan deposits, and colluviums. Whatever the specific soil mix, the rapid rainfall apparently saturated this granular soil, reducing its frictional strength, and causing the soil mass to begin moving downslope as a flowing mass.[1085]

Late on the afternoon of July 12, a large convective cell centered over Oak Creek Canyon produced a brief period of intense rainfall. While debris flows are sometimes triggered by a prolonged rain or snowmelt event, the Oak Creek debris flow was triggered by this short, intense rainfall event. The debris flow started within an hour or so of the onset of that cloudburst. The rain started sometime after 4:00 p.m. We don't have any direct precipitation data for that thunderstorm because the RAWS automated weather station located upstream from the South Fork Oak Creek junction was swept away by the debris flow just before its scheduled transmission at 4:42 p.m., less than an hour after the onset of the rain.

The Oak Creek debris flow was initiated in the headwaters of the North and South Forks of Oak Creek at about elevation 10,825 feet (3,300 m). It then traveled about 11 miles (18 km) before coming to rest on the floor of Owens Valley at about elevation 3,840 feet (1,170 m).

The debris flow that came down the North Fork of Oak Creek deeply eroded the existing creek channel through the Oak Creek Campground and severely damaged the nearby road. In the campground, a camper escaped his collapsing motor home when it became entangled in a grove of trees. He jumped into the muddy debris flow and "surfed" several hundred yards until he eventually reached stable ground where he could walk to help. This fortunate individual reported experiencing three distinct waves or surges of material with the larger one being between 6–12 feet (2–4 m) high. The velocity of that larger wave would have been about 12.5 mph (20 km/hr).

The debris flow that came down the South Fork of Oak Creek destroyed the Bright Ranch, including the main house. Fortunately the inhabitants were away at the time that the debris flow struck. That ranch had been occupied since 1872 and had never been impacted by a debris flow. That suggests that this debris flow had a long recurrence interval.

The watersheds serving as the source for the Oak Creek debris flow were burned during the Inyo Complex Fire in July 2007. There is circumstantial evidence that this might have created the conditions necessary for this debris flow. Presumably equally severe thunderstorms had occurred in the Oak Creek Basin since 1872. Yet this was the first time that the Bright Ranch had been impacted by a debris flood. That raises the possibility that the effect of the Inyo Complex Fire on this watershed contributed to this unusual event.[1086]

The South and North Forks converge at about 5,250 feet (1,600 m). Just east of the junction of these two tributaries, the debris flow struck the historic Mt. Whitney Fish Hatchery. The buildings survived, but all the fish were wiped out.

In the reach below the fish hatchery, 17 homes were damaged or destroyed. By about 5:30 p.m., most people along Oak Creek had abandoned their homes and only one person had to be rescued from a rooftop. Fortunately no one was seriously injured or killed.

The debris flow blocked the drainage structure under U.S. Highway 395, forcing the flow to mobilize over the top of the roadway. This resulted in closure of the highway for five hours, from 6:30 p.m. – 11:45 p.m. It took Caltrans nearly a week to fully restore the highway.

After crossing the highway, the debris flow damaged another 25 homes on the Ft. Independence Indian Reservation.

The debris flow had a total length of about 11 miles and a drop of 6,985 feet in elevation. It had an average gradient of 635 feet/mile. That averages the steeper mountain section and the flatter section lower down. The peak velocity measured was about 12.5 mph (20 km/hr).

The debris flow had a total estimated volume of about 2.04 million cubic yards (1.56 million cubic meters). This is the largest Sierra debris flow ever definitively measured.

### *Debris Flow: Erskine Creek*

There were multiple bouts of flash flooding in the Kern County foothills and mountains during the July 12–15 time period.[1087]

The Piute Fire was a major wildfire that began south of Lake Isabella on June 28, 2008. Precipitation information was available for this area from RAWS automated weather stations because the Lake Isabella area had experienced flash floods in the past and also because it is near a major reservoir. In addition, the firefighting efforts directed at the Piute Fire placed a temporary RAWS at Piute Peak to assist with fire weather forecasting. An NWS meteorologist was assigned to the firefighting incident, giving us a good record of the events that would unfold.

Heavy rain hit the Piute Fire area for three days in a row: July 12–14. On July 15, heavy rain occurred just south of the fire area. The rain was sometimes quite intense. For example, on July 15, a Claraville weather station (due south of Lake Isabella) reported 2.15 inches of rain in 90 minutes (a rate of 1.43 inches per hour).

The intense rain caused at least seven debris flows and flash floods in the area plus one in the town of Tehachapi. The most impressive of those was the July 12 Erskine Creek debris flow. Erskine Creek would also experience what the NWS characterized as debris flows on July 13 and July 14.[1088] Erskine Creek is within the Sequoia National Forest. The July 12 Erskine Creek debris flow was analyzed by Jerry DeGraff and others.[1089]

When thunderstorms began to develop late on the afternoon of July 12, the NWS weather specialists on the fire team and at the NWS forecast office in Hanford recognized the high likelihood of flash flooding. The Piute Fire Unified Command issued a flash flood warning and recommended evacuation notice for Erskine Creek. As the storm developed, a helicopter operated by the Kern County Fire Department was dispatched for aerial observation.

The July 12 debris flow didn't start from a prolonged rain or snowmelt event. It was triggered by a short, intense rainfall event. The debris flow started within an hour or so of the onset of that cloudburst.

Whenever a debris flow occurs after a fire, there is a tendency to assume that the debris flow was caused primarily by the fire. That is, to assume the fire was necessary to create the conditions for the debris flow to occur. The Piute Fire did contribute to the progressive bulking of the Erskine Creek debris flow. But otherwise, there was no clear-cut relationship between that fire and the debris flow.[1090]

Debris flows are rarely seen, let alone photographed. However, this one happened to occur where an incident command post was already established; one that had both good meteorological data and helicopter support. It also occurred in daylight. That combination made all the difference.

The debris surge had multiple surges. The first surge was the largest and darkest one, containing abundant ash from the burned slopes. The helicopter spotted the debris flow in the upper watershed and followed the leading edge through the town of Lake Isabella, capturing dramatic footage (video on file in the national parks).[1091]

The Erskine Creek debris flow was generated by flows from within the South, Middle and East Forks of Erskine Creek.

Like Lewis Creek and Oak Creek, the bedrock in the Erskine Creek Basin is granitic. Perhaps there are some soils in the upper elevations that are composed largely of granitic sands. However, most of the bedrock is apparently overlaid with deposits of landslide deposits, colluviums, and alluvial fan deposits. The town of Lake Isabella is built on the alluvial fan deposits.

The longest tributary, the South Fork, flows about 6.5 miles (10.5 km) from its headwaters at about 8,185 feet (2,495 m) to the junction with the other two tributaries at an elevation of 4,395 feet (1,340 m). From there, Erskine Creek flows another 8.3 miles (13.4 km) to its junction with the Kern River at an elevation of 2,445 feet (744 m).

The debris flow had a total length (including the South Fork tributary) of 14.8 miles (23.9 km) and a drop of 5,740 feet in elevation. It had an average gradient of 388 feet/mile. That averages the steeper mountain section and the flatter section lower down.

The portion of the debris flow that was on the South Fork tributary had a length of 6.5 miles and a drop of 3,790 feet in elevation. That section had an average gradient of 583 feet/mile.

The gradient along the flow path was about 23%. The peak velocity measured was about 12 mph (19 km/hr).

Because the debris flow entered and dispersed within the Kern River, there is no accurate estimate of its volume. On the basis of the affected area, it would seem comparable in size to the Oak Creek debris flow which had a total estimated volume of about 2.04 million cubic yards (1.56 cubic meters).

There was an initial and natural temptation to associate the three Erskine Creek debris flows / flash floods with the Piute Fire. That is, to assume that the debris flows were caused in some way by the fire. However, as Jerry DeGraff and others showed in their analysis, that was not the case with the Erskine Creek debris flows.

However, in 1984 the Lake Isabella area did experience a major debris flow that might have been due in part to a wildfire. That was in the Goat Ranch Canyon and Long Canyon areas (see the section of this document that describes the 1984 flood).

Erskine Creek flash flooded on three consecutive days, July 12–14.[1092, 1093] When the July 12 debris flow swept into the town of Lake Isabella at about 6:30 p.m., Erskine Creek overflowed Lake Isabella Blvd (video on file in the national parks).[1094] Because of the advance notice, crowds had gathered on each side of the barricaded area to watch this dramatic event. This was probably the most viewed and photographed debris flow ever in the Tulare Lake Basin.

Erskine Creek would overflow Lake Isabella Blvd. again on July 13 and briefly on July 14. Each time, the Piute Fire Unified Command issued a flash flood warning and recommended evacuation notice far enough in advance so that the road could be closed to traffic.

The debris flood of July 13 was very powerful. In places, the floodwaters were 100 yards wide and 18–24 inches deep. Many people had heeded the flood warning and evacuation notice. Up to 80 homes in the Erskine Canyon area were evacuated. But still, despite the warning — and the experience of the previous day — some people were caught unprepared. Helicopter crews from the

Kern County Fire Department rescued two families, consisting of a total of seven people and two dogs, from their homes along Erskine Creek. One of those families and their dog had to be plucked off their roof.[1095]

The three debris flows / flash floods in the town of Lake Isabella resulted in $1.5 million in property damage.[1096]

Erskine Creek empties into the Kern River; so much of the debris from the three days of debris flows / flash floods became dispersed into the flow of the Kern and carried downstream. At a point 17 miles (27 km) downstream is Democrat Dam, a low diversion dam to divert water to the KR1 hydroelectric plant operated by SCE. The KR1 powerhouse is located another 10.2 miles (16 km) downstream. Near that powerhouse is the Kern Canyon diversion dam to direct water to the PG&E-operated Kern Canyon powerhouse at the mouth of the Kern River Canyon. Both of those plants were taken off-line to avoid damage to generating equipment and remained closed for a number of days. Additionally, habitat supporting a genetically pure strain of Kern River rainbow trout was severely damaged within the Erskine Creek watershed.[1097]

On July 15, the Bakersfield water supply was threatened by dirt and silt washing down the Kern River. A portion of the water treatment facility had to be shut down. The city had only a three-day supply of clean emergency water, and that was beginning to run out at one of the treatment plants.[1098]

### Other Debris Flows and Flash Floods: Kern County
In addition to the three debris flows / flash floods on Erskine Creek, there were a number of other flash floods and debris flows in the Kern Mountains during the July 12–15 period. The ones that we are aware of were:

- On July 12, a debris flow passed down Thompson Creek, a tributary of Walker Basin Creek. This event was analyzed by Jerry DeGraff.[1099] Thompson Creek Basin adjoins that of Erskine Creek. Like the South Fork of Erskine Creek, the head of this debris flow was near Piute Peak where the thunderstorm cell was centered. Sediment from the Thompson Creek debris flow was evident a distance of about 12.5 miles (20 km) downstream where Kern County Highway 483 crosses Walker Basin Creek. A number of residential structures were impacted by this debris flow.
- On July 12, a debris flow passed down Clear Creek, a tributary of Havilah Canyon Creek. This event was analyzed by Jerry DeGraff.[1100] The Clear Creek Basin adjoins that of Erskine Creek. The head of the debris flow was near Piute Peak where the thunderstorm cell was centered. While the Clear Creek debris flow did not appear to pass the entire 10½ miles (17 km) to where the creek is crossed by Kern County Highway 483, the sediment from this event was visible from that point. There are no roads or other infrastructure in the bottom of Clear Creek, so there was no damage from this event.
- On the afternoon of July 14, a flash flood / debris flow occurred on Johns Rd between Caliente Creek and Walsher Rd.[1101]
- On July 14, flooding occurred in the town of Tehachapi. An apartment complex in that town sustained significant damage.[1102]
- On the afternoon of July 15, Thompson Creek Road was washed out, stranding 40 homes about 10 miles south of Lake Isabella.[1103]

### Debris Flow: Charlotte Lake

This debris flow resulted from a cloudburst that occurred on July 14, 2008. A total of 1.73 inches of rain fell between 2:00–6:00 p.m. The intense part of the storm began at about 3:45 p.m. The first 20 minutes or more was heavy hail followed by heavy rain. George Durkee, the wilderness ranger at Charlotte Lake, looked up at about 4:20 after hearing the rain intensify for about 15 minutes and saw a flood of brown water and debris (small rocks, mud, and pine needles) coming down the hillside behind the ranger station.

The cloudburst resulted in several small debris flows and numerous gullies up to a foot deep in a 30 square mile area between Charlotte Lake and Bullfrog Lake (UTM 372965E 4072095N NAD83 Zone 11) (map and multiple photographs on file in the national parks). George attributed the huge runoff to the heavy warmer rain melting the recent hail and the combined water coming down all at once.

### Debris Flow: Lewis Creek

According to Mel Manley and Ken Hires, Kings Canyon experienced a lot of rain over the weekend of July 12–13. A couple of those storms dropped 1½ inches of rain in 1 hour.[1104] The storms continued into the following two days as well. These were evidently localized storms. The Cedar Grove gage (CGR) recorded only 0.44 inches of rain for the July 13–16 period.

On July 14, a particularly intense thunderstorm cell appears to have been centered near Kennedy Mountain along the Monarch Divide. This date is based on a ranger weekly report which recorded the storm as occurring on July 14. There is some confusion about this date. A photograph taken of the ensuing flood in the Kings River has the date stamp of July 15. In addition, Bill Templin recorded two weeks later that Ken Hire, the Cedar Grove lead interpretive ranger, said that the storm happened on July 15.

Possibly some of the July 12–13 storms had included the Lewis Creek Basin; we have no way of knowing. The ground may have been relatively dry. In any case, the July 14 thunderstorm triggered a small- to moderate-size debris flow on the north side of Kennedy Mountain. We know about that debris flow because it crossed the Kennedy Pass Trail.

In addition, that storm triggered a major debris flow on the south flank of Kennedy Mountain, just west of Kennedy Pass at about 10,200 feet elevation (UTM 351797E 4081928N NAD83 Zone 11). The following description refers to this southern debris flow.

No soils map exists for this hillside. However, judging from the material that came off in this debris flow, the soil appears to be largely sand with some silt and perhaps some clay. Such granular soils depend on grain-to-grain friction for soil strength. Rapid rainfall can exceed the infiltration capacity of the soil so that it becomes saturated. This results in water filling the pore spaces of the soil and reducing the contact between individual grains. The decreased frictional contact temporarily reduces soil strength at which point the soil mass can begin to move downslope as a flowing mass. Debris flows in the Sierra are initiated when these conditions arise from intense rainfall, rain-on-snow events, or rapid snowmelt.[1105] A large debris flow can be triggered in as little as one hour from the onset of an intense rainfall event.

This southern debris flow started on an unburned slope, well above any of the fires that have burned in that area, including the 2005 Comb Fire. From there, the debris flow traveled down Lewis Creek

5–6 miles (9 km) to the valley floor. Several smaller tributary debris flows joined the main flow along the way (PowerPoint on file in the national parks).

Whenever a debris flow occurs after a fire, there is a tendency to assume that the debris flow was caused primarily by the fire. That is, to assume the fire was necessary to create the conditions for the debris flow to occur. That was the case with the 2008 Lewis Creek debris flow; it was initially attributed to the 2005 Comb Fire.

However, by using a set of pre-fire and post-fire/post-debris flow aerial images, the parks' fire ecologist Tony Caprio was able to ascertain that the flow started above and outside the area burned by the fire. That is clearly shown in the PowerPoint that Tony created to document his research (see the PowerPoint referenced above).

The debris flow scoured the hillside, removing vegetation and eroding deeply into the ground (multiple photographs on file in the national parks). Although bedrock was exposed in some places, that doesn't necessarily mean that this event had a long recurrence interval. Jerry DeGraff, a geologist for the USFS, said that sometimes debris can fill in pretty fast after one of these events.

In any case, the debris flow scoured out rock, decomposed rock (granitic sand), and ash from the 2005 Comb Fire. It crossed the Kennedy Pass Trail in two places, causing major damage (multiple photographs on file in the national parks). A California Conservation Crew worked for 1½ months the following summer repairing that damage at a cost of $80,000.

This dramatic event was similar to a debris flow that occurred in an unnamed tributary of Lewis Creek on this hillside on May 27, 1983. The high energy debris flows that occur in the Lewis Creek Basin are somewhat reminiscent of the power and ferocity of avalanches.

No one witnessed the 2008 debris flow as it came off the hillside, so we don't know how fast it was moving. Debris flows in the Sierra have an average peak velocity of 12.4 mph (20 km/hr).[1106] From looking at the aftermath of the 2008 event, it's tempting to think that it was a much faster than average debris flow. In areas of very steep slopes, debris flows can reach speeds of over 100 mph.[1107]

All of the material that was scoured off the hillside in the July 2008 event was delivered to the valley floor. When the debris flow reached the gentler slopes at the canyon mouth, the event changed from scouring mode to depositional. This transition point occurred at about elevation 5,200 feet, about one mile upstream from the Lewis Creek highway bridge. That was 5–6 horizontal miles (9 km) from where the debris flow had begun. By this point, the debris flow had dropped about 5,000 feet in elevation.

A log jam / debris dam 20–30 feet high formed about 200–300 feet downstream from the canyon mouth (multiple photographs on file in the national parks).

Bill Templin, Rick Hartley, and the Kaweah Flyfishers visited the lower Lewis Creek channel on July 26, 2008. Bill noted that the event had covered up all of the benthic invertebrate habitat in Lewis Creek with a heavy deposit of sediment. This material was presumably a mixture of granitic sand, silt, and possibly clay.

By the time of their visit, bear tracks had been pressed into a soft, ashen-colored sediment in the bed of Lewis Creek that appeared to be silt or clay (photograph on file in the national parks).

Bill also observed that there was a layer of fine, gray material that seemed to have been blasted onto the sides of the creek channel. It clung to the rocks and trees like a coating of shotcrete would adhere to the sides of a swimming pool (photograph on file in the national parks). Jerry DeGraff said that such fine, gray material is a common feature of debris flows when you are there soon enough after occurrences — before rainfall and wind essentially scrub it off the trees and rocks. It is not ash from fires because it appears in both burned and unburned areas. It is mineral in nature, presumably finely ground rock.

David Karplus, Kings Canyon National Park's trails supervisor, recalled that Lewis Creek experienced some very big channel changes as a result of this debris flow. The park has no record of what the channel was like prior to the debris flow.

David and others noticed how black or chocolate colored the flow in Lewis Creek and the Kings River was when this event occurred. Ned Kelleher (Kings Canyon district ranger) captured this in a time-lapse sequence that he took of the Kings River during the event (multiple photographs on file in the national parks). The color was presumably due to the ground rock and the mysterious silt/clay/ash component of the debris.

The 2005 Comb Fire occurred three years earlier, and the remaining ash would have been a relatively small component of the total debris that washed off the hillside. However, that ash could very well have been mixed in the early parts of the flow; it is floatable material and tends to be part of the leading edge of a debris flow.

Ned's log recorded the debris flow on Lewis Creek and the Kings as occurring between about 3:00–5:00 p.m. The lower portion of Lewis Creek came up 3–4 feet and flushed an enormous amount of debris and sediment into the Kings River. The Kings came up over a foot within 10 minutes.

A large amount of the debris (some cobblestones, but consisting primarily of granitic sand and silt/clay/ash) washed down the Kings River as part of the debris flow. Bill Templin recalled that a thick covering of sediment completely covered the benthic invertebrate habitat as far as the old USGS gaging station below Grizzly Falls. Silt deposits were observed all the way down to Boyden Bridge and presumably continued below that. Jeff, an employee at Boyden Cave, told Bill that a large amount of woody debris floated downstream past the cave.

David Karplus recalled that deep holes in the Kings River were filled with up to about 4 feet of sediment, and a thick covering was spread over the entire riverbed. The effect was to generally level out the entire riverbed.

Rick Hartley said that the bottom of the Kings was covered with roughly 8 inches of sandy sediment, and that there were many new sandbars (multiple photographs on file in the national parks). In addition to the sand, the sediment in the riverbed contained a fine gray material that could easily be stirred up. It wasn't clear whether that was silt or clay (photograph on file in the national parks). Since the ash is floatable, it wouldn't have settled out with the sand and silt in the riverbed; it would have continued on downstream.

Bill Templin and others caught several fish in the Kings River above Deer Cove Creek. Those fish were ashen-colored and full of this fine, gray material inside. They contained no benthic invertebrates, only terrestrial insects in their stomachs. The female's eggs were brown instead of orange.

Bill estimated that the sand and silt/clay debris covered the Kings riverbed thickly for about 10 miles from the junction with Lewis Creek to just below the old USGS gaging station at about elevation 4,100 feet. That would be 11 miles below where the debris flow had changed from its scouring to depositional phase. The debris flow had dropped about 1,100 feet in elevation in the 11 miles since it changed to its depositional phase.

Bill documented the effects that he observed in a July 28, 2008 email to Sequoia National Park, Sequoia National Forest, California Department of Fish and Game, and others. He also wrote it up in a newsletter for the Kaweah Flyfishers.[1108]

The total length of this debris flow was about 17 miles, 6 miles in the scouring phase and 11 miles in the depositional phase. It dropped a total of 6,100 feet in elevation, most of which was in the high-energy scouring phase. The gradient during the scouring phase (5,000 feet in 5–6 miles) was between 833–1,000 feet/mile. No debris flow like this had previously been recorded in the Tulare Lake Basin.

For comparison, the nearest equivalent debris flow that we are aware of was the Oak Creek debris flow that occurred just two days earlier on July 12, 2008. The Oak Creek debris flow had a total length of about 11 miles and a drop of 6,985 feet in elevation. It had an average gradient of 635 feet/mile. That averages the steeper mountain section and the flatter section lower down. The Oak Creek debris flow occurred just east of the Lewis Creek debris flow and was caused by the same storm system. It had an equivalent length and drop. However, it had a much greater volume.

By the end of the spring 2009 runoff, the Kings River depths and cobble sizes apparent to a casual swimmer (David Karplus) had returned to the pre-event 2008 levels.

While the big thunderstorm was triggering the Lewis Creek debris flow in the north part of the national parks in 2008, another storm event was occurring farther south. At 4 p.m. on the afternoon of July 14, the Atwell / Cold Springs area in Mineral King was receiving heavy rain, resulting in flood damage to the Mineral King Road. Mud and rocks washed onto that road, forming deposits up to three feet deep, leaving the road impassable.

Lewis Creek forms a delta where it exits the canyon; that delta extends down to the Kings River. The main highway cuts across the lower portion of that delta. The primary channel of Lewis Creek flows from the mouth of the canyon straight across that delta, under the highway bridge, and into the river.

The July 2008 debris flow left an unstable channel above the highway bridge. A large amount of sediment and debris was deposited on the delta, especially along the general course of the primary channel. In addition, a log jam formed on the primary channel a short distance below where Lewis Creek exits the canyon. Within the next year or two, about 100 feet of the primary channel above the log jam filled in. During roughly this same period, an overflow channel developed along the northwest side of the delta, along the road to the parks' wastewater treatment plant. That channel gradually came to be called the Lewis Creek overflow channel.

On June 6, 2010, a small debris jam formed in the primary channel of Lewis Creek upstream of the highway bridge. That debris jam resulted in diverting water flow outside the primary channel into the overflow channel. The 16 inch culvert where the overflow channel) goes under the highway could not carry the resulting flow, so Lewis Creek (via its overflow channel) overflowed the highway for a couple of days and caused some minor road damage. It also caused some erosion along the road to the wastewater treatment plant. This channel-shifting happened again from about June 20–30, 2011.

In the fall of 2011, the national parks' road crew installed two more culverts (an 18 inch and a 24 inch) to help carry the flow that results when a portion of Lewis Creek moves into the overflow channel.

Where Lewis Creek emerges from its canyon, there is very little impediment to keep it from changing course from the primary channel to the overflow channel. If the primary flow of Lewis Creek were to move to the overflow channel, then a different solution would have to be found. The three highway culverts on the overflow channel are not sufficient to carry the full flow of Lewis Creek. Either a bridge would be required, or the creek would have to be pushed back into its primary channel as was done after the May 1983 debris flow.

## 2009 Flood

Flooding in 2009 occurred in October. This flood occurred during what might be considered the ending months of the 2007–2009 drought.

On October 8, Super-typhoon Melor struck Japan. Four days later, rain from the remnants of that powerful typhoon encountered the Sierra. Melor itself didn't travel across the Pacific, but water vapor from the typhoon moved via an atmospheric river.[1109] Figure 21 shows that atmospheric river channeling water vapor from the decaying Typhoon Melor over the western North Pacific, across nearly the entire width of the ocean basin to the Sierra on October 14, 2009.

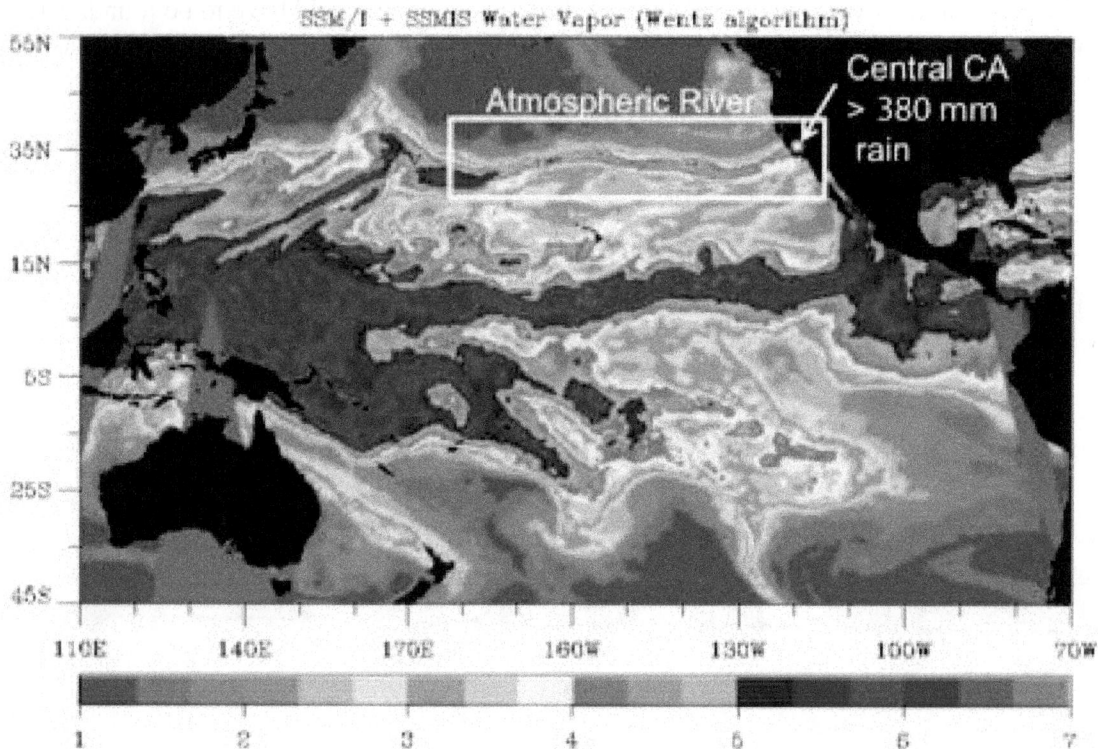

Figure 21. An atmospheric river channeling water vapor from the decaying Typhoon Melor across the Pacific Ocean to the Sierra on October 14, 2009.
Source: Michael Dettinger, USGS/Scripps.

Moisture from the remnants of Melor arrived over the Central California interior, bringing high-elevation snowfall and large amounts of rainfall. A deep low pressure trough tapped into the moisture from the remnants of Melor on October 13–14. The impacts of the storm and resulting flood were documented by the NWS forecast office in Hanford.[1110] Many valley and Sierra locations set new record-high precipitation amounts.

At the onset of the storm during the morning of October 13, quite a bit of snow fell over the crest of the Sierra. However, once the precipitation began in earnest by the afternoon of October 13, snow levels rose to elevations of over 10,000 feet. By the end of the event on October 14, most of the precipitation was falling as rain.

Maximum rainfall totals were near 19 inches in 24 hours along the Central Coast and greater than 10 inches along the Southern Sierra. Several roads and highways along the Central Coast were closed due to flooding. Landslides were reported in the Santa Cruz Mountains.

The Dinkey Creek RAWS automated weather station southwest of Shaver Lake received the most precipitation of any Sierra site. That station received a storm total of just over 13 inches, about 9 inches of which fell in 12 hours on October 13. Dinkey Creek experienced a flash flood.

Grant Grove received 7.7 inches of rain during the October 12–13 storm event. That was significantly more precipitation than had been recorded during any two-day period of October at Grant Grove or Giant Forest in the previous 80 years. The resulting flooding caused significant erosion in the newly restored Halstead Meadow and elsewhere in the national parks.

At Three Rivers (the TRR gaging station), the Kaweah River had been flowing at 27 cfs. Within a matter of a few hours, it peaked at 20,937 cfs. (That was the peak hourly flow; the peak average daily flow was 7,360 cfs.) One source said that Lake Kaweah rose 30 feet in less than 24 hours. René Ardesch captured footage of the flood as the Kaweah River passed under the Pumpkin Hollow Bridge (video on file in the national parks).[1111]

This fall flood was reminiscent of the floods that occurred in September, 1976, September 1978, September 1982, and November 2002. All of those floods were caused by the remnants of Pacific hurricanes.

The October 2009 storm caused many small-scale debris flows in the Mineral King Valley which clogged a number of the culverts on the Mineral King Road.

The storm caused significant trail damage, including to the High Sierra Trail west of Bearpaw. For example, the bank gave way under a large boulder at the Buck Creek crossing, causing that boulder to fall into Buck Creek, leaving a lot of rocks and debris on the trail.

The flood also washed out several hundred yards of the Cliff Creek Trail below Pinto Lake. The damaged sections were mainly below the area known as the Waterfalls where the trail follows the creek, and also at the trail crossing to Timber Gap. The damage was caused in large part by floodwaters running across and beside the trail. The flood deposited rocks and debris on the trail and caused some bank erosion. In some areas, the trail crew had to dig into the cut bank to reestablish the trail. The amount of trail damage from this flood was rather impressive. The trail crew repaired this section of trail in 2010.

### Debris Flow Complex: Sequoia National Park
The worst damage on the High Sierra Trail occurred at Hamilton Gorge (UTM 359334E 4047933N NAD83 Zone 11). That section of trail was the target of one of the largest of several debris flows that impacted the Sequoia National Park wilderness during this storm event. The trail in that area passes through a granite gorge. The debris flow picked up a large mass of rocks and deposited them on the trail several hundred yards downstream. It took a trail crew a good bit of effort to repair the damage to the trail. That is the largest debris flow to occur in the Sequoia National Park wilderness in recent memory.

The Middle Fork Trail in the Kaweah River Basin was also damaged in a number of places during this storm event. Moderate-sized debris flows occurred in several places along the Middle Fork Trail.

### Debris Flow: Black Rock Pass

The following year (2010), Tony Caprio, the national parks' fire ecologist, observed where several large high-elevation debris flows had occurred on the west-facing slope of Black Rock Pass (from UTM 360400E 4038080N NAD83 Zone 11 south on the slope to UTM 360530E 4037950N NAD83 Zone 11). Those flows were about 15–20 feet across and had scoured from 4–6 feet deep. They were deep enough that they were an obstacle, and it took some looking to find the right place to cross. These flows looked fresh and were presumed to have been from the October 2009 storm event.

The coordinates listed above were for the points where Tony observed the debris flows, relatively near the bottom. The 2010 NAIP aerial imagery shows that the flows began roughly ½ mile upslope from there. The imagery also shows that there are several other old debris flows in the area (multiple photographs and aerial imagery on file in the national parks).

### Debris Flow: Tablelands

Tony also observed where a large high-elevation debris flow had occurred on the Tablelands above Pear Lake (UTM 351900E 4052080N NAD83 Zone 11). The size of this debris flow was hard to judge since it just lifted mats of soil and vegetation, which were a foot or so thick, off of the bedrock and deposited them down on flatter areas (multiple photographs and video on file in the national parks).

## 2010–11 Floods (4)

There were at least four periods of flooding in 2010–11:
1. January 2010 (localized flooding from five back-to-back storms, treated as 1 flood)
2. December 2010 (2 severe storms)
3. July 2011 (multiple severe storms caused by monsoonal moisture, treated as 1 flood)

The winter of 2009–10 was a moderate- to strong-El Niño event. This coincided with the January 2010 flood.

The winter of 2010–11 experienced one of the strongest La Niña events ever.[1112] In addition, December experienced the most extreme jet stream pattern on record for that month.[1113] These conditions coincided with the December 2010 flood and the heavy snowpack that accumulated during the remainder of the season.

As a result of winter storms, flooding, and debris flows, a major federal disaster (DR-1952) was declared on January 26, 2011 for the period December 17, 2010 – January 4, 2011. It covered 10 Southern California counties including Kings, Tulare, and Kern. This was just one of 99 major federal disasters declared in the U.S. in 2011, breaking the record of 88 disasters that had been set in 2010.[1114]

The Tulare Lake Basin experienced a major precipitation event in January that consisted of five distinct storms. The storm series started on January 17 and continued for a week. This line of storms had similarities to the atmospheric river event of January–February 1998 that brought flooding to the central and southern San Joaquin Valley, but there were some significant differences.

While both events occurred during El Niño/La Niña events, the 1998 event brought warm, subtropical moisture that eroded the mountain snowpack, which appreciably added to the runoff. Although the event of January 2010 was similarly moisture-laden, surges of cold air kept snow levels low, and snow even fell on the Southern Sierra foothills on January 21–22. In addition, the spacing between the storms in the 2010 event allowed for several-hour breaks between the first few storms, enabling the ground to absorb some of the moisture before the runoff from the next storm hit. There was still some flooding in the Tulare Lake Basin in 2010, but most of the flooding events were either due to clogged storm drains or occurred in normally flood-prone areas.

The first storm moved rapidly through the area late on January 17. It brought rain but no flooding.

The second storm arrived on January 18. Strong winds over the Tehachapi Mountains ahead of that storm felled a tree onto a house, killing the occupant. This would be the first of two fatalities in this week-long storm event. Later that day, a brief tornado formed southwest of Fresno.

The third storm arrived on January 19, and an upper-level disturbance rotating around the low moved into Southern California, bringing snow to the Tehachapi Mountains. Snow levels with this storm were down to about 4,000 feet. Runoff resulted in some road flooding, and creeks in rural areas ran high. The second storm-related fatality occurred that evening when a man drove around barriers in an attempt to cross a flooded road near the Merced County-Stanislaus County line. That road had been flooded by Orestimba Creek, and the driver was swept away by the fast current.

The heavy rain associated with the third storm brought flooding to portions of Kings, Fresno, and Kern Counties on the afternoon of January 19. Among the flooded roads were:
- Highway 33 and Merced Avenue southeast of Coalinga
- The southbound lanes of Interstate 5 from Highway 269 to the Fresno County line
- The Herring Road exit on Highway 99
- Highway 33 from Highway 46 to the Lerdo Highway

The fourth storm passed through the region late on January 20. It dropped snow on the Grapevine and triggered isolated early evening thunderstorms.

The fifth and final storm of the series arrived on January 21, bringing very cold air to the region. Snow levels dropped to about 2,200 feet, with snow falling in the towns of Mariposa and Oakhurst. Both the Grapevine and Tehachapi Pass were closed for several hours by the snow. This was a very deep low pressure system, and all-time low pressure records were set in both Bakersfield and Fresno; both reported a barometric pressure of 28.94. This storm also brought very strong winds to the Kern County Mountains where 10-inch diameter tree branches were downed in the Grapevine area due to wind gusts approaching 70 mph. Arroyo Pasajero flooded, closing Highway 269 (Lassen Avenue) north of Huron at 8 p.m. on January 21.[1115]

A cold pool of air moved over the area on the afternoon of January 22, triggering strong convective showers, one of which blanketed parts of the city of Clovis with about 2–3 inches of pea-size hail on the ground. A funnel cloud was reported near Clovis Ave and Highway 168 in Clovis.

By the time the last storm moved east of the region, the total rainfall amounts in the central and southern San Joaquin Valley were mostly between 1.5 and 2.5 inches, with a few locations around 3 inches. Snowfall amounts in the Southern Sierra and Tehachapi Mountains were measured in feet, with the heaviest snowfalls reaching around 10 feet of new snow.[1116]

Winter 2009–10 was much colder than average for the U.S., and it delivered a string of record-breaking snowstorms that began on the winter solstice. (This was touted by some as proof that global climate warming was a myth.) The snow and cold didn't linger far into the spring, however. By the end of April 2010, North American snow cover had retreated to the lowest extent since satellite records began in 1967.

However, conditions were very different in the Southern Sierra. Here a snowy winter and a cool spring contributed to a snowpack that lasted later than normal. This was followed by an extended period of high water that lasted from the spring into the summer.

The winter of 2009–10 was a moderate- to strong-El Niño event. Because of this, Hawaii experienced severe drought, but Sequoia and Kings Canyon National Parks had a moderately wet year. Cedar Grove experienced the highest water levels since the 1997 flood. Flow in the South Fork of the Kings River peaked on June 6–8, 2010.

That high-water event brought ponding/flooding and lapping of water along roads at various locations throughout Kings Canyon, including the Motor Nature Trail, the North Side Road, and the main highway.

A large (50± inch dbh) ponderosa pine fell perpendicular to the flow of the South Fork Kings, creating an obstruction in the river just east of the Zumwalt Meadow parking lot. The resulting obstruction elevated the water level upstream, creating ponding along both sides of the main highway. Water was nearly up to the shoulder on both sides of the main road immediately upstream of the Zumwalt Meadow parking lot. Immediately adjacent to the root wad where the tree once stood, an eddy developed in the river, causing scouring of the river bank. About three days after the failure, that eddy location increased in size moving toward the parking lot. Scouring/erosion of the bank slowed by the end of June, but remained vulnerable.

On June 6, 2010, a small debris jam formed in Lewis Creek upstream of the main highway bridge. This resulted in diverting water flow outside the primary channel into the overflow channel. The 16-inch highway culvert for the overflow channel could not carry the resulting flow, so Lewis Creek overflowed the highway for a couple of days and caused some minor road damage. There was concern that heavy equipment might have to be used to push Lewis Creek back into its primary channel. (A bulldozer had been used for that purpose on Memorial Day Weekend, 1983.) However, the debris jam broke on the evening of June 6–7, redirecting all water flow back into the primary channel.

This debris jam was presumably caused in large part by the July 15, 2008 flood which resulted in a huge debris flow that deposited a large amount of sediment and debris on the Lewis Creek delta, created a log jam in the primary channel, and generally destabilized that channel.

This channel-shifting happened again from about June 23–30, 2011. In the fall of 2011, the national parks' road crew installed two more culverts (an 18 inch and a 24 inch) to carry the flow that results when Lewis Creek moves into the overflow channel. Where Lewis Creek emerges from its canyon, there is very little impediment to keeping it from changing course from the primary channel to the overflow channel. Should the main flow change to the overflow channel, a bridge would be required to handle the flow under the highway. Otherwise, the creek would have to be pushed back into its primary channel as was done after the May 1983 debris flow.

387

Flow in the East Fork Kaweah peaked about June 6–7, 2010. The river was very close to the bottom of the Disney Bridge. Monarch Creek jumped its banks for a few days, but resulted in no road damage (photograph on file in the national parks). This appeared to be due at least in part to debris resulting from an avalanche. (Monarch Creek would do exactly the same thing in June 2011.)

The mainstem of the Kaweah crested in Three Rivers (the TRR gaging station) on June 6, 2010 at 5,129 cfs. Total inflow into Lake Kaweah during the snowmelt period peaked on June 6 when the combined flow from all forks of the Kaweah reached 7,546 cfs. It had been years since the Kaweah had been this high during snowmelt at either of those reporting stations.

The Southern Sierra stayed unusually wet throughout the summer of 2010. Even at the end of August, meadows throughout both Sequoia and Kings Canyon National Parks continued to be reported as quite wet.

Contrary to long-term forecasts that called for the winter of 2010–11 to be a relatively dry La Niña-pattern winter, December 2010 turned out to be quite wet in the San Joaquin Valley and Southern California. Precipitation records were broken and flooding occurred. Although moderate La Niña events are often dry in our area, strong La Niña conditions are generally wet. The winter of 2010–11 developed into one of the strongest La Niña events ever.

The December 2010 flood was marked in the Tulare Lake Basin by two pulses of moisture; the first came on December 16–20 followed by a smaller one on December 28–29. The first pulse was very intense, and the media quickly dubbed it the "Wallop".

Virtually the entire state was affected by that first pulse, from coastal cities to the Central Valley, the Sierra, and the southern deserts. Large snowfalls occurred in the Sierra. Very heavy rainfalls hit Southern California.

The stormy weather began hitting the northern part of the state late on Thursday, December 16, and the southern areas on Friday after a large storm front moving out of the Gulf of Alaska met with moist subtropical air coming across the Pacific Ocean.

Writing in the *Visalia Times-Delta*, Bill Tweed described the uncommon combination of events that brought so much precipitation to the Tulare Lake Basin.[1117] First, a powerful low-pressure zone stalled off the Oregon/Washington coast and remained stationary for several days. Low-pressure areas circulate internally in a counter-clockwise manner, and since we were geographically at the bottom of the clock, this set up and held in place a strong southwesterly flow aimed at Central California.

The second critical factor in our big rain was that this southwesterly flow into Central California tapped into a huge mass of very moist tropical air from the central Pacific Ocean beyond Hawaii. The effect was to turn on a strong stream of water-saturated air, aim it directly at the Tulare Lake Basin, and then hold it in place for several days.

Finally, and this was also significant, the flow of wet air arrived from a direction that pushed it straight at the Sierra and forced the moisture to rise over the mountains at just the right angle. Put another way, our mountains run from northwest to southeast, and the wind in this storm blew at the mountains from the southwest, a perfect fit for maximum precipitation.

Many valley floor weather stations recorded 3 or 4 inches of rain during the period that began Friday, December 17, and ended Monday morning, December 20. (The storm continued through the 22nd in the southern part of the San Joaquin Valley and through the 23rd in Southern California.)

The foothills received about twice as much rain as the valley floor. Three Rivers received about 8.5 inches of rain during the storm event.

Sunday, December 19, was the date of the Christmas Bird Count in the national parks. The participants in that count were all willing to testify to the intensity of the downpour.

On December 19, flooding, landslides, rock falls, and debris flows forced the national parks to close the inbound lane of the Generals Highway at the Ash Mountain Entrance Station. To oversimplify the situation, rocks and debris were falling on the road faster than the road crews could remove them. So the entrance station was closed as a safety precaution to avoid potential accidents. In addition, the road between the two national parks had to be closed that day with more than 10 feet of snow on the roadway.

On December 21, two soil scientists from the U.S. Department of Agriculture, Kerry Arroues and Phil Smith, documented about 15 small landslides that had occurred on the Generals Highway for several miles below Hospital Rock. They concluded that most were probably related to the angle of repose. Kerry and Phil had planned to work their way farther uphill to where the big problem areas had been, but the road was still blocked by landslides and mass wasting events of various sorts.

The USACE automatic weather station in Giant Forest at 6,600 feet elevation received a total of over 15 inches of moisture during the December 17–20 storm event. Satellite data from an automatic sensor south of Mineral King at Wet Meadows (elevation 8,900 feet) also reported 15 inches, while another automatic station above Mineral King recorded almost 19 inches of precipitation.

Above 7,000 feet elevation, almost all of the precipitation fell as snow. At least 200 inches (nearly 17 feet) of snow was received at the automatic snow sensor at 9,600 feet elevation at Farewell Gap. That sensor abruptly flat-lined on December 19, apparently because it was hit by an avalanche. (When spring came, the tower was not to be seen.)

That represented an amazing amount of precipitation. In just three days, upwards of 1.5 feet of water had fallen in the Sierra above 8,000 feet elevation. For comparison, the total average annual rainfall in Visalia is about 0.75 foot.

Because the moisture flow was tropical in origin, the snowline remained above 7,000 feet elevation during most of the storm, and a great deal of foothills and middle altitude precipitation ran off into streams and rivers.

Considering the huge amount of moisture delivered by that storm event, there was relatively little runoff in the rivers of the Tulare Lake Basin. Because nearly all of the precipitation that fell above 7,000 feet elevation was captured as snow, a heavy runoff event did not occur. Had the snowline been one or two thousand feet higher, as had been forecast, mountain rivers would have risen as they did in similar storm events in 1955 and 1966.

At Three Rivers (the TRR gaging station), the mainstem of the Kaweah River had been flowing at about 229 cfs. The river began rising just before midnight on December 17. It climbed at a dramatic rate in the middle of the day on Sunday, December 19, peaking at 3:00 p.m. at 15,831 cfs, bringing the river to near flood stage in the town.

Carole Combs recalled that the North Fork of the Kaweah peaked at about the same time as the mainstem. She said that the North Fork was terrifying. Their driveway flooded so that they couldn't have escaped if they had wanted to, which she did.

Terminus Dam did the job that it was designed to do, catching the floodwaters of the Kaweah River. Valerie McKay said that Lake Kaweah rose 40 feet in 48 hours during that storm event. The USACE staff was kept busy moving stuff out of the way of the rapidly encroaching waters.

Because Terminus Dam caught the entire flood on the Kaweah, Visalia was expecting only minor flooding impacts from that storm. However, that's not exactly how events played out.

The biggest challenge Visalia officials faced during the December 17–20 storm event was overflowing ponding basins.[1118] Basins built to accept stormwater filled to capacity and spilled over, flooding nearby areas. That flooding prompted officials to call a local emergency on the evening of December 19. Areas that were flooded by overflowing ponding basins were:

- A park at Pinkham Street and Mary Avenue, near the Annie R. Mitchell Elementary School in southeast Visalia
- Pinkham Street at Cherry Avenue in southeast Visalia
- Constitution Park near Tulare Avenue and Akers
- Mooney Boulevard and Cameron Avenue in south Visalia
- Sierra Village in west Visalia
- Walnut at Roeben Road in west Visalia

Those were comparatively minor overflow situations, relatively easily dealt with. However, there was one overflowing ponding basin that proved much more consequential.[1119]

Highway 198 is below grade through much of Visalia. During rainstorms, water collects in the low spots, and Caltrans' pumps lift that water out and transfer it to city ponding basins to prevent highway flooding. Those pumps worked as designed, but the rain was so intense that the ponding basin at Linwood Street and Mineral King Avenue began overflowing about 3 a.m. on December 20, threatening adjacent homes and businesses. At the city's request, Caltrans turned off their pumps to protect those properties.

With the pumps turned off on that section of highway, water eventually started flooding the lanes (photograph on file in the national parks). As a result, the freeway was closed between Akers and Demaree starting at 10:30 a.m. on December 20. Later that day, city crews were able to fix the problem by digging a channel to divert water from the overflowing ponding basin into the Persian Ditch. Caltrans then used two pumps to drain the highway and reopened it to traffic at 10:00 p.m. that night.

The December 17–20 weekend storm caused only minor problems in Fresno and Fresno County. The brunt of the storm was felt in the southern end of the valley. As detailed in Table 86, Visalia and Bakersfield received almost twice as much moisture as Fresno did. In the Central Valley, the storm event lasted from December 17–20. In Southern California, the storm continued through December 23.

Bakersfield shattered several precipitation records during this storm event. Among those were:

- December 18 — record rainfall for the date of 1.37 inches. The old record was 0.30 inch, set in 1921.
- December 19 — record rainfall for the date of 1.53 inches. The old record was 0.48 inch, set in 1984. This was also the wettest day on record for December at Bakersfield. The previous wettest day in December was December 27, 1936 with 1.02 inches of rain.

The December 17–20 storm caused numerous, mostly minor, flooding problems across the valley, including:

- Several Fresno streets flooded. A portion of Palm Avenue near Clinton Avenue flooded. Jameson Avenue, south of Church Avenue, flooded. The northbound lane of Reed Avenue, north of Floral, flooded.
- Several streets in and around Visalia flooded. Road 64 at Avenue 308 was closed due to flooding.
- Highway 180 was closed about 7 miles west of the junction with Highway 63 for several hours due to a mud or rock slide.
- Many of the ponding basins in Tulare filled or overflowed, including Live Oak Park, Del Lago Park, and the Mission Oak High School ponding basin.
- Several streets and roads in the Tulare area flooded. The westbound lanes of Prosperity Avenue between Laspina Street and Mooney Boulevard flooded. Laspina south of Prosperity flooded. San Joaquin Avenue was closed between J Street and I Street. Highway 137 (the Tulare-Lindsay Highway) flooded west of Road 168.
- Several streets and roads in the Porterville area flooded. Highway 190 flooded at Bourbon Drive and at Westwood Street. Lots of water flooded Highway 190 at Road 284. The Eagle Mountain parking lot at Avenue 136 and Westwood Street flooded with vehicles almost submerged.
- About 30 roads in Tulare County were closed by flooding, and 1,000 acres were inundated, including farmland planted in wheat or barley.
- Ten people were evacuated on December 20 from three homes in Weldon (near Lake Isabella) due to creek flooding.
- Extensive areas of farmland in the Lamont area (southeast of Bakersfield) flooded, possibly due to the failure of a dike.
- The California Highway Patrol reported flooding, rocks, and mud on various foothill and mountain roads into the Sierra.
- Highway 59 between Merced and Los Banos was closed when Mariposa Creek overflowed it. That highway didn't reopen until December 23. Mariposa Creek overflowed it again on December 30.

Deer Creek flooded in south Tulare County. Among the roads it flooded were:

- A segment of Avenue 56 near Road 88 between Earlimart and Alpaugh
- Several segments of Highway 43, just west of Road 88

In McFarland, high water in Poso Creek caused the evacuation of about 2,000 people on December 20. Between 400 and 500 homes were in danger of flooding. Santa Fe Railway crews worked to keep that creek free of debris, helping to ensure that it didn't overflow.

The storm brought very heavy rains to Southern California. Some locations received more than 12 inches of rain during the December 17–23 event. A number of rainfall records were broken. For example, 3.45 inches of rain fell in Pasadena on December 19, shattering the old record of 1.5 inches

391

set on the same date in 1987. Los Angeles received 70% of its annual rainfall in just seven days. It was the most rainfall from one storm event since 2005.[1120]

The storm also brought very heavy rain to the Southern California deserts. The normally dry Mojave River flooded portions of the Apple Valley / Victorville area, peaking at 17 feet deep on December 21 (truly "ginormous" for those who had never seen that river in its magnificence).

The rain in the Mojave Desert was so intense that the resulting flooding wasn't restricted to the riverbeds. On the evening of December 22, Shauna Austin encountered a flash flood flowing across seemingly open desert, flooding U.S. Highway 395 a few miles north of Adelanto. The water was so deep that passenger cars that tried to push through it were being swamped.

Table 86 gives the total precipitation for that storm event for selected reporting stations.

Table 86. Precipitation during the December 17–23, 2010 storm event.

| Reporting Station | Storm Total (inches of rain) |
|---|---|
| Nature Point near Bass Lake | 18.56 |
| Wishon Dam (near Shaver Lake) | 18.89 |
| Fresno | 2.47 |
| Visalia | 4.49 |
| Three Rivers | 8.5 |
| Tulare | 2.8 |
| Bakersfield | 4.02 |
| Camp Nelson | 18.6 |
| Wofford Heights (Isabella Lake) | 15.78 |
| Near Crestline (west of Lake Arrowhead) | 26.16 |
| Tanbark Flats north of Pomona | 19.22 |

December 2010 still had one more pulse of moisture in store for the Tulare Lake Basin. As detailed in Table 87, a fast-moving storm swept through the region on December 28–29.[1121]

Table 87. Precipitation during the December 28–29, 2010 storm event.

| Reporting Station | Storm Total (inches of rain) |
|---|---|
| Fresno | 1.54 |
| Visalia | 1.24 |
| Three Rivers | 2.63 |

The brunt of this storm's effects was felt in Tulare County. The community of Seville (northwest of Woodlake) was hit particularly hard by flooding on the night of December 28. By the next morning, the surrounding area was described as looking like the Nile River.

Yokohl Creek broke through a levee, flooding dozens of orange groves and causing the closure of Avenue 304 south of Woodlake. The sight of the overflowing creek caused people to stop and stare. It's not too uncommon for Yokohl Creek to flow over the road at that location, but this particular flood caused it to cover the road to an unusual depth.

A section of the Mineral King Road ¼ mile above the Hammond Fire Station collapsed on the morning of December 29 due to erosion and undermining. That road was closed for several hours until the roadbed could be rebuilt. Anne Birkholz recalled that she couldn't get to work on the 29th because of flooding on several small tributaries along the South Fork of the Kaweah.

The wet winter of 2010–2011 caused a number of debris flows in the Kings River Special Management Area along the Garnet Dike road. There was both a December 2010 and a March 2011 event along that road. Some of those debris flows rivaled the 1937 event in the Big Creek Basin.[1122]

December 2010 was one of the wettest on record in the Tulare Lake Basin, in the Southern Sierra, and in Southern California. The December 28–29 storm pushed several communities into record-setting territory:

- Fresno received a total of 5.92 inches for the month, making it the second-wettest December in that city's history. The wettest was December 1955 with 6.73 inches.[1123]
- Three Rivers received more than 13.5 inches for the month, making it one of the wettest Decembers on record for that community.
- It was the wettest December in Visalia's history. The previous record had been 6.06, set in December 1955.
- The total precipitation for Bakersfield for the month was 5.82 inches — nearly eight times the average amount (0.76 inches) for December. That was the most rain recorded in any month since record-keeping began in 1889, and broke the record set in February 1998 during a very strong El Niño.[1124]

The winter of 2010–2011 was a La Niña season, and the long-range forecast had been for less-than-average precipitation. However, as of December 30, the season total for Three Rivers was 18.52 inches of rainfall, well above average.

At the end of December 2010, the Southern Section Sierra snowpack was reported at a phenomenal 284% of average for the date. Three of the snow sensors in the Kern River Basin — Pascoes, Tunnel Guard Station, and Casa Vieja Meadows — had recorded more snow in three months than the average amount for the entire six-month season.

To put the amount of snow in perspective, there was substantially more snow at Pascoes at the end of December 2010 than there was at the beginning of January during either the big El Niño events of 1982–83 or 1997–98. Mammoth Mountain had 208 inches of snow at the end of the year, the greatest since record-keeping began in 1968. (The previous highs at Mammoth Mountain were set in 1971 (139.8 inches) and 2002 (134.4 inches).

The snowpack continued to build and lasted through the spring. As shown in Table 88, the snowpack in the Tulare Lake Basin at the end of April 2011 was well above average.[1125]

Table 88. May 1, 2011 snowpack in the Tulare Lake Basin.

| Watershed | Predicted Runoff April 1 – July 31 (acre-feet) | % of average (1956–2005) |
|---|---|---|
| Kings | 2,050,000 | 167% |
| Kaweah | 490,000 | 171% |
| Tule | 115,000 | 180% |
| Kern | 860,000 | 187% |
| Total | 3,515,000 | 173% |

As detailed in Table 89, the winter of 2010–11 broke the snowfall record at Lodgepole that had been set in the winter of 1951–52. Snowfall in the winter of 1905–06 was even bigger than this, but that was before a weather station had been established at Lodgepole.

Table 89. Lodgepole snowfall during winter 2010–11.

| Month | Snowfall (inches of snow) |
|---|---|
| September 2010 | 0.0 |
| October 2011 | 4.0 |
| November 2010 | 52.5 |
| December 2010 | 116.7 |
| January 2011 | 44.7 |
| February 2011 | 89.0 |
| March 2011 | 127.4 |
| April 2011 | 23.5 |
| May 2011 | 17.0 |
| June 2011 | 1.8 |
| Total | 476.6 |

2010 didn't just set records in the Sierra. The Global Historical Climatology Network announced that 2010 was the wettest year that the world has seen since at least 1900. The La Niña conditions that brought so much precipitation to the Tulare Lake Basin and Southern California in December 2010 were also responsible for catastrophic flooding in Australia that month. All in all, it was a bang-up way to close out the year.

The large amount of snow caused havoc on some of the national parks' trails. The parks' trail crews expect to spend the first part of each spring clearing (logging) trees that have fallen during the preceding winter. However, significantly more trees came down in parts of the parks during the winter of 2010–11 than average.

Over 1,000 downed trees had to be cleared from trails in the Kaweah and Kern River Basins in Sequoia National Park; that's roughly five times the average. The areas that were most affected were:
- Giant Forest/Wolverton in the Kaweah River Basin
- High Sierra Trail to Buck Creek in the Kaweah River Basin
- Redwood Meadow/Cliff Creek in the Kaweah River Basin
- Tar Gap/Hockett Plateau in the Kaweah River Basin
- Lower Kern Canyon in the Kern River Basin (this area was particularly hard hit)
- Chagoopa Plateau in the Kern River Basin

Many of the above trees were tree-top failures. Tree-top failures are typically caused by a combination of heavy snow loads and wind.

There were also many up-rooted trees. Up-rooted trees are an indication of ground saturation. Based on his experience, Tyler Johnson, Sequoia National Park's trails supervisor, thinks that these events seem to occur when winter precipitation exceeds about 150% of average. Tyler recalled that up-rooted trees were also a significant problem on the parks' trails in 2006. In that year, the May 1 snowpack for the Kern was 152% of the long-term (1956–2005) average.[1126] As shown in Table 88, precipitation in the Kern in the winter of 2010–11 was 187% of average, which helps to explain why the fallen tree count was so much higher.

Kings Canyon National Park also experienced about five times the average number of trees falling across trails during the winter of 2010–11. Some trails had many trees down, scattered along the entire trail (Roaring River Basin, Woods and Bubbs Creek Basins, and South Fork Kings below Upper Paradise Valley). Some trails had about the average amount (Middle Fork Kings River Basin and Monarch Divide). Some had an average amount interspersed with large avalanches that had a lot of trees (San Joaquin River Basin). Upper Basin between Pinchot and Mather had no trees down at all — although there are typically one or two there.

One very noteworthy event was that the Middle Fork of the Kings Trail was closed to stock travel into the first half of September 2011 due to a large avalanche snow deposit that had not melted. That trail was closed just below the 7,200 foot elevation level. That was approximately mid-way between Grouse Meadow and Simpson Meadow in the Devils Washbowl area. When last reported in late September, there was still a large snow patch there, but the trail had melted out and was passable. Based on the experience of the Kings Canyon trail crew, this is the first time since at least the mid-1960s that a snow patch has lasted until late September at such a low elevation. Partly this was because the big winter of 2010–11 created the conditions necessary for a big avalanche. However, equally important was the cool spring and summer of 2011 that allowed that snow to persist into the fall. It's tempting to think that conditions such as this haven't existed since at least the winter of 1951–52.

The Mineral King Road was unusually late melting out in June 2011 due to the heavy winter of 2010–11.

On September 30, 2011, two adjoined giant sequoias failed along the Trail of 100 Giants in Giant Sequoia National Monument. One of those trees was 17 feet in diameter and 300 feet tall. Upon investigation, a forest pathologist found no rot in either tree. A suspected primary cause of the failure was lingering wet soil due to the winter of 2010–11.

In the Sierra, the majority of the snowpack usually melts in May so that there is little snow remaining by June 1. However, in 2011 the snowpack lingered well into June. This was the result of an above-average amount of snowfall during the winter, followed by an exceptionally cool spring, which helped keep the snow in place much long than normal. As of the first of June, the amount of snow still on the Sierra was nearly six times greater than average.[1127]

The snowpack at the Central Sierra Snow Laboratory monitoring site near Donner Pass lasted until June 30, 2011. That was the latest date for melt-off observed at that site since record-keeping began in 1946. It tied the record set in 1967, another big winter in the Tahoe area. (Older Southern Pacific Railroad records suggest that this might have been the latest melt-off dating back to 1879. However, those measurements were taken at a slightly different location and were not recorded in a rigorous manner, so no reliable conclusion can be drawn.) In any case, the June 30 date was a rather astounding five weeks later than the average May 23 date for melt-off at this location.

Because of the near-record amount of snowfall, there was a large amount of runoff. Since the spring was cool, that runoff didn't result in peak flooding events, just large amounts of water delivered to the valley floor. Lake Kaweah reached peak storage (714.83 feet elevation, equivalent to 185,264 acre-feet) on July 7, 2011.[1128]

As a measure of the size of the runoff, the plan for the operation of Pine Flat Reservoir was to end the irrigation season with a full reservoir.[1129] Normally Pine Flat would be drawn down to low-pool by then.

Flooding in 2011 occurred in July. It was caused by a series of individual storm events, but could be thought of as one event that occurred in multiple locations.

On July 28, southeast winds aloft began to bring mid-level moisture from northern Mexico and the Desert Southwest. Isolated thunderstorms developed over the Sierra crest around Kings Canyon and points just to the north and east. The surge of monsoonal moisture continued through the end of the month, with the strongest thunderstorm activity on July 30 and the morning of the 31st. The storm system extended south as far as Edwards Air Force Base in Kern County. A few thunderstorms over the Southern Sierra had rain rates of an inch or more in an hour.[1130]

Thunderstorms generated by this storm system deluged the Rock Creek Basin in the national parks from noon on July 29 through the evening of July 30. It was a severe and sustained event with flash floods on the afternoon of both July 29 and 30. We know about this storm event because the Rock Creek wilderness ranger station was staffed in 2011 by ranger Dave Alexander and Elizabeth Curry, a volunteer in the parks (VIP).

July 29 was one of Dave's scheduled days off. (Wilderness rangers are always on call in case of an emergency.) Dave and Elizabeth were near the ranger station on July 29 because it was cloudy and threatening to rain.

It began to rain heavily around noon on July 29 with periods of thunder and lightning. (There was only light hail that day; storms the next day would bring heavy hail.) After a few hours, it cleared somewhat. At about 3:00 p.m., Rock Creek, which runs just in front of the ranger station, turned from clear to muddy and rose rapidly over the next two hours even though it was only raining lightly (multiple pictures and video on file in the national parks, also see cover photograph of this document).

Concerned for the safety of park visitors, Dave and Elizabeth hiked down to the commonly used camp area where the Pacific Crest Trail (PCT) crosses Rock Creek. On the way there, it was obvious from the rumble of unseen boulders being swept along and large, unearthed logs floating by, that Rock Creek had become impassable. The meadow below the ranger station was flooded, and the ranger station trail was under water.

Dave and Elizabeth contacted one commercial group of pack-supported hikers who decided to stay on the ranger station side of Rock Creek. (That group spent the following two nights there without being able to safely cross. They decided to forgo their trip to Mt. Whitney and returned the way that they had come.) There was also a group of commercial packers that was stuck on the opposite side of Rock Creek. It was impossible to shout over the roar of the creek, but they indicated with hand gestures that they had decided to wait overnight and would try to cross the following morning.

Early the next morning, July 30, Rock Creek had receded somewhat from the high of the previous afternoon, but it was beginning to rain again. The commercial pack group from the opposite side was able to cross the creek downstream where it split around a small island, but the water was above the horses' bellies, and the packers were very relieved that they were able to get across without incident.

Throughout the early afternoon, there were a series of thunderstorms that covered the Rock Creek Basin with an inch of hail and heavy rain. The area around the ranger station was completely covered with hail; it looked like it had snowed. Mt. Langley, at the head of the watershed, was left blanketed with hail and snow (multiple photographs on file in the national parks). Rock Creek rose considerably higher than the previous day. It again became impassable to stock and hiker parties. The creek was so loud that Dave and Elizabeth had to yell to be heard when talking near the ranger station.

The rains ended on the evening of July 30, and Rock Creek receded to near-normal levels by late on the afternoon of July 31.

The flash flood did significant damage to the trails in the Rock Creek Basin. It swept away most of the upstream log crossings, depositing some of them in a large snag near the regular crossing below the ranger station. The section of the Rock Creek trail from Soldier Lakes to the Rock Creek ranger station was particularly hard hit. Along that section, Rock Creek jumped its banks and seriously eroded the trail, leaving it obscured, difficult to follow, and impassable to stock. The flood also caused considerable erosion of the trail near the Army Creek crossing (east end of the Rock Creek Basin, near Lower Soldier Lake). This left the trail too deep to walk in, so hikers created new trails parallel to the old trail.

Managing the logjam of stock and hiking parties unable to cross Rock Creek due to the flooding was a dangerous situation that required a great deal of intervention by Dave and Elizabeth. The heavy downpours and cold temperatures caught a number of hikers unprepared. Some who were suffering from near-hypothermia came to the ranger station to get warm and dry out. Two of them used the station as shelter for the night.

The flooding was so intense that it made significant changes to the channel structure of the creek. Most noticeable, a sand and log dam was created below the ranger station which split Rock Creek and formed what appeared to be a permanent additional channel through the meadow.

A somewhat similar storm event occurred in August 2001. In that event, an intense thunderstorm struck a large portion of the Rock Creek Basin, causing Rock Creek to flash flood and send a large quantity of water out onto the Kern valley floor about two miles north of Kern Hot Spring. The 2011 flood may have been a larger flood, but Erik Frenzel (national park meadow monitor) reported that it did not put any debris onto the High Sierra Trail. We don't know why these two floods differed in this way. Tony Caprio speculated that the 2001 flood might have cleared accumulated material from the stream channel, and there had been insufficient time to accumulate a similar quantity of material before the 2011 flood.

The 2011 Rock Creek storm and flood lasted for nearly three days, from July 29 through the morning of July 31. On the second day of that event, July 30, a number of isolated strong thunderstorms developed at various points elsewhere in the Southern Sierra and in the desert region of Kern County. The July 30 storm events that we know about included:[1131]
- Mono Hot Springs (30 miles northeast of Shaver Lake in Fresno County) received 0.89 inches of heavy rain from a thunderstorm at 1:00 p.m.
- Whiterock Creek (5 miles northeast of Tehachapi Pass in Kern County) received 1.44 inches of rainfall from a heavy thunderstorm at 4:30 p.m.

- A RAWS automated weather station located 10.5 miles south of Onyx in Kern County received 0.56 inches of rain in only 33 minutes (a rate of 1 inch per hour) from a heavy thunderstorm at 4:00 p.m.
- Several storms caused small-scale debris flows that flowed onto mountain highways. One near Johnsondale in Tulare County caused damage to Salmon Creek Highway and Mountain Highway 99.[1132]
- A severe thunderstorm moved over Edwards Air Force Base in Kern County during the afternoon.

Another of the July 30 thunderstorms occurred in the Cedar Grove / Canyon View area of Kings Canyon National Park. This was some 25 miles northwest of the Rock Creek Basin in Sequoia National Park. The Cedar Grove storm began at about 4:00 p.m. and lasted only an hour or so. The 24-hour rainfall total was 1¾ inches.

A number of small to moderate debris flows occurred toward the latter part of the storm. This storm struck in the same area where the October 2008 Cedar Bluffs prescribed burn had occurred. A small part of the storm extended into the area where the 2010 Sheep Fire occurred.

Whenever a debris flow occurs after a fire, there is a tendency to assume that the debris flow was caused primarily by the fire. That is, to assume the fire was necessary to create the conditions for the debris flow to occur. See for example when the 2008 Lewis Creek debris flow was initially attributed to a fire that had occurred three years earlier.

That was the case with the 2011 Canyon View debris flows; they were initially attributed to the Cedar Bluffs burn. However, investigation by Tony Caprio, the national parks' fire ecologist, showed that the Cedar Bluffs fire was incidental to the Canyon View debris flows. It was a low-severity burn that had occurred three years prior to the debris flows. It had contributed little to creating the conditions necessary for those debris flows.

Damage from the Canyon View runoff, erosion, and debris flows included:
- The Heliport Road suffered one washout. Otherwise, the damage to that road was mainly a lot of material and debris deposited on top of the roadbed. A lot of that is now new road grade.
- There were 7 or 8 culverts overwhelmed and plugged on Highway 180 and the Heliport Road. One of the debris flows that came out onto Highway 180 was 4–18 inches deep and prevented vehicle traffic movement until it could be cleared. There were 7–8 significantly smaller debris flows that only blocked one lane of traffic.
- There were punctures to several vehicle tires as a result of driving over the debris flows.
- Lots of debris including rocks and trees was washed down the hillside. The main water line feeding Moraine and Canyon View Campgrounds runs along the side of the highway and is buried at least four feet deep. At one spot, about 100 yards before entering Canyon View Campground, this line crosses a natural gully. At that point, the line had originally been buried, but it had been exposed prior to the storm and had not been reburied. There was nothing to protect the line from the onrushing force of the floodwater and rock and wood debris. The combination of the debris and the rushing water broke the line, draining much of the water out of the main Cedar Grove water tank before the valve could be shut off. After the flash flood, the gully was about three times the size of what it was prior to the event. The drainage that caused the pipe break flows to the southeast of the Cedar Grove Bridge; it does not flow through Canyon View Campground.

- A moderate-sized debris flow came into Canyon View Campground. When the storm started, children along with their driver jumped into a school bus to seek shelter. Rumor (probably unfounded) was that the school bus was moved sideways by the debris flow. What we do know is that this debris flow made a really big mess of the campground with the decomposed granite (DG) and other debris. This debris flow came from a drainage just east of the one that broke the main water line. In addition to the debris flow, this drainage caused erosion within the campground.
- One of the bridges on the foot/bike path to Canyon View Campground between the highway and the river was pretty well buried with some large rocks washed up to it.
- Sheep Creek deposited a moderate amount of sand and ash in the waterhead for the Sheep Creek water plant, filling it to the top of the dam. This sediment had to be removed by hand because no backhoe could access the area. This problem was compounded by the fact that the waterhead had not been cleaned out in recent years, so there were layers of old sediments that also needed to be removed. A total of 20 people worked for two days to flush all this material through the system.
- When the Sheep Creek water plant later tried to produce water from this surface intake, the turbidity was too high. That was because there was so much fine ash in the water from the 2010 Sheep Fire, and the ash couldn't be filtered out.

The storm cell that caused the July 30 Cedar Grove / Canyon View event extended south into the Roaring River area. Cindy Wood was the Roaring River wilderness ranger and was caught up in that storm event. When the storm hit, she was riding out from her station leading a string of four pack animals. Intense rain and lightning continued from Ferguson Creek through the Sugarloaf Valley (about 1½ hours by horseback).

As Cindy rode, she listened on the park radio to all of the happenings in Cedar Grove. She kept riding through the storm because stopping under a tree was not really an option due to the lightning. The rain was so intense that Cindy was soaked through in five minutes, even with a coat and long rain slicker. The Sugarloaf Creek trail crossing was very high, but she made it across okay. It was just a very long, cold ride out to the trailhead.

Total flow for water year 2011 was 197% of the 1894–2011 average for the Kings, 200% for the Kaweah, 198% for the Tule, and 199% for the Kern.

This document ends with the fall of 2011, the end of water year 2011. It covers what we know about the floods and droughts that have occurred within the Tulare Lake Basin over the preceding 2,000 years or so. For a summary and conclusions regarding this material, see the Summary section of this document.

.

# Endnotes and Literature Cited

[1] U.S. Geological Survey (USGS). 2011. Hydrologic Definitions website.
http://water.usgs.gov/wsc/glossary.html#Drainagebasin (accessed 22 July 2012).

[2] Galloway, D.L., and F.S. Riley. 1999, San Joaquin Valley, California — Largest human alteration of the Earth's surface. Pages 23–34 *in* Galloway, D.L., D.R. Jones, and S.E Ingebritsen, editors. Land Subsidence in the United States: U.S. Geological Survey Circular 1182. U.S. Geological Survey, Washington, D.C. Available at http://pubs.usgs.gov/circ/circ1182/pdf/06SanJoaquinValley.pdf (accessed 1 September 2012).

[3] U.S. Geological Survey (USGS). 2011. HUC Boundary Descriptions and Names of Regions, Subregions, Accounting Units and Cataloging Units website.
http://water.usgs.gov/GIS/huc_name.html (accessed 17 August 2012).

[4] U.S. Geological Survey (USGS). 2011. HUC Boundary Descriptions and Names of Regions, Subregions, Accounting Units and Cataloging Units website.
http://water.usgs.gov/GIS/huc_name.html (accessed 17 August 2012).

[5] California Department of Water Resources (DWR). 2009. California water plan update 2009. California Department of Water Resources. Sacramento, California. Available at
http://www.waterplan.water.ca.gov/cwpu2009/index.cfm/ (accessed 1 February 2011).

[6] Lee, C.H. 1907. The Possibility of the Permanent Reclamation of Tulare Lake Basin, California. Engineering News 57:27–30.

[7] Harding, S.T. 1949. Inflow to Tulare Lake from its tributary streams. Tulare Lake Basin Water Storage District open file report, Hanford, California.

[8] Grunsky, C.E. 1898. Irrigation near Fresno, California. Water Supply Paper No. 18. U.S. Geological Survey (USGS), Washington, D.C.

[9] ECORP Consulting Inc. 2007. Tulare Lake Basin hydrology and hydrography: a summary of the movement of water and aquatic species. Prepared for U.S. Environmental Protection Agency. Document number 909R07002. Available at
http://nepis.epa.gov/Exe/ZyPURL.cgi?Dockey=P1002E2I.txt. (accessed 1 February 2011).

[10] Preston, W. L. 1981. Vanishing Landscapes: Land and Life in the Tulare Lake Basin. University of California Press, Berkeley, California.

[11] Meyer, J., and J.S. Rosenthal; Far Western Anthropological Research Group, Inc., Davis, California. 2010. A Geoarchaeological Overview and Assessment of Caltrans Districts 6 and 9 — Cultural Resources Inventory of Caltrans District 6/9, Rural Conventional Highways. Prepared for California Department of Transportation.

[12] May 6, 2011 issue of the Kaweah Commonwealth.

[13] U.S. Bureau of Reclamation (USBR). 1970. A summary of hydrologic data for the test case on acreage limitation in Tulare Lake. U.S. Bureau of Reclamation, Sacramento, California.

[14] ECORP Consulting Inc. 2007. Tulare Lake Basin hydrology and hydrography: a summary of the movement of water and aquatic species. Prepared for U.S. Environmental Protection Agency. Document number 909R07002. Available at
http://nepis.epa.gov/Exe/ZyPURL.cgi?Dockey=P1002E2I.txt. (accessed 1 February 2011).

[15] U.S. Geological Survey (USGS). 2008. Flood Definitions website.
http://ks.water.usgs.gov/waterwatch/flood/definition.html (accessed 30 July 2011).

[16] June 8, 2011 issue of the *Fresno Bee*. Available at
http://www.fresnobee.com/2011/06/08/2420219/despite-large-snowpack-little.html (accessed 25
June 2011).

[17] Muir, J. 1894. The Mountains of California, The Century Co. New York. Available at
http://www.sierraclub.org/john_muir_exhibit/writings/the_mountains_of_california/chapter_11.aspx
(accessed 26 January 2012)

[18] DeGraff, J.V., D.L. Wagner, A.J. Gallegos, M. DeRose, C. Shannon, and T. Ellsworth. 2011. The
remarkable occurrence of large rainfall-induced debris flows at two different locations on July 12,
2008, Southern Sierra Nevada, CA. Landslides 8(2) 343–353. Available at
http://www.springerlink.com/content/f88568301m244650/ (accessed 19 August 2011).

[19] U.S. Geological Survey (USGS). 2008. Flood Definitions website.
http://ks.water.usgs.gov/waterwatch/flood/definition.html (accessed 30 July 2011).

[20] December 17, 2011 issue of the Visalia Times-Delta.

[21] Dettinger, M.D. et al. 2011. Atmospheric rivers, floods, and the water resources of California:
Water, 3 (Special Issue on Managing Water Resources and Development in a Changing
Climate):455-478. Available at http://www.mdpi.com/2073-4441/3/2/445 (accessed 24 March 2011).

[22] U.S. Geological Survey (USGS). 2011. Multi-Hazard West Coast Winter Storm Project website.
http://urbanearth.gps.caltech.edu/winter-storm/ (accessed 1 February 2011).

[23] National Oceanic and Atmospheric Administration (NOAA). n.d. Top ten published atmospheric
river events website. http://www.esrl.noaa.gov/psd/atmrivers/events/ (accessed 1 February 2011).

[24] U.S. Geological Survey (USGS). 2011. Multi-Hazard West Coast Winter Storm Project website.
http://urbanearth.gps.caltech.edu/winter-storm/ (accessed 1 February 2011).

[25] U.S. Army Corps of Engineers (USACE). 1999. Post-flood assessment for 1983, 1986, 1995, and
1997. U.S. Army Corps of Engineers, Sacramento, California. Chapter 2 available at
http://snugharbor.net/images2010/misc/2002_usace_sac_flood_history.pdf (accessed 28 May 2012).

[26] U.S. Geological Survey (USGS). 2011. Multi-Hazard West Coast Winter Storm Project website.
http://urbanearth.gps.caltech.edu/winter-storm/ (accessed 1 February 2011).

[27] Transcript of the March 25, 2011 meeting of the Central Valley Flood Protection Board, pages
150–163. Available at http://www.cvfpb.ca.gov/transcripts/2011/03-25-2011transcript.pdf (accessed
(27 June 2012)

[28] December 18, 2001 issue of The Valley Voice. Available at
http://www.valleyvoicenewspaper.com/vvarc/2001/december182001.htm (accessed (27 June 2012)

[29] U.S. Army Corps of Engineers (USACE). 1999. Post-flood assessment for 1983, 1986, 1995, and
1997. U.S. Army Corps of Engineers, Sacramento, California. Chapter 2 available at
http://snugharbor.net/images2010/misc/2002_usace_sac_flood_history.pdf (accessed 28 May 2012).

[30] National Park Service (NPS). 2009. Environmental assessment: Replacement of the Cedar Grove
Bridge, Kings Canyon. National Park Service, Three Rivers, California.

[31] National Park Service (NPS). 2009. Environmental assessment: Replacement of the Cedar Grove
Bridge, Kings Canyon. National Park Service, Three Rivers, California.

[32] Paulsen, C.G. 1953. Floods of November-December, 1950 in the Central Valley Basin, California.
Pages 505–789 *in* U.S. Geological Survey Water–Supply Paper 1137-F. U.S. Geological Survey,
Washington, D.C.

[33] U.S. Geological Survey (USGS). n.d. National Water Information System: Web Interface.
http://waterdata.usgs.gov/nwis/dv/?site_no=11212500&agency_cd=USGS&referred_module=sw
(accessed 4 August 2011).

[34] U.S. Geological Survey (USGS). 2010. National Streamflow Information Program website.
http://water.usgs.gov/nsip/definition9.html (accessed 3 August 2011).

[35] U.S. Geological Survey (USGS). 2011. National Streamflow Information Program website. http://nwis.waterdata.usgs.gov/ca/nwis/peak?site_no=11212500&agency_cd=USGS&format=html (accessed 9 August 2011).

[36] U.S. Geological Survey (USGS). 2011. National Streamflow Information Program website. http://nwis.waterdata.usgs.gov/ca/nwis/peak?site_no=11212450&agency_cd=USGS&format=html (accessed 24 November 2012).

[37] Thiros, S.A. 2010. Conceptual understanding and groundwater quality of the basin-fill aquifer in the Central Valley, California. Section 13 *in* Thiros, S.A., et al. editors. Conceptual understanding and groundwater quality of selected basin-fill aquifers in the Southwestern United States. U.S. Geological Survey Professional Paper 1781. U.S. Geological Survey, Washington, D.C. Available at http://pubs.usgs.gov/pp/1781/pdf/pp1781_section13.pdf (accessed 21 September 2012).

[38] Mayfield, T. 1929. San Joaquin primeval: Uncle Jeff's story, a tale of a San Joaquin Valley pioneer and his life with the Yokuts Indians. Tulare Times Press, Tulare, California.

[39] Griggs, F.T. 1983. Creighton Ranch Preserve — A relict of Tulare Lake. *Fremontia* 10:3-8.

[40] Preston, W. L. 1981. Vanishing Landscapes: Land and Life in the Tulare Lake Basin. University of California Press, Berkeley, California.

[41] August 14, 1898 issue of The San Francisco Call. Available at http://chroniclingamerica.loc.gov/lccn/sn85066387/1898-08-14/ed-1/seq-19/ (accessed 24 April 2012).

[42] Latta, F.F. 1937. The flood of 1937: Little journeys in the San Joaquin #20. Article.

[43] August 14, 1898 issue of The San Francisco Call. Available at http://chroniclingamerica.loc.gov/lccn/sn85066387/1898-08-14/ed-1/seq-19/ (accessed 24 April 2012).

[44] Thompson, T.H. 1892. Official Historical Atlas Map of Tulare County, California, Author, Tulare, California.

[45] Latta, F.F. 1937. Territory covered by old Tulare Lake. August 12/13, 1937 issue of the Visalia Times-Delta.

[46] Harding, S.T. 1949. Inflow to Tulare Lake from its tributary streams. Tulare Lake Basin Water Storage District open file report, Hanford, California.

[47] U.S. Bureau of Reclamation (USBR). 1970. A summary of hydrologic data for the test case on acreage limitation in Tulare Lake. U.S. Bureau of Reclamation, Sacramento, California.

[48] Harding, S.T. 1949. Inflow to Tulare Lake from its tributary streams. Tulare Lake Basin Water Storage District open file report, Hanford, California.

[49] Haslam, G.W. 1990. The Lake That Will Not Die. Capra Press, Santa Barbara, California.

[50] Heizer, R.F. 1978. Handbook of North American Indians: Volume 8, California. Smithsonian Institution, Washington, D.C.

[51] Preston, W. L. 1981. Vanishing Landscapes: Land and Life in the Tulare Lake Basin. University of California Press, Berkeley, California.

[52] Preston, W. L. 1981. Vanishing Landscapes: Land and Life in the Tulare Lake Basin. University of California Press, Berkeley, California.

[53] Heizer, R.F. 1978. Handbook of North American Indians: Volume 8, California. Smithsonian Institution, Washington, D.C.

[54] March 1997 *Los Tulares*, quarterly bulletin of the Tulare County Historical Society, Tulare, California.

[55] November 2, 1971 issue of Visalia Times-Delta.

[56] Latta, F.F. 1977. Handbook of Yokuts Indians. Bear State Books, Santa Cruz, California.

[57] Hurtado, Albert L. 1990. Indian Survival on the California Frontier. Yale University Press, New Haven, Connecticut.

[58] Preston, W. L. 1981. Vanishing Landscapes: Land and Life in the Tulare Lake Basin. University of California Press, Berkeley, California.

[59] Heizer, R.F. 1978. Handbook of North American Indians: Volume 8, California. Smithsonian Institution, Washington, D.C.

[60] The Bay Institute. 1998. From the Sierra to the sea: The ecological history of the San Francisco Bay-Delta Watershed. The Bay Institute, San Francisco, California. Available at http://www.bay.org/publications/from-the-sierra-to-the-sea-the-ecological-history-of-the-san-francisco-bay-delta-waters (accessed 1 February 2011).

[61] The Bay Institute. 1998. From the Sierra to the sea: The ecological history of the San Francisco Bay-Delta Watershed. The Bay Institute, San Francisco, California. Available at http://www.bay.org/publications/from-the-sierra-to-the-sea-the-ecological-history-of-the-san-francisco-bay-delta-waters (accessed 1 February 2011).

[62] Tulare Basin Wildlife Partners. 2006. Tulare Lake Basin regional conservation plan: Sand Ridge — Tulare Lake. Unpublished report, Three Rivers, California.

[63] Derby, Lt. G. H. 1850. A report on the Tulare Valley. U.S. House of Representatives, Washington, D.C. Reprinted as Browning, P., editor. 1991. Bright gem of the western seas: California 1846–1852. Great West Books, Lafayette, California.

[64] Lee, C.H. 1907. The Possibility of the Permanent Reclamation of Tulare Lake Basin, California. Engineering News 57:27–30.

[65] June 1956 *Los Tulares*, quarterly bulletin of the Tulare County Historical Society, Tulare, California

[66] Tulare Basin Wildlife Partners. 2009. Tulare Basin riparian and wildlife corridor report. Unpublished report, Three Rivers, California.

[67] Tulare Basin Wildlife Partners. 2009. Tulare Basin riparian and wildlife corridor report. Unpublished report, Three Rivers, California.

[68] Tulare Basin Wildlife Partners. 2006. Tulare Lake Basin regional conservation plan: Sand Ridge — Tulare Lake. Unpublished report, Three Rivers, California.

[69] Preston, W. L. 1981. Vanishing Landscapes: Land and Life in the Tulare Lake Basin. University of California Press, Berkeley, California.

[70] Atwater, B.F. et al. 1986. A fan dam for Tulare Lake, California, and implications for the Wisconsin glacial history of the Sierra Nevada. Geological Society of America Bulletin 97:97–109.

[71] Latta, F.F. 1977. Handbook of Yokuts Indians. Bear State Books, Santa Cruz, California.

[72] Meyer, J., and J.S. Rosenthal; Far Western Anthropological Research Group, Inc., Davis, California. 2010. A Geoarchaeological Overview and Assessment of Caltrans Districts 6 and 9 — Cultural Resources Inventory of Caltrans District 6/9, Rural Conventional Highways. Prepared for California Department of Transportation.

[73] Thompson, T.H. 1892. Official Historical Atlas Map of Tulare County, California, Author, Tulare, California.

[74] Goodridge, J. 1996. Data on California's Extreme Rainfall from 1862–1995. Proceedings of the 1996 California Extreme Precipitation Symposium. Sierra College Science Center, Rocklin, California. Available at http://www.cepsym.info/proceedings_1996.php (accessed 1 July 2011).

[75] U.S. Bureau of Reclamation (USBR). 1970. A summary of hydrologic data for the test case on acreage limitation in Tulare Lake. U.S. Bureau of Reclamation, Sacramento, California.

[76] The Bay Institute. 1998. From the Sierra to the sea: The ecological history of the San Francisco Bay-Delta Watershed. The Bay Institute, San Francisco, California. Available at http://www.bay.org/publications/from-the-sierra-to-the-sea-the-ecological-history-of-the-san-francisco-bay-delta-waters (accessed 1 February 2011).

[77] U.S. Bureau of Reclamation (USBR). 1970. A summary of hydrologic data for the test case on acreage limitation in Tulare Lake. U.S. Bureau of Reclamation, Sacramento, California.

[78] ECORP Consulting Inc. 2007. Tulare Lake Basin hydrology and hydrography: a summary of the movement of water and aquatic species. Prepared for U.S. Environmental Protection Agency. Document number 909R07002. Available at http://nepis.epa.gov/Exe/ZyPURL.cgi?Dockey=P1002E2I.txt. (accessed 1 February 2011).

[79] February 6, 1864 issue of the Porterville Messenger.

[80] Boyd, W.H. 1972. A California Middle Border, the Kern River Country, 1772–1880. The Havilah Press, Richardson, Texas.

[81] February 6, 1864 issue of the Porterville Messenger.

[82] Latta, F.F. 1977. Handbook of Yokuts Indians. Bear State Books, Santa Cruz, California.

[83] Latta, F.F. 1977. Handbook of Yokuts Indians. Bear State Books, Santa Cruz, California.

[84] Linton, C.B. 1908. Notes from Buena Vista Lake, May 20 to June 16, 1907. Condor 10:196–198.

[85] Tulare Basin Wildlife Partners. 2006. Tulare Lake Basin regional conservation plan: Goose Lake conservation plan. Unpublished report, Three Rivers, California.

[86] Latta, F.F. 1977. Handbook of Yokuts Indians. Bear State Books, Santa Cruz, California.

[87] Menefee, E.L and F.A. Dodge. 1913. History of Tulare and Kings Counties, California, with biographical sketches of the leading men and women of the counties who have been identified with their growth and development from the early days to the present. Historic Record Co., Los Angeles, California. Available at http://archive.org/stream/historyoftularek00mene/historyoftularek00mene_djvu.txt (accessed 16 April 2012).

[88] July 29, 1886 issue of the Tulare County Times.

[89] Thompson, T.H. 1892. Official Historical Atlas Map of Tulare County, California. Author, Tulare, California.

[90] Brewer, C. 2004. Historic Tulare County: A Sesquicentennial History, 1852–2002. Historical Publishing Network, San Antonio, Texas. Preview available at: http://books.google.com/books?id=pZGPLnBTDuwC&pg=PP1&lpg=PP1&dq=Historic+Tulare+County+%E2%80%94+A+Sesquicentennial+History,+1852+%E2%80%932002.+Chris+Brewer.+2004&source=bl&ots=_ERU3vhynt&sig=ZItNArBTXoonHm-wNP_kf-934-0&hl=en&sa=X&ei=02qIT8uCB-fu0gG7ypzJCQ&sqi=2&ved=0CC4Q6AEwAA#v=onepage&q=Historic%20Tulare%20County%20%E2%80%94%20A%20Sesquicentennial%20History%2C%201852%20%E2%80%932002.%20Chris%20Brewer.%202004&f=false (accessed 13 April 2012).

[91] City of Woodlake. n.d. Woodlake history website. http://www.cityofwoodlake.com/home.asp?icat=history# (accessed 12 April 2012).

[92] Jan 30, 1913 issue of the Tulare County Times.

[93] Carson, J.H. 1852. Early recollections of the mines, Tulare Plains, and Life in California. San Joaquin Republican, Stockton, California. Reprinted as Browning, P., editor. 1991. Bright gem of the western seas: California 1846–1852. Great West Books, Lafayette, California.

[94] Preston, W. L. 1981. Vanishing Landscapes: Land and Life in the Tulare Lake Basin. University of California Press, Berkeley, California.

[95] Cleland, R.G. 1950. This Reckless Breed of Men: The Trappers and Fur Traders of the Southwest. Alfred A. Knopf, New York.

[96] Wikipedia. n.d. Beaver in the Sierra Nevada web article. http://en.wikipedia.org/wiki/Beaver_in_the_Sierra_Nevada (accessed 20 January 2012).

[97] December 1968 *Los Tulares*, quarterly bulletin of the Tulare County Historical Society, Tulare, California.

[98] Dale, H.C. 1918. The Ashley-Smith Explorations and the Discovery of a Central Route to the Pacific 1822–1829. A.H. Clark Co., Cleveland, Ohio.

[99] Farquhar, F.P. 1965. Pages 23–29 *in* History of the Sierra Nevada. University of California Press, Berkeley and Los Angeles, California.

[100] Dilsaver, L. and W. Tweed. 1990. Chapter Three: Exploration and Exploitation (1850–1885) Arrival of the Anglo-Americans. *in* Challenge of the Big Trees: A Resource History of Sequoia and Kings Canyon National Parks. Sequoia Natural History Association, Three Rivers, California. Available at http://www.cr.nps.gov/history/online_books/dilsaver-tweed/chap3a.htm (accessed 20 January 2012).

[101] Townsend, W.R. 1979. Beaver in the upper Kern Canyon, Sequoia National Park. Thesis. University of California, Fresno, California.

[102] Farquhar, F.P. 1965. Pages 23–29 *in* History of the Sierra Nevada. University of California Press, Berkeley and Los Angeles, California.

[103] Dilsaver, L. and W. Tweed. 1990. Chapter Three: Exploration and Exploitation (1850–1885) Arrival of the Anglo-Americans. *in* Challenge of the Big Trees: A Resource History of Sequoia and Kings Canyon National Parks. Sequoia Natural History Association, Three Rivers, California. Available at http://www.cr.nps.gov/history/online_books/dilsaver-tweed/chap3a.htm (accessed 20 January 2012).

[104] December 1968 *Los Tulares*, quarterly bulletin of the Tulare County Historical Society, Tulare, California.

[105] Maloney, A.B. 1940 Peter Skene Ogden's trapping expedition to the Gulf of California 1829–1830. California Historical Society Quarterly 19(4) 308–316.

[106] Tappe, D.T. 1942. The Status of Beavers in California. Game Bulletin No. 3. California Division of Fish and Game, Sacramento, California. Available at http://www.martinezbeavers.org/wordpress/wp-content/docs/The%20Status%20of%20Beavers%20in%20California%20Tappe%20DT%20Game%20Bullletin%20_3%20California%20DFG%201942.pdf (accessed 19 January 2012).

[107] Townsend, W.R. 1979. Beaver in the upper Kern Canyon, Sequoia National Park. Thesis. University of California, Fresno, California.

[108] Farquhar, F.P. 1965. Pages 23–29 *in* History of the Sierra Nevada. University of California Press, Berkeley and Los Angeles, California.

[109] Leonard Z. 1839. Narrative of the adventures of Zenas Leonard. D. W. Moore, Clearfield, Pennsylvania. Available at http://user.xmission.com/~drudy/mtman/html/leonintr.html (accessed 4 September 2012).

[110] Meek, S.H. April, 1885. The Autobiography of Stephen Hall Meek. *in* A Sketch of the Life of the First Pioneer *in* the April 1885 issue of the Golden Era. Available at http://www.xmission.com/~drudy/mtman/html/smeek.html (accessed 20 January 2012)

[111] Colorado Department of Natural Resources. 2012. Beaver website. http://wildlife.state.co.us/WildlifeSpecies/Profiles/Mammals/Pages/Beaver.aspx (accessed 12 September 2012).

[112] Grinnell, J., J.S. Dixon, and J.M. Linsdale. 1937. Page 636 *in* Fur-Bearing Mammals of California: Their Natural History, Systematic Status and Relations to Man. University of California Press, Berkeley, California.

[113] Tappe, D.T. 1942. The Status of Beavers in California. Game Bulletin No. 3. California Division of Fish and Game, Sacramento, California. Available at http://www.martinezbeavers.org/wordpress/wp-content/docs/The%20Status%20of%20Beavers%20in%20California%20Tappe%20DT%20Game%20Bullletin%20_3%20California%20DFG%201942.pdf (accessed 19 January 2012).

[114] Tappe, D.T. 1942. The Status of Beavers in California. Game Bulletin No. 3. California Division of Fish and Game, Sacramento, California. Available at http://www.martinezbeavers.org/wordpress/wp-content/docs/The%20Status%20of%20Beavers%20in%20California%20Tappe%20DT%20Game%20Bullletin%20_3%20California%20DFG%201942.pdf (accessed 19 January 2012).

[115] Warner, Col. J.J. 1907. Reminiscences of early California from 1831 to 1846. Page 187 *in* Southern California Quarterly, Vol. 7. Los Angeles County Pioneers of Southern California, Historical Society of Southern California, Los Angeles, California. Available at http://books.google.com/books?id=sKeiqjSIUTAC&pg=PA176&lpg=PA176&dq=reminiscences+of+early+california+warner&hl=en#v=onepage&q=reminiscences%20of%20early%20california%20warner&f=false (accessed 19 January 2012).

[116] Wikipedia. n.d. Beaver in the Sierra Nevada web article. http://en.wikipedia.org/wiki/Beaver_in_the_Sierra_Nevada (accessed 20 January 2012).

[117] Williams, E.E. 1973. Tales of Old San Joaquin City. San Joaquin Historian: 9(2):9. Available at http://www.sanjoaquinhistory.org/documents/HistorianOS9-2.pdf (accessed 19 January 2012).

[118] Farquhar, F.P. 1965. Pages 23–29 *in* History of the Sierra Nevada. University of California Press, Berkeley and Los Angeles, California.

[119] Williams, E.E. 1973. Tales of Old San Joaquin City. San Joaquin Historian: 9(2):9. Available at http://www.sanjoaquinhistory.org/documents/HistorianOS9-2.pdf (accessed 19 January 2012).

[120] Tappe, D.T. 1942. The Status of Beavers in California. Game Bulletin No. 3. California Division of Fish and Game, Sacramento, California. Available at http://www.martinezbeavers.org/wordpress/wp-content/docs/The%20Status%20of%20Beavers%20in%20California%20Tappe%20DT%20Game%20Bullletin%20_3%20California%20DFG%201942.pdf (accessed 19 January 2012).

[121] Townsend, W.R. 1979. Beaver in the upper Kern Canyon, Sequoia National Park. Thesis. University of California, Fresno, California.

[122] Latta, F.F. 1977. Handbook of Yokuts Indians. Bear State Books, Santa Cruz, California.

[123] Sumner, L. and J. S. Dixon. 1953. Birds and Mammals of the Sierra Nevada. University California Press, Berkeley, California.

[124] Sumner, L. and J. S. Dixon. 1953. Birds and Mammals of the Sierra Nevada. University California Press, Berkeley, California.

[125] Preston, W. L. 1981. Vanishing Landscapes: Land and Life in the Tulare Lake Basin. University of California Press, Berkeley, California.

[126] Townsend, W.R. 1979. Beaver in the upper Kern Canyon, Sequoia National Park. Thesis. University of California, Fresno, California.

[127] Carson, J.H. 1852. Early recollections of the mines, Tulare Plains, and Life in California. San Joaquin Republican, Stockton, California. Reprinted as Browning, P., editor. 1991. Bright gem of the western seas: California 1846–1852. Great West Books, Lafayette, California.

[128] Heizer, R.F. 1978. Handbook of North American Indians: Volume 8, California. Smithsonian Institution, Washington, D.C.

[129] McCullough, D.R. 1969. The Tule Elk: Its History, Behavior, and Ecology. University of California Publications in Zoology no. 88.

[130] The Phantom Antelope from the May 20, 1904 issue of the Bakersfield Daily Californian

[131] Schmidt, R.H. 1991. Gray wolves in California: their presence and absence. California Fish and Game 77(2):79-85. Available at http://www.archive.org/stream/californiafishga77_2cali/californiafishga77_2cali_djvu.txt (accessed 21 April 2011).

[132] Grinnell, J., J.S. Dixon, and J. M. Linsdale. 1937. Page 529 *in* Fur-bearing mammals of California. Vol. II. University of California Press, Berkeley.

[133] Fremont, John C. 1885. The daring adventures of Kit Carson and Fremont: among buffaloes, grizzlies and Indians: being a spirited diary of the most difficult and wonderful explorations ever made: opening through yawning chasms and over perilous peaks, the great pathway to the Pacific. J.W. Lovell, New York.

[134] Audubon, John W. 1906. Audubon's Western Journal, 1849–1850: Being the MS Record of a Trip from New York to Texas, and an Overland Journey through New Mexico and Arizona to the Gold-fields of California. Arthur H. Clark, Cleveland. Available at http://www.archive.org/stream/audubonswesternj017578mbp/audubonswesternj017578mbp_djvu.txt (accessed 14 April 2011).

[135] Fry, W. 1924. The discovery of Sequoia National Park and the sequoia groves of big trees it contains: Sequoia Nature Guide Service Bulletin no. 1. Sequoia National Park, Three Rivers, California.

[136] Jurek, R.M. 1994. The former distribution of gray wolves in California. California Fish and Game, Sacramento, California. Available at nrm.dfg.ca.gov/FileHandler.ashx?DocumentVersionID=44334 (accessed 4 January 2012)

[137] Fry, W. 1932. A twenty-five year survey of the animals of Sequoia National Park — 1906–1931. Pages 129-31 (described and heavily quoted) *in* G.M. Wright, J.S. Dixon and B.H. Thompson. Fauna of the national parks of the United States, Fauna Series No. 1. U.S. Dept. of Interior, Washington, D.C. Available at http://www.nps.gov/history/history/online_books/fauna1/fauna4c4.htm (accessed 1 March 2012)

[138] Sumner, L. and J.S. Dixon. 1953. Page 464 *in* Birds and Mammals of the Sierra Nevada. University California Press, Berkeley, California.

[139] Sequoia and Kings Canyon National Park (SEKI). n.d. Walter Fry: ambassador of nature website. http://www.nps.gov/seki/historyculture/fry.htm (accessed 2 March 2012)

[140] Fry, W. 1932. A twenty-five year survey of the animals of Sequoia National Park — 1906–1931. Pages 129-31 (described and heavily quoted) *in* G.M. Wright, J.S. Dixon and B.H. Thompson. Fauna of the national parks of the United States, Fauna Series No. 1. U.S. Dept. of Interior, Washington, D.C. Available at http://www.nps.gov/history/history/online_books/fauna1/fauna4c4.htm (accessed 1 March 2012)

[141] Jurek, R.M. 1994. The former distribution of gray wolves in California. California Fish and Game, Sacramento, California. Available at nrm.dfg.ca.gov/FileHandler.ashx?DocumentVersionID=44334 (accessed 4 January 2012)

[142] California Department of Fish and Game (CDFG). n.d. Gray wolf website. http://www.dfg.ca.gov/wildlife/nongame/wolf/ (accessed 28 January 2012).

[143] Haslam, G.W. 1990. The Lake That Will Not Die. Capra Press, Santa Barbara, California.

[144] Lewis, J.C. 1995. Introduction of non-native red foxes in California: implications for the Sierra Nevada red fox. Transactions of the Western Section of the Wildlife Society 31:29–32. Available at http://www.tws-west.org/transactions/Lewis%20Golightly%20Jurek.pdf (accessed 29 January 2012).

[145] California Fish and Game. 1920. Game in the San Joaquin Valley in 1853. California Fish and Game, Sacramento, California.

[146] August 14, 1898 issue of The San Francisco Call. Available at http://chroniclingamerica.loc.gov/lccn/sn85066387/1898-08-14/ed-1/seq-19/ (accessed 24 April 2012).

[147] Fry, W. 1932. A twenty-five year study of the bird life of Sequoia National Park — 1906–1931: Bulletin no. 5. Sequoia National Park, Three Rivers, California.

[148] Latta, F.F. 1937. Tule boats on Tulare Lake: Little journeys in the San Joaquin #13. Article.

[149] Weis, Mae. 1938. Development of this fertile area from primeval state covers comparatively short span. The Corcoran Journal article, Corcoran, California.

[150] Latta, F.F. 1977. Handbook of Yokuts Indians. Bear State Books, Santa Cruz, California.

[151] February 10, 2004 issue of San Francisco Chronicle. Available at http://www.sfgate.com/cgi-bin/article.cgi?f=/c/a/2004/02/10/MNG7T4SRV31.DTL (accessed 3 February 2012)

[152] February 10, 2004 issue of the Los Angeles Times. Available at http://articles.latimes.com/2004/feb/10/local/me-sealion10 (accessed 3 February 2012)

[153] The Marine Mammal Center. 2012. Patient success story website for Chippy. http://www.marinemammalcenter.org/patients/success-stories/chippy-the-media-darling.html (accessed 3 February 2012)

[154] Haslam, G.W. 1990. The Lake That Will Not Die. Capra Press, Santa Barbara, California.

[155] U.S. Bureau of Reclamation (USBR). 1970. A summary of hydrologic data for the test case on acreage limitation in Tulare Lake. U.S. Bureau of Reclamation, Sacramento, California.

[156] ECORP Consulting Inc. 2007. Tulare Lake Basin hydrology and hydrography: a summary of the movement of water and aquatic species. Prepared for U.S. Environmental Protection Agency. Document number 909R07002. Available at http://nepis.epa.gov/Exe/ZyPURL.cgi?Dockey=P1002E2I.txt. (accessed 1 February 2011).

[157] U.S. Army Corps of Engineers (USACE). 1999. Post-flood assessment for 1983, 1986, 1995, and 1997. U.S. Army Corps of Engineers, Sacramento, California. Chapter 2 available at http://snugharbor.net/images2010/misc/2002_usace_sac_flood_history.pdf (accessed 28 May 2012).

[158] April 6, 2011 issue of The Bakersfield Californian. Available at http://www.bakersfieldcalifornian.com/local/x529881015/What-will-we-do-with-all-this-dam-water (accessed 26 August 2012).

[159] ECORP Consulting Inc. 2007. Tulare Lake Basin hydrology and hydrography: a summary of the movement of water and aquatic species. Prepared for U.S. Environmental Protection Agency. Document number 909R07002. Available at http://nepis.epa.gov/Exe/ZyPURL.cgi?Dockey=P1002E2I.txt. (accessed 1 February 2011).

[160] Mitchell, A.R. May 3, 1952. Untitled article filed in the Tulare County Historical Library.

[161] March 1956 *Los Tulares*, quarterly bulletin of the Tulare County Historical Society, Tulare, California.

[162] Farquhar, F.P. 1965. Pages 59–60 *in* History of the Sierra Nevada. University of California Press, Berkeley and Los Angeles, California.

[163] Derby, Lt. G. H. 1850. A report on the Tulare Valley. U.S. House of Representatives, Washington, D.C. Reprinted as Browning, P., editor. 1991. Bright gem of the western seas: California 1846–1852. Great West Books, Lafayette, California.

[164] August 12, 1937 issue of the Visalia Times-Delta, reprinted in December 2000 *Los Tulares*, quarterly bulletin of the Tulare County Historical Society, Tulare, California.

[165] February 6, 1864 issue of the Porterville Messenger.

[166] Harding, S.T. 1949. Inflow to Tulare Lake from its tributary streams. Tulare Lake Basin Water Storage District open file report, Hanford, California.

[167] February 6, 1864 issue of the Porterville Messenger.

[168] Boyd, W.H. 1972. A California Middle Border, the Kern River Country, 1772–1880. The Havilah Press, Richardson, Texas.

[169] Arax, M. and R. Wartman. 2005. The King of California: J.G. Boswell and the Making of a Secret American Empire. PublicAffairs, Jackson, Tennessee.

[170] Kings River Water Conservation District and Kings River Water Association. 2003. The Kings River Handbook. Available at http://www.centralvalleywater.org/_pdf/KingsRiverHandbook-03final.pdf (accessed 20 April 2011).

[171] August 14, 1898 issue of The San Francisco Call. Available at http://chroniclingamerica.loc.gov/lccn/sn85066387/1898-08-14/ed-1/seq-19/ (accessed 24 April 2012).

[172] March 1954 and June 1976 *Los Tulares*, quarterly bulletin of the Tulare County Historical Society, Tulare, California.

[173] March 1954 *Los Tulares*, quarterly bulletin of the Tulare County Historical Society, Tulare, California.

[174] April 22, 1875 issue of the Tulare Times.

[175] August 14, 1898 issue of The San Francisco Call. Available at http://chroniclingamerica.loc.gov/lccn/sn85066387/1898-08-14/ed-1/seq-19/ (accessed 24 April 2012).

[176] January 4, 1978 issue of the Farmersville Herald.

[177] Grunsky, C.E. 1930. Tulare Lake — A contribution to long-time weather history. Monthly Weather Review 58(7): 288–290. Available at http://docs.lib.noaa.gov/rescue/mwr/058/mwr-058-07-0288.pdf (accessed 25 June 2012.)

[178] August 14, 1898 issue of The San Francisco Call. Available at http://chroniclingamerica.loc.gov/lccn/sn85066387/1898-08-14/ed-1/seq-19/ (accessed 24 April 2012).

[179] Thompson, T.H. 1892. Official Historical Atlas Map of Tulare County, California, Author, Tulare, California.

[180] July 15, 1981 issue of the Pixley Enterprise.

[181] August 24, 2007 issue of The South Valley Bee.

[182] March 1997 *Los Tulares*, quarterly bulletin of the Tulare County Historical Society, Tulare, California.

[183] Latta, F.F. 1937. Tulare Lake boats and steamboats: Little journeys in the San Joaquin #12. Article.

[184] January 12, 1884 issue of the New York Times.

[185] 1887 Hanford Sentinel, reprinted in June 1958 *Los Tulares*, quarterly bulletin of the Tulare County Historical Society, Tulare, California.

[186] July 15, 1981 issue of the Pixley Enterprise.

[187] August 14, 1898 issue of The San Francisco Call. Available at http://chroniclingamerica.loc.gov/lccn/sn85066387/1898-08-14/ed-1/seq-19/ (accessed 24 April 2012).

[188] August 14, 1898 issue of The San Francisco Call. Available at http://chroniclingamerica.loc.gov/lccn/sn85066387/1898-08-14/ed-1/seq-19/ (accessed 24 April 2012).

[189] December 20, 1888 issue of the Tulare County Times, reprinted in the September 1997 *Los Tulares*, quarterly bulletin of the Tulare County Historical Society, Tulare, California.

[190] August 11, 1889 of the New York Times.

[191] October 24, 1889 issue of Visalia Weekly Delta.

[192] August 15, 1898 issue of the New York Times.

[193] August 14, 1898 issue of The San Francisco Call. Available at http://chroniclingamerica.loc.gov/lccn/sn85066387/1898-08-14/ed-1/seq-19/ (accessed 24 April 2012).

[194] August 14, 1898 issue of The San Francisco Call. Available at http://chroniclingamerica.loc.gov/lccn/sn85066387/1898-08-14/ed-1/seq-19/ (accessed 24 April 2012).

[195] August 14, 1898 issue of The San Francisco Call. Available at http://chroniclingamerica.loc.gov/lccn/sn85066387/1898-08-14/ed-1/seq-19/ (accessed 24 April 2012).

[196] U.S. Bureau of Reclamation (USBR). 1970. A summary of hydrologic data for the test case on acreage limitation in Tulare Lake. U.S. Bureau of Reclamation, Sacramento, California.

[197] U.S. Bureau of Reclamation (USBR). 1970. A summary of hydrologic data for the test case on acreage limitation in Tulare Lake. U.S. Bureau of Reclamation, Sacramento, California.

[198] June 28, 1952 issue of the Tulare Advance Register.

[199] Latta, F.F. 1977. Handbook of Yokuts Indians. Bear State Books, Santa Cruz, California.

[200] February 3, 1945 issue of the Visalia Times-Delta.

[201] Arax, M and R. Wartman. 2005. The King of California: J.G. Boswell and the making of a secret American empire. PublicAffairs, Jackson, Tennessee.

[202] April 2, 1969 issue of the Visalia Times-Delta.

[203] May 11, 1969 issue of the Fresno Bee.

[204] May 11, 1969 issue of the Fresno Bee.

[205] ECORP Consulting Inc. 2007. Tulare Lake Basin hydrology and hydrography: a summary of the movement of water and aquatic species. Prepared for U.S. Environmental Protection Agency. Document number 909R07002. Available at http://nepis.epa.gov/Exe/ZyPURL.cgi?Dockey=P1002E2I.txt. (accessed 1 February 2011).

[206] Dill, W. A. and A.J. Cordone. 1997. History and Status of Introduced Fishes in California, 1871–1996; Fish Bulletin 178. California Department of Fish and Game, Sacramento, California. Available at http://content.cdlib.org/view?docId=kt8p30069f;NAAN=13030&doc.view=frames&chunk.id=0&toc.id=d0e445&brand=calisphere (accessed 21 June 2011).

[207] ECORP Consulting Inc. 2007. Tulare Lake Basin hydrology and hydrography: a summary of the movement of water and aquatic species. Prepared for U.S. Environmental Protection Agency. Document number 909R07002. Available at http://nepis.epa.gov/Exe/ZyPURL.cgi?Dockey=P1002E2I.txt. (accessed 1 February 2011).

[208] U.S. Bureau of Reclamation (USBR). 1970. A summary of hydrologic data for the test case on acreage limitation in Tulare Lake. U.S. Bureau of Reclamation, Sacramento, California.

[209] U.S. Army Corps of Engineers (USACE). n.d. Water control manual for Pine Flat Dam: pertinent data sheets. U.S. Army Corps of Engineers, Sacramento, California.

[210] U.S. Army Corps of Engineers (USACE). n.d. Water control manual for Pine Flat Dam: pertinent data sheets. U.S. Army Corps of Engineers, Sacramento, California.

[211] U.S. Army Corps of Engineers (USACE). n.d. Water control manual for Pine Flat Dam: pertinent data sheets. U.S. Army Corps of Engineers, Sacramento, California.

[212] U.S. Army Corps of Engineers (USACE). n.d. Water control manual for Pine Flat Dam: pertinent data sheets. U.S. Army Corps of Engineers, Sacramento, California.

[213] U.S. Army Corps of Engineers (USACE). 1999. Post-flood assessment for 1983, 1986, 1995, and 1997: Chapter 3. U.S. Army Corps of Engineers, Sacramento, California. Available at http://www.auburndamcouncil.org/pages/pdf-files/3-cv_floodmgmt_system.pdf (accessed 1 June 2012).

[214] U.S. Army Corps of Engineers (USACE). n.d. Water control manual for Pine Flat Dam: pertinent data sheets. U.S. Army Corps of Engineers, Sacramento, California.

[215] U.S. Army Corps of Engineers (USACE). n.d. Water control manual for Pine Flat Dam: pertinent data sheets. U.S. Army Corps of Engineers, Sacramento, California.

[216] Kings River Water Conservation District and Kings River Water Association. 2003. The Kings River Handbook. Available at http://www.centralvalleywater.org/_pdf/KingsRiverHandbook-03final.pdf (accessed 20 April 2011).

[217] ECORP Consulting Inc. 2007. Tulare Lake Basin hydrology and hydrography: a summary of the movement of water and aquatic species. Prepared for U.S. Environmental Protection Agency. Document number 909R07002. Available at http://nepis.epa.gov/Exe/ZyPURL.cgi?Dockey=P1002E2I.txt. (accessed 1 February 2011).

[218] ECORP Consulting Inc. 2007. Tulare Lake Basin hydrology and hydrography: a summary of the movement of water and aquatic species. Prepared for U.S. Environmental Protection Agency. Document number 909R07002. Available at http://nepis.epa.gov/Exe/ZyPURL.cgi?Dockey=P1002E2I.txt. (accessed 1 February 2011).

[219] U.S. Army Corps of Engineers (USACE). n.d. Water control manual for Terminus Dam: pertinent data sheets. U.S. Army Corps of Engineers, Sacramento, California.

[220] U.S. Bureau of Reclamation (USBR). 1970. A summary of hydrologic data for the test case on acreage limitation in Tulare Lake. U.S. Bureau of Reclamation, Sacramento, California.

[221] Wikipedia. 2012. Terminus Dam web article. http://en.wikipedia.org/wiki/Terminus_dam (accessed 17 August 2012).

[222] U.S. Army Corps of Engineers (USACE). n.d. Water control manual for Pine Flat Dam: pertinent data sheets. U.S. Army Corps of Engineers, Sacramento, California.

[223] U.S. Army Corps of Engineers (USACE). n.d. Water control manual for Pine Flat Dam: pertinent data sheets. U.S. Army Corps of Engineers, Sacramento, California.

[224] U.S. Army Corps of Engineers (USACE). June 1989. Kaweah River Sediment Investigation: Hydraulic Design. Unpublished report, Sacramento, California.

[225] U.S. Army Corps of Engineers (USACE). n.d. Water control manual for Pine Flat Dam: pertinent data sheets. U.S. Army Corps of Engineers, Sacramento, California.

[226] U.S. Geological Survey (USGS). 2011. HUC Boundary Descriptions and Names of Regions, Subregions, Accounting Units and Cataloging Units website. http://water.usgs.gov/GIS/huc_name.html (accessed 17 August 2012).

[227] U.S. Army Corps of Engineers (USACE). June 1989. Kaweah River Sediment Investigation: Hydraulic Design. Unpublished report, Sacramento, California.

[228] U.S. Bureau of Reclamation (USBR). 1970. A summary of hydrologic data for the test case on acreage limitation in Tulare Lake. U.S. Bureau of Reclamation, Sacramento, California.

[229] U.S. Army Corps of Engineers (USACE). 1999. Post-flood assessment for 1983, 1986, 1995, and 1997: Chapter 3. U.S. Army Corps of Engineers, Sacramento, California. Available at http://www.auburndamcouncil.org/pages/pdf-files/3-cv_floodmgmt_system.pdf (accessed 1 June 2012).

[230] Transcript of the March 25, 2011 meeting of the Central Valley Flood Protection Board, pages 150–163. Available at http://www.cvfpb.ca.gov/transcripts/2011/03-25-2011transcript.pdf (accessed (27 June 2012)

[231] December 18, 2001 issue of The Valley Voice. Available at http://www.valleyvoicenewspaper.com/vvarc/2001/december182001.htm (accessed (27 June 2012)

[232] Austin, J.T. Forthcoming. Erosion and mass wasting. Appendix 7 *in*: National Park Service (NPS). Natural resource condition assessment for Sequoia and Kings Canyon National Parks. To be published in the National Park Service Natural Resource Report Series. Panek, J.A. and C.A. Sydoriak, editors. National Park Service, Fort Collins, Colorado.

[233] Austin, J.T. Forthcoming. Erosion and mass wasting. Appendix 7 *in*: National Park Service (NPS). Natural resource condition assessment for Sequoia and Kings Canyon National Parks. To be published in the National Park Service Natural Resource Report Series. Panek, J.A. and C.A. Sydoriak, editors. National Park Service, Fort Collins, Colorado.

[234] U.S. Army Corps of Engineers (USACE). June 1989. Kaweah River Sediment Investigation: Hydraulic Design. Unpublished report, Sacramento, California.

[235] U.S. Army Corps of Engineers (USACE). May 1989. Kaweah River, California: Hydrology report (draft). U.S. Army Corps of Engineers, Sacramento, California.

[236] U.S. Army Corps of Engineers (USACE). June 1989. Kaweah River Sediment Investigation: Hydraulic Design. Unpublished report, Sacramento, California.

[237] Austin, J.T. Forthcoming. Erosion and mass wasting. Appendix 7 *in*: National Park Service (NPS). Natural resource condition assessment for Sequoia and Kings Canyon National Parks. To be published in the National Park Service Natural Resource Report Series. Panek, J.A. and C.A. Sydoriak, editors. National Park Service, Fort Collins, Colorado.

[238] U.S. Army Corps of Engineers (USACE). 1999. Post-flood assessment for 1983, 1986, 1995, and 1997: Chapter 3. U.S. Army Corps of Engineers, Sacramento, California. Available at http://www.auburndamcouncil.org/pages/pdf-files/3-cv_floodmgmt_system.pdf (accessed 1 June 2012).

[239] Transcript of the March 25, 2011 meeting of the Central Valley Flood Protection Board, pages 150–163. Available at http://www.cvfpb.ca.gov/transcripts/2011/03-25-2011transcript.pdf (accessed (27 June 2012)

[240] December 18, 2001 issue of The Valley Voice. Available at http://www.valleyvoicenewspaper.com/vvarc/2001/december182001.htm (accessed (27 June 2012)

[241] U.S. Army Corps of Engineers (USACE). May 1989. Kaweah River, California: Hydrology report (draft). U.S. Army Corps of Engineers, Sacramento, California.

[242] U.S. Army Corps of Engineers (USACE). 1999. Post-flood assessment for 1983, 1986, 1995, and 1997: Chapter 3. U.S. Army Corps of Engineers, Sacramento, California. Available at http://www.auburndamcouncil.org/pages/pdf-files/3-cv_floodmgmt_system.pdf (accessed 1 June 2012).

[243] U.S. Army Corps of Engineers (USACE). n.d. Water control manual for Terminus Dam: pertinent data sheets. U.S. Army Corps of Engineers, Sacramento, California.

[244] Transcript of the March 25, 2011 meeting of the Central Valley Flood Protection Board, pages 150–163. Available at http://www.cvfpb.ca.gov/transcripts/2011/03-25-2011transcript.pdf (accessed (27 June 2012)

[245] December 18, 2001 issue of The Valley Voice. Available at http://www.valleyvoicenewspaper.com/vvarc/2001/december182001.htm (accessed (27 June 2012)

[246] U.S. Army Corps of Engineers (USACE). n.d. Water control manual for Terminus Dam: pertinent data sheets. U.S. Army Corps of Engineers, Sacramento, California.

[247] U.S. Army Corps of Engineers (USACE). n.d. Water control manual for Terminus Dam: pertinent data sheets. U.S. Army Corps of Engineers, Sacramento, California.

[248] U.S. Army Corps of Engineers (USACE). 1999. Post-flood assessment for 1983, 1986, 1995, and 1997: Chapter 3. U.S. Army Corps of Engineers, Sacramento, California. Available at http://www.auburndamcouncil.org/pages/pdf-files/3-cv_floodmgmt_system.pdf (accessed 1 June 2012).

[249] U.S. Army Corps of Engineers (USACE). n.d. Water control manual for Terminus Dam: pertinent data sheets. U.S. Army Corps of Engineers, Sacramento, California.

[250] U.S. Army Corps of Engineers (USACE). n.d. Water control manual for Success Dam: pertinent data sheets. U.S. Army Corps of Engineers, Sacramento, California.

[251] U.S. Army Corps of Engineers (USACE). n.d. Water control manual for Success Dam: pertinent data sheets. U.S. Army Corps of Engineers, Sacramento, California.

[252] Dean, W.W. 1971. Floods of December, 1966 in the Kern-Kaweah Area, Kern and Tulare Counties, California. U.S. Geological Survey Water–Supply Paper 1870-C. U.S. Geological Survey, Washington, D.C.

[253] U.S. Bureau of Reclamation (USBR). 1970. A summary of hydrologic data for the test case on acreage limitation in Tulare Lake. U.S. Bureau of Reclamation, Sacramento, California.

[254] U.S. Army Corps of Engineers (USACE). n.d. Water control manual for Success Dam: pertinent data sheets. U.S. Army Corps of Engineers, Sacramento, California.

[255] U.S. Army Corps of Engineers (USACE). n.d. Water control manual for Success Dam: pertinent data sheets. U.S. Army Corps of Engineers, Sacramento, California.

[256] U.S. Army Corps of Engineers (USACE). 1999. Post-flood assessment for 1983, 1986, 1995, and 1997: Chapter 3. U.S. Army Corps of Engineers, Sacramento, California. Available at http://www.auburndamcouncil.org/pages/pdf-files/3-cv_floodmgmt_system.pdf (accessed 1 June 2012).

[257] U.S. Army Corps of Engineers (USACE). n.d. Water control manual for Success Dam: pertinent data sheets. U.S. Army Corps of Engineers, Sacramento, California.

[258] May 13, 2012 issue of the Visalia Times-Delta. Available at http://www.visaliatimesdelta.com/article/20120514/NEWS01/205140307 (accessed 17 May 2012).

[259] U.S. Army Corps of Engineers (USACE). n.d. Water control manual for Isabella Dam: pertinent data sheets. U.S. Army Corps of Engineers, Sacramento, California.

[260] U.S. Army Corps of Engineers (USACE). n.d. Water control manual for Isabella Dam: pertinent data sheets. U.S. Army Corps of Engineers, Sacramento, California.

[261] U.S. Army Corps of Engineers (USACE). n.d. Water control manual for Isabella Dam: pertinent data sheets. U.S. Army Corps of Engineers, Sacramento, California.

[262] U.S. Geological Survey (USGS). 2011. HUC Boundary Descriptions and Names of Regions, Subregions, Accounting Units and Cataloging Units website. http://water.usgs.gov/GIS/huc_name.html (accessed 17 August 2012).

[263] U.S. Army Corps of Engineers (USACE). n.d. Water control manual for Isabella Dam: pertinent data sheets. U.S. Army Corps of Engineers, Sacramento, California.

[264] U.S. Army Corps of Engineers (USACE). n.d. Water control manual for Isabella Dam: pertinent data sheets. U.S. Army Corps of Engineers, Sacramento, California.

[265] U.S. Army Corps of Engineers (USACE). 1999. Post-flood assessment for 1983, 1986, 1995, and 1997: Chapter 3. U.S. Army Corps of Engineers, Sacramento, California. Available at http://www.auburndamcouncil.org/pages/pdf-files/3-cv_floodmgmt_system.pdf (accessed 1 June 2012).

[266] U.S. Army Corps of Engineers (USACE). n.d. Water control manual for Isabella Dam: pertinent data sheets. U.S. Army Corps of Engineers, Sacramento, California.

[267] U.S. Army Corps of Engineers (USACE). n.d. Water control manual for Isabella Dam: pertinent data sheets. U.S. Army Corps of Engineers, Sacramento, California.

[268] California Data Exchange Center. n.d. Monthly data by water year website for Sensor #65 and Station ID KGF, KWT, SCC, and KRB. http://cdec.water.ca.gov/cgi-progs/queryWY (accessed 30 October 2012).

[269] U.S. Bureau of Reclamation (USBR). 1970. A summary of hydrologic data for the test case on acreage limitation in Tulare Lake. U.S. Bureau of Reclamation, Sacramento, California.

[270] California Department of Water Resources (DWR). 2011. Water year hydrologic classification indices website. http://cdec.water.ca.gov/cgi-progs/iodir/WSIHIST (accessed 26 June 2012).

[271] Earle, C.J. and H.C. Fritts. 1986. Reconstructing riverflow in the Sacramento Basin since 1560. Report to California Department of Water Resources, Agreement No. DWR B–55398. Laboratory of Tree-ring Research, University of Arizona, Tucson.

[272] Meko, D.M., M.D. Therrell, C.H. Baisan, and M.K. Hughes. 2001. Sacramento River flow reconstructed to A.D. 869 from tree-rings. Journal of the American Water Resources Association 37:1029–1040.

[273] Graumlich, L.J. 1993. A 1,000-year record of temperature and precipitation in the Sierra Nevada Quaternary Research 39:249–255.

[274] Cook, E.R., D.M. Meko, D.W. Stahle, and M.K. Cleaveland. 1999. Drought reconstructions for the continental United States. Journal of Climate 12:1145–1162.

[275] Cook, E.R., C.A. Woodhouse, C.M. Eakin, D.M. Meko, and D.W. Stahle. 2004. Long-term aridity changes in the western United States. Science 306:1015–1018.

[276] Swetnam, T.W., C.H. Baisan, A.C. Caprio, P.M. Brown, R. Touchan, R.S. Anderson, and D.J. Hallett. 2009. Multi-millennial fire history from the Giant Forest, Sequoia National Park. California Fire Ecology 5:120–150. Available at http://www.rmtrr.org/data/Swetnametal_2009.pdf (accessed 1 February 2011).

[277] Mensing, S.A., Benson, L.V., Kashgarian, M. and Lund, S. 2004. A Holocene pollen record of persistent droughts from Pyramid Lake, Nevada, USA. Quaternary Research 62:29-38.

[278] Davis, O.K. 1999. Pollen analysis of Tulare Lake, California: Great Basinlike vegetation in Central California during the full-glacial and early Holocene. Review of Palaeobotany and Playnology 107:249–257. Abstract available at http://www.sciencedirect.com/science/article/pii/S0034666799000202 (accessed 23 November 2012).

[279] Negrini, R.M., et al. 2006. The Rambla highstand shoreline and the Holocene lake-level history of Tulare Lake, California, USA. Quaternary Science Reviews 25:1599–1618.

[280] Intergovernmental Panel on Climate Change (IPCC). 2007. Climate Change 2007. Working Group I: The Physical Science Basis. Available at http://www.ipcc.ch/publications_and_data/ar4/wg1/en/ch6s6-6.html (accessed 29 March 2011)

[281] Freedman, A. 2012. Historic heat wave marches on as drought expands: Climate Central blog. http://www.climatecentral.org/news/historic-heat-wave-marches-on-as-drought-expands/ (accessed 14 August 2012).

[282] National Oceanic and Atmospheric Administration (NOAA) National Climatic Data Center. 2012. State of the climate report for July 2012. Available at http://www.ncdc.noaa.gov/sotc/global/2012/7 (accessed 14 August 2012).

[283] National Oceanic and Atmospheric Administration (NOAA) National Climatic Data Center. 2012. State of the climate global analysis for October 2012. Available at http://www.ncdc.noaa.gov/sotc/global/2012/10 (accessed 24 November 2012).

[284] Breschini, G.S. 1996. Sebastian Vizcaíno's exploration of Monterey in 1602–1603. Available at http://www.mchsmuseum.com/vizcaino.html (accessed 15 March 2011).

[285] Breschini, G.S. 2000. The Portolá Expedition of 1769. Available at http://www.mchsmuseum.com/portola1769.html (accessed 15 March 2011).

[286] Dale, R.H., J.J. French, and H.D. Wilson Jr. 1964. The story of groundwater in the San Joaquin Valley. U.S. Geological Survey Circular 459. U.S. Geological Survey, Washington, D.C. Available at http://pubs.usgs.gov/circ/1964/0459/report.pdf (accessed 1 September 2012).

[287] Galloway, D.L., and F.S. Riley. 1999, San Joaquin Valley, California — Largest human alteration of the Earth's surface. Pages 23–34 in Galloway, D.L., D.R. Jones, and S.E Ingebritsen, editors. Land Subsidence in the United States: U.S. Geological Survey Circular 1182. U.S. Geological Survey,

Washington, D.C. Available at http://pubs.usgs.gov/circ/circ1182/pdf/06SanJoaquinValley.pdf (accessed 1 September 2012).

[288] Gronberg, J.A., et al. 1998. Environmental setting of the San Joaquin–Tulare Basins, California. U.S. Geological Survey Water-Resources Investigations Report 97-4205. U.S. Geological Survey, Sacrament, California. Available at http://ca.water.usgs.gov/sanj/pub/usgs/wrir97-4205/wrir97-4205.pdf (accessed 21 September 2012).

[289] Thiros, S.A. 2010. Conceptual understanding and groundwater quality of the basin-fill aquifer in the Central Valley, California. Section 13 *in* Thiros, S.A., et al. editors. Conceptual understanding and groundwater quality of selected basin-fill aquifers in the Southwestern United States. U.S. Geological Survey Professional Paper 1781. U.S. Geological Survey, Washington, D.C. Available at http://pubs.usgs.gov/pp/1781/pdf/pp1781_section13.pdf (accessed 21 September 2012).

[290] Gronberg, J.A., et al. 1998. Environmental setting of the San Joaquin–Tulare Basins, California. U.S. Geological Survey Water-Resources Investigations Report 97-4205. U.S. Geological Survey, Sacrament, California. Available at http://ca.water.usgs.gov/sanj/pub/usgs/wrir97-4205/wrir97-4205.pdf (accessed 21 September 2012).

[291] Thiros, S.A. 2010. Conceptual understanding and groundwater quality of the basin-fill aquifer in the Central Valley, California. Section 13 *in* Thiros, S.A., et al. editors. Conceptual understanding and groundwater quality of selected basin-fill aquifers in the Southwestern United States. U.S. Geological Survey Professional Paper 1781. U.S. Geological Survey, Washington, D.C. Available at http://pubs.usgs.gov/pp/1781/pdf/pp1781_section13.pdf (accessed 21 September 2012).

[292] Galloway, D.L., and F.S. Riley. 1999, San Joaquin Valley, California — Largest human alteration of the Earth's surface. Pages 23–34 *in* Galloway, D.L., D.R. Jones, and S.E Ingebritsen, editors. Land Subsidence in the United States: U.S. Geological Survey Circular 1182. U.S. Geological Survey, Washington, D.C. Available at http://pubs.usgs.gov/circ/circ1182/pdf/06SanJoaquinValley.pdf (accessed 1 September 2012).

[293] Kaweah Delta Water Conservation District. n.d. Kaweah Delta Water Conservation District: about us website. http://www.kdwcd.com/kdwcdweb_003.htm (accessed 29 September 2012)

[294] Parker, J. n.d. Information and testimony provided to the Little Hoover Commission about the Kern Water Bank. Available at http://www.lhc.ca.gov/studies/201/watergovernance/ParkerJan10.pdf (accessed 29 September 2012).

[295] Kern Water Bank Authority. n.d. Kern Water Bank website. http://www.kwb.org/index.cfm/fuseaction/Pages.Page/id/330 (accessed 29 September 2012)

[296] Segrest, M. 2012. The History of Pumps: Through the Years: Page 2 of 4. http://www.pump-zone.com/topics/pumps/pumps/history-pumps-through-years/page/0/1 (accessed 20 September 2012).

[297] Galloway, D.L., and F.S. Riley. 1999, San Joaquin Valley, California — Largest human alteration of the Earth's surface. Pages 23–34 *in* Galloway, D.L., D.R. Jones, and S.E Ingebritsen, editors. Land Subsidence in the United States: U.S. Geological Survey Circular 1182. U.S. Geological Survey, Washington, D.C. Available at http://pubs.usgs.gov/circ/circ1182/pdf/06SanJoaquinValley.pdf (accessed 1 September 2012).

[298] June 1969 *Los Tulares*, quarterly bulletin of the Tulare County Historical Society, Tulare, California.

[299] Dale, R.H., J.J. French, and H.D. Wilson Jr. 1964. The story of groundwater in the San Joaquin Valley. U.S. Geological Survey Circular 459. U.S. Geological Survey, Washington, D.C. Available at http://pubs.usgs.gov/circ/1964/0459/report.pdf (accessed 1 September 2012).

[300] Faunt, C.C., editor. 2009. Groundwater availability of the Central Valley aquifer, California: U.S. Geological Survey Professional Paper 1766. U.S. Geological Survey, Washington, D.C. Available at http://pubs.usgs.gov/pp/1766/PP_1766.pdf (accessed 2 September 2012).

[301] Galloway, D.L., and F.S. Riley. 1999, San Joaquin Valley, California — Largest human alteration of the Earth's surface. Pages 23–34 *in* Galloway, D.L., D.R. Jones, and S.E Ingebritsen, editors. Land Subsidence in the United States: U.S. Geological Survey Circular 1182. U.S. Geological Survey, Washington, D.C. Available at http://pubs.usgs.gov/circ/circ1182/pdf/06SanJoaquinValley.pdf (accessed 1 September 2012).

[302] Wikipedia. 2012. California State Water Project. http://en.wikipedia.org/wiki/California_State_Water_Project (accessed 20 January 2012).

[303] Wikipedia. 2012. California State Water Project. http://en.wikipedia.org/wiki/California_State_Water_Project (accessed 20 January 2012).

[304] Wikipedia. 2012. Central Valley Project. http://en.wikipedia.org/wiki/Central_Valley_Project (accessed 20 January 2012).

[305] Gronberg, J.A., et al. 1998. Environmental setting of the San Joaquin–Tulare Basins, California. U.S. Geological Survey Water-Resources Investigations Report 97-4205. U.S. Geological Survey, Sacrament, California. Available at http://ca.water.usgs.gov/sanj/pub/usgs/wrir97-4205/wrir97-4205.pdf (accessed 21 September 2012).

[306] Aasen, G. 2012. Of interest to managers. Interagency Ecological Program for the San Francisco Estuary Newsletter edition. Interagency Ecological Program for the San Francisco Estuary Newsletter, 25(1). Available at http://www.water.ca.gov/iep/newsletters/2012/IEPNewsletter_FinalWINTER2012.pdf (accessed 22 September 2012).

[307] Revive the San Joaquin shrinking San Joaquin valley aquifer website. Available at http://www.revivethesanjoaquin.org/content/feds-document-shrinking-san-joaquin-valley-aquifer (accessed 1 September 12).

[308] Thiros, S.A. 2010. Conceptual understanding and groundwater quality of the basin-fill aquifer in the Central Valley, California. Section 13 *in* Thiros, S.A., et al. editors. Conceptual understanding and groundwater quality of selected basin-fill aquifers in the Southwestern United States. U.S. Geological Survey Professional Paper 1781. U.S. Geological Survey, Washington, D.C. Available at http://pubs.usgs.gov/pp/1781/pdf/pp1781_section13.pdf (accessed 21 September 2012).

[309] ECORP Consulting Inc. 2007. Tulare Lake Basin hydrology and hydrography: a summary of the movement of water and aquatic species. Prepared for U.S. Environmental Protection Agency. Document number 909R07002. Available at http://nepis.epa.gov/Exe/ZyPURL.cgi?Dockey=P1002E2I.txt. (accessed 1 February 2011).

[310] Wikipedia. 2012. California State Water Project. http://en.wikipedia.org/wiki/California_State_Water_Project (accessed 20 January 2012).

[311] Cohen, R.B., G. Wolff, and B. Nelson. 2004. Energy down the drain: the hidden costs of California's water supply. Natural Resources Defense Council and Pacific Institute. Available at http://www.nrdc.org/water/conservation/edrain/edrain.pdf. (accessed 8 July 2011).

[312] Trask, M. 2005. California's Water-Energy Relationship. California Energy Commission. Available at http://www.energy.ca.gov/2005publications/CEC-700-2005-011/CEC-700-2005-011.PDF (accessed 16 July 2011)

[313] Cohen, R.B., G. Wolff, and B. Nelson. 2004. Energy down the drain: the hidden costs of California's water supply. Natural Resources Defense Council and Pacific Institute. Available at http://www.nrdc.org/water/conservation/edrain/edrain.pdf. (accessed 8 July 2011).

[314] ECORP Consulting Inc. 2007. Tulare Lake Basin hydrology and hydrography: a summary of the movement of water and aquatic species. Prepared for U.S. Environmental Protection Agency. Document number 909R07002. Available at http://nepis.epa.gov/Exe/ZyPURL.cgi?Dockey=P1002E2I.txt. (accessed 1 February 2011).

[315] ECORP Consulting Inc. 2007. Tulare Lake Basin hydrology and hydrography: a summary of the movement of water and aquatic species. Prepared for U.S. Environmental Protection Agency. Document number 909R07002. Available at http://nepis.epa.gov/Exe/ZyPURL.cgi?Dockey=P1002E2I.txt. (accessed 1 February 2011).

[316] UC Irvine. 2009. California's troubled waters: satellite-based findings by UCI, NASA reveal significant groundwater loss in Central Valley. http://today.uci.edu/news/2009/12/nr_centralvalleywater_091214.html (accessed 23 April 2012)

[317] January 30, 2010 issue of the Visalia Times-Delta.

[318] Scanlon, B.R., et al. 2012. Groundwater depletion and sustainability of irrigation in the US High Plains and Central Valley. Proc. Natl. Acad. Sci. Available at http://www.pnas.org/content/early/2012/05/24/1200311109.full.pdf+html (accessed 24 July 2012).

[319] July 22, 2012 issue of the Sacramento Bee. Available at http://www.sacbee.com/2012/07/22/4648322/valley-groundwater-threatened.html (accessed 24 July 2012).

[320] Bertoldi, G.L. 1991. Subsidence and consolidation in alluvial aquifer systems. Proceedings of the 18th Biennial Conference on Groundwater. U.S. Geological Survey, Sacramento, California.

[321] Galloway, D.L., and F.S. Riley. 1999, San Joaquin Valley, California — Largest human alteration of the Earth's surface. Pages 23–34 in Galloway, D.L., D.R. Jones, and S.E Ingebritsen, editors. Land Subsidence in the United States: U.S. Geological Survey Circular 1182. U.S. Geological Survey, Washington, D.C. Available at http://pubs.usgs.gov/circ/circ1182/pdf/06SanJoaquinValley.pdf (accessed 1 September 2012).

[322] May 6, 2011 issue of the Kaweah Commonwealth.

[323] The Aquifer of the San Joaquin Valley web site. Available at http://academic.emporia.edu/schulmem/hydro/TERM%20PROJECTS/2008/Trump/mineral%20resources%20of%20Kansas.html (accessed 1 September 2012).

[324] Galloway, D.L., and F.S. Riley. 1999, San Joaquin Valley, California — Largest human alteration of the Earth's surface. Pages 23–34 in Galloway, D.L., D.R. Jones, and S.E Ingebritsen, editors. Land Subsidence in the United States: U.S. Geological Survey Circular 1182. U.S. Geological Survey, Washington, D.C. Available at http://pubs.usgs.gov/circ/circ1182/pdf/06SanJoaquinValley.pdf (accessed 1 September 2012).

[325] November 2009 issue of the California Farmer. Available at http://magissues.farmprogress.com/CLF/CF11Nov09/clf030.pdf (accessed 1 September 2012)

[326] Paulsen, C.G. 1953. Floods of November-December, 1950 in the Central Valley Basin, California. Pages 505–789 in U.S. Geological Survey Water–Supply Paper 1137-F. U.S. Geological Survey, Washington, D.C.

[327] U.S. Army Corps of Engineers (USACE). 1999. Post-flood assessment for 1983, 1986, 1995, and 1997. U.S. Army Corps of Engineers, Sacramento, California. Chapter 2 available at http://snugharbor.net/images2010/misc/2002_usace_sac_flood_history.pdf (accessed 28 May 2012).

[328] California Department of Water Resources (DWR). 2000. Preparing for California's next drought-Changes since 1987-92. California Department of Water Resources, Sacramento, California. Available at http://www.water.ca.gov/drought/nextdrought.cfm (accessed 1 February 2011).

[329] Cook, E.R., D.M. Meko, D.W. Stahle, and M.K. Cleaveland. 1999. Drought reconstructions for the continental United States. Journal of Climate 12:1145–1162.

[330] Cook, E.R., C.A. Woodhouse, C.M. Eakin, D.M. Meko, and D.W. Stahle. 2004. Long-term aridity changes in the western United States. Science 306:1015–1018.

[331] Stine, S. 1994. Extreme and persistent drought in California and Patagonia during medieval time. Nature 369:546-549.

[332] Stine, S. 2001. The great drought of Y1K. Sierra Nature Notes volume 1. Available at http://www.sierranaturenotes.com/naturenotes/paleodrought1.htm (accessed 1 February 2011).

[333] Stine, S. 2001. The great drought of Y1K. Sierra Nature Notes volume 1. Available at http://www.sierranaturenotes.com/naturenotes/paleodrought1.htm (accessed 1 February 2011).

[334] Stine, S. 2001. The great drought of Y1K. Sierra Nature Notes volume 1. Available at http://www.sierranaturenotes.com/naturenotes/paleodrought1.htm (accessed 1 February 2011).

[335] California Department of Water Resources (DWR). 2000. Preparing for California's next drought-Changes since 1987-92. California Department of Water Resources, Sacramento, California. Available at http://www.water.ca.gov/drought/nextdrought.cfm (accessed 1 February 2011).

[336] Ingram, B. L., Ingle, J. C., and Conrad, M. E. 1996. A 2,000-yr record of San Joaquin and Sacramento River inflow to San Francisco Bay California. Geology 24:331-334.

[337] Meko, D.M., M.D. Therrell, C.H. Baisan, and M.K. Hughes. 2001. Sacramento River flow reconstructed to A.D. 869 from tree-rings. Journal of the American Water Resources Association 37:1029–1040.

[338] Caprio, A.C., L.S. Mutch, T.W. Swetnam, and C.H. Baisan. 1994. Temporal and spatial patterns of giant sequoia radial growth response to a high severity fire in A.D. 1297. Contract report to the California Department of Forestry and Fire Protection, Mountain Home State Forest.

[339] Meko, D.M., M.D. Therrell, C.H. Baisan, and M.K. Hughes. 2001. Sacramento River flow reconstructed to A.D. 869 from tree-rings. Journal of the American Water Resources Association 37:1029–1040.

[340] Graumlich, L.J. 1993. A 1,000-year record of temperature and precipitation in the Sierra Nevada. Quaternary Research 39:249–255.

[341] Earle, C.J. and H.C. Fritts. 1986. Reconstructing riverflow in the Sacramento Basin since 1560. Report to California Department of Water Resources, Agreement No. DWR B–55398. Laboratory of Tree-ring Research, University of Arizona, Tucson.

[342] Earle, C.J. and H.C. Fritts. 1986. Reconstructing riverflow in the Sacramento Basin since 1560. Report to California Department of Water Resources, Agreement No. DWR B–55398. Laboratory of Tree-ring Research, University of Arizona, Tucson.

[343] Schimmelmann, A, M. Zhao, C. Harvey, and C. Lange. 1998. A Large California Flood and Correlative Global Climatic Events 400 Years Ago. Quaternary Research 49:51–61.

[344] Schimmelmann, A, C.B. Lange, and B.J. Meggers. 2003. Paleoclimatic and archaeological evidence for a 200-yr recurrence of floods and droughts linking California, Mesoamerica and South America over the past 2,000 years. The Holocene, 13(5):763–778.

[345] Bradley, R.S., and P.D. Jones. 1993. 'Little Ice Age' summer temperature variations: Their nature and relevance to recent global warming tends. The Holocene 3(4):367–376.

[346] Scuderi, L.A. 1990. Tree-ring evidence for climatically effective volcanic eruptions. Quaternary Research 34:67–85.

[347] Earle, C.J. and H.C. Fritts. 1986. Reconstructing riverflow in the Sacramento Basin since 1560. Report to California Department of Water Resources, Agreement No. DWR B–55398. Laboratory of Tree-ring Research, University of Arizona, Tucson.

[348] Graumlich, L.J. 1993. A 1,000-year record of temperature and precipitation in the Sierra Nevada. Quaternary Research 39:249–255.

[349] Stine, S. 1990. Late Holocene fluctuations of Mono Lake, eastern California. Paleogeography, Paleoclimatology, Paleoecology 78:333–381

[350] Haston, L. and J. Michaelsen. 1994. Long-term central coastal California precipitation variability and relationships to El Niño -Southern Oscillation. Journal of Climate 7:1373–1387.

[351] Earle, C.J. and H.C. Fritts. 1986. Reconstructing riverflow in the Sacramento Basin since 1560. Report to California Department of Water Resources, Agreement No. DWR B–55398. Laboratory of Tree-ring Research, University of Arizona, Tucson.

[352] Earle, C.J. and H.C. Fritts. 1986. Reconstructing riverflow in the Sacramento Basin since 1560. Report to California Department of Water Resources, Agreement No. DWR B–55398. Laboratory of Tree-ring Research, University of Arizona, Tucson.

[353] Earle, C.J. and H.C. Fritts. 1986. Reconstructing riverflow in the Sacramento Basin since 1560. Report to California Department of Water Resources, Agreement No. DWR B–55398. Laboratory of Tree-ring Research, University of Arizona, Tucson.

[354] Earle, C.J. and H.C. Fritts. 1986. Reconstructing riverflow in the Sacramento Basin since 1560. Report to California Department of Water Resources, Agreement No. DWR B–55398. Laboratory of Tree-ring Research, University of Arizona, Tucson.

[355] Earle, C.J. and H.C. Fritts. 1986. Reconstructing riverflow in the Sacramento Basin since 1560. Report to California Department of Water Resources, Agreement No. DWR B–55398. Laboratory of Tree-ring Research, University of Arizona, Tucson.

[356] Earle, C.J. and H.C. Fritts. 1986. Reconstructing riverflow in the Sacramento Basin since 1560. Report to California Department of Water Resources, Agreement No. DWR B–55398. Laboratory of Tree-ring Research, University of Arizona, Tucson.

[357] Earle, C.J. and H.C. Fritts. 1986. Reconstructing riverflow in the Sacramento Basin since 1560. Report to California Department of Water Resources, Agreement No. DWR B–55398. Laboratory of Tree-ring Research, University of Arizona, Tucson.

[358] Paulsen, C.G. 1953. Floods of November-December, 1950 in the Central Valley Basin, California. Pages 505–789 *in* U.S. Geological Survey Water–Supply Paper 1137-F. U.S. Geological Survey, Washington, D.C.

[359] U.S. Army Corps of Engineers (USACE). 1999. Post-flood assessment for 1983, 1986, 1995, and 1997. U.S. Army Corps of Engineers, Sacramento, California. Chapter 2 available at http://snugharbor.net/images2010/misc/2002_usace_sac_flood_history.pdf (accessed 28 May 2012).

[360] Thompson and West. 1879. History of Yuba County, California. Thompson and West, Oakland, California.

[361] Kattelmann, R., N. Berg, and B. McGurk. 1991. A history of rain-on-snow floods in the Sierra Nevada. Western Snow Conference. Juneau, Alaska, April 1991:138–141. Available at http://www.westernsnowconference.org/proceedings/pdf_Proceedings/1991%20WEB/Kattelmann,Rain-On-SnowFloodsSierraNevada.pdf (accessed 14 June 2011).

[362] Paulsen, C.G. 1953. Floods of November-December, 1950 in the Central Valley Basin, California. Pages 505–789 *in* U.S. Geological Survey Water–Supply Paper 1137-F. U.S. Geological Survey, Washington, D.C.

[363] Thompson and West. 1879. History of Yuba County, California. Thompson and West, Oakland, California.

[364] U.S. Army Corps of Engineers (USACE). 1999. Post-flood assessment for 1983, 1986, 1995, and 1997. U.S. Army Corps of Engineers, Sacramento, California. Chapter 2 available at http://snugharbor.net/images2010/misc/2002_usace_sac_flood_history.pdf (accessed 28 May 2012).

[365] Mitchell, A.R. May 3, 1952. Untitled article filed in the Tulare County Historical Library.

[366] Paulsen, C.G. 1953. Floods of November-December, 1950 in the Central Valley Basin, California. Pages 505–789 *in* U.S. Geological Survey Water–Supply Paper 1137-F. U.S. Geological Survey, Washington, D.C.

[367] Earle, C.J. and H.C. Fritts. 1986. Reconstructing riverflow in the Sacramento Basin since 1560. Report to California Department of Water Resources, Agreement No. DWR B–55398. Laboratory of Tree-ring Research, University of Arizona, Tucson.

[368] Earle, C.J. and H.C. Fritts. 1986. Reconstructing riverflow in the Sacramento Basin since 1560. Report to California Department of Water Resources, Agreement No. DWR B–55398. Laboratory of Tree-ring Research, University of Arizona, Tucson.

[369] Paulsen, C.G. 1953. Floods of November-December, 1950 in the Central Valley Basin, California. Pages 505–789 *in* U.S. Geological Survey Water–Supply Paper 1137-F. U.S. Geological Survey, Washington, D.C.

[370] U.S. Army Corps of Engineers (USACE). 1999. Post-flood assessment for 1983, 1986, 1995, and 1997. U.S. Army Corps of Engineers, Sacramento, California. Chapter 2 available at http://snugharbor.net/images2010/misc/2002_usace_sac_flood_history.pdf (accessed 28 May 2012).

[371] U.S. Army Corps of Engineers (USACE). 1999. Post-flood assessment for 1983, 1986, 1995, and 1997. U.S. Army Corps of Engineers, Sacramento, California. Chapter 2 available at http://snugharbor.net/images2010/misc/2002_usace_sac_flood_history.pdf (accessed 28 May 2012).

[372] Paulsen, C.G. 1953. Floods of November-December, 1950 in the Central Valley Basin, California. Pages 505–789 *in* U.S. Geological Survey Water–Supply Paper 1137-F. U.S. Geological Survey, Washington, D.C.

[373] Thompson, T.H. 1892. Official Historical Atlas Map of Tulare County, California. Author, Tulare, California.

[374] Thompson and West. 1879. History of Yuba County, California. Thompson and West, Oakland, California.

[375] Derby, Lt. G. H. 1850. A report on the Tulare Valley. U.S. House of Representatives, Washington, D.C. Reprinted as Browning, P., editor. 1991. Bright gem of the western seas: California 1846–1852. Great West Books, Lafayette, California.

[376] January 6, 2011 issue of the Valley Voice.

[377] U.S. Bureau of Reclamation (USBR). 1970. A summary of hydrologic data for the test case on acreage limitation in Tulare Lake. U.S. Bureau of Reclamation, Sacramento, California.

[378] McCulley, Frick & Gilman and William Lettis & Associates. 1998. Panoche/Silver Creek Watershed Assessment. Prepared for Panoche/Silver Creek Watershed Coordinated Resource Management and Planning Group and the City of Mendota, California. Available at http://www.water.ca.gov/pubs/environment/watersheds/panoche_silver_creek_watershed_assessment_final_report/psc-assessment.pdf (accessed 23 August 2011).

[379] Paulsen, C.G. 1953. Floods of November-December, 1950 in the Central Valley Basin, California. Pages 505–789 *in* U.S. Geological Survey Water–Supply Paper 1137-F. U.S. Geological Survey, Washington, D.C.

[380] California Department of Water Resources (DWR). 2010. Fact Sheet, Sacramento River Flood Control System Weirs and Flood Relief Structures. California Department of Water Resources. Sacramento, California. Available at http://www.water.ca.gov/newsroom/docs/WeirsReliefStructures.pdf (accessed 29 April 2012).

[381] Holiday, J.S. 1999. Page 192 *in* Rush for Riches Gold Fever and the Making of California. University of California Press, Berkeley, California.

[382] January 29, 1853 issue of the New York Times.

[383] February 14, 1853 issue of the New York Times.

[384] U.S. Army Corps of Engineers (USACE). 1999. Post-flood assessment for 1983, 1986, 1995, and 1997. U.S. Army Corps of Engineers, Sacramento, California. Chapter 2 available at http://snugharbor.net/images2010/misc/2002_usace_sac_flood_history.pdf (accessed 28 May 2012).

[385] Undated January 1853 article from the San Francisco Alta California newspaper, reprinted in the January 29 1853 issue of the New York Times.

[386] Holiday, J.S. 1999. Page 192 *in* Rush for Riches Gold Fever and the Making of California. University of California Press, Berkeley, California.

[387] Thompson and West. 1879. History of Yuba County, California. Thompson and West, Oakland, California.

[388] January 29 1853 issue of the New York Times.

[389] Article from the December 31, 1852 Sacramento Delta reprinted in the January 29 issue of the New York Times.

[390] Undated article from the Alta California newspaper reprinted in the January 29 1853 issue of the New York Times.

[391] Perlot, J-N. 1985. Page 154 *in* Gold Seeker: Adventures of a Belgian Argonaut during the Gold Rush Years. Yale University Press, New Haven, Connecticut.

[392] McCulley, Frick & Gilman and William Lettis & Associates. 1998. Panoche/Silver Creek Watershed Assessment. Prepared for Panoche/Silver Creek Watershed Coordinated Resource Management and Planning Group and the City of Mendota, California. Available at http://www.water.ca.gov/pubs/environment/watersheds/panoche_silver_creek_watershed_assessment _final_report/psc-assessment.pdf (accessed 23 August 2011).

[393] Thompson, T.H. 1892. Official Historical Atlas Map of Tulare County, California. Author, Tulare, California.

[394] February 6, 1862 issue of the Porterville Messenger.

[395] Gia, Gilbert. 2010. Gordon's Ferry and Other Crossings Of The Kern River, 1852–1937.

[396] Latta, F.F. 1977. Handbook of Yokuts Indians. Bear State Books, Santa Cruz, California.

[397] Otter, F.L. 1963. The Men of Mammoth Forest. Edwards Bros. Inc., Ann Arbor, Michigan.

[398] Otter, F.L. 1963. The Men of Mammoth Forest. Edwards Bros. Inc., Ann Arbor, Michigan.

[399] Preston, W. L. 1981. Vanishing Landscapes: Land and Life in the Tulare Lake Basin. University of California Press, Berkeley, California.

[400] November 7, 1861 issue of the Red Bluff Independent.

[401] Goodridge, J. 1996. Data on California's Extreme Rainfall from 1862–1995. Proceedings of the 1996 California Extreme Precipitation Symposium. Sierra College Science Center, Rocklin, California. Available at http://www.cepsym.info/proceedings_1996.php (accessed 1 July 2011).

[402] U.S. Army Corps of Engineers (USACE). 1999. Post-flood assessment for 1983, 1986, 1995, and 1997. U.S. Army Corps of Engineers, Sacramento, California. Chapter 2 available at http://snugharbor.net/images2010/misc/2002_usace_sac_flood_history.pdf (accessed 28 May 2012).

[403] Null, J. and J. Hulbert. 2007. California washed away, the great flood of 1862. Weatherwise 60:26–30. Available at http://ggweather.com/1862flood.pdf (accessed 22 June 2011).

[404] U.S. Geological Survey (USGS). 2011. Multi-Hazard West Coast Winter Storm Project website. http://urbanearth.gps.caltech.edu/winter-storm/ (accessed 1 February 2011).

[405] Brewer, W.H. 1930. Up and Down California in 1860–1864; the Journal of William H. Brewer. Yale University Press, New Haven, Connecticut. Available at http://www.yosemite.ca.us/library/up_and_down_california/3-1.html (accessed 21 June 2011).

[406] Goodridge, J. 1996. Data on California's Extreme Rainfall from 1862–1995. Proceedings of the 1996 California Extreme Precipitation Symposium. Sierra College Science Center, Rocklin, California. Available at http://www.cepsym.info/proceedings_1996.php (accessed 1 July 2011).

[407] Wells, E.L. 1947, Notes on the winter of 1861–1862 in the Pacific Northwest. Northwest Science v XXI. Available at http://www.vetmed.wsu.edu/org_nws/NWSci%20journal%20articles/1940-1949/1947%20vol%2021/21-2/v21%20p76%20Wells.PDF (accessed 1 February 2011).

[408] Perlot, J-N. 1985. Page 373 *in* Gold Seeker: Adventures of a Belgian Argonaut during the Gold Rush Years. Yale University Press, New Haven, Connecticut.

[409] December 7, 1861 issue of the Oregonian as reprinted in the January 2, 1862 issue of the Porterville Messenger.

[410] U.S. Army Corps of Engineers (USACE). 1999. Post-flood assessment for 1983, 1986, 1995, and 1997. U.S. Army Corps of Engineers, Sacramento, California. Chapter 2 available at http://snugharbor.net/images2010/misc/2002_usace_sac_flood_history.pdf (accessed 28 May 2012).

[411] December 10, 1861 issue of the Red Bluff Independent.

[412] Goodridge, J. 1996. Data on California's Extreme Rainfall from 1862–1995. Proceedings of the 1996 California Extreme Precipitation Symposium. Sierra College Science Center, Rocklin, California. Available at http://www.cepsym.info/proceedings_1996.php (accessed 1 July 2011).

[413] Null, J. and J. Hulbert. 2007. California washed away, the great flood of 1862. Weatherwise 60:26–30. Available at http://ggweather.com/1862flood.pdf (accessed 22 June 2011).

[414] Paulsen, C.G. 1953. Floods of November-December, 1950 in the Central Valley Basin, California. Pages 505–789 *in* U.S. Geological Survey Water–Supply Paper 1137-F. U.S. Geological Survey, Washington, D.C.

[415] Brewer, W.H. 1930. Up and Down California in 1860–1864; the Journal of William H. Brewer. Yale University Press, New Haven, Connecticut. Available at http://www.yosemite.ca.us/library/up_and_down_california/ (accessed 21 June 2011).

[416] Goodridge, J. 1996. Data on California's Extreme Rainfall from 1862–1995. Proceedings of the 1996 California Extreme Precipitation Symposium. Sierra College Science Center, Rocklin, California. Available at http://www.cepsym.info/proceedings_1996.php (accessed 1 July 2011).

[417] Brewer, W.H. 1930. Up and Down California in 1860–1864; the Journal of William H. Brewer. Yale University Press, New Haven, Connecticut. Available at http://www.yosemite.ca.us/library/up_and_down_california/3-1.html (accessed 28 February 2012).

[418] Brewer, W.H. 1930. Up and Down California in 1860–1864; the Journal of William H. Brewer. Yale University Press, New Haven, Connecticut. Available at http://www.yosemite.ca.us/library/up_and_down_california/3-1.html (accessed 28 February 2012).

[419] Goodridge, J. 1996. Data on California's Extreme Rainfall from 1862–1995. Proceedings of the 1996 California Extreme Precipitation Symposium. Sierra College Science Center, Rocklin, California. Available at http://www.cepsym.info/proceedings_1996.php (accessed 1 July 2011).

[420] Thompson and West. 1879. History of Yuba County, California. Thompson and West, Oakland, California.

[421] December 10, 1861 issue of the Marysville Appeal.

[422] January 26, 1862 issue of the Mariposa Gazette as reprinted in the February 6, 1862 issue of the Porterville Messenger.

[423] U.S. Army Corps of Engineers (USACE). 1999. Post-flood assessment for 1983, 1986, 1995, and 1997. U.S. Army Corps of Engineers, Sacramento, California. Chapter 2 available at http://snugharbor.net/images2010/misc/2002_usace_sac_flood_history.pdf (accessed 28 May 2012).

[424] U.S. Army Corps of Engineers (USACE). 1999. Post-flood assessment for 1983, 1986, 1995, and 1997. U.S. Army Corps of Engineers, Sacramento, California. Chapter 2 available at http://snugharbor.net/images2010/misc/2002_usace_sac_flood_history.pdf (accessed 28 May 2012).

[425] McCulley, Frick & Gilman and William Lettis & Associates. 1998. Panoche/Silver Creek Watershed Assessment. Prepared for Panoche/Silver Creek Watershed Coordinated Resource Management and Planning Group and the City of Mendota, California. Available at http://www.water.ca.gov/pubs/environment/watersheds/panoche_silver_creek_watershed_assessment_final_report/psc-assessment.pdf (accessed 23 August 2011).

[426] Paulsen, C.G. 1953. Floods of November-December, 1950 in the Central Valley Basin, California. Pages 505–789 *in* U.S. Geological Survey Water–Supply Paper 1137-F. U.S. Geological Survey, Washington, D.C.

[427] January 20, 1862 issue of the Porterville Messenger.

[428] U.S. Army Corps of Engineers (USACE). 1999. Post-flood assessment for 1983, 1986, 1995, and 1997. U.S. Army Corps of Engineers, Sacramento, California. Chapter 2 available at http://snugharbor.net/images2010/misc/2002_usace_sac_flood_history.pdf (accessed 28 May 2012).

[429] Latta, F.F. 1937. Tulare Lake boats and steamboats: Little journeys in the San Joaquin #12. Article.

[430] June 1976 and March 1997 newsletters of *Los Tulares*.

[431] Thompson, T.H. 1892. Official Historical Atlas Map of Tulare County, California. Author, Tulare, California.

[432] October 24, 1899 issue of the Tulare County Times.

[433] January 23, 1862 issue of the Porterville Messenger.

[434] Grunsky, C.E. 1898. Irrigation near Fresno, California. Water Supply Paper No. 18. U.S. Geological Survey, Washington, D.C.

[435] March 1956 *Los Tulares*, quarterly bulletin of the Tulare County Historical Society, Tulare, California.

[436] March 24, 2001 issue of the Visalia Times-Delta.

[437] January 23, 1862 issue of the Visalia Delta.

[438] March 24, 2001 issue of the Visalia Times-Delta.

[439] Small, K.E. 1926. History of Tulare County, Vol. I, S.J. Clarke Publishing Co., Chicago, Illinois.

[440] August 3, 1905 issue of the Tulare County Times.

[441] February 5, 1862 issue of the Porterville Messenger.

[442] January 23, 1862 issue of the Visalia Delta.

[443] January 30, 1862 issue of the Visalia Delta.

[444] February 13, 1862 issue of the Visalia Delta.

[445] January 30, 1862, February 6, 1862, and February 13, 1862 issues of the Porterville Messenger.

[446] March 1971 *Los Tulares*, quarterly bulletin of the Tulare County Historical Society, Tulare, California.

[447] March 1956 *Los Tulares*, quarterly bulletin of the Tulare County Historical Society, Tulare, California.

[448] March 1953 *Los Tulares*, quarterly bulletin of the Tulare County Historical Society, Tulare, California.

[449] September 1958 *Los Tulares*, quarterly bulletin of the Tulare County Historical Society, Tulare, California.

[450] U.S. Army Corps of Engineers (USACE). 1999. Post-flood assessment for 1983, 1986, 1995, and 1997. U.S. Army Corps of Engineers, Sacramento, California. Chapter 2 available at http://snugharbor.net/images2010/misc/2002_usace_sac_flood_history.pdf (accessed 28 May 2012).

[451] March 1956 *Los Tulares*, quarterly bulletin of the Tulare County Historical Society, Tulare, California.

[452] January 23, 1862 and February 6, 1862 issues of Porterville Messenger.

[453] February 6, 1862 issue of the Porterville Messenger.

[454] January 30, 1862, February 6, 1864 and February 20, 1864 issues of Porterville Messenger.

[455] February 6, 1864 issue of the Visalia Delta.

[456] Paulsen, C.G. 1953. Floods of November-December, 1950 in the Central Valley Basin, California. Pages 505–789 *in* U.S. Geological Survey Water–Supply Paper 1137-F. U.S. Geological Survey, Washington, D.C.

[457] February 6, 1864 issue of the Porterville Messenger.

[458] February 6, 1864 issue of the Porterville Messenger.

[459] Boyd, William H. 1972. A California Middle Border, the Kern River Country, 1772–1880. The Havilah Press, Richardson, Texas.

[460] February 20, 1862 issue of the Visalia Delta

[461] Harding, S.T. 1949. Inflow to Tulare Lake from its tributary streams. Tulare Lake Basin Water Storage District open file report, Hanford, California.

[462] Taylor, W.L., and R.W. Taylor. 2007. The great California flood of 1862. The Fortnightly Club of Redlands, California, Redlands, California. Available at http://www.redlandsfortnightly.org/papers/Taylor06.htm (accessed 1 February 2011).

[463] National Weather Service (NWS). 2010. A History of Significant Weather Events in Southern California. http://www.wrh.noaa.gov/sgx/document/weatherhistory.pdf (accessed 17 June 2011).

[464] California State University San Bernardino. n.d. Alluvial Fan Task Force (AFTF) Study Area Flood History. Available at http://aftf.csusb.edu/documents/AFTF%20Study%20Area%20Flood%20History_ALL.pdf (accessed 17 June 2011).

[465] Rose, G. 1992. San Joaquin: A River Betrayed. Linrose Publishing Co., Fresno, California.

[466] March 1971 *Los Tulares*, quarterly bulletin of the Tulare County Historical Society, Tulare, California.

[467] Porterville Messenger.

[468] Dilsaver, L. and W. Tweed. 1990. Challenge of the Big Trees: A Resource History of Sequoia and Kings Canyon National Parks. Sequoia Natural History Association, Three Rivers, California.

[469] Small, K.E. 1926. History of Tulare County, Vol. I, S.J. Clarke Publishing Co., Chicago, Illinois.

[470] Otter, F.L. 1963. The Men of Mammoth Forest, Edwards Bros. Inc., Ann Arbor, Michigan.

[471] Porter, Samuel Thomas. n.d. The Silver Rush at Mineral King, California, 1873–1882. Author.

[472] Paulsen, C.G. 1953. Floods of November-December, 1950 in the Central Valley Basin, California. Pages 505–789 *in* U.S. Geological Survey Water–Supply Paper 1137-F. U.S. Geological Survey, Washington, D.C.

[473] Paulsen, C.G. 1953. Floods of November-December, 1950 in the Central Valley Basin, California. Pages 505–789 *in* U.S. Geological Survey Water–Supply Paper 1137-F. U.S. Geological Survey, Washington, D.C.

[474] U.S. Army Corps of Engineers (USACE). 1999. Post-flood assessment for 1983, 1986, 1995, and 1997. U.S. Army Corps of Engineers, Sacramento, California. Chapter 2 available at http://snugharbor.net/images2010/misc/2002_usace_sac_flood_history.pdf (accessed 28 May 2012).

[475] Paulsen, C.G. 1953. Floods of November-December, 1950 in the Central Valley Basin, California. Pages 505–789 *in* U.S. Geological Survey Water–Supply Paper 1137-F. U.S. Geological Survey, Washington, D.C.

[476] Kings River Water Conservation District and Kings River Water Association. 2003. The Kings River Handbook. Available at http://www.centralvalleywater.org/_pdf/KingsRiverHandbook-03final.pdf (accessed 20 April 2011).

[477] Paulsen, C.G. 1953. Floods of November-December, 1950 in the Central Valley Basin, California. Pages 505–789 *in* U.S. Geological Survey Water–Supply Paper 1137-F. U.S. Geological Survey, Washington, D.C.

[478] Paulsen, C.G. 1953. Floods of November-December, 1950 in the Central Valley Basin, California. Pages 505–789 *in* U.S. Geological Survey Water–Supply Paper 1137-F. U.S. Geological Survey, Washington, D.C.

[479] U.S. Army Corps of Engineers (USACE). 1999. Post-flood assessment for 1983, 1986, 1995, and 1997. U.S. Army Corps of Engineers, Sacramento, California. Chapter 2 available at http://snugharbor.net/images2010/misc/2002_usace_sac_flood_history.pdf (accessed 28 May 2012).

[480] March 1956 and March 1988 newsletters of *Los Tulares*.

[481] Tilchen, M., editor. 2009. Floods of the Kaweah. Sequoia Natural History Association, Three Rivers, California.

[482] September 1985 *Los Tulares*, quarterly bulletin of the Tulare County Historical Society, Tulare, California.

[483] The Kaweah Commonwealth. n.d. The history of Woodlake, CA. website. http://www.kaweahcommonwealth.com/woodlakehistory.html (accessed 13 April 2012).

[484] Porterville Messenger.

[485] Menefee, E.L and F.A. Dodge. 1913. History of Tulare and Kings Counties, California, with biographical sketches of the leading men and women of the counties who have been identified with their growth and development from the early days to the present. Historic Record Co., Los Angeles, California. Available at

http://archive.org/stream/historyoftularek00mene/historyoftularek00mene_djvu.txt (accessed 16 April 2012).

[486] Paulsen, C.G. 1953. Floods of November-December, 1950 in the Central Valley Basin, California. Pages 505–789 *in* U.S. Geological Survey Water–Supply Paper 1137-F. U.S. Geological Survey, Washington, D.C.

[487] U.S. Army Corps of Engineers (USACE). 1999. Post-flood assessment for 1983, 1986, 1995, and 1997. U.S. Army Corps of Engineers, Sacramento, California. Chapter 2 available at http://snugharbor.net/images2010/misc/2002_usace_sac_flood_history.pdf (accessed 28 May 2012).

[488] Boyd, William H. 1972. A California Middle Border, the Kern River Country, 1772–1880. The Havilah Press, Richardson, Texas.

[489] Latta, F.F. 1937. The flood of 1937: Little journeys in the San Joaquin #20. Article.

[490] U.S. Army Corps of Engineers (USACE). 1999. Post-flood assessment for 1983, 1986, 1995, and 1997. U.S. Army Corps of Engineers, Sacramento, California. Chapter 2 available at http://snugharbor.net/images2010/misc/2002_usace_sac_flood_history.pdf (accessed 28 May 2012).

[491] Arax, M and R. Wartman. 2005. The King of California: J.G. Boswell and the making of a secret American empire. PublicAffairs, Jackson, Tennessee.

[492] Costa, J.E., R.L. Schuster. 1991. Documented historical landslide dams from around the world. U.S. Geological Survey. Open-file report 91-239:486.

[493] Fry, W. 1931. The Great Sequoia Avalanche: Sequoia Nature Guide Service Bulletin no. 8. Sequoia National Park, Three Rivers, California.

[494] Harp, E.L., M.E. Reid, J.W. Godt, J.V. DeGraff, and A.J. Gallegos. 2008. Ferguson rock slide buries California state highway near Yosemite National Park. Landslides 5:331–337. Available at http://www.fs.usda.gov/Internet/FSE_DOCUMENTS/stelprdb5238394.pdf (accessed 13 August 2011).

[495] February 12, 1945 issue of the Visalia Times-Delta.

[496] June 17, 1808 issue of the Visalia Weekly Delta.

[497] March 1956 *Los Tulares*, quarterly bulletin of the Tulare County Historical Society, Tulare, California.

[498] Winchell, L.A. 1920. History of Fresno County. Sierra Publishing Co.

[499] Rose, G. 1992. San Joaquin: A River Betrayed. Linrose Publishing Co., Fresno, California.

[500] Otter, F.L. 1963. The Men of Mammoth Forest. Edwards Bros. Inc., Ann Arbor, Michigan.

[501] Rose, G. 1992. San Joaquin: A River Betrayed. Linrose Publishing Co., Fresno, California.

[502] July 16, 1881 issue of the Visalia Weekly Delta.

[503] Otter, F.L. 1963. The Men of Mammoth Forest. Edwards Bros. Inc., Ann Arbor, Michigan.

[504] Lawson, A.C. 1904. The geomorphogeny of the Upper Kern Basin. California Univ. Dept. Geology Bulletin 3:291-376.

[505] Moore, J.G., 2000. Exploring the Highest Sierra, Stanford University Press, Stanford, California.

[506] Lawson, A.C. 1904. The geomorphogeny of the Upper Kern Basin. California Univ. Dept. Geology Bulletin 3:291-376.

[507] Lawson, A.C. 1906. The geomorphic features of the Middle Kern. California Univ. Dept. Geology Bulletin 4:397-409.

[508] U.S. Army Corps of Engineers (USACE). 1999. Post-flood assessment for 1983, 1986, 1995, and 1997. U.S. Army Corps of Engineers, Sacramento, California. Chapter 2 available at http://snugharbor.net/images2010/misc/2002_usace_sac_flood_history.pdf (accessed 28 May 2012).

[509] U.S. Army Corps of Engineers (USACE). 1999. Post-flood assessment for 1983, 1986, 1995, and 1997. U.S. Army Corps of Engineers, Sacramento, California. Chapter 2 available at http://snugharbor.net/images2010/misc/2002_usace_sac_flood_history.pdf (accessed 28 May 2012).

[510] June 1953 *Los Tulares*, quarterly bulletin of the Tulare County Historical Society, Tulare, California.

[511] Porterville Messenger.

[512] October 3, 2001 issue of the Fresno Bee. Available at http://historical.fresnobeehive.com/2011/09/blackhorse-jones-ship-of-life/ (accessed 3 February 2012)

[513] The California Gen Web Project. n.d. Grangeville Cemetery website. http://www.cagenweb.com/archives/Cemetery/Fresno/GrangevilleCemeteryPart1.htm (accessed 4 February 2012)

[514] Rose, G. 1992. San Joaquin: A River Betrayed. Linrose Publishing Co., Fresno, California.

[515] Preston, W. L. 1981. Vanishing Landscapes: Land and Life in the Tulare Lake Basin. University of California Press, Berkeley, California.

[516] Porterville Messenger.

[517] Kings River Water Conservation District and Kings River Water Association. 2003. The Kings River Handbook. Available at http://www.centralvalleywater.org/_pdf/KingsRiverHandbook-03final.pdf (accessed 20 April 2011).

[518] Muir, J. 1894. The Mountains of California, The Century Co. New York. Available at http://www.sierraclub.org/john_muir_exhibit/writings/our_national_parks/chapter_9.aspx (accessed 28 February 2012)

[519] July 29, 1875 issue of the Visalia Weekly Delta.

[520] Tulare County Historical Society, 1958.

[521] Otter, F.L. 1963. The Men of Mammoth Forest. Edwards Bros. Inc., Ann Arbor, Michigan.

[522] Thompson, T.H. 1892. Official Historical Atlas Map of Tulare County, California. Author, Tulare, California.

[523] March 1989 *Los Tulares*, quarterly bulletin of the Tulare County Historical Society, Tulare, California.

[524] Goodridge, J. 1996. Data on California's Extreme Rainfall from 1862–1995. Proceedings of the 1996 California Extreme Precipitation Symposium. Sierra College Science Center, Rocklin, California. Available at http://www.cepsym.info/proceedings_1996.php (accessed 1 July 2011).

[525] Thompson and West. 1879. History of Yuba County, California. Thompson and West, Oakland, California.

[526] Muir, J. 1875. Flood-Storm in the Sierra. Overland Monthly and Out West magazine 14:489–496. Available at http://quod.lib.umich.edu/cgi/t/text/pageviewer-idx?c=moajrnl;cc=moajrnl;g=moagrp;xc=1;q1=Flood-Storm%20in%20the%20Sierra;rgn=full%20text;view=image;seq=0485;idno=ahj1472.1-14.006;node=ahj1472.1-14.006%3A1 (accessed 25 March 2011).

[527] January 21, 1875 issue of the Visalia Weekly Delta.

[528] March 1956 *Los Tulares*, quarterly bulletin of the Tulare County Historical Society, Tulare, California.

[529] Davis, M. 2001. Late Victorian Holocaust: El Niño Famines and the Making of the Third World. Verso, London.

[530] U.S. Army Corps of Engineers (USACE). 1999. Post-flood assessment for 1983, 1986, 1995, and 1997. U.S. Army Corps of Engineers, Sacramento, California. Chapter 2 available at http://snugharbor.net/images2010/misc/2002_usace_sac_flood_history.pdf (accessed 28 May 2012).

[531] March 1997 *Los Tulares*, quarterly bulletin of the Tulare County Historical Society, Tulare, California.

[532] Spring 1994 newsletter of the Kings County Historical Review.

[533] March 1956 issue of the *Los Tulares*.

[534] Goodridge, J. 1996. Data on California's Extreme Rainfall from 1862–1995. Proceedings of the 1996 California Extreme Precipitation Symposium. Sierra College Science Center, Rocklin, California. Available at http://www.cepsym.info/proceedings_1996.php (accessed 1 July 2011).

[535] Paulsen, C.G. 1953. Floods of November-December, 1950 in the Central Valley Basin, California. Pages 505–789 *in* U.S. Geological Survey Water–Supply Paper 1137-F. U.S. Geological Survey, Washington, D.C.

[536] U.S. Army Corps of Engineers (USACE). 1999. Post-flood assessment for 1983, 1986, 1995, and 1997. U.S. Army Corps of Engineers, Sacramento, California. Chapter 2 available at http://snugharbor.net/images2010/misc/2002_usace_sac_flood_history.pdf (accessed 28 May 2012).

[537] U.S. Army Corps of Engineers (USACE). 1999. Post-flood assessment for 1983, 1986, 1995, and 1997. U.S. Army Corps of Engineers, Sacramento, California. Chapter 2 available at http://snugharbor.net/images2010/misc/2002_usace_sac_flood_history.pdf (accessed 28 May 2012).

[538] February 13, 1945 issue of the Visalia Times-Delta.

[539] Goodridge, J. 1996. Data on California's Extreme Rainfall from 1862–1995. Proceedings of the 1996 California Extreme Precipitation Symposium. Sierra College Science Center, Rocklin, California. Available at http://www.cepsym.info/proceedings_1996.php (accessed 1 July 2011).

[540] McCulley, Frick & Gilman and William Lettis & Associates. 1998. Panoche/Silver Creek Watershed Assessment. Prepared for Panoche/Silver Creek Watershed Coordinated Resource Management and Planning Group and the City of Mendota, California. Available at http://www.water.ca.gov/pubs/environment/watersheds/panoche_silver_creek_watershed_assessment_final_report/psc-assessment.pdf (accessed 23 August 2011).

[541] October 1950 *Los Tulares*, quarterly bulletin of the Tulare County Historical Society, Tulare, California.

[542] U.S. Army Corps of Engineers (USACE). 1999. Post-flood assessment for 1983, 1986, 1995, and 1997. U.S. Army Corps of Engineers, Sacramento, California. Chapter 2 available at http://snugharbor.net/images2010/misc/2002_usace_sac_flood_history.pdf (accessed 28 May 2012).

[543] Harding, S.T. 1949. Inflow to Tulare Lake from its tributary streams. Tulare Lake Basin Water Storage District open file report, Hanford, California.

[544] U.S. Army Corps of Engineers (USACE). 1999. Post-flood assessment for 1983, 1986, 1995, and 1997. U.S. Army Corps of Engineers, Sacramento, California. Chapter 2 available at http://snugharbor.net/images2010/misc/2002_usace_sac_flood_history.pdf (accessed 28 May 2012).

[545] U.S. Army Corps of Engineers (USACE). 1999. Post-flood assessment for 1983, 1986, 1995, and 1997. U.S. Army Corps of Engineers, Sacramento, California. Chapter 2 available at http://snugharbor.net/images2010/misc/2002_usace_sac_flood_history.pdf (accessed 28 May 2012).

[546] Goodridge, J. 1996. Data on California's Extreme Rainfall from 1862–1995. Proceedings of the 1996 California Extreme Precipitation Symposium. Sierra College Science Center, Rocklin, California. Available at http://www.cepsym.info/proceedings_1996.php (accessed 1 July 2011).

[547] Paulsen, C.G. 1953. Floods of November-December, 1950 in the Central Valley Basin, California. Pages 505–789 *in* U.S. Geological Survey Water–Supply Paper 1137-F. U.S. Geological Survey, Washington, D.C.

[548] U.S. Army Corps of Engineers (USACE). 1999. Post-flood assessment for 1983, 1986, 1995, and 1997. U.S. Army Corps of Engineers, Sacramento, California. Chapter 2 available at http://snugharbor.net/images2010/misc/2002_usace_sac_flood_history.pdf (accessed 28 May 2012).

[549] Rose, G. 1992. San Joaquin: A River Betrayed. Linrose Publishing Co., Fresno, California.

[550] U.S. Army Corps of Engineers (USACE). 1999. Post-flood assessment for 1983, 1986, 1995, and 1997. U.S. Army Corps of Engineers, Sacramento, California. Chapter 2 available at http://snugharbor.net/images2010/misc/2002_usace_sac_flood_history.pdf (accessed 28 May 2012).

[551] January 1, 1891 issue of the Visalia Weekly Delta.

[552] U.S. Army Corps of Engineers (USACE). 1999. Post-flood assessment for 1983, 1986, 1995, and 1997. U.S. Army Corps of Engineers, Sacramento, California. Chapter 2 available at http://snugharbor.net/images2010/misc/2002_usace_sac_flood_history.pdf (accessed 28 May 2012).

[553] National Weather Service (NWS). 2009. The Interior Central California climate calendar. Available at http://www.wrh.noaa.gov/hnx/WXCALENDER.pdf (accessed 1 February 2011).

[554] January 5, 1924 issue of the Visalia Morning Delta.

[555] Otter, F.L. 1963. The Men of Mammoth Forest. Edwards Bros. Inc., Ann Arbor, Michigan.

[556] August 15, 1898 issue of the New York Times.

[557] Goodridge, J. 1996. Data on California's Extreme Rainfall from 1862–1995. Proceedings of the 1996 California Extreme Precipitation Symposium. Sierra College Science Center, Rocklin, California. Available at http://www.cepsym.info/proceedings_1996.php (accessed 1 July 2011).

[558] Stafford, H.M. 1956. Snowmelt Flood of 1952 in Kern River, Tulare Lake, and San Joaquin River Basins. Pages 562–575 *in* U.S. Geological Survey Water–Supply Paper 1260-D. U.S. Geological Survey, Washington, D.C. Available at http://books.google.com/books?id=RrAZAQAAIAAJ&pg=PA530-IA7&lpg=PA530-IA7&dq=stafford+%22snowmelt+flood+of+1952+in+kern+river%22+%22Water%E2%80%93Supply+Paper+1260%22&source=bl&ots=l5-ZwdsnFO&sig=JSX80Q6x22tFJvlPKf2yo-l5rcg&hl=en&sa=X&ei=ULx3T_-hMILUiAKT87WnDg&ved=0CFcQ6AEwBw#v=onepage&q=stafford%20%22snowmelt%20flood%20of%201952%20in%20kern%20river%22%20%22Water%E2%80%93Supply%20Paper%201260%22&f=false (accessed 30 March 2012).

[559] January 16, 1906 issue of the Daily Visalia Delta.

[560] Fry, W., and J.R. White. 1930. Big Trees. Stanford University Press, Stanford, California.

[561] January 16, 1906 issue of the Daily Visalia Delta.

[562] January 16, 1906 issue of the Daily Visalia Delta.

[563] March 1975 *Los Tulares*, quarterly bulletin of the Tulare County Historical Society, Tulare, California.

[564] December 14, 2007 issue of The Kaweah Commonwealth. Available at http://www.kaweahcommonwealth.com/12-14-07features.htm (accessed 2 July 2012).

[565] September 1982 *Los Tulares*, quarterly bulletin of the Tulare County Historical Society, Tulare, California.

[566] July 11, 1906 issue of the Visalia Daily Times.

[567] Latta, F.F. 1937. The flood of 1937: Little journeys in the San Joaquin #20. Article.

[568] U.S. Bureau of Reclamation (USBR). 1970. A summary of hydrologic data for the test case on acreage limitation in Tulare Lake. U.S. Bureau of Reclamation, Sacramento, California.

[569] July 19, 1906 issue of the Hanford Weekly Sentinel.

[570] Goodridge, J. 1996. Data on California's Extreme Rainfall from 1862–1995. Proceedings of the 1996 California Extreme Precipitation Symposium. Sierra College Science Center, Rocklin, California. Available at http://www.cepsym.info/proceedings_1996.php (accessed 1 July 2011).

[571] National Weather Service (NWS). 2009. The Interior Central California climate calendar. Available at http://www.wrh.noaa.gov/hnx/WXCALENDER.pdf (accessed 1 February 2011).

[572] Goodridge, J. 1996. Data on California's Extreme Rainfall from 1862–1995. Proceedings of the 1996 California Extreme Precipitation Symposium. Sierra College Science Center, Rocklin, California. Available at http://www.cepsym.info/proceedings_1996.php (accessed 1 July 2011).

[573] U.S. Army Corps of Engineers (USACE). 1999. Post-flood assessment for 1983, 1986, 1995, and 1997. U.S. Army Corps of Engineers, Sacramento, California. Chapter 2 available at http://snugharbor.net/images2010/misc/2002_usace_sac_flood_history.pdf (accessed 28 May 2012).

[574] U.S. Army Corps of Engineers (USACE). 1999. Post-flood assessment for 1983, 1986, 1995, and 1997. U.S. Army Corps of Engineers, Sacramento, California. Chapter 2 available at http://snugharbor.net/images2010/misc/2002_usace_sac_flood_history.pdf (accessed 28 May 2012).

[575] September 1979 *Los Tulares*, quarterly bulletin of the Tulare County Historical Society, Tulare, California.

[576] October 3, 2001 issue of the Fresno Bee. Available at http://historical.fresnobeehive.com/2011/09/blackhorse-jones-ship-of-life/ (accessed 3 February 2012)

[577] Goodridge, J. 1996. Data on California's Extreme Rainfall from 1862–1995. Proceedings of the 1996 California Extreme Precipitation Symposium. Sierra College Science Center, Rocklin, California. Available at http://www.cepsym.info/proceedings_1996.php (accessed 1 July 2011).

[578] U.S. Army Corps of Engineers (USACE). 1999. Post-flood assessment for 1983, 1986, 1995, and 1997. U.S. Army Corps of Engineers, Sacramento, California. Chapter 2 available at http://snugharbor.net/images2010/misc/2002_usace_sac_flood_history.pdf (accessed 28 May 2012).

[579] February 13, 1945 issue of the Visalia Times-Delta.

[580] McCulley, Frick & Gilman and William Lettis & Associates. 1998. Panoche/Silver Creek Watershed Assessment. Prepared for Panoche/Silver Creek Watershed Coordinated Resource Management and Planning Group and the City of Mendota, California. Available at http://www.water.ca.gov/pubs/environment/watersheds/panoche_silver_creek_watershed_assessment_final_report/psc-assessment.pdf (accessed 23 August 2011).

[581] McCulley, Frick & Gilman and William Lettis & Associates. 1998. Panoche/Silver Creek Watershed Assessment. Prepared for Panoche/Silver Creek Watershed Coordinated Resource Management and Planning Group and the City of Mendota, California. Available at http://www.water.ca.gov/pubs/environment/watersheds/panoche_silver_creek_watershed_assessment_final_report/psc-assessment.pdf (accessed 23 August 2011).

[582] Paulsen, C.G. 1953. Floods of November-December, 1950 in the Central Valley Basin, California. Pages 505–789 *in* U.S. Geological Survey Water–Supply Paper 1137-F. U.S. Geological Survey, Washington, D.C.

[583] Paulsen, C.G. 1953. Floods of November-December, 1950 in the Central Valley Basin, California. Pages 505–789 *in* U.S. Geological Survey Water–Supply Paper 1137-F. U.S. Geological Survey, Washington, D.C.

[584] January 29, 1914 issue of the Corcoran Journal as retold in the August 10, 1989 issue of the Corcoran Journal.

[585] Various 1914 issues of the Corcoran Journal as retold in the September 30, 1982 issue of the Corcoran Journal.

[586] Goodridge, J. 1996. Data on California's Extreme Rainfall from 1862–1995. Proceedings of the 1996 California Extreme Precipitation Symposium. Sierra College Science Center, Rocklin, California. Available at http://www.cepsym.info/proceedings_1996.php (accessed 1 July 2011).

[587] Paulsen, C.G. 1953. Floods of November-December, 1950 in the Central Valley Basin, California. Pages 505–789 *in* U.S. Geological Survey Water–Supply Paper 1137-F. U.S. Geological Survey, Washington, D.C.

[588] Paulsen, C.G. 1953. Floods of November-December, 1950 in the Central Valley Basin, California. Pages 505–789 *in* U.S. Geological Survey Water–Supply Paper 1137-F. U.S. Geological Survey, Washington, D.C.

[589] U.S. Bureau of Reclamation (USBR). 1970. A summary of hydrologic data for the test case on acreage limitation in Tulare Lake. U.S. Bureau of Reclamation, Sacramento, California.

[590] Pourade, R.F. 1965. The History of San Diego; Gold in the Sun (1900–1919), Chapter 11. Union-Tribune Publishing Co., San Diego, California. Available at http://www.sandiegohistory.org/books/pourade/gold/goldchapter11.htm (accessed 18 June 2011).

[591] California State University San Bernardino. n.d. Alluvial Fan Task Force (AFTF) Study Area Flood History. Available at http://aftf.csusb.edu/documents/AFTF%20Study%20Area%20Flood%20History_ALL.pdf (accessed 17 June 2011).

[592] Goodridge, J. 1996. Data on California's Extreme Rainfall from 1862–1995. Proceedings of the 1996 California Extreme Precipitation Symposium. Sierra College Science Center, Rocklin, California. Available at http://www.cepsym.info/proceedings_1996.php (accessed 1 July 2011).

[593] Goodridge, J. 1996. Data on California's Extreme Rainfall from 1862–1995. Proceedings of the 1996 California Extreme Precipitation Symposium. Sierra College Science Center, Rocklin, California. Available at http://www.cepsym.info/proceedings_1996.php (accessed 1 July 2011).

[594] California State University San Bernardino. n.d. Alluvial Fan Task Force (AFTF) Study Area Flood History. Available at http://aftf.csusb.edu/documents/AFTF%20Study%20Area%20Flood%20History_ALL.pdf (accessed 17 June 2011).

[595] California State University San Bernardino. n.d. Alluvial Fan Task Force (AFTF) Study Area Flood History. Available at http://aftf.csusb.edu/documents/AFTF%20Study%20Area%20Flood%20History_ALL.pdf (accessed 17 June 2011).

[596] California State University San Bernardino. n.d. Alluvial Fan Task Force (AFTF) Study Area Flood History. Available at http://aftf.csusb.edu/documents/AFTF%20Study%20Area%20Flood%20History_ALL.pdf (accessed 17 June 2011).

[597] California State University San Bernardino. n.d. Alluvial Fan Task Force (AFTF) Study Area Flood History. Available at http://aftf.csusb.edu/documents/AFTF%20Study%20Area%20Flood%20History_ALL.pdf (accessed 17 June 2011).

[598] Giese, B.S., N.C. Slowey, R. Sulagna, G.P. Compo, P.D. Sardeshmukh, J.A. Carton, and J.S. Whitaker. 2010. The 1918/19 El Niño. Bull. Amer. Meteor. Soc. 91:177–183.

[599] Goodridge, J. 1996. Data on California's Extreme Rainfall from 1862–1995. Proceedings of the 1996 California Extreme Precipitation Symposium. Sierra College Science Center, Rocklin, California. Available at http://www.cepsym.info/proceedings_1996.php (accessed 1 July 2011).

[600] Goodridge, J. 1996. Data on California's Extreme Rainfall from 1862–1995. Proceedings of the 1996 California Extreme Precipitation Symposium. Sierra College Science Center, Rocklin, California. Available at http://www.cepsym.info/proceedings_1996.php (accessed 1 July 2011).

[601] Goodridge, J. 1996. Data on California's Extreme Rainfall from 1862–1995. Proceedings of the 1996 California Extreme Precipitation Symposium. Sierra College Science Center, Rocklin, California. Available at http://www.cepsym.info/proceedings_1996.php (accessed 1 July 2011).

[602] U.S. Bureau of Reclamation (USBR). 1970. A summary of hydrologic data for the test case on acreage limitation in Tulare Lake. U.S. Bureau of Reclamation, Sacramento, California.

[603] Weis, Mae. 1938. Development of this fertile area from primeval state covers comparatively short span. The Corcoran Journal article, Corcoran, California.

[604] National Weather Service (NWS). 2009. The Interior Central California climate calendar. Available at http://www.wrh.noaa.gov/hnx/WXCALENDER.pdf (accessed 1 February 2011).

[605] Earle, C.J. and H.C. Fritts. 1986. Reconstructing riverflow in the Sacramento Basin since 1560. Report to California Department of Water Resources, Agreement No. DWR B–55398. Laboratory of Tree-ring Research, University of Arizona, Tucson.

[606] Kings River Water Conservation District and Kings River Water Association. 2003. The Kings River Handbook. Available at http://www.centralvalleywater.org/_pdf/KingsRiverHandbook-03final.pdf (accessed 20 April 2011).

[607] January 5, 1924 issue of the Visalia Morning Delta.

[608] January 5, 1924 issue of the Visalia Morning Delta.

[609] Weis, Mae. 1938. Development of this fertile area from primeval state covers comparatively short span. The Corcoran Journal article, Corcoran, California.

[610] Barry, J. M. 1997. Rising Tide: The Great Mississippi Flood of 1927 and How it Changed America. Simon & Schuster, New York, N.Y.

[611] May 8, 2011 broadcast of Weekend Edition Saturday, National Public Radio. Available at http://www.npr.org/player/v2/mediaPlayer.html?action=1&t=1&islist=false&id=136304625&m=136304679 (accessed 1 February 2011). (accessed 8 May 2011).

[612] Galloway, D.L., and F.S. Riley. 1999, San Joaquin Valley, California — Largest human alteration of the Earth's surface. Pages 23–34 in Galloway, D.L., D.R. Jones, and S.E Ingebritsen, editors. Land Subsidence in the United States: U.S. Geological Survey Circular 1182. U.S. Geological Survey, Washington, D.C. Available at http://pubs.usgs.gov/circ/circ1182/pdf/06SanJoaquinValley.pdf (accessed 1 September 2012).

[613] June 1969 Los Tulares, quarterly bulletin of the Tulare County Historical Society, Tulare, California.

[614] The Weather Service. 2012.. 2012 Drought Rivals Dust Bowl website. http://www.weather.com/news/drought-disaster-new-data-20120715 (accessed 17 July 2012).

[615] January 5, 1924 issue of the Visalia Morning Delta.

[616] Paulsen, C.G. 1953. Floods of November-December, 1950 in the Central Valley Basin, California. Pages 505–789 in U.S. Geological Survey Water–Supply Paper 1137-F. U.S. Geological Survey, Washington, D.C.

[617] February 1March 1045 issue of the Visalia Times-Delta.

[618] National Weather Service (NWS). 2009. The Interior Central California climate calendar. Available at http://www.wrh.noaa.gov/hnx/WXCALENDER.pdf (accessed 1 February 2011).

[619] National Weather Service (NWS). 2009. The Interior Central California climate calendar. Available at http://www.wrh.noaa.gov/hnx/WXCALENDER.pdf (accessed 1 February 2011).

[620] October 1, 1932 issue of the San Mateo California Times.

[621] October 2, 1932 issue of the Ogden Utah Standard Examiner.

[622] California State University San Bernardino. n.d. Alluvial Fan Task Force (AFTF) Study Area Flood History: Kern County Flood History. Available at http://aftf.csusb.edu/documents/AFTF%20Study%20Area%20Flood%20History_ALL.pdf (accessed 17 June 2011).

[623] Goodridge, J. 1996. Data on California's Extreme Rainfall from 1862–1995. Proceedings of the 1996 California Extreme Precipitation Symposium. Sierra College Science Center, Rocklin, California. Available at http://www.cepsym.info/proceedings_1996.php (accessed 1 July 2011).

[624] Paulsen, C.G. 1953. Floods of November-December, 1950 in the Central Valley Basin, California. Pages 505–789 *in* U.S. Geological Survey Water–Supply Paper 1137-F. U.S. Geological Survey, Washington, D.C.

[625] February 13, 1945 issue of the Visalia Times-Delta.

[626] Paulsen, C.G. 1953. Floods of November-December, 1950 in the Central Valley Basin, California. Pages 505–789 *in* U.S. Geological Survey Water–Supply Paper 1137-F. U.S. Geological Survey, Washington, D.C.

[627] Paulsen, C.G. 1953. Floods of November-December, 1950 in the Central Valley Basin, California. Pages 505–789 *in* U.S. Geological Survey Water–Supply Paper 1137-F. U.S. Geological Survey, Washington, D.C.

[628] California State University San Bernardino. n.d. Alluvial Fan Task Force (AFTF) Study Area Flood History: Kern County Flood History. Available at http://aftf.csusb.edu/documents/AFTF%20Study%20Area%20Flood%20History_ALL.pdf (accessed 17 June 2011).

[629] Paulsen, C.G. 1953. Floods of November-December, 1950 in the Central Valley Basin, California. Pages 505–789 *in* U.S. Geological Survey Water–Supply Paper 1137-F. U.S. Geological Survey, Washington, D.C.

[630] DeGraff, J.V. 1994. The geomorphology of some debris flows in the southern Sierra Nevada, California. Geomorphology 10:231–252.

[631] Goodridge, J. 1996. Data on California's Extreme Rainfall from 1862–1995. Proceedings of the 1996 California Extreme Precipitation Symposium. Sierra College Science Center, Rocklin, California. Available at http://www.cepsym.info/proceedings_1996.php (accessed 1 July 2011).

[632] Paulsen, C.G. 1953. Floods of November-December, 1950 in the Central Valley Basin, California. Pages 505–789 *in* U.S. Geological Survey Water–Supply Paper 1137-F. U.S. Geological Survey, Washington, D.C.

[633] Paulsen, C.G. 1953. Floods of November-December, 1950 in the Central Valley Basin, California. Pages 505–789 *in* U.S. Geological Survey Water–Supply Paper 1137-F. U.S. Geological Survey, Washington, D.C.

[634] December 14, 2007 issue of The Kaweah Commonwealth. Available at http://www.kaweahcommonwealth.com/12-14-07features.htm (accessed 2 July 2012).

[635] Paulsen, C.G. 1953. Floods of November-December, 1950 in the Central Valley Basin, California. Pages 505–789 *in* U.S. Geological Survey Water–Supply Paper 1137-F. U.S. Geological Survey, Washington, D.C.

[636] December 14, 2007 issue of The Kaweah Commonwealth. Available at http://www.kaweahcommonwealth.com/12-14-07features.htm (accessed 2 July 2012).

[637] December 14, 2007 issue of The Kaweah Commonwealth. Available at http://www.kaweahcommonwealth.com/12-14-07features.htm (accessed 2 July 2012).

[638] December 14, 2007 issue of The Kaweah Commonwealth. Available at http://www.kaweahcommonwealth.com/12-14-07features.htm (accessed 2 July 2012).

[639] December 14, 2007 issue of The Kaweah Commonwealth. Available at http://www.kaweahcommonwealth.com/12-14-07features.htm (accessed 2 July 2012).

[640] December 14, 2007 issue of The Kaweah Commonwealth. Available at http://www.kaweahcommonwealth.com/12-14-07features.htm (accessed 2 July 2012).

[641] December 14, 2007 issue of The Kaweah Commonwealth. Available at http://www.kaweahcommonwealth.com/12-14-07features.htm (accessed 2 July 2012).

[642] February 13, 1945 issue of the Visalia Times-Delta.

[643] Paulsen, C.G. 1953. Floods of November-December, 1950 in the Central Valley Basin, California. Pages 505–789 *in* U.S. Geological Survey Water–Supply Paper 1137-F. U.S. Geological Survey, Washington, D.C.

[644] Paulsen, C.G. 1953. Floods of November-December, 1950 in the Central Valley Basin, California. Pages 505–789 *in* U.S. Geological Survey Water–Supply Paper 1137-F. U.S. Geological Survey, Washington, D.C.

[645] DeGraff, J.V. 1994. The geomorphology of some debris flows in the southern Sierra Nevada, California. Geomorphology 10:231–252.

[646] U.S. Army Corps of Engineers (USACE). 1999. Post-flood assessment for 1983, 1986, 1995, and 1997. U.S. Army Corps of Engineers, Sacramento, California. Chapter 2 available at http://snugharbor.net/images2010/misc/2002_usace_sac_flood_history.pdf (accessed 28 May 2012).

[647] Goodridge, J. 1996. Data on California's Extreme Rainfall from 1862–1995. Proceedings of the 1996 California Extreme Precipitation Symposium. Sierra College Science Center, Rocklin, California. Available at http://www.cepsym.info/proceedings_1996.php (accessed 1 July 2011).

[648] McCulley, Frick & Gilman and William Lettis & Associates. 1998. Panoche/Silver Creek Watershed Assessment. Prepared for Panoche/Silver Creek Watershed Coordinated Resource Management and Planning Group and the City of Mendota, California. Available at http://www.water.ca.gov/pubs/environment/watersheds/panoche_silver_creek_watershed_assessment_final_report/psc-assessment.pdf (accessed 23 August 2011).

[649] California State University San Bernardino. n.d. Alluvial Fan Task Force (AFTF) Study Area Flood History. Available at http://aftf.csusb.edu/documents/AFTF%20Study%20Area%20Flood%20History_ALL.pdf (accessed 17 June 2011).

[650] Paulsen, C.G. 1953. Floods of November-December, 1950 in the Central Valley Basin, California. Pages 505–789 *in* U.S. Geological Survey Water–Supply Paper 1137-F. U.S. Geological Survey, Washington, D.C.

[651] U.S. Bureau of Reclamation (USBR). 1970. A summary of hydrologic data for the test case on acreage limitation in Tulare Lake. U.S. Bureau of Reclamation, Sacramento, California.

[652] June 28, 1952 issue of the Tulare Advance Register.

[653] U.S. Bureau of Reclamation (USBR). 1970. A summary of hydrologic data for the test case on acreage limitation in Tulare Lake. U.S. Bureau of Reclamation, Sacramento, California.

[654] Latta, F.F. 1977. Handbook of Yokuts Indians. Bear State Books, Santa Cruz, California.

[655] June 28, 1952 issue of the Tulare Advance Register.

[656] Stanchelski, C., and G. Sanger. 2008: The climate of Fresno, California. National Oceanic and Atmospheric Administration Technical Memorandum NWS WR-280. Available at http://www.wrh.noaa.gov/wrh/techMemos/TM-280.pdf (accessed 1 February 2011).

[657] National Weather Service (NWS). 2009. The Interior Central California climate calendar. Available at http://www.wrh.noaa.gov/hnx/WXCALENDER.pdf (accessed 1 February 2011).

[658] National Weather Service (NWS). 2009. The Interior Central California climate calendar. Available at http://www.wrh.noaa.gov/hnx/WXCALENDER.pdf (accessed 1 February 2011).

[659] Stanchelski, C., and G. Sanger, 2008: The climate of Fresno, California. NOAA Tech. Memo., NWS WR-280. Available at http://www.wrh.noaa.gov/wrh/techMemos/TM-280.pdf (accessed 1 February 2011).

[660] Goodridge, J. 1996. Data on California's Extreme Rainfall from 1862–1995. Proceedings of the 1996 California Extreme Precipitation Symposium. Sierra College Science Center, Rocklin, California. Available at http://www.cepsym.info/proceedings_1996.php (accessed 1 July 2011).

[661] National Weather Service (NWS). 2011. Storm data and unusual weather phenomenon: January 2012. Available at http://www.wrh.noaa.gov/hnx/stormdat/2012jan.pdf (accessed 15 June 2012).

[662] Wikipedia. n.d. Katabatic wind web article. http://en.wikipedia.org/wiki/Katabatic_wind (accessed 13 June 2012).

[663] Null, J. 2000. Winds of the World. Weatherwise, 53(2). Available from http://ggweather.com/windsoftheworld.htm (accessed 7 June 2012).

[664] National Weather Service (NWS). 2011. Storm data and unusual weather phenomenon: November 2004. Available at http://www.wrh.noaa.gov/hnx/stormdat/2004nov.pdf (accessed 16 June 2012).

[665] Fosberg, M.A. 1986. Windthrown trees on the Kings River Ranger District, Sierra National Forest: meteorological aspects. Research Note PSW-RN-381. U.S. Department of Agriculture, Forest Service, Pacific Southwest Forest and Range Experiment Station, Berkeley, California. Available at http://www.fs.fed.us/psw/publications/documents/psw_rn381/psw_rn381.pdf (accessed 13 June 2012).

[666] National Park Service (NPS). 2003. Fire and Fuels Management Plan. National Park Service, Three Rivers, California. Available a http://www.nps.gov/seki/naturescience/upload/2011-FFMP-final_reduce-file-size.pdf (accessed 16 June, 2012).

[667] Devils Postpile Windstorm Lecture. 2012. Mammoth Lakes, California web video. Available at http://www.youtube.com/watch?v=52wFKqU8VUo&feature=plcp (accessed 10 July 2012).

[668] National Weather Service (NWS). 2011. Storm data and unusual weather phenomenon: December 2011. Available at http://www.wrh.noaa.gov/hnx/stormdat/2011dec.pdf (accessed 6 June 2012).

[669] National Weather Service (NWS). 2011. Storm data and unusual weather phenomenon: December 2011. Available at http://www.wrh.noaa.gov/hnx/stormdat/2011dec.pdf (accessed 6 June 2012).

[670] Whiteman, C.D. 2000. Page 149 *in* Mountain Meteorology; Fundamentals and Applications. Oxford University Press, New York. Available at http://books.google.com/books?id=Mz_7qLK5hQcC&pg=PA149&lpg=PA149&dq=mountain+lee+wave+blowdown+park+range&source=bl&ots=6V3luzpLUa&sig=bnEf-WmU1q3nVoi7dt68A9tvvPg&hl=en&sa=X&ei=OWbVT5f6EomO2wXq9LSVDw&ved=0CEYQ6AEwAA#v=onepage&q=mountain%20lee%20wave%20blowdown%20park%20range&f=false (accessed 12 June 2012).

[671] National Oceanic and Atmospheric Administration (NOAA). n.d. Changes to the Oceanic Niño Index (ONI) website. http://www.cpc.ncep.noaa.gov/products/analysis_monitoring/ensostuff/ensoyears.shtml (accessed 10 July 2012).

[672] February 13, 1945 issue of the Visalia Times-Delta.

[673] Goodridge, J. 1996. Data on California's Extreme Rainfall from 1862–1995. Proceedings of the 1996 California Extreme Precipitation Symposium. Sierra College Science Center, Rocklin, California. Available at http://www.cepsym.info/proceedings_1996.php (accessed 1 July 2011).

[674] Paulsen, C.G. 1953. Floods of November-December, 1950 in the Central Valley Basin, California. Pages 505–789 *in* U.S. Geological Survey Water–Supply Paper 1137-F. U.S. Geological Survey, Washington, D.C.

[675] February 13, 1945issue of the Visalia Times-Delta.

[676] Dean, W.W. 1971. Floods of December, 1966 in the Kern-Kaweah Area, Kern and Tulare Counties, California. U.S. Geological Survey Water–Supply Paper 1870-C. U.S. Geological Survey, Washington, D.C.

[677] Paulsen, C.G. 1953. Floods of November-December, 1950 in the Central Valley Basin, California. Pages 505–789 *in* U.S. Geological Survey Water–Supply Paper 1137-F. U.S. Geological Survey, Washington, D.C.

[678] Paulsen, C.G. 1953. Floods of November-December, 1950 in the Central Valley Basin, California. Pages 505–789 *in* U.S. Geological Survey Water–Supply Paper 1137-F. U.S. Geological Survey, Washington, D.C.

[679] Special to the February 4, 1945 issue of the New York Times.

[680] February 17, 1945 issue of the Visalia Times-Delta.

[681] February 5, 1945 issue of the Visalia Morning Delta.

[682] February 6, 1945 issue of the Visalia Times-Delta.

[683] February 7, 1945 issue of the Visalia Times-Delta.

[684] February 8, 1945 issue of the Visalia Times-Delta.

[685] February 12, 1945 issue of the Visalia Times-Delta.

[686] February 13, 1945 issue of the Visalia Times-Delta.

[687] March 2000 *Los Tulares*, quarterly bulletin of the Tulare County Historical Society, Tulare, California

[688] Paulsen, C.G. 1953. Floods of November-December, 1950 in the Central Valley Basin, California. Pages 505–789 *in* U.S. Geological Survey Water–Supply Paper 1137-F. U.S. Geological Survey, Washington, D.C.

[689] February 3, 1945 issue of the Visalia Times-Delta.

[690] United States Weather Bureau. 1945. Two reports on October flooding in the Central Valley of California. Monthly Weather Review 73:172–177. Available at http://docs.lib.noaa.gov/rescue/mwr/073/mwr-073-10-0172.pdf (accessed 1 February 2011).

[691] United States Weather Bureau. 1945. Two reports on October flooding in the Central Valley of California. Monthly Weather Review 73:177. Available at http://docs.lib.noaa.gov/rescue/mwr/073/mwr-073-10-0172.pdf (accessed 1 February 2011).

[692] February 7, 1945 and February 8, 1945 issues of Visalia Times-Delta.

[693] Tulare County Civil Grand Jury. n.d. Tulare County Grand Jury Final Report: 2005–2006. Superior Court of the State of California, County of Tulare, Visalia, California. Available at http://www.co.tulare.ca.us/civica/filebank/blobdload.asp?BlobID=4002 (accessed 1 July 2011).

[694] National Weather Service (NWS). 2009. The Interior Central California climate calendar. Available at http://www.wrh.noaa.gov/hnx/WXCALENDER.pdf (accessed 1 February 2011).

[695] Paulsen, C.G. 1953. Floods of November-December, 1950 in the Central Valley Basin, California. Pages 505–789 *in* U.S. Geological Survey Water–Supply Paper 1137-F. U.S. Geological Survey, Washington, D.C.

[696] Goodridge, J. 1996. Data on California's Extreme Rainfall from 1862–1995. Proceedings of the 1996 California Extreme Precipitation Symposium. Sierra College Science Center, Rocklin, California. Available at http://www.cepsym.info/proceedings_1996.php (accessed 1 July 2011).

[697] November 24, 1950 issue of the Three Rivers Current.

[698] National Weather Service (NWS). 2009. The Interior Central California climate calendar. Available at http://www.wrh.noaa.gov/hnx/WXCALENDER.pdf (accessed 1 February 2011).

[699] Paulsen, C.G. 1953. Floods of November-December, 1950 in the Central Valley Basin, California. Pages 505–789 *in* U.S. Geological Survey Water–Supply Paper 1137-F. U.S. Geological Survey, Washington, D.C.

[700] November 24, 1950 issue of the Three Rivers Current.

[701] Paulsen, C.G. 1953. Floods of November-December, 1950 in the Central Valley Basin, California. Pages 505–789 *in* U.S. Geological Survey Water–Supply Paper 1137-F. U.S. Geological Survey, Washington, D.C.

[702] November 20, 1950 issue of the New York Times.

[703] November 24, 1950 issue of the New York Times.

[704] Paulsen, C.G. 1953. Floods of November-December, 1950 in the Central Valley Basin, California. Pages 505–789 *in* U.S. Geological Survey Water–Supply Paper 1137-F. U.S. Geological Survey, Washington, D.C.

[705] Paulsen, C.G. 1953. Floods of November-December, 1950 in the Central Valley Basin, California. Pages 505–789 *in* U.S. Geological Survey Water–Supply Paper 1137-F. U.S. Geological Survey, Washington, D.C.

[706] Paulsen, C.G. 1953. Floods of November-December, 1950 in the Central Valley Basin, California. Pages 505–789 *in* U.S. Geological Survey Water–Supply Paper 1137-F. U.S. Geological Survey, Washington, D.C.

[707] Paulsen, C.G. 1953. Floods of November-December, 1950 in the Central Valley Basin, California. Pages 505–789 *in* U.S. Geological Survey Water–Supply Paper 1137-F. U.S. Geological Survey, Washington, D.C.

[708] Paulsen, C.G. 1953. Floods of November-December, 1950 in the Central Valley Basin, California. Pages 505–789 *in* U.S. Geological Survey Water–Supply Paper 1137-F. U.S. Geological Survey, Washington, D.C.

[709] Paulsen, C.G. 1953. Floods of November-December, 1950 in the Central Valley Basin, California. Pages 505–789 *in* U.S. Geological Survey Water–Supply Paper 1137-F. U.S. Geological Survey, Washington, D.C.

[710] November 24, 1950 issue of the Three Rivers Current.

[711] Paulsen, C.G. 1953. Floods of November-December, 1950 in the Central Valley Basin, California. Pages 505–789 *in* U.S. Geological Survey Water–Supply Paper 1137-F. U.S. Geological Survey, Washington, D.C.

[712] Stafford, H.M. 1956. Snowmelt Flood of 1952 in Kern River, Tulare Lake, and San Joaquin River Basins. Pages 562–575 *in* U.S. Geological Survey Water–Supply Paper 1260. U.S. Geological Survey, Washington, D.C.

[713] Dean, W.W. 1971. Floods of December, 1966 in the Kern-Kaweah Area, Kern and Tulare Counties, California. U.S. Geological Survey Water–Supply Paper 1870-C. U.S. Geological Survey, Washington, D.C.

[714] November 24, 1950 issue of the Three Rivers Current.

[715] Paulsen, C.G. 1953. Floods of November-December, 1950 in the Central Valley Basin, California. Pages 505–789 *in* U.S. Geological Survey Water–Supply Paper 1137-F. U.S. Geological Survey, Washington, D.C.

[716] Paulsen, C.G. 1953. Floods of November-December, 1950 in the Central Valley Basin, California. Pages 505–789 *in* U.S. Geological Survey Water–Supply Paper 1137-F. U.S. Geological Survey, Washington, D.C.

[717] Paulsen, C.G. 1953. Floods of November-December, 1950 in the Central Valley Basin, California. Pages 505–789 *in* U.S. Geological Survey Water–Supply Paper 1137-F. U.S. Geological Survey, Washington, D.C.

[718] Paulsen, C.G. 1953. Floods of November-December, 1950 in the Central Valley Basin, California. Pages 505–789 *in* U.S. Geological Survey Water–Supply Paper 1137-F. U.S. Geological Survey, Washington, D.C.

[719] Dean, W.W. 1971. Floods of December, 1966 in the Kern-Kaweah Area, Kern and Tulare Counties, California. U.S. Geological Survey Water–Supply Paper 1870-C. U.S. Geological Survey, Washington, D.C.

[720] Dean, W.W. 1971. Floods of December, 1966 in the Kern-Kaweah Area, Kern and Tulare Counties, California. U.S. Geological Survey Water–Supply Paper 1870-C. U.S. Geological Survey, Washington, D.C.

[721] Paulsen, C.G. 1953. Floods of November-December, 1950 in the Central Valley Basin, California. Pages 505–789 *in* U.S. Geological Survey Water–Supply Paper 1137-F. U.S. Geological Survey, Washington, D.C.

[722] Dean, W.W. 1971. Floods of December, 1966 in the Kern-Kaweah Area, Kern and Tulare Counties, California. U.S. Geological Survey Water–Supply Paper 1870-C. U.S. Geological Survey, Washington, D.C.

[723] Dean, W.W. 1971. Floods of December, 1966 in the Kern-Kaweah Area, Kern and Tulare Counties, California. U.S. Geological Survey Water–Supply Paper 1870-C. U.S. Geological Survey, Washington, D.C.

[724] Paulsen, C.G. 1953. Floods of November-December, 1950 in the Central Valley Basin, California. Pages 505–789 *in* U.S. Geological Survey Water–Supply Paper 1137-F. U.S. Geological Survey, Washington, D.C.

[725] Dean, W.W. 1971. Floods of December, 1966 in the Kern-Kaweah Area, Kern and Tulare Counties, California. U.S. Geological Survey Water–Supply Paper 1870-C. U.S. Geological Survey, Washington, D.C.

[726] Dean, W.W. 1971. Floods of December, 1966 in the Kern-Kaweah Area, Kern and Tulare Counties, California. U.S. Geological Survey Water–Supply Paper 1870-C. U.S. Geological Survey, Washington, D.C.

[727] California State University San Bernardino. n.d. Alluvial Fan Task Force (AFTF) Study Area Flood History: Kern County Flood History. Available at http://aftf.csusb.edu/documents/AFTF%20Study%20Area%20Flood%20History_ALL.pdf (accessed 17 June 2011).

[728] Paulsen, C.G. 1953. Floods of November-December, 1950 in the Central Valley Basin, California. Pages 505–789 *in* U.S. Geological Survey Water–Supply Paper 1137-F. U.S. Geological Survey, Washington, D.C.

[729] Selvidge, J.B. n.d. Tracy Ranch History website. http://www.tracyranch.com/history.htm (accessed 14 June 2011).

[730] Paulsen, C.G. 1953. Floods of November-December, 1950 in the Central Valley Basin, California. Pages 505–789 *in* U.S. Geological Survey Water–Supply Paper 1137-F. U.S. Geological Survey, Washington, D.C.

[731] Paulsen, C.G. 1953. Floods of November-December, 1950 in the Central Valley Basin, California. Pages 505–789 *in* U.S. Geological Survey Water–Supply Paper 1137-F. U.S. Geological Survey, Washington, D.C.

[732] U.S. Bureau of Reclamation (USBR). 1970. A summary of hydrologic data for the test case on acreage limitation in Tulare Lake. U.S. Bureau of Reclamation, Sacramento, California.

[733] Paulsen, C.G. 1953. Floods of November-December, 1950 in the Central Valley Basin, California. Pages 505–789 *in* U.S. Geological Survey Water–Supply Paper 1137-F. U.S. Geological Survey, Washington, D.C.

[734] Stafford, H.M. 1956. Snowmelt Flood of 1952 in Kern River, Tulare Lake, and San Joaquin River Basins. Pages 562–575 *in* U.S. Geological Survey Water–Supply Paper 1260-D. U.S. Geological Survey, Washington, D.C. Available at http://books.google.com/books?id=RrAZAQAAIAAJ&pg=PA530-IA7&lpg=PA530-IA7&dq=stafford+%22snowmelt+flood+of+1952+in+kern+river%22+%22Water%E2%80%93Supply+Paper+1260%22&source=bl&ots=l5-ZwdsnFO&sig=JSX80Q6x22tFJvlPKf2yo-l5rcg&hl=en&sa=X&ei=ULx3T_-hMILUiAKT87WnDg&ved=0CFcQ6AEwBw#v=onepage&q=stafford%20%22snowmelt%20flood%20of%201952%20in%20kern%20river%22%20%22Water%E2%80%93Supply%20Paper%201260%22&f=false (accessed 30 March 2012).

[735] National Weather Service (NWS). 2009. The Interior Central California climate calendar. Available at http://www.wrh.noaa.gov/hnx/WXCALENDER.pdf (accessed 1 February 2011).

[736] National Weather Service (NWS). 2009. The Interior Central California climate calendar. Available at http://www.wrh.noaa.gov/hnx/WXCALENDER.pdf (accessed 1 February 2011).

[737] Stafford, H.M. 1956. Snowmelt Flood of 1952 in Kern River, Tulare Lake, and San Joaquin River Basins. Pages 562–575 *in* U.S. Geological Survey Water–Supply Paper 1260-D. U.S. Geological Survey, Washington, D.C. Available at http://books.google.com/books?id=RrAZAQAAIAAJ&pg=PA530-IA7&lpg=PA530-IA7&dq=stafford+%22snowmelt+flood+of+1952+in+kern+river%22+%22Water%E2%80%93Supply+Paper+1260%22&source=bl&ots=l5-ZwdsnFO&sig=JSX80Q6x22tFJvlPKf2yo-l5rcg&hl=en&sa=X&ei=ULx3T_-hMILUiAKT87WnDg&ved=0CFcQ6AEwBw#v=onepage&q=stafford%20%22snowmelt%20flood%20of%201952%20in%20kern%20river%22%20%22Water%E2%80%93Supply%20Paper%201260%22&f=false (accessed 30 March 2012).

[738] Stafford, H.M. 1956. Snowmelt Flood of 1952 in Kern River, Tulare Lake, and San Joaquin River Basins. Pages 562–575 *in* U.S. Geological Survey Water–Supply Paper 1260-D. U.S. Geological Survey, Washington, D.C. Available at http://books.google.com/books?id=RrAZAQAAIAAJ&pg=PA530-IA7&lpg=PA530-IA7&dq=stafford+%22snowmelt+flood+of+1952+in+kern+river%22+%22Water%E2%80%93Supply+Paper+1260%22&source=bl&ots=l5-ZwdsnFO&sig=JSX80Q6x22tFJvlPKf2yo-l5rcg&hl=en&sa=X&ei=ULx3T_-hMILUiAKT87WnDg&ved=0CFcQ6AEwBw#v=onepage&q=stafford%20%22snowmelt%20flood%20of%201952%20in%20kern%20river%22%20%22Water%E2%80%93Supply%20Paper%201260%22&f=false (accessed 30 March 2012).

[739] Stafford, H.M. 1956. Snowmelt Flood of 1952 in Kern River, Tulare Lake, and San Joaquin River Basins. Pages 562–575 *in* U.S. Geological Survey Water–Supply Paper 1260-D. U.S. Geological Survey, Washington, D.C. Available at http://books.google.com/books?id=RrAZAQAAIAAJ&pg=PA530-IA7&lpg=PA530-IA7&dq=stafford+%22snowmelt+flood+of+1952+in+kern+river%22+%22Water%E2%80%93Supply+Paper+1260%22&source=bl&ots=l5-ZwdsnFO&sig=JSX80Q6x22tFJvlPKf2yo-l5rcg&hl=en&sa=X&ei=ULx3T_-hMILUiAKT87WnDg&ved=0CFcQ6AEwBw#v=onepage&q=stafford%20%22snowmelt%20flood%20of%201952%20in%20kern%20river%22%20%22Water%E2%80%93Supply%20Paper%201260%22&f=false (accessed 30 March 2012).

[740] U.S. Geological Survey (USGS). 2011. National Streamflow Information Program website. http://nwis.waterdata.usgs.gov/ca/nwis/peak?site_no=11212500&agency_cd=USGS&format=html (accessed 9 August 2011).

[741] U.S. Bureau of Reclamation (USBR). 1970. A summary of hydrologic data for the test case on acreage limitation in Tulare Lake. U.S. Bureau of Reclamation, Sacramento, California.

[742] Selvidge, J.B. n.d. Tracy Ranch History website. http://www.tracyranch.com/history.htm (accessed 14 June 2011).

[743] Stafford, H.M. 1956. Snowmelt Flood of 1952 in Kern River, Tulare Lake, and San Joaquin River Basins. Pages 562–575 *in* U.S. Geological Survey Water–Supply Paper 1260-D. U.S. Geological Survey, Washington, D.C. Available at http://books.google.com/books?id=RrAZAQAAIAAJ&pg=PA530-IA7&lpg=PA530-IA7&dq=stafford+%22snowmelt+flood+of+1952+in+kern+river%22+%22Water%E2%80%93Supply+Paper+1260%22&source=bl&ots=l5-ZwdsnFO&sig=JSX80Q6x22tFJvlPKf2yo-l5rcg&hl=en&sa=X&ei=ULx3T_-

hMILUiAKT87WnDg&ved=0CFcQ6AEwBw#v=onepage&q=stafford%20%22snowmelt%20flood%20of%201952%20in%20kern%20river%22%20%22Water%E2%80%93Supply%20Paper%201260%22&f=false (accessed 30 March 2012).

[744] June 4, 1952 issue of the New York Times.

[745] U.S. Bureau of Reclamation (USBR). 1970. A summary of hydrologic data for the test case on acreage limitation in Tulare Lake. U.S. Bureau of Reclamation, Sacramento, California.

[746] J. G. Boswell Company and J. G. Boswell Company (Successor by Merger to Tulare Lake Land Company) v. Commissioner of Internal Revenue, U.S. Court of Appeals Ninth Circuit. April 17, 1962. Available at http://ca.findacase.com/research/wfrmDocViewer.aspx/xq/fac.%2FFCT%2FC09%2F1962%2F19620417_0002.C09.htm/qx (accessed 30 May 2011).

[747] Goodridge, J. 1996. Data on California's Extreme Rainfall from 1862–1995. Proceedings of the 1996 California Extreme Precipitation Symposium. Sierra College Science Center, Rocklin, California. Available at http://www.cepsym.info/proceedings_1996.php (accessed 1 July 2011).

[748] U.S. Army Corps of Engineers (USACE). 1999. Post-flood assessment for 1983, 1986, 1995, and 1997. U.S. Army Corps of Engineers, Sacramento, California. Chapter 2 available at http://snugharbor.net/images2010/misc/2002_usace_sac_flood_history.pdf (accessed 28 May 2012).

[749] U.S. Army Corps of Engineers (USACE). 1999. Post-flood assessment for 1983, 1986, 1995, and 1997. U.S. Army Corps of Engineers, Sacramento, California. Chapter 2 available at http://snugharbor.net/images2010/misc/2002_usace_sac_flood_history.pdf (accessed 28 May 2012).

[750] National Weather Service (NWS). 2009. The Interior Central California climate calendar. Available at http://www.wrh.noaa.gov/hnx/WXCALENDER.pdf (accessed 1 February 2011).

[751] National Weather Service (NWS). 2009. The Interior Central California climate calendar. Available at http://www.wrh.noaa.gov/hnx/WXCALENDER.pdf (accessed 1 February 2011).

[752] November 20 issue of the New York Times.

[753] December 26 issue of the New York Times.

[754] December 27 issue of the New York Times.

[755] Donahue, M. n.d. Yosemite Valley spring runoff and flooding website. http://faculty.deanza.edu/donahuemary/stories/storyReader$2113 (accessed 1 February 2011).

[756] California Disaster Office. 1956. The Big Flood, California 1955. California Disaster Office. Sacramento, California.

[757] California Disaster Office. 1956. The Big Flood, California 1955. California Disaster Office. Sacramento, California.

[758] California Disaster Office. 1956. The Big Flood, California 1955. California Disaster Office. Sacramento, California.

[759] National Weather Service (NWS). 2009. The Interior Central California climate calendar. Available at http://www.wrh.noaa.gov/hnx/WXCALENDER.pdf (accessed 1 February 2011).

[760] McCulley, Frick & Gilman and William Lettis & Associates. 1998. Panoche/Silver Creek Watershed Assessment. Prepared for Panoche/Silver Creek Watershed Coordinated Resource Management and Planning Group and the City of Mendota, California. Available at http://www.water.ca.gov/pubs/environment/watersheds/panoche_silver_creek_watershed_assessment_final_report/psc-assessment.pdf (accessed 23 August 2011).

[761] U.S. Geological Survey (USGS). 2011. National Streamflow Information Program website. http://nwis.waterdata.usgs.gov/ca/nwis/peak?site_no=11212500&agency_cd=USGS&format=html (accessed 9 August 2011).

[762] January 7, 1956 issue of the Three Rivers Current.

[763] January 7, 1956 issue of the Three Rivers Current.

[764] January 7, 1956 issue of the Three Rivers Current.

[765] January 7, 1956 issue of the Three Rivers Current.

[766] Dean, W.W. 1971. Floods of December, 1966 in the Kern-Kaweah Area, Kern and Tulare Counties, California. U.S. Geological Survey Water–Supply Paper 1870-C. U.S. Geological Survey, Washington, D.C.

[767] January 7, 1956 issue of the Three Rivers Current.

[768] U.S. Army Corps of Engineers (USACE). June 1989. Kaweah River Sediment Investigation: Hydraulic Design. Unpublished report, Sacramento, California.

[769] California Disaster Office. 1956. The Big Flood, California 1955. California Disaster Office. Sacramento, California.

[770] U.S. Bureau of Reclamation (USBR). 1970. A summary of hydrologic data for the test case on acreage limitation in Tulare Lake. U.S. Bureau of Reclamation, Sacramento, California.

[771] Dean, W.W. 1971. Floods of December, 1966 in the Kern-Kaweah Area, Kern and Tulare Counties, California. U.S. Geological Survey Water–Supply Paper 1870-C. U.S. Geological Survey, Washington, D.C.

[772] December 24, 2010 issue of the Kaweah Commonwealth.

[773] National Weather Service (NWS). 2009. The Interior Central California climate calendar. Available at http://www.wrh.noaa.gov/hnx/WXCALENDER.pdf (accessed 1 February 2011).

[774] California Disaster Office. 1956. The Big Flood, California 1955. California Disaster Office. Sacramento, California.

[775] California Disaster Office. 1956. The Big Flood, California 1955. California Disaster Office. Sacramento, California.

[776] California Disaster Office. 1956. The Big Flood, California 1955. California Disaster Office. Sacramento, California.

[777] California Disaster Office. 1956. The Big Flood, California 1955. California Disaster Office. Sacramento, California.

[778] California Disaster Office. 1956. The Big Flood, California 1955. California Disaster Office. Sacramento, California.

[779] California Disaster Office. 1956. The Big Flood, California 1955. California Disaster Office. Sacramento, California.

[780] California Disaster Office. 1956. The Big Flood, California 1955. California Disaster Office. Sacramento, California.

[781] January 31, 1956 issue of the Visalia Times-Delta.

[782] California Disaster Office. 1956. The Big Flood, California 1955. California Disaster Office. Sacramento, California.

[783] California Disaster Office. 1956. The Big Flood, California 1955. California Disaster Office. Sacramento, California.

[784] California Disaster Office. 1956. The Big Flood, California 1955. California Disaster Office. Sacramento, California.

[785] California State University San Bernardino. n.d. Alluvial Fan Task Force (AFTF) Study Area Flood History: Kern County Flood History. Available at http://aftf.csusb.edu/documents/AFTF%20Study%20Area%20Flood%20History_ALL.pdf (accessed 17 June 2011).

[786] California Disaster Office. 1956. The Big Flood, California 1955. California Disaster Office. Sacramento, California.

[787] California Disaster Office. 1956. The Big Flood, California 1955. California Disaster Office. Sacramento, California.

[788] National Weather Service (NWS). 2009. The Interior Central California climate calendar. Available at http://www.wrh.noaa.gov/hnx/WXCALENDER.pdf (accessed 1 February 2011).

[789] January 26, 1956 issue of the New York Times.

[790] Dean, W.W. 1971. Floods of December, 1966 in the Kern-Kaweah Area, Kern and Tulare Counties, California. U.S. Geological Survey Water–Supply Paper 1870-C. U.S. Geological Survey, Washington, D.C.

[791] U.S. Bureau of Reclamation (USBR). 1970. A summary of hydrologic data for the test case on acreage limitation in Tulare Lake. U.S. Bureau of Reclamation, Sacramento, California.

[792] U.S. Geological Survey (USGS). 2011. National Streamflow Information Program website. http://nwis.waterdata.usgs.gov/ca/nwis/peak?site_no=11212500&agency_cd=USGS&format=html (accessed 9 August 2011).

[793] National Weather Service (NWS). 2009. The Interior Central California climate calendar. Available at http://www.wrh.noaa.gov/hnx/WXCALENDER.pdf (accessed 1 February 2011).

[794] Goodridge, J. 1996. Data on California's Extreme Rainfall from 1862–1995. Proceedings of the 1996 California Extreme Precipitation Symposium. Sierra College Science Center, Rocklin, California. Available at http://www.cepsym.info/proceedings_1996.php (accessed 1 July 2011).

[795] McCulley, Frick & Gilman and William Lettis & Associates. 1998. Panoche/Silver Creek Watershed Assessment. Prepared for Panoche/Silver Creek Watershed Coordinated Resource Management and Planning Group and the City of Mendota, California. Available at http://www.water.ca.gov/pubs/environment/watersheds/panoche_silver_creek_watershed_assessment_final_report/psc-assessment.pdf (accessed 23 August 2011).

[796] U.S. Bureau of Reclamation (USBR). 1970. A summary of hydrologic data for the test case on acreage limitation in Tulare Lake. U.S. Bureau of Reclamation, Sacramento, California.

[797] National Weather Service (NWS). 2009. The Interior Central California climate calendar. Available at http://www.wrh.noaa.gov/hnx/WXCALENDER.pdf (accessed 1 February 2011).

[798] Goodridge, J. 1996. Data on California's Extreme Rainfall from 1862–1995. Proceedings of the 1996 California Extreme Precipitation Symposium. Sierra College Science Center, Rocklin, California. Available at http://www.cepsym.info/proceedings_1996.php (accessed 1 July 2011).

[799] U.S. Geological Survey (USGS). 2011. National Streamflow Information Program website. http://nwis.waterdata.usgs.gov/ca/nwis/peak?site_no=11212500&agency_cd=USGS&format=html (accessed 9 August 2011).

[800] Goodridge, J. 1996. Data on California's Extreme Rainfall from 1862–1995. Proceedings of the 1996 California Extreme Precipitation Symposium. Sierra College Science Center, Rocklin, California. Available at http://www.cepsym.info/proceedings_1996.php (accessed 1 July 2011).

[801] Goodridge, J. 1996. Data on California's Extreme Rainfall from 1862–1995. Proceedings of the 1996 California Extreme Precipitation Symposium. Sierra College Science Center, Rocklin, California. Available at http://www.cepsym.info/proceedings_1996.php (accessed 1 July 2011).

[802] U.S. Geological Survey (USGS). 2011. National Streamflow Information Program website. http://nwis.waterdata.usgs.gov/ca/nwis/peak?site_no=11212450&agency_cd=USGS&format=html (accessed 24 November 2012).

[803] California State University San Bernardino. n.d. Alluvial Fan Task Force (AFTF) Study Area Flood History: Kern County Flood History. Available at http://aftf.csusb.edu/documents/AFTF%20Study%20Area%20Flood%20History_ALL.pdf (accessed 17 June 2011).

[804] Dean, W.W. 1971. Floods of December, 1966 in the Kern-Kaweah Area, Kern and Tulare Counties, California. U.S. Geological Survey Water–Supply Paper 1870-C. U.S. Geological Survey, Washington, D.C.

[805] Goodridge, J. 1996. Data on California's Extreme Rainfall from 1862–1995. Proceedings of the 1996 California Extreme Precipitation Symposium. Sierra College Science Center, Rocklin, California. Available at http://www.cepsym.info/proceedings_1996.php (accessed 1 July 2011).

[806] Goodridge, J. 1996. Data on California's Extreme Rainfall from 1862–1995. Proceedings of the 1996 California Extreme Precipitation Symposium. Sierra College Science Center, Rocklin, California. Available at http://www.cepsym.info/proceedings_1996.php (accessed 1 July 2011).

[807] U.S. Army Corps of Engineers (USACE). 1999. Post-flood assessment for 1983, 1986, 1995, and 1997. U.S. Army Corps of Engineers, Sacramento, California. Chapter 2 available at http://snugharbor.net/images2010/misc/2002_usace_sac_flood_history.pdf (accessed 28 May 2012).

[808] U.S. Army Corps of Engineers (USACE). 1999. Post-flood assessment for 1983, 1986, 1995, and 1997. U.S. Army Corps of Engineers, Sacramento, California. Chapter 2 available at http://snugharbor.net/images2010/misc/2002_usace_sac_flood_history.pdf (accessed 28 May 2012).

[809] U.S. Geological Survey (USGS). 2011. National Streamflow Information Program website. http://nwis.waterdata.usgs.gov/ca/nwis/peak?site_no=11212500&agency_cd=USGS&format=html (accessed 9 August 2011).

[810] National Weather Service (NWS). 2009. The Interior Central California climate calendar. Available at http://www.wrh.noaa.gov/hnx/WXCALENDER.pdf (accessed 1 February 2011).

[811] U.S. Geological Survey (USGS). 2011. National Streamflow Information Program website. http://nwis.waterdata.usgs.gov/ca/nwis/peak?site_no=11212450&agency_cd=USGS&format=html (accessed 24 November 2012).

[812] Goodridge, J. 1996. Data on California's Extreme Rainfall from 1862–1995. Proceedings of the 1996 California Extreme Precipitation Symposium. Sierra College Science Center, Rocklin, California. Available at http://www.cepsym.info/proceedings_1996.php (accessed 1 July 2011).

[813] Dean, W.W. 1971. Floods of December, 1966 in the Kern-Kaweah Area, Kern and Tulare Counties, California. U.S. Geological Survey Water–Supply Paper 1870-C. U.S. Geological Survey, Washington, D.C.

[814] December 9, 1966 issue of Visalia Times-Delta.

[815] Goodridge, J. 1996. Data on California's Extreme Rainfall from 1862–1995. Proceedings of the 1996 California Extreme Precipitation Symposium. Sierra College Science Center, Rocklin, California. Available at http://www.cepsym.info/proceedings_1996.php (accessed 1 July 2011).

[816] U.S. Army Corps of Engineers (USACE). 1999. Post-flood assessment for 1983, 1986, 1995, and 1997. U.S. Army Corps of Engineers, Sacramento, California. Chapter 2 available at http://snugharbor.net/images2010/misc/2002_usace_sac_flood_history.pdf (accessed 28 May 2012).

[817] U.S. Army Corps of Engineers (USACE). 1999. Post-flood assessment for 1983, 1986, 1995, and 1997. U.S. Army Corps of Engineers, Sacramento, California. Chapter 2 available at http://snugharbor.net/images2010/misc/2002_usace_sac_flood_history.pdf (accessed 28 May 2012).

[818] McCulley, Frick & Gilman and William Lettis & Associates. 1998. Panoche/Silver Creek Watershed Assessment. Prepared for Panoche/Silver Creek Watershed Coordinated Resource Management and Planning Group and the City of Mendota, California. Available at http://www.water.ca.gov/pubs/environment/watersheds/panoche_silver_creek_watershed_assessment_final_report/psc-assessment.pdf (accessed 23 August 2011).

[819] U.S. Geological Survey (USGS). 2011. National Streamflow Information Program website. http://nwis.waterdata.usgs.gov/ca/nwis/peak?site_no=11212500&agency_cd=USGS&format=html (accessed 9 August 2011).

[820] Dean, W.W. 1971. Floods of December, 1966 in the Kern-Kaweah Area, Kern and Tulare Counties, California. U.S. Geological Survey Water–Supply Paper 1870-C. U.S. Geological Survey, Washington, D.C.

[821] Dean, W.W. 1971. Floods of December, 1966 in the Kern-Kaweah Area, Kern and Tulare Counties, California. U.S. Geological Survey Water–Supply Paper 1870-C. U.S. Geological Survey, Washington, D.C.

[822] Dean, W.W. 1971. Floods of December, 1966 in the Kern-Kaweah Area, Kern and Tulare Counties, California. U.S. Geological Survey Water–Supply Paper 1870-C. U.S. Geological Survey, Washington, D.C.

[823] Dean, W.W. 1971. Floods of December, 1966 in the Kern-Kaweah Area, Kern and Tulare Counties, California. U.S. Geological Survey Water–Supply Paper 1870-C. U.S. Geological Survey, Washington, D.C.

[824] U.S. Geological Survey (USGS). 2010. Water-data Report 2010: 11208730 East Fork Kaweah River near Three Rivers, CA. Water-data report website. http://wdr.water.usgs.gov/wy2010/pdfs/11208730.2010.pdf (accessed 5 March 2012).

[825] December 1966 issue (date unknown) of the Visalia Times-Delta.

[826] December 1966 issue (date unknown) of the Visalia Times-Delta.

[827] December 8, 1966 issue of the Visalia Times-Delta.

[828] December 8, 1966 issue of the Visalia Times-Delta.

[829] December 7, 1966 issue of the Visalia Times-Delta.

[830] December 8, 1966 issue of the Visalia Times-Delta.

[831] December 7, 1966 issue of the Visalia Times-Delta.

[832] Dean, W.W. 1971. Floods of December, 1966 in the Kern-Kaweah Area, Kern and Tulare Counties, California. U.S. Geological Survey Water–Supply Paper 1870-C. U.S. Geological Survey, Washington, D.C.

[833] U.S. Bureau of Reclamation (USBR). 1970. A summary of hydrologic data for the test case on acreage limitation in Tulare Lake. U.S. Bureau of Reclamation, Sacramento, California.

[834] December 8, 1966 issue of the Visalia Times-Delta.

[835] U.S. Bureau of Reclamation (USBR). 1970. A summary of hydrologic data for the test case on acreage limitation in Tulare Lake. U.S. Bureau of Reclamation, Sacramento, California.

[836] December 8, 1966 issue of the Visalia Times-Delta.

[837] U.S. Army Corps of Engineers (USACE). June 1989. Kaweah River Sediment Investigation: Hydraulic Design. Unpublished report, Sacramento, California.

[838] December 7, 1966 issue of the Visalia Times-Delta.

[839] December 6, 1966 issue of the Visalia Times-Delta.

[840] December 8, 1966 issue of the Visalia Times-Delta.

[841] December 6, 1966 issue of the Visalia Times-Delta.

[842] December 7, 1966 issue of the Visalia Times-Delta.

[843] Dean, W.W. 1971. Floods of December, 1966 in the Kern-Kaweah Area, Kern and Tulare Counties, California. U.S. Geological Survey Water–Supply Paper 1870-C. U.S. Geological Survey, Washington, D.C.

[844] December 6, 1966 issue of the Visalia Times-Delta.

[845] December 9, 1966 issue of the Visalia Times-Delta.

[846] Dean, W.W. 1971. Floods of December, 1966 in the Kern-Kaweah Area, Kern and Tulare Counties, California. U.S. Geological Survey Water–Supply Paper 1870-C. U.S. Geological Survey, Washington, D.C.

[847] Dean, W.W. 1971. Floods of December, 1966 in the Kern-Kaweah Area, Kern and Tulare Counties, California. U.S. Geological Survey Water–Supply Paper 1870-C. U.S. Geological Survey, Washington, D.C.

[848] Dean, W.W. 1971. Floods of December, 1966 in the Kern-Kaweah Area, Kern and Tulare Counties, California. U.S. Geological Survey Water–Supply Paper 1870-C. U.S. Geological Survey, Washington, D.C.

[849] U.S. Army Corps of Engineers (USACE). n.d. Water control manual for Success Dam: pertinent data sheets. U.S. Army Corps of Engineers, Sacramento, California.

[850] Dean, W.W. 1971. Floods of December, 1966 in the Kern-Kaweah Area, Kern and Tulare Counties, California. U.S. Geological Survey Water–Supply Paper 1870-C. U.S. Geological Survey, Washington, D.C.

[851] December 12, 1966 issue of the Visalia Times-Delta.

[852] Dean, W.W. 1971. Floods of December, 1966 in the Kern-Kaweah Area, Kern and Tulare Counties, California. U.S. Geological Survey Water–Supply Paper 1870-C. U.S. Geological Survey, Washington, D.C.

[853] December 8, 1966 issue of the Visalia Times-Delta.

[854] Dean, W.W. 1971. Floods of December, 1966 in the Kern-Kaweah Area, Kern and Tulare Counties, California. U.S. Geological Survey Water–Supply Paper 1870-C. U.S. Geological Survey, Washington, D.C.

[855] Dean, W.W. 1971. Floods of December, 1966 in the Kern-Kaweah Area, Kern and Tulare Counties, California. U.S. Geological Survey Water–Supply Paper 1870-C. U.S. Geological Survey, Washington, D.C.

[856] California State University San Bernardino. n.d. Alluvial Fan Task Force (AFTF) Study Area Flood History: Kern County Flood History. Available at http://aftf.csusb.edu/documents/AFTF%20Study%20Area%20Flood%20History_ALL.pdf (accessed 17 June 2011).

[857] Dean, W.W. 1971. Floods of December, 1966 in the Kern-Kaweah Area, Kern and Tulare Counties, California. U.S. Geological Survey Water–Supply Paper 1870-C. U.S. Geological Survey, Washington, D.C.

[858] Dean, W.W. 1971. Floods of December, 1966 in the Kern-Kaweah Area, Kern and Tulare Counties, California. U.S. Geological Survey Water–Supply Paper 1870-C. U.S. Geological Survey, Washington, D.C.

[859] Dean, W.W. 1971. Floods of December, 1966 in the Kern-Kaweah Area, Kern and Tulare Counties, California. U.S. Geological Survey Water–Supply Paper 1870-C. U.S. Geological Survey, Washington, D.C.

[860] Dean, W.W. 1971. Floods of December, 1966 in the Kern-Kaweah Area, Kern and Tulare Counties, California. U.S. Geological Survey Water–Supply Paper 1870-C. U.S. Geological Survey, Washington, D.C.

[861] Dean, W.W. 1971. Floods of December, 1966 in the Kern-Kaweah Area, Kern and Tulare Counties, California. U.S. Geological Survey Water–Supply Paper 1870-C. U.S. Geological Survey, Washington, D.C.

[862] U.S. Army Corps of Engineers (USACE). n.d. Water control manual for Isabella Dam: pertinent data sheets. U.S. Army Corps of Engineers, Sacramento, California.

[863] Dean, W.W. 1971. Floods of December, 1966 in the Kern-Kaweah Area, Kern and Tulare Counties, California. U.S. Geological Survey Water–Supply Paper 1870-C. U.S. Geological Survey, Washington, D.C.

[864] California State University San Bernardino. n.d. Alluvial Fan Task Force (AFTF) Study Area Flood History: Kern County Flood History. Available at http://aftf.csusb.edu/documents/AFTF%20Study%20Area%20Flood%20History_ALL.pdf (accessed 17 June 2011).

[865] Dean, W.W. 1971. Floods of December, 1966 in the Kern-Kaweah Area, Kern and Tulare Counties, California. U.S. Geological Survey Water–Supply Paper 1870-C. U.S. Geological Survey, Washington, D.C.

[866] California State University San Bernardino. n.d. Alluvial Fan Task Force (AFTF) Study Area Flood History: Kern County Flood History. Available at http://aftf.csusb.edu/documents/AFTF%20Study%20Area%20Flood%20History_ALL.pdf (accessed 17 June 2011).

[867] December 9, 1966 issue of the Visalia Times-Delta.

[868] Dean, W.W. 1971. Floods of December, 1966 in the Kern-Kaweah Area, Kern and Tulare Counties, California. U.S. Geological Survey Water–Supply Paper 1870-C. U.S. Geological Survey, Washington, D.C.

[869] California State University San Bernardino. n.d. Alluvial Fan Task Force (AFTF) Study Area Flood History: Kern County Flood History. Available at http://aftf.csusb.edu/documents/AFTF%20Study%20Area%20Flood%20History_ALL.pdf (accessed 17 June 2011).

[870] December 7, 1966 issue of the New York Times.

[871] Dean, W.W. 1971. Floods of December, 1966 in the Kern-Kaweah Area, Kern and Tulare Counties, California. U.S. Geological Survey Water–Supply Paper 1870-C. U.S. Geological Survey, Washington, D.C.

[872] U.S. Army Corps of Engineers (USACE). 1969. Flood Plain Information, Kern River, Bakersfield, California. U.S. Army Corps of Engineers, Sacramento, California.

[873] December 9, 1966 issue of the Visalia Times-Delta.

[874] December 16, 1966 issue of the Visalia Times-Delta.

[875] U.S. Bureau of Reclamation (USBR). 1970. A summary of hydrologic data for the test case on acreage limitation in Tulare Lake. U.S. Bureau of Reclamation, Sacramento, California.

[876] December 13, 1966 issue of the Visalia Times-Delta.

[877] U.S. Army Corps of Engineers (USACE). 1999. Post-flood assessment for 1983, 1986, 1995, and 1997. U.S. Army Corps of Engineers, Sacramento, California. Chapter 2 available at http://snugharbor.net/images2010/misc/2002_usace_sac_flood_history.pdf (accessed 28 May 2012).

[878] U.S. Army Corps of Engineers (USACE). 1999. Post-flood assessment for 1983, 1986, 1995, and 1997. U.S. Army Corps of Engineers, Sacramento, California. Chapter 2 available at http://snugharbor.net/images2010/misc/2002_usace_sac_flood_history.pdf (accessed 28 May 2012).

[879] U.S. Geological Survey (USGS). 2011. National Streamflow Information Program website. http://nwis.waterdata.usgs.gov/ca/nwis/peak?site_no=11212450&agency_cd=USGS&format=html (accessed 24 November 2012).

[880] U.S. Army Corps of Engineers (USACE). 1999. Post-flood assessment for 1983, 1986, 1995, and 1997. U.S. Army Corps of Engineers, Sacramento, California. Chapter 2 available at http://snugharbor.net/images2010/misc/2002_usace_sac_flood_history.pdf (accessed 28 May 2012).

[881] U.S. Bureau of Reclamation (USBR). 1970. A summary of hydrologic data for the test case on acreage limitation in Tulare Lake. U.S. Bureau of Reclamation, Sacramento, California.

[882] Goodridge, J. 1996. Data on California's Extreme Rainfall from 1862–1995. Proceedings of the 1996 California Extreme Precipitation Symposium. Sierra College Science Center, Rocklin, California. Available at http://www.cepsym.info/proceedings_1996.php (accessed 1 July 2011).

[883] U.S. Army Corps of Engineers (USACE). 1999. Post-flood assessment for 1983, 1986, 1995, and 1997. U.S. Army Corps of Engineers, Sacramento, California. Chapter 2 available at http://snugharbor.net/images2010/misc/2002_usace_sac_flood_history.pdf (accessed 28 May 2012).

[884] National Weather Service (NWS). 2009. The Interior Central California climate calendar. Available at http://www.wrh.noaa.gov/hnx/WXCALENDER.pdf (accessed 1 February 2011).

[885] National Weather Service (NWS). 2009. The Interior Central California climate calendar. Available at http://www.wrh.noaa.gov/hnx/WXCALENDER.pdf (accessed 1 February 2011).

[886] National Weather Service (NWS). 2009. The Interior Central California climate calendar. Available at http://www.wrh.noaa.gov/hnx/WXCALENDER.pdf (accessed 1 February 2011).

[887] National Weather Service (NWS). 2009. The Interior Central California climate calendar. Available at http://www.wrh.noaa.gov/hnx/WXCALENDER.pdf (accessed 1 February 2011).

[888] McCulley, Frick & Gilman and William Lettis & Associates. 1998. Panoche/Silver Creek Watershed Assessment. Prepared for Panoche/Silver Creek Watershed Coordinated Resource Management and Planning Group and the City of Mendota, California. Available at http://www.water.ca.gov/pubs/environment/watersheds/panoche_silver_creek_watershed_assessment_final_report/psc-assessment.pdf (accessed 23 August 2011).

[889] U.S. Bureau of Reclamation (USBR). 1970. A summary of hydrologic data for the test case on acreage limitation in Tulare Lake. U.S. Bureau of Reclamation, Sacramento, California.

[890] Jensen, D. n.d. Hockett Trail notes website. http://igneousrange.wordpress.com/tag/hiking/ (accessed 12 July 2011).

[891] McCulley, Frick & Gilman and William Lettis & Associates. 1998. Panoche/Silver Creek Watershed Assessment. Prepared for Panoche/Silver Creek Watershed Coordinated Resource Management and Planning Group and the City of Mendota, California. Available at http://www.water.ca.gov/pubs/environment/watersheds/panoche_silver_creek_watershed_assessment_final_report/psc-assessment.pdf (accessed 23 August 2011).

[892] U.S. Bureau of Reclamation (USBR). 1970. A summary of hydrologic data for the test case on acreage limitation in Tulare Lake. U.S. Bureau of Reclamation, Sacramento, California.

[893] U.S. Bureau of Reclamation (USBR). 1970. A summary of hydrologic data for the test case on acreage limitation in Tulare Lake. U.S. Bureau of Reclamation, Sacramento, California.

[894] April 2, 1969 issue of the Visalia Times-Delta.

[895] Goodridge, J. 1996. Data on California's Extreme Rainfall from 1862–1995. Proceedings of the 1996 California Extreme Precipitation Symposium. Sierra College Science Center, Rocklin, California. Available at http://www.cepsym.info/proceedings_1996.php (accessed 1 July 2011).

[896] U.S. Bureau of Reclamation (USBR). Upper San Joaquin River Basin Storage Investigation: Raise Pine Flat Dam. 2003. U.S. Bureau of Reclamation, Mid-Pacific Regional Office, Sacramento, California. Available at http://www.usbr.gov/mp/sccao/storage/docs/phase1_rpt_fnl/tech_app/09_pine_flat.pdf (accessed 24 May 2012).

[897] U.S. Army Corps of Engineers (USACE). n.d. Water control manual for Pine Flat Dam: pertinent data sheets. U.S. Army Corps of Engineers, Sacramento, California.

[898] U.S. Army Corps of Engineers (USACE). 1976. Pine Flat Lake Master Plan Design Memorandum No. 7. U.S. Army Corps of Engineers, Sacramento, California.

[899] U.S. Bureau of Reclamation (USBR). Upper San Joaquin River Basin Storage Investigation: Raise Pine Flat Dam. 2003. U.S. Bureau of Reclamation, Mid-Pacific Regional Office, Sacramento, California. Available at http://www.usbr.gov/mp/sccao/storage/docs/phase1_rpt_fnl/tech_app/09_pine_flat.pdf (accessed 24 May 2012).

[900] U.S. Geological Survey (USGS). 2012. Water-data Report 2011: 11253500 James Bypass near San Joaquin, CA. Water-data report website. http://wdr.water.usgs.gov/wy2011/pdfs/11253500.2011.pdf (accessed 21 September 2012).

[901] May 11, 1969 issue of the Fresno Bee.

[902] Salyer Land Co. vs. Tulare Lake Basin Water Storage District. 410 U.S. 719 1973. Available at http://supreme.justia.com/us/410/719/case.html (accessed 25 May 2011).

[903] May 11, 1969 issue of the Fresno Bee.

[904] U.S. Bureau of Reclamation (USBR). 1970. A summary of hydrologic data for the test case on acreage limitation in Tulare Lake. U.S. Bureau of Reclamation, Sacramento, California.

[905] U.S. Bureau of Reclamation (USBR). 1970. A summary of hydrologic data for the test case on acreage limitation in Tulare Lake. U.S. Bureau of Reclamation, Sacramento, California.

[906] California State University San Bernardino. n.d. Alluvial Fan Task Force (AFTF) Study Area Flood History. Available at http://aftf.csusb.edu/documents/AFTF%20Study%20Area%20Flood%20History_ALL.pdf (accessed 17 June 2011).

[907] California State University San Bernardino. n.d. Alluvial Fan Task Force (AFTF) Study Area Flood History. Available at http://aftf.csusb.edu/documents/AFTF%20Study%20Area%20Flood%20History_ALL.pdf (accessed 17 June 2011).

[908] National Weather Service (NWS). 2009. The Interior Central California climate calendar. Available at http://www.wrh.noaa.gov/hnx/WXCALENDER.pdf (accessed 1 February 2011).

[909] California State University San Bernardino. n.d. Alluvial Fan Task Force (AFTF) Study Area Flood History: Kern County Flood History. Available at http://aftf.csusb.edu/documents/AFTF%20Study%20Area%20Flood%20History_ALL.pdf (accessed 17 June 2011).

[910] California State University San Bernardino. n.d. Alluvial Fan Task Force (AFTF) Study Area Flood History: Kern County Flood History. Available at http://aftf.csusb.edu/documents/AFTF%20Study%20Area%20Flood%20History_ALL.pdf (accessed 17 June 2011).

[911] National Weather Service (NWS). 2009. The Interior Central California climate calendar. Available at http://www.wrh.noaa.gov/hnx/WXCALENDER.pdf (accessed 1 February 2011).

[912] California State University San Bernardino. n.d. Alluvial Fan Task Force (AFTF) Study Area Flood History: Kern County Flood History. Available at http://aftf.csusb.edu/documents/AFTF%20Study%20Area%20Flood%20History_ALL.pdf (accessed 17 June 2011).

[913] U.S. Army Corps of Engineers (USACE). n.d. Water control manual for Pine Flat Dam: pertinent data sheets. U.S. Army Corps of Engineers, Sacramento, California.

[914] National Weather Service (NWS). 2009. The Interior Central California climate calendar. Available at http://www.wrh.noaa.gov/hnx/WXCALENDER.pdf (accessed 1 February 2011).

[915] National Weather Service (NWS). 2010. A History of Significant Weather Events in Southern California. http://www.wrh.noaa.gov/sgx/document/weatherhistory.pdf (accessed 17 June 2011).

[916] National Weather Service (NWS). 2009. The Interior Central California climate calendar. Available at http://www.wrh.noaa.gov/hnx/WXCALENDER.pdf (accessed 1 February 2011).

[917] National Weather Service (NWS). 2009. The Interior Central California climate calendar. Available at http://www.wrh.noaa.gov/hnx/WXCALENDER.pdf (accessed 1 February 2011).

[918] National Weather Service (NWS). 2009. The Interior Central California climate calendar. Available at http://www.wrh.noaa.gov/hnx/WXCALENDER.pdf (accessed 1 February 2011).

[919] National Weather Service (NWS). 2009. The Interior Central California climate calendar. Available at http://www.wrh.noaa.gov/hnx/WXCALENDER.pdf (accessed 1 February 2011).

[920] Goodridge, J. 1996. Data on California's Extreme Rainfall from 1862–1995. Proceedings of the 1996 California Extreme Precipitation Symposium. Sierra College Science Center, Rocklin, California. Available at http://www.cepsym.info/proceedings_1996.php (accessed 1 July 2011).

[921] National Weather Service (NWS). 2009. The Interior Central California climate calendar. Available at http://www.wrh.noaa.gov/hnx/WXCALENDER.pdf (accessed 1 February 2011).

[922] National Weather Service (NWS). n.d. Bakersfield Normals, Means, and Extremes website. http://www.wrh.noaa.gov/hnx/bflmain.php (accessed 3 May 2011).

[923] California State University San Bernardino. n.d. Alluvial Fan Task Force (AFTF) Study Area Flood History: Kern County Flood History. Available at http://aftf.csusb.edu/documents/AFTF%20Study%20Area%20Flood%20History_ALL.pdf (accessed 17 June 2011).

[924] McCulley, Frick & Gilman and William Lettis & Associates. 1998. Panoche/Silver Creek Watershed Assessment. Prepared for Panoche/Silver Creek Watershed Coordinated Resource Management and Planning Group and the City of Mendota, California. Available at http://www.water.ca.gov/pubs/environment/watersheds/panoche_silver_creek_watershed_assessment_final_report/psc-assessment.pdf (accessed 23 August 2011).

[925] Goodridge, J. 1996. Data on California's Extreme Rainfall from 1862–1995. Proceedings of the 1996 California Extreme Precipitation Symposium. Sierra College Science Center, Rocklin, California. Available at http://www.cepsym.info/proceedings_1996.php (accessed 1 July 2011).

[926] Goodridge, J. 1996. Data on California's Extreme Rainfall from 1862–1995. Proceedings of the 1996 California Extreme Precipitation Symposium. Sierra College Science Center, Rocklin, California. Available at http://www.cepsym.info/proceedings_1996.php (accessed 1 July 2011).

[927] National Weather Service (NWS). 2009. The Interior Central California climate calendar. Available at http://www.wrh.noaa.gov/hnx/WXCALENDER.pdf (accessed 1 February 2011).

[928] National Weather Service (NWS). 2009. The Interior Central California climate calendar. Available at http://www.wrh.noaa.gov/hnx/WXCALENDER.pdf (accessed 1 February 2011).

[929] Goodridge, J. 1996. Data on California's Extreme Rainfall from 1862–1995. Proceedings of the 1996 California Extreme Precipitation Symposium. Sierra College Science Center, Rocklin, California. Available at http://www.cepsym.info/proceedings_1996.php (accessed 1 July 2011).

[930] National Weather Service (NWS). 2009. The Interior Central California climate calendar. Available at http://www.wrh.noaa.gov/hnx/WXCALENDER.pdf (accessed 1 February 2011).

[931] Wikipedia. n.d. 1982 Pacific hurricane season web article. http://en.wikipedia.org/wiki/1982_Pacific_hurricane_season#Hurricane_Olivia (accessed 31 July 2012).

[932] National Weather Service (NWS). 2009. The Interior Central California climate calendar. Available at http://www.wrh.noaa.gov/hnx/WXCALENDER.pdf (accessed 1 February 2011).

[933] Goodridge, J. 1996. Data on California's Extreme Rainfall from 1862–1995. Proceedings of the 1996 California Extreme Precipitation Symposium. Sierra College Science Center, Rocklin, California. Available at http://www.cepsym.info/proceedings_1996.php (accessed 1 July 2011).

[934] U.S. Army Corps of Engineers (USACE). 1999. Post-flood assessment for 1983, 1986, 1995, and 1997. U.S. Army Corps of Engineers, Sacramento, California. Chapter 2 available at http://snugharbor.net/images2010/misc/2002_usace_sac_flood_history.pdf (accessed 28 May 2012).

[935] Goodridge, J. 1996. Data on California's Extreme Rainfall from 1862–1995. Proceedings of the 1996 California Extreme Precipitation Symposium. Sierra College Science Center, Rocklin, California. Available at http://www.cepsym.info/proceedings_1996.php (accessed 1 July 2011).

[936] U.S. Army Corps of Engineers (USACE). 1999. Post-flood assessment for 1983, 1986, 1995, and 1997. U.S. Army Corps of Engineers, Sacramento, California. Chapter 2 available at http://snugharbor.net/images2010/misc/2002_usace_sac_flood_history.pdf (accessed 28 May 2012).

[937] U.S. Army Corps of Engineers (USACE). 1999. Post-flood assessment for 1983, 1986, 1995, and 1997. U.S. Army Corps of Engineers, Sacramento, California. Chapter 2 available at http://snugharbor.net/images2010/misc/2002_usace_sac_flood_history.pdf (accessed 28 May 2012).

[938] U.S. Army Corps of Engineers (USACE). 1999. Post-flood assessment for 1983, 1986, 1995, and 1997. U.S. Army Corps of Engineers, Sacramento, California. Chapter 2 available at http://snugharbor.net/images2010/misc/2002_usace_sac_flood_history.pdf (accessed 28 May 2012).

[939] National Weather Service (NWS). 2009. The Interior Central California climate calendar. Available at http://www.wrh.noaa.gov/hnx/WXCALENDER.pdf (accessed 1 February 2011).

[940] U.S. Army Corps of Engineers (USACE). 1999. Post-flood assessment for 1983, 1986, 1995, and 1997. U.S. Army Corps of Engineers, Sacramento, California. Chapter 2 available at http://snugharbor.net/images2010/misc/2002_usace_sac_flood_history.pdf (accessed 28 May 2012).

[941] National Weather Service (NWS). 2009. The Interior Central California climate calendar. Available at http://www.wrh.noaa.gov/hnx/WXCALENDER.pdf (accessed 1 February 2011).

[942] U.S. Army Corps of Engineers (USACE). 1999. Post-flood assessment for 1983, 1986, 1995, and 1997. U.S. Army Corps of Engineers, Sacramento, California. Chapter 2 available at http://snugharbor.net/images2010/misc/2002_usace_sac_flood_history.pdf (accessed 28 May 2012).

[943] DeGraff, J.V. 1994. The geomorphology of some debris flows in the southern Sierra Nevada, California. Geomorphology 10:231–252.

[944] National Weather Service (NWS). 2009. The Interior Central California climate calendar. Available at http://www.wrh.noaa.gov/hnx/WXCALENDER.pdf (accessed 1 February 2011).

[945] California State University San Bernardino. n.d. Alluvial Fan Task Force (AFTF) Study Area Flood History: Kern County Flood History. Available at http://aftf.csusb.edu/documents/AFTF%20Study%20Area%20Flood%20History_ALL.pdf (accessed 17 June 2011).

[946] National Weather Service (NWS). 2009. The Interior Central California climate calendar. Available at http://www.wrh.noaa.gov/hnx/WXCALENDER.pdf (accessed 1 February 2011).

[947] National Weather Service (NWS). 2009. The Interior Central California climate calendar. Available at http://www.wrh.noaa.gov/hnx/WXCALENDER.pdf (accessed 1 February 2011).

[948] U.S. Army Corps of Engineers (USACE). 1999. Post-flood assessment for 1983, 1986, 1995, and 1997. U.S. Army Corps of Engineers, Sacramento, California. Chapter 2 available at http://snugharbor.net/images2010/misc/2002_usace_sac_flood_history.pdf (accessed 28 May 2012).

[949] U.S. Army Corps of Engineers (USACE). 1999. Post-flood assessment for 1983, 1986, 1995, and 1997. U.S. Army Corps of Engineers, Sacramento, California. Chapter 2 available at http://snugharbor.net/images2010/misc/2002_usace_sac_flood_history.pdf (accessed 28 May 2012).

[950] Sydnor, R.H., 1997, Reconnaissance engineering geology of the Mill Creek landslide of January 24, 1997. California Geology 50:74–83. Available at http://www.consrv.ca.gov/CGS/rghm/landslides/cal_geology/1997%20Hwy%2050%20Cal%20Geology.pdf (accessed 12 August 2011).

[951] Wieczorek, G.F. 2002. Catastrophic rockfalls and rockslides in the Sierra Nevada, USA. Pages 165–190 *in* S.G. Evans and J.V. DeGraff, editors. Catastrophic landslides: effects, occurrence, and mechanisms. Geological Society of America Reviews in Engineering Geology XV, Boulder, Colorado. Available at http://landslides.usgs.gov/docs/wieczorek/reg015-08.pdf (accessed 13 August 2011).

[952] Wieczorek, G.F. 2002. Catastrophic rockfalls and rockslides in the Sierra Nevada, USA. Pages 165–190 *in* S.G. Evans and J.V. DeGraff, editors. Catastrophic landslides: effects, occurrence, and mechanisms. Geological Society of America Reviews in Engineering Geology XV, Boulder, Colorado. Available at http://landslides.usgs.gov/docs/wieczorek/reg015-08.pdf (accessed 13 August 2011).

[953] DeGraff, J.V. 1994. The geomorphology of some debris flows in the southern Sierra Nevada, California. Geomorphology 10:231–252.

[954] DeGraff, J.V. 1994. The geomorphology of some debris flows in the southern Sierra Nevada, California. Geomorphology 10:231–252.

[955] DeGraff, J.V. 1994. The geomorphology of some debris flows in the southern Sierra Nevada, California. Geomorphology 10:231–252.

[956] DeGraff, J.V. 1994. The geomorphology of some debris flows in the southern Sierra Nevada, California. Geomorphology 10:231–252.

[957] Austin, J.T. Forthcoming. Erosion and mass wasting. Appendix 7 *in*: National Park Service (NPS). Natural resource condition assessment for Sequoia and Kings Canyon National Parks. To be published in the National Park Service Natural Resource Report Series. Panek, J.A. and C.A. Sydoriak, editors. National Park Service, Fort Collins, Colorado.

[958] California State University San Bernardino. n.d. Alluvial Fan Task Force (AFTF) Study Area Flood History: Kern County Flood History. Available at http://aftf.csusb.edu/documents/AFTF%20Study%20Area%20Flood%20History_ALL.pdf (accessed 17 June 2011).

[959] Robert Olson Associates and AMEC Earth and Environment, Inc. 2005. Kern County, California, Multi-Hazard Mitigation Plan. Prepared for Kern County Fire Department Office of Emergency Services. Available at http://hazardmitigation.calema.ca.gov/docs/lhmp/Kern_County_LHMP.pdf. (accessed 18 June 2011).

[960] California State University San Bernardino. n.d. Alluvial Fan Task Force (AFTF) Study Area Flood History: Kern County Flood History. Available at http://aftf.csusb.edu/documents/AFTF%20Study%20Area%20Flood%20History_ALL.pdf (accessed 17 June 2011).

[961] National Weather Service (NWS). 2009. The Interior Central California climate calendar. Available at http://www.wrh.noaa.gov/hnx/WXCALENDER.pdf (accessed 1 February 2011).

[962] National Weather Service (NWS). 2009. The Interior Central California climate calendar. Available at http://www.wrh.noaa.gov/hnx/WXCALENDER.pdf (accessed 1 February 2011).

[963] National Oceanic and Atmospheric Administration (NOAA). n.d. Top ten published atmospheric river events website. http://www.esrl.noaa.gov/psd/atmrivers/events/ (accessed 1 February 2011).

[964] U.S. Army Corps of Engineers (USACE). 1999. Post-flood assessment for 1983, 1986, 1995, and 1997. U.S. Army Corps of Engineers, Sacramento, California. Chapter 2 available at http://snugharbor.net/images2010/misc/2002_usace_sac_flood_history.pdf (accessed 28 May 2012).

[965] U.S. Army Corps of Engineers (USACE). 1999. Post-flood assessment for 1983, 1986, 1995, and 1997. U.S. Army Corps of Engineers, Sacramento, California. Chapter 2 available at http://snugharbor.net/images2010/misc/2002_usace_sac_flood_history.pdf (accessed 28 May 2012).

[966] U.S. Army Corps of Engineers (USACE). 1999. Post-flood assessment for 1983, 1986, 1995, and 1997. U.S. Army Corps of Engineers, Sacramento, California. Chapter 2 available at http://snugharbor.net/images2010/misc/2002_usace_sac_flood_history.pdf (accessed 28 May 2012).

[967] U.S. Army Corps of Engineers (USACE). 1999. Post-flood assessment for 1983, 1986, 1995, and 1997. U.S. Army Corps of Engineers, Sacramento, California. Chapter 2 available at http://snugharbor.net/images2010/misc/2002_usace_sac_flood_history.pdf (accessed 28 May 2012).

[968] U.S. Army Corps of Engineers (USACE). 1999. Post-flood assessment for 1983, 1986, 1995, and 1997. U.S. Army Corps of Engineers, Sacramento, California. Chapter 2 available at http://snugharbor.net/images2010/misc/2002_usace_sac_flood_history.pdf (accessed 28 May 2012).

[969] Goodridge, J. 1996. Data on California's Extreme Rainfall from 1862–1995. Proceedings of the 1996 California Extreme Precipitation Symposium. Sierra College Science Center, Rocklin, California. Available at http://www.cepsym.info/proceedings_1996.php (accessed 1 July 2011).

[970] U.S. Army Corps of Engineers (USACE). 1999. Post-flood assessment for 1983, 1986, 1995, and 1997. U.S. Army Corps of Engineers, Sacramento, California. Chapter 2 available at http://snugharbor.net/images2010/misc/2002_usace_sac_flood_history.pdf (accessed 28 May 2012).

[971] U.S. Army Corps of Engineers (USACE). 1999. Post-flood assessment for 1983, 1986, 1995, and 1997. U.S. Army Corps of Engineers, Sacramento, California. Chapter 2 available at http://snugharbor.net/images2010/misc/2002_usace_sac_flood_history.pdf (accessed 28 May 2012).

[972] U.S. Army Corps of Engineers (USACE). 1999. Post-flood assessment for 1983, 1986, 1995, and 1997. U.S. Army Corps of Engineers, Sacramento, California. Chapter 2 available at http://snugharbor.net/images2010/misc/2002_usace_sac_flood_history.pdf (accessed 28 May 2012).

[973] National Weather Service (NWS). 2009. The Interior Central California climate calendar. Available at http://www.wrh.noaa.gov/hnx/WXCALENDER.pdf (accessed 1 February 2011).

[974] National Weather Service (NWS). 2009. The Interior Central California climate calendar. Available at http://www.wrh.noaa.gov/hnx/WXCALENDER.pdf (accessed 1 February 2011).

[975] U.S. Bureau of Reclamation (USBR). Upper San Joaquin River Basin Storage Investigation: Raise Pine Flat Dam. 2003. U.S. Bureau of Reclamation, Mid-Pacific Regional Office, Sacramento, California. Available at http://www.usbr.gov/mp/sccao/storage/docs/phase1_rpt_fnl/tech_app/09_pine_flat.pdf (accessed 24 May 2012).

[976] DeGraff, J.V. 1994. The geomorphology of some debris flows in the southern Sierra Nevada, California. Geomorphology 10:231–252.

[977] DeGraff, J.V. 1994. The geomorphology of some debris flows in the southern Sierra Nevada, California. Geomorphology 10:231–252.

[978] California Department of Water Resources (DWR). 1992. The hydrology of the 1987-92 California drought: Technical Information Paper. California Department of Water Resources, Sacramento, California. Available at http://www.water.ca.gov/drought/docs/hydrology_drought-1987-1992.pdf (accessed 1 February 2011).

[979] Bostock, M. and S. Carter. 2012. Drought and deluge in the Lower 48 interactive website. http://www.nytimes.com/interactive/2012/08/11/sunday-review/drought-history.html?ref=sunday-review (accessed 15 August 2012).

[980] Wikipedia. 2012. Drought in the United States web article. http://en.wikipedia.org/wiki/Drought_in_the_United_States (accessed 14 August 2012).

[981] California Department of Water Resources (DWR). 1992. The hydrology of the 1987-92 California drought: Technical Information Paper. California Department of Water Resources, Sacramento, California. Available at http://www.water.ca.gov/drought/docs/hydrology_drought-1987-1992.pdf (accessed 1 February 2011).

[982] National Weather Service (NWS). 2009. The Interior Central California climate calendar. Available at http://www.wrh.noaa.gov/hnx/WXCALENDER.pdf (accessed 1 February 2011).

[983] U.S. Fish and Wildlife Service (USFWS). 2005. Comprehensive conservation plan. U.S. Fish and Wildlife Service Region 1, California/Nevada Refuge Planning Office, Sacramento, California and Kern National Wildlife Refuge Complex, Delano, California. Available at http://www.fws.gov/cno/refuges/kern/Final_CCP.pdf (accessed 30 August 2012).

[984] Goodridge, J. 1996. Data on California's Extreme Rainfall from 1862–1995. Proceedings of the 1996 California Extreme Precipitation Symposium. Sierra College Science Center, Rocklin, California. Available at http://www.cepsym.info/proceedings_1996.php (accessed 1 July 2011).

[985] Goodridge, J. 1996. Data on California's Extreme Rainfall from 1862–1995. Proceedings of the 1996 California Extreme Precipitation Symposium. Sierra College Science Center, Rocklin, California. Available at http://www.cepsym.info/proceedings_1996.php (accessed 1 July 2011).

[986] DeGraff, J.V. 1994. The geomorphology of some debris flows in the southern Sierra Nevada, California. Geomorphology 10:231–252.

[987] National Weather Service (NWS). 2009. The Interior Central California climate calendar. Available at http://www.wrh.noaa.gov/hnx/WXCALENDER.pdf (accessed 1 February 2011).

[988] Goodridge, J. 1996. Data on California's Extreme Rainfall from 1862–1995. Proceedings of the 1996 California Extreme Precipitation Symposium. Sierra College Science Center, Rocklin, California. Available at http://www.cepsym.info/proceedings_1996.php (accessed 1 July 2011).

[989] National Weather Service (NWS). 2009. The Interior Central California climate calendar. Available at http://www.wrh.noaa.gov/hnx/WXCALENDER.pdf (accessed 1 February 2011).

[990] Goodridge, J. 1996. Data on California's Extreme Rainfall from 1862–1995. Proceedings of the 1996 California Extreme Precipitation Symposium. Sierra College Science Center, Rocklin, California. Available at http://www.cepsym.info/proceedings_1996.php (accessed 1 July 2011).

[991] Goodridge, J. 1996. Data on California's Extreme Rainfall from 1862–1995. Proceedings of the 1996 California Extreme Precipitation Symposium. Sierra College Science Center, Rocklin, California. Available at http://www.cepsym.info/proceedings_1996.php (accessed 1 July 2011).

[992] Goodridge, J. 1996. Data on California's Extreme Rainfall from 1862–1995. Proceedings of the 1996 California Extreme Precipitation Symposium. Sierra College Science Center, Rocklin, California. Available at http://www.cepsym.info/proceedings_1996.php (accessed 1 July 2011).

[993] U.S. Army Corps of Engineers (USACE). 1999. Post-flood assessment for 1983, 1986, 1995, and 1997. U.S. Army Corps of Engineers, Sacramento, California. Chapter 2 available at http://snugharbor.net/images2010/misc/2002_usace_sac_flood_history.pdf (accessed 28 May 2012).

[994] Goodridge, J. 1996. Data on California's Extreme Rainfall from 1862–1995. Proceedings of the 1996 California Extreme Precipitation Symposium. Sierra College Science Center, Rocklin, California. Available at http://www.cepsym.info/proceedings_1996.php (accessed 1 July 2011).

[995] Goodridge, J. 1996. Data on California's Extreme Rainfall from 1862–1995. Proceedings of the 1996 California Extreme Precipitation Symposium. Sierra College Science Center, Rocklin, California. Available at http://www.cepsym.info/proceedings_1996.php (accessed 1 July 2011).

[996] National Weather Service (NWS). 2009. The Interior Central California climate calendar. Available at http://www.wrh.noaa.gov/hnx/WXCALENDER.pdf (accessed 1 February 2011).

[997] National Weather Service (NWS). 2009. The Interior Central California climate calendar. Available at http://www.wrh.noaa.gov/hnx/WXCALENDER.pdf (accessed 1 February 2011).

[998] National Weather Service (NWS). 2009. The Interior Central California climate calendar. Available at http://www.wrh.noaa.gov/hnx/WXCALENDER.pdf (accessed 1 February 2011).

[999] McCulley, Frick & Gilman and William Lettis & Associates. 1998. Panoche/Silver Creek Watershed Assessment. Prepared for Panoche/Silver Creek Watershed Coordinated Resource Management and Planning Group and the City of Mendota, California. Available at http://www.water.ca.gov/pubs/environment/watersheds/panoche_silver_creek_watershed_assessment_final_report/psc-assessment.pdf (accessed 23 August 2011).

[1000] U.S. Army Corps of Engineers (USACE). 1999. Post-flood assessment for 1983, 1986, 1995, and 1997. U.S. Army Corps of Engineers, Sacramento, California. Chapter 2 available at http://snugharbor.net/images2010/misc/2002_usace_sac_flood_history.pdf (accessed 28 May 2012).

[1001] U.S. Army Corps of Engineers (USACE). 1999. Post-flood assessment for 1983, 1986, 1995, and 1997. U.S. Army Corps of Engineers, Sacramento, California. Chapter 2 available at http://snugharbor.net/images2010/misc/2002_usace_sac_flood_history.pdf (accessed 28 May 2012).

[1002] U.S. Army Corps of Engineers (USACE). 1999. Post-flood assessment for 1983, 1986, 1995, and 1997. U.S. Army Corps of Engineers, Sacramento, California. Chapter 2 available at http://snugharbor.net/images2010/misc/2002_usace_sac_flood_history.pdf (accessed 28 May 2012).

[1003] U.S. Army Corps of Engineers (USACE). 1999. Post-flood assessment for 1983, 1986, 1995, and 1997. U.S. Army Corps of Engineers, Sacramento, California. Chapter 2 available at http://snugharbor.net/images2010/misc/2002_usace_sac_flood_history.pdf (accessed 28 May 2012).

[1004] U.S. Army Corps of Engineers (USACE). 1999. Post-flood assessment for 1983, 1986, 1995, and 1997. U.S. Army Corps of Engineers, Sacramento, California. Chapter 2 available at http://snugharbor.net/images2010/misc/2002_usace_sac_flood_history.pdf (accessed 28 May 2012).

[1005] Goodridge, J. 1996. Data on California's Extreme Rainfall from 1862–1995. Proceedings of the 1996 California Extreme Precipitation Symposium. Sierra College Science Center, Rocklin, California. Available at http://www.cepsym.info/proceedings_1996.php (accessed 1 July 2011).

[1006] National Weather Service (NWS). 2009. The Interior Central California climate calendar. Available at http://www.wrh.noaa.gov/hnx/WXCALENDER.pdf (accessed 1 February 2011).

[1007] U.S. Army Corps of Engineers (USACE). 1999. Post-flood assessment for 1983, 1986, 1995, and 1997. U.S. Army Corps of Engineers, Sacramento, California. Chapter 2 available at http://snugharbor.net/images2010/misc/2002_usace_sac_flood_history.pdf (accessed 28 May 2012).

[1008] Goodridge, J. 1996. Data on California's Extreme Rainfall from 1862–1995. Proceedings of the 1996 California Extreme Precipitation Symposium. Sierra College Science Center, Rocklin, California. Available at http://www.cepsym.info/proceedings_1996.php (accessed 1 July 2011).

[1009] U.S. Army Corps of Engineers (USACE). 1999. Post-flood assessment for 1983, 1986, 1995, and 1997. U.S. Army Corps of Engineers, Sacramento, California. Chapter 2 available at http://snugharbor.net/images2010/misc/2002_usace_sac_flood_history.pdf (accessed 28 May 2012).

[1010] National Weather Service (NWS). 2009. The Interior Central California climate calendar. Available at http://www.wrh.noaa.gov/hnx/WXCALENDER.pdf (accessed 1 February 2011).

[1011] McCulley, Frick & Gilman and William Lettis & Associates. 1998. Panoche/Silver Creek Watershed Assessment. Prepared for Panoche/Silver Creek Watershed Coordinated Resource Management and Planning Group and the City of Mendota, California. Available at http://www.water.ca.gov/pubs/environment/watersheds/panoche_silver_creek_watershed_assessment_final_report/psc-assessment.pdf (accessed 23 August 2011).

[1012] Goodridge, J. 1996. Data on California's Extreme Rainfall from 1862–1995. Proceedings of the 1996 California Extreme Precipitation Symposium. Sierra College Science Center, Rocklin, California. Available at http://www.cepsym.info/proceedings_1996.php (accessed 1 July 2011).

[1013] U.S. Army Corps of Engineers (USACE). 1999. Post-flood assessment for 1983, 1986, 1995, and 1997. U.S. Army Corps of Engineers, Sacramento, California. Chapter 2 available at http://snugharbor.net/images2010/misc/2002_usace_sac_flood_history.pdf (accessed 28 May 2012).

[1014] National Weather Service (NWS). 2009. The Interior Central California climate calendar. Available at http://www.wrh.noaa.gov/hnx/WXCALENDER.pdf (accessed 1 February 2011).

[1015] National Weather Service (NWS). 2009. The Interior Central California climate calendar. Available at http://www.wrh.noaa.gov/hnx/WXCALENDER.pdf (accessed 1 February 2011).

[1016] Donahue, M. n.d. Yosemite Valley spring runoff and flooding website. http://faculty.deanza.edu/donahuemary/stories/storyReader$2113 (accessed 1 February 2011).

[1017] Federal Emergency Management Agency (FEMA). n.d. Declared Disasters by Year or State website. http://www.fema.gov/news/disaster_totals_annual.fema (accessed 15 January 2012).

[1018] National Oceanic and Atmospheric Administration (NOAA). n.d. Top ten published atmospheric river events website. http://www.esrl.noaa.gov/psd/atmrivers/events/ (accessed 1 February 2011).

[1019] U.S. Army Corps of Engineers (USACE). 1999. Post-flood assessment for 1983, 1986, 1995, and 1997. U.S. Army Corps of Engineers, Sacramento, California. Chapter 2 available at http://snugharbor.net/images2010/misc/2002_usace_sac_flood_history.pdf (accessed 28 May 2012).

[1020] U.S. Army Corps of Engineers (USACE). 1999. Post-flood assessment for 1983, 1986, 1995, and 1997. U.S. Army Corps of Engineers, Sacramento, California. Chapter 2 available at http://snugharbor.net/images2010/misc/2002_usace_sac_flood_history.pdf (accessed 28 May 2012).

[1021] U.S. Army Corps of Engineers (USACE). 1999. Post-flood assessment for 1983, 1986, 1995, and 1997. U.S. Army Corps of Engineers, Sacramento, California. Chapter 2 available at http://snugharbor.net/images2010/misc/2002_usace_sac_flood_history.pdf (accessed 28 May 2012).

[1022] U.S. Army Corps of Engineers (USACE). 1999. Post-flood assessment for 1983, 1986, 1995, and 1997. U.S. Army Corps of Engineers, Sacramento, California. Chapter 2 available at http://snugharbor.net/images2010/misc/2002_usace_sac_flood_history.pdf (accessed 28 May 2012).

[1023] U.S. Army Corps of Engineers (USACE). 1999. Post-flood assessment for 1983, 1986, 1995, and 1997. U.S. Army Corps of Engineers, Sacramento, California. Chapter 2 available at http://snugharbor.net/images2010/misc/2002_usace_sac_flood_history.pdf (accessed 28 May 2012).

[1024] National Weather Service (NWS). 2009. The Interior Central California climate calendar. Available at http://www.wrh.noaa.gov/hnx/WXCALENDER.pdf (accessed 1 February 2011).

[1025] National Weather Service (NWS). 2009. The Interior Central California climate calendar. Available at http://www.wrh.noaa.gov/hnx/WXCALENDER.pdf (accessed 1 February 2011).

[1026] Donahue, M. n.d. Yosemite Valley spring runoff and flooding website. http://faculty.deanza.edu/donahuemary/stories/storyReader$2113 (accessed 1 February 2011).

[1027] U.S. Army Corps of Engineers (USACE). 1999. Post-flood assessment for 1983, 1986, 1995, and 1997. U.S. Army Corps of Engineers, Sacramento, California. Chapter 2 available at http://snugharbor.net/images2010/misc/2002_usace_sac_flood_history.pdf (accessed 28 May 2012).

[1028] National Weather Service (NWS). 2009. The Interior Central California climate calendar. Available at http://www.wrh.noaa.gov/hnx/WXCALENDER.pdf (accessed 1 February 2011).

[1029] National Weather Service (NWS). 2009. Storm data and unusual weather phenomenon: January 1997. Available at http://www.wrh.noaa.gov/hnx/stormdat/1997jan.pdf (accessed 1 February 2011).

[1030] Kings River Water Conservation District and Kings River Water Association. 2003. The Kings River Handbook. Available at http://www.centralvalleywater.org/_pdf/KingsRiverHandbook-03final.pdf (accessed 20 April 2011).

[1031] U.S. Army Corps of Engineers (USACE). 1999. Post-flood assessment for 1983, 1986, 1995, and 1997: Chapter 3. U.S. Army Corps of Engineers, Sacramento, California. Available at http://www.auburndamcouncil.org/pages/pdf-files/3-cv_floodmgmt_system.pdf (accessed 1 June 2012).

[1032] U.S. Army Corps of Engineers (USACE). 1999. Post-flood assessment for 1983, 1986, 1995, and 1997: Chapter 3. U.S. Army Corps of Engineers, Sacramento, California. Available at http://www.auburndamcouncil.org/pages/pdf-files/3-cv_floodmgmt_system.pdf (accessed 1 June 2012).

[1033] U.S. Army Corps of Engineers (USACE). 1999. Post-flood assessment for 1983, 1986, 1995, and 1997. U.S. Army Corps of Engineers, Sacramento, California. Chapter 2 available at http://snugharbor.net/images2010/misc/2002_usace_sac_flood_history.pdf (accessed 28 May 2012).

[1034] National Weather Service (NWS). 2009. Storm data and unusual weather phenomenon: January 1997. Available at http://www.wrh.noaa.gov/hnx/stormdat/1997jan.pdf (accessed 1 February 2011).

[1035] National Weather Service (NWS). 2009. Storm data and unusual weather phenomenon: July 1997. Available at http://www.wrh.noaa.gov/hnx/stormdat/1997jul.pdf (accessed 5 October 2012).

[1036] National Weather Service (NWS). 2009. Storm data and unusual weather phenomenon: September 1997. Available at http://www.wrh.noaa.gov/hnx/stormdat/1997sep.pdf (accessed 1 February 2011).

[1037] National Weather Service (NWS). 2009. The Interior Central California climate calendar. Available at http://www.wrh.noaa.gov/hnx/WXCALENDER.pdf (accessed 1 February 2011).

[1038] California State University San Bernardino. n.d. Alluvial Fan Task Force (AFTF) Study Area Flood History: Kern County Flood History. Available at http://aftf.csusb.edu/documents/AFTF%20Study%20Area%20Flood%20History_ALL.pdf (accessed 17 June 2011).

[1039] National Weather Service (NWS). 2009. The Interior Central California climate calendar. Available at http://www.wrh.noaa.gov/hnx/WXCALENDER.pdf (accessed 1 February 2011).

[1040] Sydnor, R.H., 1997, Reconnaissance engineering geology of the Mill Creek landslide of January 24, 1997. California Geology 50:74–83. Available at http://www.consrv.ca.gov/CGS/rghm/landslides/cal_geology/1997%20Hwy%2050%20Cal%20Geology.pdf (accessed 12 August 2011).

[1041] Harp, E.L., M.E. Reid, J.W. Godt, J.V. DeGraff, and A.J. Gallegos. 2008. Ferguson rock slide buries California state highway near Yosemite National Park. Landslides 5:331–337. Available at http://www.fs.usda.gov/Internet/FSE_DOCUMENTS/stelprdb5238394.pdf (accessed 13 August 2011).

[1042] DeGraff, J.V. 2001. Sourgrass debris flow — a landslide triggered in the Sierra Nevada by the 1997 New Year Storm. Pages 69–76 *in* H. Ferriz and R. Anderson, editors. Engineering Geology Practice in Northern California. Jointly published as Association of Engineering Geologists Special Publication 12 and California Division of Mines and Geology Bulletin 210. California Division of Mines and Geology, Sacramento, California. Available at http://ia700305.us.archive.org/23/items/minesbulletin210calirich/minesbulletin210calirich.pdf (accessed 10 May 2012).

[1043] DeGraff, J.V. 1997. Briefing paper — Sourgrass debris flow. Available at http://virtual.yosemite.cc.ca.us/jtolhurst/ESGIS/Earth_Science/Sourgrass_Debris_Flow/Sourgrass_Dorrington%20Slide.htm (accessed 12 August 2011).

[1044] U.S. Forest Service (USFS). n.d. The Sourgrass Slide of 1997. Available at http://www.scenic4.org/documents/sourgrass_slide_brochure_47.pdf (accessed 13 August 2011).

[1045] U.S. Geological Survey (USGS). 2008. Aerial reconnaissance of potential landslide activity in the Central Sierra Nevada Mountains. Landslide Hazards Program website. http://landslides.usgs.gov/recent/archives/1997sierra.php (accessed 12 August 2011).

[1046] National Oceanic and Atmospheric Administration (NOAA). n.d. Top ten published atmospheric river events website. http://www.esrl.noaa.gov/psd/atmrivers/events/ (accessed 1 February 2011).

[1047] National Weather Service (NWS). 2009. The Interior Central California climate calendar. Available at http://www.wrh.noaa.gov/hnx/WXCALENDER.pdf (accessed 1 February 2011).

[1048] National Weather Service (NWS). 2009. Storm data and unusual weather phenomenon: February 1998. Available at http://www.wrh.noaa.gov/hnx/stormdat/1998feb.pdf (accessed 1 February 2011).

[1049] National Weather Service (NWS). 2009. The Interior Central California climate calendar. Available at http://www.wrh.noaa.gov/hnx/WXCALENDER.pdf (accessed 1 February 2011).

[1050] Federal Emergency Management Agency (FEMA). n.d. Declared Disasters by Year or State website. http://www.fema.gov/news/disaster_totals_annual.fema (accessed 15 January 2012).

[1051] National Weather Service (NWS). 2009. The Interior Central California climate calendar. Available at http://www.wrh.noaa.gov/hnx/WXCALENDER.pdf (accessed 1 February 2011).

[1052] McCulley, Frick & Gilman and William Lettis & Associates. 1998. Panoche/Silver Creek Watershed Assessment. Prepared for Panoche/Silver Creek Watershed Coordinated Resource Management and Planning Group and the City of Mendota, California. Available at http://www.water.ca.gov/pubs/environment/watersheds/panoche_silver_creek_watershed_assessment_final_report/psc-assessment.pdf (accessed 23 August 2011).

[1053] National Weather Service (NWS). 2009. The Interior Central California climate calendar. Available at http://www.wrh.noaa.gov/hnx/WXCALENDER.pdf (accessed 1 February 2011).

[1054] National Weather Service (NWS). 2009. The Interior Central California climate calendar. Available at http://www.wrh.noaa.gov/hnx/WXCALENDER.pdf (accessed 1 February 2011).

[1055] National Weather Service (NWS). 2009. The Interior Central California climate calendar. Available at http://www.wrh.noaa.gov/hnx/WXCALENDER.pdf (accessed 1 February 2011).

[1056] National Weather Service (NWS). 2009. The Interior Central California climate calendar. Available at http://www.wrh.noaa.gov/hnx/WXCALENDER.pdf (accessed 1 February 2011).

[1057] National Weather Service (NWS). 2009. The Interior Central California climate calendar. Available at http://www.wrh.noaa.gov/hnx/WXCALENDER.pdf (accessed 1 February 2011).

[1058] National Weather Service (NWS). 2009. The Interior Central California climate calendar. Available at http://www.wrh.noaa.gov/hnx/WXCALENDER.pdf (accessed 1 February 2011).

[1059] National Weather Service (NWS). 2009. Storm data and unusual weather phenomenon: February 2001. Available at http://www.wrh.noaa.gov/hnx/stormdat/2001feb.pdf (accessed 29 March 2011).

[1060] National Weather Service (NWS). 2009. The Interior Central California climate calendar. Available at http://www.wrh.noaa.gov/hnx/WXCALENDER.pdf (accessed 1 February 2011).

[1061] National Weather Service (NWS). 2009. The Interior Central California climate calendar. Available at http://www.wrh.noaa.gov/hnx/WXCALENDER.pdf (accessed 1 February 2011).

[1062] National Weather Service (NWS). 2009. Storm data and unusual weather phenomenon: November 2002. Available at http://www.wrh.noaa.gov/hnx/stormdat/2002nov.pdf (accessed 1 February 2011).

[1063] National Weather Service (NWS). 2009. The Interior Central California climate calendar. Available at http://www.wrh.noaa.gov/hnx/WXCALENDER.pdf (accessed 1 February 2011).

[1064] National Weather Service (NWS). 2009. Storm data and unusual weather phenomenon: December 2003. Available at http://www.wrh.noaa.gov/hnx/stormdat/2003dec.pdf (accessed 24 August 2011).

[1065] Debris flow event, December 25, 2003, Devore, California web video. Available at http://www.youtube.com/watch?v=k3W-wDIR-Os (accessed 19 August 2011).

[1066] National Weather Service (NWS). 2009. The Interior Central California climate calendar. Available at http://www.wrh.noaa.gov/hnx/WXCALENDER.pdf (accessed 1 February 2011).

[1067] National Weather Service (NWS). 2009. The Interior Central California climate calendar. Available at http://www.wrh.noaa.gov/hnx/WXCALENDER.pdf (accessed 1 February 2011).

[1068] National Weather Service (NWS). 2009. The Interior Central California climate calendar. Available at http://www.wrh.noaa.gov/hnx/WXCALENDER.pdf (accessed 1 February 2011).

[1069] National Weather Service (NWS). 2009. Storm data and unusual weather phenomenon: May 2005. Available at http://www.wrh.noaa.gov/hnx/stormdat/2005may.pdf (accessed 23 May 2011).

[1070] Dettinger, Michael, Jessica Lundquist, Dan Cayan, and Joe Meyer. n.d. The 16 May 2005 Flood in Yosemite — A glimpse into high-country flood generation in the Sierra Nevada Available at http://tenaya.ucsd.edu/~dettinge/Yosemite_flood_poster.pdf (accessed 23 May 2011).

[1071] National Weather Service (NWS). 2009. Storm data and unusual weather phenomenon: August 2005. Available at http://www.wrh.noaa.gov/hnx/stormdat/2005aug.pdf (accessed 1 February 2011).

[1072] National Weather Service (NWS). 2009. Storm data and unusual weather phenomenon: October 2005. Available at http://www.wrh.noaa.gov/hnx/stormdat/2005oct.pdf (accessed 1 February 2011).

[1073] Federal Emergency Management Agency (FEMA). n.d. Declared Disasters by Year or State website. http://www.fema.gov/news/disaster_totals_annual.fema (accessed 15 January 2012).

[1074] National Weather Service (NWS). 2009. Storm data and unusual weather phenomenon: January 2006. Available at http://www.wrh.noaa.gov/hnx/stormdat/2006jan.pdf (accessed 1 February 2011).

[1075] January 20, 2006 issue of The Kaweah Commonwealth. Available at http://www.kaweahcommonwealth.com/01-20-06features.htm (accessed 15 June 2012).

[1076] National Weather Service (NWS). 2009. Storm data and unusual weather phenomenon: January 2006. Available at http://www.wrh.noaa.gov/hnx/stormdat/2006jan.pdf (accessed 1 February 2011).

[1077] Federal Emergency Management Agency (FEMA). n.d. Declared Disasters by Year or State website. http://www.fema.gov/news/disaster_totals_annual.fema (accessed 15 January 2012).

[1078] National Weather Service (NWS). 2009. The Interior Central California climate calendar. Available at http://www.wrh.noaa.gov/hnx/WXCALENDER.pdf (accessed 1 February 2011).

[1079] May/Jun 2011 Preserving Quantity and Quality: Groundwater Management in California, bimonthly newsletter of the Water Education Foundation. Available at https://www.watereducation.org/doc.asp?id=2144 (accessed 1 September 2012).

[1080] July 22, 2012 issue of the Sacramento Bee. Available at http://www.sacbee.com/2012/07/22/4648322/valley-groundwater-threatened.html (accessed 24 July 2012).

[1081] DeGraff, J.V., D.L. Wagner, A.J. Gallegos, M. DeRose, C. Shannon, and T. Ellsworth. 2011. The remarkable occurrence of large rainfall-induced debris flows at two different locations on July 12, 2008, Southern Sierra Nevada, CA. Landslides 8(2) 343–353. Available at http://www.springerlink.com/content/f88568301m244650/ (accessed 19 August 2011).

[1082] National Weather Service (NWS). 2009. Storm data and unusual weather phenomenon: July 2008. Available at http://www.wrh.noaa.gov/hnx/stormdat/2008jul.pdf (accessed 27 July 2011).

[1083] National Weather Service (NWS). 2009. Storm data and unusual weather phenomenon: July 2008. Available at http://www.wrh.noaa.gov/hnx/stormdat/2008jul.pdf (accessed 27 July 2011).

[1084] DeGraff, J.V., D.L. Wagner, A.J. Gallegos, M. DeRose, C. Shannon, and T. Ellsworth. 2011. The remarkable occurrence of large rainfall-induced debris flows at two different locations on July 12, 2008, Southern Sierra Nevada, CA. Landslides 8(2) 343–353. Available at http://www.springerlink.com/content/f88568301m244650/ (accessed 19 August 2011).

[1085] DeGraff, J.V. 1994. The geomorphology of some debris flows in the southern Sierra Nevada, California. Geomorphology 10:231–252.

[1086] DeGraff, J.V., D.L. Wagner, A.J. Gallegos, M. DeRose, C. Shannon, and T. Ellsworth. 2011. The remarkable occurrence of large rainfall-induced debris flows at two different locations on July 12, 2008, Southern Sierra Nevada, CA. Landslides 8(2) 343–353. Available at http://www.springerlink.com/content/f88568301m244650/ (accessed 19 August 2011).

[1087] National Weather Service (NWS). 2009. Storm data and unusual weather phenomenon: July 2008. Available at http://www.wrh.noaa.gov/hnx/stormdat/2008jul.pdf (accessed 27 July 2011).

[1088] National Weather Service (NWS). 2008. July 2008 Weather Summary. Available at http://www.wrh.noaa.gov/hnx/wxsummaries/2008/JULY%202008%20WEATHER%20SUMMARY.pdf (accessed 23 August 2011).

[1089] DeGraff, J.V., D.L. Wagner, A.J. Gallegos, M. DeRose, C. Shannon, and T. Ellsworth. 2011. The remarkable occurrence of large rainfall-induced debris flows at two different locations on July 12, 2008, Southern Sierra Nevada, CA. Landslides 8(2) 343–353. Available at http://www.springerlink.com/content/f88568301m244650/ (accessed 19 August 2011).

[1090] DeGraff, J.V., D.L. Wagner, A.J. Gallegos, M. DeRose, C. Shannon, and T. Ellsworth. 2011. The remarkable occurrence of large rainfall-induced debris flows at two different locations on July 12, 2008, Southern Sierra Nevada, CA. Landslides 8(2) 343–353. Available at http://www.springerlink.com/content/f88568301m244650/ (accessed 19 August 2011).

[1091] Lake Isabella Day 1 — Kern County helicopter video of July 12, 2008, Erskine Creek debris flow web video. Available at http://www.youtube.com/watch?v=GX4TFBAuL3s (accessed 19 August 2011).

[1092] July 17, 2008 issue of the Kern Valley Sun. Available at
http://www.kvsun.com/articles/2008/08/14/news/doc487fb1180e622918637247.txt (accessed 19
August 2011).

[1093] National Weather Service (NWS). 2008. July 2008 Weather Summary. Available at
http://www.wrh.noaa.gov/hnx/wxsummaries/2008/JULY%202008%20WEATHER%20SUMMARY.
pdf (accessed 23 August 2011).

[1094] Flash Flood in Lake Isabella 2 — Video of July 12, 2008, Erskine Creek debris flow overflowing
Lake Isabella Blvd web video. Available at http://www.youtube.com/watch?v=XHbysXUYDdc
(accessed 19 August 2011).

[1095] Hourly news story on July 14, 2008 from The Bakersfield Californian. Available at
http://www.bakersfield.com/hourly_news/story/496430.html (no longer available online)

[1096] National Weather Service (NWS). 2009. Storm data and unusual weather phenomenon: July
2008. Available at http://www.wrh.noaa.gov/hnx/stormdat/2008jul.pdf (accessed 27 July 2011).

[1097] DeGraff, J.V., D.L. Wagner, A.J. Gallegos, M. DeRose, C. Shannon, and T. Ellsworth. 2011. The
remarkable occurrence of large rainfall-induced debris flows at two different locations on July 12,
2008, Southern Sierra Nevada, CA. Landslides 8(2) 343–353. Available at
http://www.springerlink.com/content/f88568301m244650/ (accessed 19 August 2011).

[1098] National Weather Service (NWS). 2009. Storm data and unusual weather phenomenon: July
2008. Available at http://www.wrh.noaa.gov/hnx/stormdat/2008jul.pdf (accessed 27 July 2011).

[1099] DeGraff, J.V., D.L. Wagner, A.J. Gallegos, M. DeRose, C. Shannon, and T. Ellsworth. 2011. The
remarkable occurrence of large rainfall-induced debris flows at two different locations on July 12,
2008, Southern Sierra Nevada, CA. Landslides 8(2) 343–353. Available at
http://www.springerlink.com/content/f88568301m244650/ (accessed 19 August 2011).

[1100] DeGraff, J.V., D.L. Wagner, A.J. Gallegos, M. DeRose, C. Shannon, and T. Ellsworth. 2011. The
remarkable occurrence of large rainfall-induced debris flows at two different locations on July 12,
2008, Southern Sierra Nevada, CA. Landslides 8(2) 343–353. Available at
http://www.springerlink.com/content/f88568301m244650/ (accessed 19 August 2011).

[1101] National Weather Service (NWS). 2009. Storm data and unusual weather phenomenon: July
2008. Available at http://www.wrh.noaa.gov/hnx/stormdat/2008jul.pdf (accessed 27 July 2011).

[1102] National Weather Service (NWS). 2009. Storm data and unusual weather phenomenon: July
2008. Available at http://www.wrh.noaa.gov/hnx/stormdat/2008jul.pdf (accessed 27 July 2011).

[1103] National Weather Service (NWS). 2009. Storm data and unusual weather phenomenon: July
2008. Available at http://www.wrh.noaa.gov/hnx/stormdat/2008jul.pdf (accessed 27 July 2011).

[1104] August 2008 newsletter of the Kaweah Flyfishers. Available at
http://www.kaweahflyfishers.org/images/stories/kaweah_fly_fishers/newsletters/nl0808/aug2008new
sletter.pdf (accessed 29 July 2011).

[1105] DeGraff, J.V. 1994. The geomorphology of some debris flows in the southern Sierra Nevada,
California. Geomorphology 10:231–252.

[1106] DeGraff, J.V. 1994. The geomorphology of some debris flows in the southern Sierra Nevada,
California. Geomorphology 10:231–252.

[1107] U.S. Geological Survey (USGS). n.d. Debris Flow Hazards website.
http://geology.com/articles/debris-flow/ (accessed 14 August 2011).

[1108] August 2008 newsletter of the Kaweah Flyfishers. Available at
http://www.kaweahflyfishers.org/images/stories/kaweah_fly_fishers/newsletters/nl0808/aug2008new
sletter.pdf (accessed 29 July 2011).

[1109] National Oceanic and Atmospheric Administration (NOAA). n.d. Top ten published atmospheric
river events website. http://www.esrl.noaa.gov/psd/atmrivers/events/ (accessed 1 February 2011).

[1110] National Weather Service (NWS). 2009. Storm data and unusual weather phenomenon: October 2009. Available at http://www.wrh.noaa.gov/hnx/stormdat/2009oct.pdf (accessed 1 February 2011).

[1111] Ardesch, R. 2009. October 2009 Flood of the Kaweah River in Three Rivers, California from the Pumpkin Hollow Bridge web video. Available at http://www.youtube.com/watch?v=B7x5mpmi-LA&feature=youtu.be (accessed 5 October 2012).

[1112] Wikipedia. n.d. El Niño-Southern Oscillation web article. http://en.wikipedia.org/wiki/El_Ni%C3%B1o-Southern_Oscillation#North_America_2 (accessed 6 March 2012).

[1113] National Air and Space Administration (NASA). 2012. Earth Observatory: Diagnosing the snow deficit of December website. http://earthobservatory.nasa.gov/IOTD/view.php?id=77076&src=eoa-iotd. (accessed 6 March 2012)

[1114] Federal Emergency Management Agency (FEMA). n.d. Declared Disasters by Year or State website. http://www.fema.gov/news/disaster_totals_annual.fema (accessed 15 January 2012).

[1115] http://www.fresnobee.com/2010/01/19/1788556/rain-may-cause-valley-flooding.html

[1116] National Weather Service (NWS). 2009. Storm data and unusual weather phenomenon: October 2009. Available at http://www.wrh.noaa.gov/hnx/stormdat/2010jan.pdf (accessed 27 March 2011).

[1117] December 22, 2010 issue of the Visalia Times-Delta.

[1118] December 20, 2011 issue of the Visalia Times-Delta.

[1119] December 20, 2011 online news feed of the Fresno Bee. No longer available online.

[1120] December 20, 2011 Associated Press online story. No longer available online.

[1121] December 28, 2010 online news feed of Fresno Bee. No longer available online.

[1122] DeGraff, J.V. 1994. The geomorphology of some debris flows in the southern Sierra Nevada, California. Geomorphology 10:231–252.

[1123] National Weather Service (NWS). n.d. Fresno Normals, Means, and Extremes website. http://www.wrh.noaa.gov/hnx/fat/normals/fath2oyrwettest.pdf (accessed 3 May 2011).

[1124] National Weather Service (NWS). n.d. Bakersfield Normals, Means, and Extremes website. http://www.wrh.noaa.gov/hnx/bflmain.php (accessed 3 May 2011).

[1125] California Department of Water Resources (DWR). 2011. Summary of Water Conditions (B120 Bulletin) for May 1, 2011. California Department of Water Resources. Sacramento, California. Available at http://cdec.water.ca.gov/snow/bulletin120/b120may11.pdf (accessed 16 May 2011).

[1126] California Department of Water Resources (DWR). 2011. Summary of Water Conditions (B120 Bulletin) for May 1, 2011. California Department of Water Resources. Sacramento, California. Available at http://cdec.water.ca.gov/snow/bulletin120/b120may11.pdf (accessed 16 May 2011).

[1127] Kenward, A. 2011. Exploring California's record Sierra Nevada snowpack: Climate Central blog. http://www.climatecentral.org/blogs/2011s-record-sierra-nevada-snowpack (accessed 18 July 2011).

[1128] July 8, 2011 issue of The Kaweah Commonwealth.

[1129] June 8, 2011 issue of the Fresno Bee. Available at http://www.fresnobee.com/2011/06/08/2420219/despite-large-snowpack-little.html (accessed 18 July 2011).

[1130] National Weather Service (NWS). 2011. Storm data and unusual weather phenomenon: July 2011. Available at http://www.wrh.noaa.gov/hnx/stormdat/2011jul.pdf (accessed 8 February 2012).

[1131] National Weather Service (NWS). 2011. Storm data and unusual weather phenomenon: July 2011. Available at http://www.wrh.noaa.gov/hnx/stormdat/2011jul.pdf (accessed 8 February 2012).

[1132] National Weather Service (NWS). 2011. Storm data and unusual weather phenomenon: July 2011. Available at http://www.wrh.noaa.gov/hnx/stormdat/2011jul.pdf (accessed 8 February 2012).

---

Do you love the experience of visiting America's public lands?
You can help protect our nation's treasures!

The Sequoia Natural History Association is the primary non-profit educational partner of the U.S. Army Corps of Engineers at Lake Kaweah and the National Park Service for Sequoia and Kings Canyon National Parks and Devils Postpile National Monument. The association is not a government agency. They depend solely on revenue from book sales, fees from educational services, membership dues, and donations. The mission of the Sequoia Natural History Association is to enhance understanding and appreciation of Sequoia and Kings Canyon National Parks and surrounding public lands. In addition to publishing books, the association:

Operates the bookstores in the park visitor centers
Operates the Sequoia Field Institute and Beetle Rock Education Center
Leads educational field seminars
Prints the parks' free visitor guide
Conducts tours of Sequoia's Crystal Cave
Operates the Pear Lake backcountry ski hut
Fund park exhibits, ranger programs, and black bear protection.

**Members receive:**
Discount on everything we sell in Sequoia and Kings Canyon
National Park visitor centers
Discount on books at most other national park visitor centers
If Educational program and Crystal Cave tour discounts
An invitation to special park and member events
The knowledge you are helping to protect America's natural treasures!
Join and invest in quality park experiences for all. We are a non-profit 501(c)3 organization.
Dues and donations are tax deductible to the extent allowed by law.

www.sequoiahistory.org